T0312282

Coulson and Richardson's Chemical Engineering

Coulson & Richardson's Chemical Engineering Series

Chemical Engineering, Volume 1A, Seventh edition
Fluid Flow: Fundamentals and Applications
Raj Chhabra and V. Shankar

Chemical Engineering, Volume 1B, Seventh edition
Heat and Mass Transfer: Fundamentals and Applications
Raj Chhabra and V. Shankar

Chemical Engineering, Volume 2A, Sixth edition
Particulate Systems and Particle Technology
Raj Chhabra and Basa M. Gurappa

Chemical Engineering, Volume 2B, Sixth edition
Separation Processes
A. K. Ray

Chemical Engineering, Volume 3A, Fourth edition
Chemical and Biochemical Reactors and Reaction Engineering
R. Ravi, R. Vinu and S. N. Gummadi

Chemical Engineering, Volume 3B, Fourth edition
Process Control
Sohrab Rohani

Chemical Engineering
Solutions to the Problems in Volume 1
J. R. Backhurst, J. H. Harker and J. F. Richardson

Chemical Engineering
Solutions to the Problems in Volumes 2 and 3
J. R. Backhurst, J. H. Harker and J. F. Richardson

Chemical Engineering, Volume 6, Third edition
Chemical Engineering Design
R. K. Sinnott

Coulson and Richardson's Chemical Engineering

Volume 1A: Fluid Flow: Fundamentals and Applications

Seventh Edition

Raj Chhabra
V. Shankar

Butterworth-Heinemann is an imprint of Elsevier
The Boulevard, Langford Lane, Kidlington, Oxford OX5 1GB, United Kingdom
50 Hampshire Street, 5th Floor, Cambridge, MA 02139, United States

Notices

Knowledge and best practice in this field are constantly changing. As new research and experience broaden our understanding, changes in research methods, professional practices, or medical treatment may become necessary.

Practitioners and researchers must always rely on their own experience and knowledge in evaluating and using any information, methods, compounds, or experiments described herein. In using such information or methods they should be mindful of their own safety and the safety of others, including parties for whom they have a professional responsibility.

To the fullest extent of the law, neither the Publisher nor the authors, contributors, or editors, assume any liability for any injury and/or damage to persons or property as a matter of products liability, negligence or otherwise, or from any use or operation of any methods, products, instructions, or ideas contained in the material herein.

Library of Congress Cataloging-in-Publication Data
A catalog record for this book is available from the Library of Congress

British Library Cataloguing-in-Publication Data
A catalogue record for this book is available from the British Library

ISBN: 978-0-08-101099-0

For information on all Butterworth-Heinemann publications
visit our website at https://www.elsevier.com/books-and-journals

Publisher: Joe Hayton
Acquisition Editor: Anita Koch
Editorial Project Manager: Ashlie Jackman
Production Project Manager: Sruthi Satheesh
Cover Designer: Victoria Pearson

Typeset by SPi Global, India

Contents

About Professor Coulson xi
About Professor Richardson xiii
Preface to Seventh Edition xv
Preface to Sixth Edition xvii
Preface to Fifth Edition xix
Preface to Fourth Edition xxi
Preface to Third Edition xxiii
Preface to Second Edition xxv
Preface to First Edition xxvii
Acknowledgements xxxi
Introduction xxxiii

Chapter 1: Units and Dimensions 1
1.1 Introduction 1
1.2 Systems of Units 2
1.2.1 The Centimetre-Gram-Second (cgs) System 4
1.2.2 The Metre–Kilogram–Second (mks) System and the *Système International d' Unités* (SI) 4
1.2.3 The Foot–Pound–Second (fps) System 5
1.2.4 The British Engineering System 6
1.2.5 Noncoherent System Employing Pound Mass and Pound Force Simultaneously 6
1.2.6 Derived Units 7
1.2.7 Thermal (Heat) Units 8
1.2.8 Molar Units 9
1.2.9 Electrical Units 10
1.3 Conversion of Units 10
1.4 Dimensional Analysis 13
1.5 Buckingham's Π Theorem 17
1.6 Scale Up 23
1.7 Redefinition of the Length and Mass Dimensions 26
1.7.1 Vector and Scalar Quantities 26

1.7.2 Quantity Mass and Inertia Mass 27
1.8 Nomenclature 28
References 29
Further reading 29

Chapter 2: Flow of Fluids—Energy and Momentum Relationships 31
2.1 Introduction 31
2.2 Internal Energy 31
2.3 Types of Fluid 35
2.3.1 The Incompressible Fluid (Liquid) 36
2.3.2 The Ideal Gas 36
2.3.3 The Nonideal Gas 40
2.4 The Fluid in Motion 46
2.4.1 Continuity 47
2.4.2 Momentum Changes in a Fluid 48
2.4.3 Energy of a Fluid in Motion 51
2.4.4 Pressure and Fluid Head 54
2.4.5 Constant Flow Per Unit Area 55
2.4.6 Separation 55
2.5 Pressure-Volume Relationships 56
2.5.1 Incompressible Fluids 56
2.5.2 Compressible Fluids 56
2.6 Rotational or Vortex Motion in a Fluid 59
2.6.1 The Forced Vortex 60
2.6.2 The Free Vortex 63
2.7 Nomenclature 64
References 66
Further Reading 66

Chapter 3: Flow of liquids in Pipes and Open Channels 67
3.1 Introduction 67
3.2 The Nature of Fluid Flow 68
3.2.1 Flow Over a Surface 70
3.2.2 Flow in a Pipe 71
3.3 Newtonian Fluids 71
3.3.1 Shearing Characteristics of a Newtonian Fluid 71
3.3.2 Pressure Drop for Flow of Newtonian Liquids Through a Pipe 73
3.3.3 Reynolds Number and Shear Stress 89
3.3.4 Velocity Distributions and Volumetric Flowrates for Streamline Flow 90
3.3.5 The Transition From Laminar to Turbulent Flow in a Pipe 98
3.3.6 Velocity Distributions and Volumetric Flowrates for Turbulent Flow 100
3.3.7 Flow Through Curved Pipes 104
3.3.8 Miscellaneous Friction Losses 104
3.3.9 Flow Over Banks of Tubes 112

3.3.10 Flow With a Free Surface 112
3.4 Non-Newtonian Fluids 124
3.4.1 Steady-State Shear-Dependent Behaviour 126
3.4.2 Time-Dependent Behaviour 135
3.4.3 Viscoelastic Behaviour 137
3.4.4 Characterisation of Non-Newtonian Fluids 140
3.4.5 Dimensionless Characterisation of Viscoelastic Flows 142
3.4.6 Relation Between Rheology and Structure of Material 143
3.4.7 Streamline Flow in Pipes and Channels of Regular Geometry 144
3.4.8 Turbulent flow 161
3.4.9 The Transition From Laminar to Turbulent Flow 164
3.5 Nomenclature 166
References 168
Further Reading 170

Chapter 4: Flow of Compressible Fluids 171
4.1 Introduction 171
4.2 Flow of Gas Through a Nozzle or Orifice 171
4.2.1 Isothermal Flow 173
4.2.2 Nonisothermal Flow 175
4.3 Velocity of Propagation of a Pressure Wave 182
4.4 Converging-Diverging Nozzles for Gas Flow 184
4.4.1 Maximum Flow and Critical Pressure Ratio 185
4.4.2 The Pressure and Area for Flow 186
4.4.3 Effect of Backpressure on Flow in Nozzle 188
4.5 Flow in a Pipe 189
4.5.1 Energy Balance for Flow of Ideal Gas 190
4.5.2 Isothermal Flow of an Ideal Gas in a Horizontal Pipe 191
4.5.3 Nonisothermal Flow of an Ideal Gas in a Horizontal Pipe 201
4.5.4 Adiabatic Flow of an Ideal Gas in a Horizontal Pipe 201
4.5.5 Flow of Nonideal Gases 206
4.6 Shock Waves 207
4.7 Nomenclature 212
References 214
Further Reading 214

Chapter 5: Flow of Multiphase Mixtures 215
5.1 Introduction 215
5.2 Two-Phase Gas (Vapour)-Liquid Flow 217
5.2.1 Introduction 217
5.2.2 Flow Regimes and Flow Patterns 218
5.2.3 Hold-Up 221
5.2.4 Pressure, Momentum, and Energy Relations 223

5.2.5 Erosion ... 231
5.3 Flow of Liquid–Liquid Mixtures ... 232
5.3.1 Introduction ... 232
5.3.2 Flow Patterns ... 233
5.3.3 Average Holdup ... 237
5.3.4 Pressure Gradient ... 238
5.4 Flow of Solids–Liquid Mixtures ... 239
5.4.1 Introduction ... 239
5.4.2 Homogeneous Nonsettling Suspensions ... 240
5.4.3 Coarse Solids ... 243
5.4.4 Coarse Solids in Horizontal Flow ... 243
5.4.5 Coarse Solids in Vertical Flow ... 256
5.5 Flow of Gas–Solids Mixtures ... 260
5.5.1 General Considerations ... 260
5.5.2 Horizontal Transport ... 261
5.5.3 Vertical Transport ... 271
5.5.4 Practical Applications ... 272
5.6 Nomenclature ... 277
References ... 279
Further Reading ... 284

Chapter 6: Flow and Pressure Measurement ... 285
6.1 Introduction ... 285
6.2 Fluid Pressure ... 286
6.2.1 Static Pressure ... 286
6.2.2 Pressure Measuring Devices ... 287
6.2.3 Pressure Signal Transmission—The Differential Pressure Cell ... 292
6.2.4 Intelligent Electronic Pressure Transmitters ... 293
6.2.5 Impact Pressure ... 295
6.3 Measurement of Fluid Flow ... 297
6.3.1 The Pitot Tube ... 298
6.3.2 Measurement by Flow Through a Constriction ... 299
6.3.3 The Orifice Meter ... 302
6.3.4 The Nozzle ... 311
6.3.5 The Venturi Meter ... 312
6.3.6 Pressure Recovery in Orifice-Type Meters ... 313
6.3.7 Variable Area Meters—Rotameters ... 315
6.3.8 The Notch or Weir ... 319
6.3.9 Other Methods of Measuring Flowrates ... 322
6.4 Nomenclature ... 330
References ... 332
Further Reading ... 332

Chapter 7: Liquid Mixing *333*
7.1 Introduction—Types of Mixing 333
7.1.1 Single-Phase Liquid Mixing 333
7.1.2 Mixing of Immiscible Liquids 334
7.1.3 Gas–Liquid Mixing 334
7.1.4 Liquid–Solids Mixing 334
7.1.5 Gas–Liquid–Solids Mixing 335
7.1.6 Solids–Solids Mixing 335
7.1.7 Miscellaneous Mixing Applications 335
7.2 Mixing Mechanisms 336
7.2.1 Laminar Mixing 336
7.2.2 Turbulent Mixing 339
7.3 Scale-Up of Stirred Vessels 340
7.4 Power Consumption in Stirred Vessels 343
7.4.1 Low Viscosity Systems 343
7.4.2 High Viscosity Systems 351
7.5 Flow Patterns in Stirred Tanks 355
7.6 Rate and Time for Mixing 360
7.7 Mixing Equipment 363
7.7.1 Mechanical Agitation 363
7.7.2 Portable Mixers 368
7.7.3 Extruders 368
7.7.4 Static Mixers 370
7.7.5 Other Types of Mixer 373
7.8 Mixing in Continuous Systems 373
7.9 Nomenclature 374
References 375
Further Reading 377

Chapter 8: Pumping of Fluids *379*
8.1 Introduction 379
8.2 Pumping Equipment for Liquids 380
8.2.1 Reciprocating Pump 381
8.2.2 Positive-Displacement Rotary Pumps 387
8.2.3 The Centrifugal Pump 396
8.3 Pumping Equipment for Gases 412
8.3.1 Fans and Rotary Compressors 412
8.3.2 Centrifugal and Turbocompressors 414
8.3.3 The Reciprocating Piston Compressor 415
8.3.4 Power Required for the Compression of Gases 416
8.4 The Use of Compressed Air for Pumping 429
8.4.1 The Air-Lift Pump 429
8.5 Vacuum Pumps 436
8.6 Power Requirements for Pumping Through Pipelines 438

8.6.1 Liquids 439
8.6.2 Gases 447
8.7 Effect of Minor Losses 449
8.8 Nomenclature 456
References 458
Further Reading 459

Appendix 461

A.1 Tables of Physical Properties 461
Table 1 Thermal conductivities of liquids 461
Table 2 Latent heats of vaporisation 463
Table 3 Specific heats of liquids 465
Table 4 Specific heats at constant pressure of gases and vapours at 101.3 kN/m^2 467
Table 5 Viscosity of water 469
Table 6 Thermal conductivities of gases and vapours 470
Table 7 Viscosities of gases 472
Table 8 Viscosities and densities of liquids 474
Table 9 Critical constants of gases 478
Table 10 Emissivities of surfaces 479
A.2 Steam Tables 483
Table 11A Properties of saturated steam (S.I. units) 484
Table 11B Properties of saturated steam (Centigrade and Fahrenheit units) 488
Table 11C Enthalpy of superheated steam, H (kJ/kg) 491
Table 11D Entropy of superheated steam, S (kJ/kg K) 492
Fig. 11A Pressure-enthalpy diagram for water and steam 493
Fig. 11B Temperature-entropy diagram for water and steam 494
A.3 Mathematical Tables 495
Table 12 Laplace transforms 495
Table 13 Error function and its derivative 501

Problems 505

Index 529

About Professor Coulson

John Coulson, who died on 6 January 1990 at the age of 79, came from a family with close involvement with education. Both he and his twin brother Charles (renowned physicist and mathematician), who predeceased him, became professors. John did his undergraduate studies at Cambridge and then moved to Imperial College where he took the postgraduate course in chemical engineering—the normal way to qualify at that time—and then carried out research on the flow of fluids through packed beds. He then became an assistant lecturer at Imperial College and, after wartime service in the Royal Ordnance Factories, returned as lecturer and was subsequently promoted to a readership. At Imperial College he initially had to run the final year of the undergraduate course almost single-handed, a very demanding assignment. During this period, he collaborated with Sir Frederick (Ned) Warner to write a model design exercise for the IChemE home paper on 'The Manufacture of Nitrotoluene'. He published research papers on heat transfer and evaporation, on distillation, and on liquid extraction and co-authored this textbook of chemical engineering. He did valiant work for the Institution of Chemical Engineers that awarded him its Davis medal in 1973 and was also a member of the advisory board for what was then a new Pergamon journal, Chemical Engineering Science.

In 1954, he was appointed to the newly established chair at Newcastle upon Tyne, where chemical engineering became a separate department and independent of mechanical engineering of which it was formerly part and remained there until his retirement in 1975. He took a period of secondment to Heriot Watt University where, following the splitting of the joint Department of Chemical Engineering with Edinburgh, he acted as adviser and de facto head of department. The Scottish university awarded him an honorary DSc in 1973.

John's first wife Dora sadly died in 1961; they had two sons, Anthony and Simon. He remarried in 1965 and is survived by Christine.

John F. Richardson

About Professor Richardson

Professor **John Francis Richardson**, Jack to all who knew him, was born at Palmers Green, North London, on 29 July 1920 and attended the Dame Alice Owen's School in Islington. Subsequently, after studying chemical engineering at Imperial College, he embarked on research into the suppression of burning liquids and of fires. This early work contributed much to our understanding of the extinguishing properties of foams, carbon dioxide, and halogenated hydrocarbons, and he spent much time during the war years on large-scale fire control experiments in Manchester and at the Llandarcy refinery in South Wales. At the end of the war, Jack returned to Imperial College as a lecturer where he focussed on research in the broad area of multiphase fluid mechanics, especially sedimentation and fluidisation, two-phase flow of a gas and a liquid in pipes. This laid the foundation for the design of industrial processes like catalytic crackers and led to a long-lasting collaboration with the Nuclear Research Laboratories at Harwell. This work also led to the publication of the famous paper, now common knowledge, the so-called Richardson-Zaki equation, which was selected as the Week's citation classic (*Current Contents*, 12 February 1979)!

After a brief spell with Boake Roberts in East London, where he worked on the development of novel processes for flavours and fragrances, he was appointed the professor of chemical engineering at the then University College of Swansea (now University of Swansea) in 1960. He remained there until his retirement in 1987 and thereafter continued as an emeritus professor until his death on 4 January 2011.

Throughout his career, his major thrust was on the wellbeing of the discipline of chemical engineering. In the early years of his teaching duties at Imperial College, he and his colleague John Coulson recognised the lack of satisfactory textbooks available in the field of chemical engineering. They set about rectifying the situation, and this is how the now well-known Coulson-Richardson series of books on chemical engineering was born. The fact that this series of books (six volumes) is as relevant today as it was at the time of their first appearance is a testimony to the foresight of John Coulson and Jack Richardson.

Throughout his entire career spanning almost 40 years, Jack contributed significantly to all facets of professional life, teaching, research in multiphase fluid mechanics, and service to the Institution of Chemical Engineers (IChemE, the United Kingdom). His professional work and long-standing public service was well recognised. Jack was the president of IChemE during the period 1975–76; he was named a fellow of the Royal Academy of Engineering in 1978 and was awarded an OBE in 1981.

In his spare time, Jack and his wife Joan were keen dancers, having been founder members of the Society of International Folk Dancing, and they also shared a love of hill walking.

Raj Chhabra

Preface to Seventh Edition

The sixth edition of this title appeared in 1999 at the dawn of the new millennium (with reprints in 2000 and 2003). There is a gap of almost two decades between the sixth edition and the seventh edition in your hands now. The fact that this title has never been out of print is a testimony to its timelessness and 'evergreen' character, both in terms of content and style. The sole reason for this unusually long gap between the sixth and seventh editions is obviously the fact that Jack Richardson passed away in 2011 and, therefore, the publishers needed to establish whether it was a worthwhile project to continue with. The question was easily answered in the affirmative by numerous independent formal reviews and by the continuous feedback from students, teachers, and working professionals from all over the world. Having established that there was a definite need for this title, the next step was to identify individuals who would have the inclination to carry forward the legacy of Coulson and Richardson. Indeed, we feel privileged to have been entrusted with this onerous task.

The basic philosophy and the objectives of this edition remain the same as articulated so very well by the previous authors of the sixth edition, except for the fact that it has been split into two sub-volumes, 1A and 1B. In essence, these two volumes continue to concentrate on the fundamentals of fluid flow (momentum transfer) in volume 1A and heat and mass transfer in volume 1B, as applied to a wide-ranging industrial settings. Most of the concepts have been illustrated by including examples of practical applications in the areas of the pumping of fluids in pipes and pipe networks, selection of pumps and of flow metering devices, two-phase flow, etc. The entire volume has been reviewed keeping in mind the feedback received from the readers and the reviewers. Wherever needed, both contents and presentation have been improved by reorganising the existing material for easier understanding, or new material has been added to provide updated and reliable information. Apart from the general revision of all the chapters, the specific changes made in this edition are summarised below:

(i) In Chapter 1, a short section on scale-up has been added.
(ii) In Chapter 5, a new section on the flow of liquid–liquid mixtures in pipes has been included (Section 5.3).
(iii) In Chapter 8, the discussion of minor losses in pipe fittings has been amplified together with two new examples (Section 8.7).

Most of these changes are based on the first author's extensive conversations and discussion with Jack Richardson over a period of 30 years.

We are grateful to the many individuals who have facilitated the publication of the seventh edition. Over the past 2 years, it has been a wonderful experience working with the staff at Butterworth–Heinemann. Each one of them has been extremely helpful, and some of these individuals deserve a mention here. First and foremost, we are grateful to Fiona Geraghty (and her successor Anita Koch) for commissioning the new edition. They not only patiently answered our endless queries but also came to our rescue on several occasions. Similarly, Maria Convoy and Ashlie Jackman went much beyond their call of duty to see this project through. Finally, Sruthi Satheesh assembled the numerous fragments in different forms and formats—ranging from handwritten notes to latex files—into the finished product in your hands. We end this preface with an appeal to our readers to please let us know as and when you spot errors or inconsistencies so that these can be rectified at the earliest opportunity.

Raj Chhabra
V. Shankar
Kanpur, September 2017

Preface to Sixth Edition

It is somewhat sobering to realise that the sixth edition of Volume 1 appears 45 years after the publication of the first edition in 1954. Over the intervening period, there have been considerable advances in both the underlying theory and the practical applications of chemical engineering, all of which are reflected in parallel developments in undergraduate courses. In successive editions, we have attempted to adapt the scope and depth of treatment in the text to meet the changes in the needs of both students and practitioners of the subject.

Volume 1 continues to concentrate on the basic processes of momentum transfer (as in fluid flow), heat transfer, and mass transfer, and it also includes examples of practical applications of these topics in areas of commercial interest such as the pumping of fluids, the design of shell and tube heat exchangers, and the operation and performance of cooling towers. In response to the many requests from the readers (and the occasional note of encouragement from our reviewers), additional examples and their solutions have now been included in the main text. The principal areas of application, particularly of the theories of mass transfer across a phase boundary, form the core material of Volume 2; however, whilst in Volume 6, material presented in other volumes is utilised in the practical design of process plant.

The more important additions and modifications that have been introduced into this sixth edition of Volume 1 are the following:

Dimensionless analysis. The idea and advantages of treating length as a vector quantity and of distinguishing between the separate role of mass in representing a quantity of matter as opposed to its inertia are introduced.

Fluid flow. The treatment of the behaviour of non-Newtonian fluids is extended, and the methods used for pumping and metering of such fluids are updated.

Heat transfer. A more detailed discussion of the problem of unsteady-state heat transfer by conduction where bodies of various shapes are heated or cooled is offered together with a more complete treatment of heat transfer by radiation and a reorientation of the introduction to the design of shell and tube heat exchangers.

Mass transfer. The section on mass transfer accompanied by chemical reaction has been considerably expanded, and it is hoped that this will provide a good basis for the understanding of the operation of both homogeneous and heterogeneous catalytic reactions.

As ever, we are grateful for a great deal of help in the preparation of this new edition from a number of people. In particular, we should like to thank Dr. D.G. Peacock for the great enthusiasm and dedication he has shown in the production of the index, a task he has undertaken for us over many years. We would also mention especially Dr. R.P. Chhabra of the Indian Institute of Technology at Kanpur for his contribution on unsteady-state heat transfer by conduction, those commercial organisations that have so generously contributed new figures and diagrams of equipment, our publishers who cope with our perhaps overwhelming number of suggestions and alterations with a never-failing patience, and, most of all, our readers who with great kindness make so many extremely useful and helpful suggestions all of which are incorporated wherever practicable. With their continued help and support, the signs are that this present work will continue to be of real value as we move into the new millennium.

Swansea, 1999
Newcastle upon Tyne, 1999

John F. Richardson
John R. Backhurst
John H. Harker

Preface to Fifth Edition

This textbook has been the subject of continual updating since it was first published in 1954. There have been numerous revised impressions, and the opportunity has been taken on the occasion of each reprinting to make corrections and revisions, many of them in response to readers who have kindly pointed out errors or who have suggested modifications and additions. When the summation of the changes has reached a sufficiently high level, a new edition has been produced. We have now reached this point again, and the fifth edition incorporates all the alterations in the 1993 revision of the fourth edition, together with new material, particularly on simultaneous mass transfer and chemical reaction for unsteady-state processes.

There have been changes in the publisher too. Since the appearance of the fourth edition in 1990, Pergamon Press has become part of Elsevier Science, and now, following a reorganisation in the Reed Elsevier group of companies, the responsibility for publishing the Chemical Engineering series has passed to Butterworth-Heinemann, another Reed Elsevier company.

We are grateful to our readers for their interest and very much hope they will continue to make suggestions for the improvement of the series.

John F. Richardson

Preface to Fourth Edition

The first edition of Volume 1 was published in 1954, and Volume 2 appeared a year later. In the intervening 35 years or so, there have been far-reaching developments in *Chemical Engineering*, and the whole approach to the subject has undergone a number of fundamental changes. The question therefore arises as to whether it is feasible to update a textbook written to meet the needs of the final-year students of an undergraduate course in the 1950s so that it can continue to fulfil a useful purpose in the last decade of the century. Perhaps, it would have been better if a new textbook had been written by an entirely new set of authors. Although at one stage this had seemed likely through the sponsorship of the Institution of Chemical Engineers, there is now no sign of any such replacement book appearing in the United Kingdom.

In producing the fourth edition, it has been necessary to consider whether to start again with a clean sheet of paper—an impossibly daunting task—or whether to retain the original basic structure with relatively small modifications. In following the latter course, the authors were guided by the results of a questionnaire sent to a wide range of university ('Do not tamper overmuch with the devil we know, in spite of all his faults!' and polytechnic) departments throughout the English-speaking world. The clear message that came back was

It was in 1971 that Volume 3 was added to the series, essentially to make good some of the more glaring omissions in the earlier volumes. Volume 3 contains a series of seven specialist chapters written by members of the staff of the Chemical Engineering Department at the University College of Swansea, with Dr. D.G. Peacock of the School of Pharmacy, London, as a joint editor. In 1977–79, as well as contributing significantly to the new editions of Volumes 1 and 2, two colleagues at the University of Newcastle upon Tyne, Dr. J.R. Backhurst and the Revd Dr. J.H. Harker, prepared Volumes 4 and 5, the solutions to the problems in Volumes 1 and 2, respectively. The final major development was the publication of Volume 6 on chemical engineering design by Mr. R.K. Sinnott in 1983. With the preparation of a fourth edition, the opportunity has presented itself for a degree of rationalisation, without introducing major changes to the structure. This has led to the following format:

Volume 1	Fluid Flow, Heat Transfer, and Mass Transfer
Volume 2	Particle Technology and Separation Processes
Volume 3	Chemical and Biochemical Reactor Engineering and Control
Volume 4/5	Solutions to the Problems in Volumes 1, 2, and 3
Volume 6	Chemical Engineering Design

The details of this new arrangement are as follows:

Volume 1 has acquired an abbreviated treatment of non-Newtonian flow, formerly in Volume 3.
Liquid Mixing appears as a new chapter in Volume 1, which incorporates the relevant material formerly in Volumes 2 and 3.
Separate chapters now appear in Volume 1 on compressible flow and on multiphase flow, the latter absorbing material previously scattered between Volumes 1 and 2.
New chapters are added to Volume 2 to cover four separation processes of increasing importance—adsorption (from Volume 3), ion exchange, chromatographic separations, and membrane separations.
Volume 3 is now devoted to various aspects of reaction engineering and control, material that is considerably expanded.
Some aspects of design, previously in the earlier volumes, are now transferred to a more appropriate home in Volume 6.

As far as Volume 1 is concerned, the opportunity has been taken to update existing material. The major changes in fluid flow include the incorporation of non-Newtonian flow, an extensive revision of compressible flow and the new chapters on multiphase flow and liquid mixing. Material for this last chapter has been contributed by Dr. R.P. Chhabra of the Indian Institute of Technology at Kanpur. There has also been a substantial revision of the presentation of material on mass transfer and momentum, heat, and mass transfer. To the Appendix have been added the tables of Laplace transform and error functions, which were formerly in Volume 3, and throughout this new edition, all the diagrams have been redrawn. Some further problems have been added at the end.

Sadly, John Coulson was not able to contribute as he had done previously, and his death in January 1990 leaves us with a gap that is difficult to fill. John Backhurst and John Harker, who made a substantial contribution to the preparation of the third edition in 1977, have taken an increased share of the burden of revising the book and contributing new material and have taken a special responsibility for those sections that originated from John Coulson, in addition to the special task of updating the illustrations. Without their continued support and willing cooperation, there would have been no fourth edition.

Finally, we would all like to thank our many readers who have made such helpful suggestions in the past and have pointed out errors, many of which the authors would never have spotted. It is hoped that readers will continue to act in this way as unseen authors.

John F. Richardson
June 1990

Preface to Third Edition

The introduction of the SI system of units by the United Kingdom and many other countries has itself necessitated the revision of this engineering text. This clear implementation of a single system of units will be welcomed not only by those already in the engineering profession but also by those who are about to join. The system that is based on the cgs and mks systems using length (**L**), mass (**M**), and time (**T**) as the three basic dimensions, as is the practice in the physical sciences, has the very great advantage that it removes any possible confusion between mass and force that arises in the engineering system from the common use of the term *pound* for both quantities. We have therefore presented the text, problems, and examples in the SI system but have arranged the tables of physical data in the Appendix to include both SI and other systems wherever possible. This we regard as important because so many of the physical data have been published in cgs units. For similar reasons, engineering units have been retained as an alternative where appropriate.

In addition to the change to the SI system of units, we have taken the opportunity to update and to clarify the text. A new section on the flow of two-phase gas–liquid mixtures has been added to reflect the increased interest in the gas and petroleum industries and in its application to the boiling of liquids in vertical tubes.

The chapter on mass transfer, the subject that is so central and specific to chemical engineering, has been considerably extended and modernised. Here, we have thought it important in presenting some of the theoretical work to stress its tentative nature and to show that, although some of the theories may often lack a full scientific basis, they provide the basis of a workable technique for solving problems. In the discussion on fluid flow, reference has been made to American methods, and the emphasis on flow measurement has been slanted more to the use of instruments as part of a control system. We have emphasised the importance of pipe-flow networks, which represent a substantial cost item in modern large-scale enterprises.

This text covers the physical basis of the three major transfer operations of fluid flow, heat transfer, and mass transfer. We feel that it is necessary to provide a thorough grounding in these operations before introducing techniques that have been developed to give workable solutions in the most convenient manner for practical application. At the same time, we have directed the attention of the reader to such invaluable design codes as TEMA and the British Standards for heat exchanger design and to other manuals for pipe-flow systems.

It is important for designers always to have in their minds the need for reliability and safety; this is likely to follow from an understanding of the basic principles involved, many of which are brought out in the text.

We would like to thank our many friends from several countries who have written with suggestions, and it is our hope that this edition will help in furthering growth and interest in the profession. We should also like to thank a number of industrialists who have made available much useful information for incorporation in this edition; this help is acknowledged at the appropriate point. Our particular thanks are due to B. Waldie for his contribution to the high temperature aspects of heat transfer and to the Kellogg International Corporation and Humphreys and Glasgow Limited for their help. In conclusion, we would like to thank J.R. Backhurst and J.H. Harker for their editorial work and for recalculating the problems in SI units and converting the charts and tables.

Since the publication of the second edition of this volume, Volume 3 of *Chemical Engineering* has been published in order to give a more complete coverage of those areas of chemical engineering that are of importance in both universities and industry in the 1970s.

John M. Coulson
John F. Richardson
January 1976

Preface to Second Edition

In presenting this second edition, we should like to thank our many friends from various parts of the world who have so kindly made suggestions for clarifying parts of the text and for additions that they have felt to be important. During the last eight years, there have been changes in the general approach to chemical engineering in the universities with a shift in emphasis towards the physical mechanisms of transport processes and with a greater interest in unsteady-state conditions. We have taken this opportunity to strengthen those sections dealing with the mechanisms of processes, particularly in Chapter 7 on mass transfer and in the chapters on fluid mechanics where we have laid greater emphasis on the use of momentum exchange. Many chemical engineers are primarily concerned with the practical design of plant, and we have tried to include a little more material of use in this field in Chapter 6 on heat transfer. An introductory section on dimensional analysis has been added, but it has been possible to do no more than outline the possibilities opened up by the use of this technique. Small changes will be found throughout the text, and we have tried to meet many readers' requests by adding some more worked examples and a further selection of problems for the student. The selection of material and its arrangement are becoming more difficult and must be to a great extent a matter of personal choice, but we hope that this new edition will provide a sound basis for the study of the fundamentals of the subject and will perhaps be of some value to practising engineers.

John M. Coulson
John F. Richardson

Preface to First Edition

The idea of treating the various processes of the chemical industry as a series of unit operations was first brought out as a basis for a new technology by Walker, Lewis, and McAdams in their book in 1923. Before this, the engineering of chemical plants had been regarded as individual to an industry, and there was little common ground between one industry and another. Since the early 1920s, chemical engineering as a separate subject has been introduced into the universities of both America and England and has expanded considerably in recent years so that there are now a number of university courses in both countries. During the past 20 years, the subject matter has been extensively increased by various researches described in a number of technical journals to which frequent reference is made in the present work.

Despite the increased attention given to the subject, there are few general books, although there have been a number of specialised books on certain sections such as distillation and heat transfer. It is the purpose of the present work to present to the student an account of the fundamentals of the subject. The physical basis of the mechanisms of many of the chemical engineering operations forms a major feature of chemical engineering technology. Before tackling the individual operations, it is important to stress the general mechanisms that are found in so many of the operations. We have therefore divided the subject matter into two volumes, the first of which contains an account of these fundamentals—diffusion, fluid flow, and heat transfer. In Volume 2, we shall show how these theoretical foundations are applied in the design of individual units such as distillation columns, filters, crystallisers, and evaporators.

Volume 1 is divided into four sections—fluid flow, heat transfer, mass transfer and humidification. Since the chemical engineer must handle fluids of all kinds, including compressible gases at high pressures, we believe that it is a good plan to consider the problem from a thermodynamic aspect and to derive general equations for flow that can be used in a wide range of circumstances. We have paid special attention to showing how the boundary layer is developed over plane surfaces and in pipes, since it is so important in controlling heat and mass transfer. At the same time, we have included a chapter on pumping since chemical engineering is an essentially practical subject, and the normal engineering texts do not cover the problem as experienced in the chemical and petroleum industries.

The chapter on heat transfer contains an account of the generally accepted techniques for calculation of film transfer coefficients for a wide range of conditions and includes a section on the general construction of tubular exchangers that form a major feature of many works. The possibilities of the newer plate type units are indicated.

In Section 3, the chapter on mass transfer introduces the mechanism of diffusion, and this is followed by an account of the common relationships between heat, mass, and momentum transfer and the elementary boundary layer theory. The final section includes the practical problem of humidification where both heat and mass transfer are taking place simultaneously.

It will be seen that in all chapters, there are sections in small print. In a subject such as this, which ranges from very theoretical and idealised systems to the practical problems with empirical or experimentally determined relations, there is much to be said for omitting the more theoretical features in a first reading, and in fact, this is frequently done in the more practical courses. For this reason, the more difficult theoretical sections have been put in small print, and the whole of Chapter 9 may be omitted by those who are more concerned with the practical utility of the subject.

In many of the derivations, we have given the mathematical analysis in more detail than is customary. It is our experience that the mathematical treatment should be given in full and that the student should then apply similar analysis to a variety of problems.

We have introduced into each chapter a number of worked examples that we believe are essential to a proper understanding of the methods of treatment given in the text. It is very desirable for a student to understand a worked example before tackling fresh practical problems himself. Chemical engineering problems require a numerical answer, and it is essential to become familiar with the different techniques so that the answer is obtained by systematic methods rather than by intuition.

In preparing this text, we have been guided by courses of lectures that we have given over a period of years and have presented an account of the subject with the major emphasis on the theoretical side. With a subject that has grown so rapidly and that extends from the physical sciences to practical techniques, the choice of material must be a matter of personal selection. It is, however, more important to give the principles than the practice, which is best acquired in the factory. We hope that the text may also prove useful to those in industry who, whilst perhaps successfully employing empirical relationships, feel that they would like to find the extent to which the fundamentals are of help.

We should like to take this opportunity of thanking a number of friends who have helped by their criticism and suggestions. We are particularly indebted to Mr F.E. Warner, to M. Guter, to D.J. Rasbash, and to L.L. Katan. We are also indebted to a number of companies who have kindly permitted us to use illustrations of their equipment. We have given a number of references to technical journals, and we are grateful to the publishers for permission to use illustrations from their works. In particular, we would thank the Institution of Chemical Engineers, the American Institute of Chemical Engineers, the American Chemical Society, the Oxford University Press, and the McGraw-Hill Book Company.

South Kensington,
London S.W.7
1953

Acknowledgements

The authors and publishers acknowledge the kind assistance of the following organisations in providing illustrative material:

Fig. 6.6A and B, Bowden Sedema Ltd
Figs 6.7 and 6.6C, Foxboro Great Britain Ltd, Redhill, Surrey
Fig. 6.17, Kent Meters Ltd
Fig. 6.21, Rotameter Manufacturing Co Ltd
Fig. 6.27, Endress and Hauser
Fig. 6.29, Baird and Tatlock Ltd
Figs 7.26, 7.27, and 7.28, Chemineer Ltd, Derby
Fig. 7.29, Sulzer Ltd, Switzerland
Fig. 8.1, Worthington-Simpson Ltd
Fig. 8.3, ECD Co Ltd
Fig. 8.9, Varley Pumps and Engineering Ltd
Fig. 8.11, Watson-Marlow Ltd
Figs 8.12, 8.13, and 8.14, Mono Pumps Ltd, Manchester
Figs 8.19 and 8.20, Sigmund Pumps Ltd
Fig. 8.29, Crane Packing Ltd, Wolverhampton, W. Midlands
Fig. 8.31, International Combustion Ltd, Derby
Fig. 8.36, Nash Europe, Manchester
Figs 8.37 and 8.38, Reavell and Co Ltd
Figs 8.45 and 8.46, Hick Hargreaves and Co Ltd.

Introduction

Welcome to the next generation of Coulson–Richardson series of books on *chemical engineering*. I would like to convey to you all my feelings about this project that have evolved over the past 30 years and are based on numerous conversations with Jack Richardson himself (1981 onwards until his death in 2011) and with some of the other contributors to previous editions including Tony Wardle, Ray Sinnott, Bill Wilkinson, and John Smith. So what follows here is the essence of these interactions combined with what the independent (solicited and unsolicited) reviewers had to say about this series of books on several occasions.

The Coulson–Richardson series of books has served the academia, students, and working professionals extremely well since their first publication more than 50 years ago. This is a testimony to their robustness and, to some extent, their timelessness. I have often heard much praise, from different parts of the world, for these volumes both for their informal and user-friendly yet authoritative style and for their extensive coverage. Therefore, there is a strong case for continuing with its present style and pedagogical approach.

On the other hand, advances in our discipline in terms of new applications (for instance, energy, bio, microfluidics, nanoscale engineering, smart materials, new control strategies, and reactor configurations) are occurring so rapidly and in such a significant manner that it will be naive, even detrimental, to ignore them. Therefore, while we have tried to retain the basic structure of this series, the contents have been thoroughly revised. Wherever the need was felt, the material has been updated, revised, and expanded as deemed appropriate. Therefore, the reader whether a student or a researcher or a working professional should feel confident that what is in the book is the most up-to-date, accurate, and reliable piece of information on the topic he/she is interested in.

Evidently, this is a massive undertaking that cannot be managed by a single individual. Therefore, we now have a team of volume editors responsible for each volume having the individual chapters being written by experts in some cases. I am most grateful to all of them for having joined us in the endeavour. Furthermore, based on extensive deliberations and feedback from a large number of individuals, some structural changes were deemed to be appropriate as detailed here. In this edition, Volumes 1 to 3 have been split into two volumes each as follows:

Volume 1A: Fluid Flow: Fundamentals and Applications.

Volume 1B: Heat and Mass Transfer: Fundamentals and Applications.

Volume 2A: Particulate Systems and Particle Technology.

Volume 2B: Separation Processes.

Volume 3A: Chemical and Biochemical Reactors and Reaction Engineering.

Volume 3B: Process Control.

Undoubtedly, the success of a project of such a vast scope and magnitude hinges on the cooperation and assistance of many individuals. In this regard, we have been extremely fortunate in working with some of the outstanding individuals at Butterworth–Heinemann, a few of whom deserve to be singled out: Jonathan Simpson, Fiona Geraghty, Maria Convey, and Ashlie Jackman who have taken personal interest in this project and have come to our rescue whenever needed, going much beyond the call of duty.

Finally, this series has had a glorious past, but I sincerely hope that its future will be even brighter by presenting the best possible books to the global chemical engineering community for the next 50 years, if not for longer. I sincerely hope that the new edition of this series will meet (if not exceed) your expectations! Lastly, a request to the readers, please continue to do the good work by letting me know if, no not if, but when you spot a mistake so that these can be corrected at the first opportunity.

Raj Chhabra
Editor-in-chief
Kanpur, September 2017

CHAPTER 1

Units and Dimensions

1.1 Introduction

Students of chemical engineering quickly discover that the data used by them are expressed in a great variety of different units, and that quantities must be converted into a common system before proceeding with calculations. Standardisation has been largely achieved with the introduction of the *Système International d' Unités* (SI)[1,2] to be discussed later, which is used throughout all the volumes of this series of books. This system is now in general use in Europe and is rapidly being adopted throughout the rest of the world, including the USA where the initial inertia is now being overcome. Most of the physical properties determined in the laboratory will originally have been expressed in the cgs system, whereas the dimensions of the full-scale plant, its throughput, design, and operating characteristics appear either in some form of general engineering units or in special (often mixed) units which have their origin in the history of the particular industry. This inconsistency is quite unavoidable and is a reflection of the fact that chemical engineering has in many cases developed as a synthesis of scientific knowledge and practical experience from diverse industrial applications. Familiarity with the various systems of units and an ability to convert from one to another are therefore essential tools, as it will frequently be necessary to access literature in which the SI system has not been used. In this chapter, the main systems of units are discussed, and the importance of understanding dimensions is emphasised. It is also shown how dimensions can be used to considerably help in the formulation of relationships between large numbers of parameters.

The magnitude of any physical quantity is expressed as the product of two quantities; one is the magnitude of the unit and the other is the number of those units. Thus the distance between two points may be expressed as 1 m or as 100 cm or as 3.28 ft. The metre, centimetre, and foot are respectively the size of the units, while 1, 100, and 3.28 are the corresponding numbers of units required to express the distance between the two points.

Since the physical properties of a system are governed by a series of interconnected mechanical and physical laws, it is convenient to regard certain quantities as basic and other quantities as derived. The choice of the basic dimensions varies from one system to another, although it is usual to take length and time as fundamental. These quantities are denoted by **L** and **T**, respectively. The dimensions of velocity, which is a rate of increase of distance with time,

Coulson and Richardson's Chemical Engineering. https://doi.org/10.1016/B978-0-08-101099-0.00001-X

may be written as $\mathbf{LT}^{-1}$, and those of acceleration, the rate of increase of velocity, are $\mathbf{LT}^{-2}$. An area has dimensions $\mathbf{L}^2$ and a volume has the dimensions $\mathbf{L}^3$.

The volume of a body does not completely define the amount of material that it contains, and therefore it is usual to define a third basic quantity, the amount of matter in the body, that is its mass **M**. Thus the density of the material, its mass per unit volume, has the dimensions $\mathbf{ML}^{-3}$. However, in the British Engineering System (Section 1.2.4) force **F** is used as the third fundamental and mass then becomes a derived dimension by using the Newton's law of motion.

Physical and mechanical laws provide a further set of relations between dimensions. The most important of these is that the force required to produce a given acceleration of a body is proportional to its mass and, similarly, the acceleration imparted to a body is proportional to the applied force.

Thus force is proportional to the product of mass and acceleration (Newton's law), or:

$$\mathbf{F} = \text{const}\ \mathbf{M}\left(\mathbf{LT}^{-2}\right) \tag{1.1}$$

The proportionality constant therefore has the dimensions:

$$\frac{\mathbf{F}}{\mathbf{M}\left(\mathbf{LT}^{-2}\right)} = \mathbf{FM}^{-1}\mathbf{L}^{-1}\mathbf{T}^{2} \tag{1.2}$$

In any set of *consistent* or *coherent* units, the proportionality constant in Eq. (1.1) is considered equal to unity, and unit force is that force which will impart unit acceleration to unit mass. When no other relationship between force and mass is used, the constant may be arbitrarily regarded as dimensionless and the dimensional relationship:

$$\mathbf{F} = \mathbf{MLT}^{-2} \tag{1.3}$$

is obtained.

If, however, some other physical law were to be introduced such that, for instance, the attractive force between two bodies would be proportional to the product of their masses, then this relation between **F** and **M** would no longer hold. It should be noted here that mass has essentially two connotations. First, it is a measure of the amount of material and appears in this role when the density of a fluid or solid is considered. Second, it is a measure of the inertia of the material when used, for example, in Eqs (1.1)–(1.3). Although mass is normally taken as the third fundamental quantity, as already mentioned, in some engineering systems force is used in place of mass, which then becomes a derived unit.

1.2 Systems of Units

Although in scientific work mass is taken as the third fundamental quantity and in engineering force is sometimes used as mentioned above, the fundamental quantities **L**, **M**, **F**, and **T** may be used interchangeably. A summary of the various systems of units and the quantities associated with them, is given in Table 1.1. In the cgs system, which has historically been used for

Table 1.1 Units in different systems

Quantity	Cgs	SI	fps	Dimensions in M, L, T, θ	Engineering system	Dimensions F, L, T, θ	Dimensions in F, M, L, T, θ
Mass	gram	kilogram	pound	M	slug	$FL^{-1}T^2$	M
Length	centimetre	metre	foot	L	foot	L	L
Time	second	second	second	T	second	T	T
Force	dyne	Newton	poundal	MLT^{-2}	pound force	F	F
Energy	erg (= 10^{-7} Joules)	Joule	foot-poundal	ML^2T^{-2}	foot-pound	FL	FL
Pressure	dyne/square centimetre	Newton/sq metre or Pascal, Pa	poundal/ square foot	$ML^{-1}T^{-2}$	pound force/ square foot	FL^{-2}	FL^{-2}
Power	erg/second	Watt	foot-poundal/ second	ML^2T^{-3}	foot-pound/ second	FLT^{-1}	FLT^{-1}
Entropy per unit mass	erg/gram °C	Joule/kilogram K	foot-poundal/ pound °C	$L^2T^{-2}\theta^{-1}$	foot-pound/ slug °F	$L^2T^{-2}\theta^{-1}$	$FM^{-1}L\theta^{-1}$
Universal gas constant	8.314×10^7 erg/mol °C	8314 J/kmol K	8.94 ft-poundal/ lb mol °C	$MN^{-1}L^2T^{-2}\theta^{-1}$	4.96×10^4 foot-pound/ slug mol °F	$MN^{-1}L^2T^{-2}\theta^{-1}$	$FN^{-1}L\theta^{-1}$

Heat Units

Quantity	Cgs	SI	British/American Engineering System	Dimensions in M, L, T, θ	Dimensions in H, M, L, T, θ
Temperature	degree centigrade	degree Kelvin	degree Fahrenheit	θ	θ
Thermal energy or heat	calorie	Joule	British thermal unit (Btu)	$M\theta$	H
Entropy per unit mass, specific heat	calorie/gram °C	Joule/kilogram K	Btu/pound °F	–	$HM^{-1}\theta^{-1}$
Mechanical equivalent of heat, J	4.18×10^7 erg/gram-°C	1 J (heat energy) = 1 J (mechanical energy)	2.50×10^4 foot-poundal/pound °F	$L^2T^{-2}\theta^{-1}$	$H^{-1}ML^2T^{-2}$
Universal gas constant **R**	1.986 calorie/mol °C	8314 J/kmol K	1.986 Btu/lb-mol °F	$MN^{-1}L^2T^{-2}\theta^{-1}$	$HN^{-1}\theta^{-1}$

scientific work, metric units are employed. From this developed the mks system which employs larger units of mass and length (kilogram in place of gram, and metre in place of centimetre); this system has been favoured by electrical engineers because the fundamental and the practical electrical units (volt, ampere and ohm) are then identical. The SI system is essentially based on the mks system of units.

1.2.1 The Centimetre-Gram-Second (cgs) System

In this system, the basic units are of length **L**, mass **M**, and time **T** with the nomenclature:

Length:	Dimension **L**:	Unit 1 centimetre	(1 cm)
Mass:	Dimension **M**:	Unit 1 gram	(1 g)
Time:	Dimension **T**:	Unit 1 second	(1 s)

The unit of force is defined as that force which will give a mass of 1 g an acceleration of 1 cm/s^2 and is known as the dyne:

Force:	Dimension $\mathbf{F} = \mathbf{MLT}^{-2}$:	Unit	1 dyne (1 dyn)
Energy:	Dimensions $\mathbf{ML}^2\mathbf{T}^{-2}$	Unit	1 erg
Power:	Dimensions $\mathbf{ML}^2\mathbf{T}^{-3}$	Unit	1 erg/s

1.2.2 The Metre–Kilogram–Second (mks) System and the Système International d'Unités (SI)

These systems are in essence modifications of the cgs system but employ larger units. The basic dimensions are again of **L**, **M**, and **T**.

Length:	Dimension **L**:	Unit 1 metre	(1 m)
Mass:	Dimension **M**:	Unit 1 kilogram	(1 kg)
Time:	Dimension **T**:	Unit 1 second	(1 s)

The unit of force, known as the *Newton*, is that force which will give an acceleration of 1 m/s^2 to a mass of one kilogram. Thus 1 N = 1 kg m/s^2 with dimensions $\mathbf{MLT}^{-2}$, and one Newton equals 10^5 dynes. The energy unit, the Newton-metre, is 10^7 ergs and is called the *Joule*; while the power unit, equal to one Joule per second, is known as the *Watt*.

Thus:	Force:	Dimensions $\mathbf{MLT}^{-2}$:	Unit 1 Newton (1 N)	or 1 kg m/s^2
	Energy:	Dimensions $\mathbf{ML}^2\mathbf{T}^{-2}$:	Unit 1 Joule (1 J)	or 1 kg m^2/s^2
	Power:	Dimensions $\mathbf{ML}^2\mathbf{T}^{-3}$:	Unit 1 Watt (1 W)	or 1 kg m^2/s^3

In many cases, the chosen unit in the SI system will be practically either too large or too small, and the following prefixes are adopted as standard. Multiples or submultiples in powers of 10^3 are preferred and thus, for example, millimetre should always be used in preference to centimetre.

10^{18}	exa	(E)	10^{-1}	Deci	(d)
10^{15}	peta	(P)	10^{-2}	centi	(c)
10^{12}	tera	(T)	10^{-3}	milli	(m)
10^{9}	giga	(G)	10^{-6}	micro	(μ)
10^{6}	mega	(M)	10^{-9}	nano	(n)
10^{3}	kilo	(k)	10^{-12}	Pico	(p)
10^{2}	hecto	(h)	10^{-15}	femto	(f)
10^{1}	deca	(da)	10^{-18}	Alto	(a)

These prefixes should be used with great care and be written immediately adjacent to the unit to be valid; furthermore only one prefix should be used at a time to precede a given unit. Thus, for example, 10^{-3} m, which is one millimetre, is written 1 mm. 10^3 kg is written as 1 Mg, not as 1 kkg. This immediately shows that the name *kilogram* is an unsuitable choice for the basic unit of mass and therefore a new name may well be given to it in the future.

Some special terms are acceptable and commonly used in the SI system, for example, a mass of 10^3 kg (1 Mg) is called a *tonne* (t); and a pressure of 100 kN/m^2 is called a *bar.*

The most important practical difference between the mks and the SI systems lies in the units used for thermal energy (heat), and this topic is discussed in Section 1.2.7.

A detailed account of the structure and implementation of the SI system is given in publications of the British Standards Institution,[1] and of Her Majesty's Stationery Office.[2]

1.2.3 The Foot–Pound–Second (fps) System

The basic units in this system are:

Length:	Dimension **L**:	Unit 1 foot	(1 ft)
Mass:	Dimension **M**:	Unit 1 pound	(1 lb)
Time:	Dimension **T**:	Unit 1 second	(1 s)

The unit of force that gives a mass of 1 lb an acceleration of 1 ft/s^2 is known as the poundal (pdl).

The unit of energy (or work) is the foot-poundal, and the unit of power is the foot-poundal per second.

Thus:	Force	Dimensions $\mathbf{MLT}^{-2}$	Unit	1 poundal (1 pdl)
	Energy	Dimensions $\mathbf{ML}^2\mathbf{T}^{-2}$	Unit	1 ft-poundal
	Power	Dimensions $\mathbf{ML}^2\mathbf{T}^{-3}$	Unit	1 foot-poundal/s

1.2.4 The British Engineering System

In an alternative form of the fps system (*Engineering system*), the units of length (ft) and time (s) are unchanged, but the third fundamental is a unit of force (**F**) instead of mass and is known as the pound force (lb_f). This is defined as the force that gives a mass of 1 lb an acceleration of 32.1740 ft/s^2, the 'standard' value of the acceleration due to gravity. It is therefore a fixed quantity and must not be confused with the pound weight which is the force exerted by the earth's gravitational field on a mass of one pound and which varies from place to place as the value of g varies. It will be noted therefore that the pound force and the pound weight have the same value only when g is 32.1740 ft^2/s.

The unit of mass in this system is known as the slug, and it is the mass which is given an acceleration of 1 ft/s^2 by a one pound force:

$$1\,\text{slug} = 1\,\text{lb}_\text{f}\,\text{ft}^{-1}\,\text{s}^2$$

Misunderstanding often arises from the fact that the pound, which is the unit of mass in the fps system has the same name as the unit of force in the engineering system. To avoid confusion, the pound mass should be written as lb or even lb_m and the unit of force is always written as lb_f.

It will be noted that:

$$1\,\text{slug} = 32.1740\text{lbmass} \quad \text{and} \quad 1\,\text{lb}_\text{f} = 32.1740\text{pdl}$$

To summarise:

The basic units are:

Length	Dimension **L**	Unit 1 foot (1 ft)
Force	Dimension **F**	Unit 1 pound-force (1 lb_f)
Time	Dimension **T**	Unit 1 second (1 s)

The derived units are:

Mass	Dimensions $\mathbf{FL^{-1}T^{-2}}$	Unit 1 slug (=32.1740 pounds)
Energy	Dimensions **FL**	Unit 1 foot pound-force (1 ft lb_f)
Power	Dimensions $\mathbf{FLT^{-1}}$	Unit 1 foot-pound force/s (1 ft-lb_f/s)
		Note: 1 horsepower is defined as 550 ft-lb_f/s

1.2.5 Noncoherent System Employing Pound Mass and Pound Force Simultaneously

Two units, which have never been popular in the last two systems of units (Sections 1.2.3 and 1.2.4) are the poundal (for force) and the slug (for mass). As a result, many writers, particularly in America, use both the pound mass and pound force as basic units in the same equation because they are the units in common use. This is an essentially an incoherent system and requires great care when using it. In this system a proportionality factor between force and mass is defined as g_c given by:

$$\text{Force (in pounds force)} = (\text{mass in pounds})\left(\text{acceleration in ft/s}^2\right)/g_c$$

Thus in terms of dimensions:

$$\mathbf{F} = (\mathbf{M})\left(\mathbf{LT}^{-2}\right)/g_c \tag{1.4}$$

From Eq. (1.4), it is seen that g_c has the dimensions $\mathbf{F}^{-1}\mathbf{MLT}^{-2}$ or, putting $\mathbf{F} = \mathbf{MLT}^{-2}$, it is seen to be dimensionless. Thus:

$$g_c = 32.1740\,\text{lb}_\text{f}/\left(\text{lb}_\text{m}\,\text{ft s}^{-2}\right)$$

or:

$$g_c = \frac{32.1740\,\text{ft s}^{-2}}{1\,\text{ft s}^{-2}} = 32.1740$$

i.e. g_c is a dimensionless quantity whose numerical value corresponds to the acceleration due to gravity expressed in the appropriate units.

(It should be noted that a force in the cgs system is sometimes expressed as a *gram force* and in the mks system as *kilogram force*, although this is not a good practice. It should also be noted that the gram force = 980.665 dyne and the kilogram force = 9.80665 N)

1.2.6 Derived Units

The three fundamental units of the SI and of the cgs systems are length, mass, and time. It has been shown that force can be regarded as having the dimensions of $\mathbf{MLT}^{-2}$, and the dimensions of many other parameters may be worked out in terms of the basic **MLT** system.

For example:

> Energy is given as the product of force and distance with dimensions $\mathbf{ML}^2\mathbf{T}^{-2}$, and pressure is the force per unit area with dimensions $\mathbf{ML}^{-1}\mathbf{T}^{-2}$.
> Viscosity is defined as the shear stress per unit velocity gradient with dimensions $\left(\mathbf{MLT}^{-2}/\mathbf{L}^2\right)/\left(\mathbf{LT}^{-1}/\mathbf{L}\right) = \mathbf{ML}^{-1}\mathbf{T}^{-1}$.
> And kinematic viscosity is the viscosity divided by the density with dimensions $\mathbf{ML}^{-1}\mathbf{T}^{-1}/\mathbf{ML}^{-3} = \mathbf{L}^2\mathbf{T}^{-1}$.

The units, dimensions, and normal form of expression for these quantities in the SI system are:

Quantity	Unit	Dimensions	Units in kg, m, s
Force	Newton	$\mathbf{MLT}^{-2}$	1 kg m/s^2
Energy or work	Joule	$\mathbf{ML}^2\mathbf{T}^{-2}$	1 kg m^2/s^2 (=1 N m = 1 J)

Power	Watt	$\mathbf{ML^{2}T^{-3}}$	1 kg m^2/s^3 (=1 J/s)
Pressure	Pascal	$\mathbf{ML^{-1}T^{-2}}$	1 kg/m s^2 (=1 N/m^2)
Viscosity	Pascal-second	$\mathbf{ML^{-1}T^{-1}}$	1 kg/m s (=1 N s/m^2)
Frequency	Hertz	$\mathbf{T^{-1}}$	1 s^{-1}

1.2.7 Thermal (Heat) Units

Heat is a form of energy and therefore its dimensions are $\mathbf{ML^{2}T^{-2}}$. In many cases, however, no account is taken of interconversion of heat and 'mechanical' energy (for example, potential and kinetic energy), and heat can be treated as a quantity which is conserved. It may then be regarded as having its own independent dimension **H** which can be used as an additional fundamental. It will be seen in Section 1.4 on dimensional analysis that increasing the number of fundamental units by one leads to an additional relation and consequently to one less dimensionless group.

Wherever heat is involved, temperature also fulfils an important role: firstly because the heat content of a body is a function of its temperature and, secondly, because temperature difference or temperature gradient determines the rate at which heat is transferred from one point to another. Temperature has the dimension θ which is independent of **M**,**L**, and **T**, provided that no resort is made to the kinetic theory of gases in which temperature is shown to be directly proportional to the square of the velocity of the molecules.

It is not *incorrect* to express heat and temperature in terms of the **M**,**L**, and **T** dimensions, although it is unhelpful in that it prevents the maximum of information being extracted from the process of dimensional analysis and reduces the insight that it affords into the physical nature of the process under consideration.

Dimensionally, the relation between **H**, **M**, and θ can be expressed in the form:

$$\mathbf{H} \propto \mathbf{M}\theta = C_p\mathbf{M}\theta \tag{1.5}$$

where C_p, the specific heat capacity has dimensions $\mathbf{HM}^{-1}\theta^{-1}$.

Eq. (1.5) is similar in nature to the relationship between force, mass, and acceleration given by Eq. (1.1) with one important exception. The proportionality constant in Eq. (1.1) is not a function of the material concerned and it has been possible arbitrarily to put it equal to unity. The constant in Eq. (1.5), the specific heat capacity C_p, differs from one material to another, and, in some cases, it varies with temperature (θ) *itself.*

In the SI system, the unit of heat is taken similar to that of mechanical energy and is therefore the Joule. For water at 298 K (the datum used for many definitions), the specific heat capacity C_p is 4186.8 J/kg K.

Prior to the almost universal adoption of the SI system of units now, the unit of heat was defined as the quantity of heat required to raise the temperature of unit mass of water by one degree. This heat quantity is designated as the calorie in the cgs system and kilocalorie in the mks system, and in both cases temperature is expressed in degrees Celsius (Centigrade). As the specific heat capacity is a function of temperature, it has been necessary to set a datum temperature that is chosen as 298 K or 25°C.

In the British system of units, the pound, but never the slug, taken as the unit of mass and temperature may be expressed either in degrees Centigrade or in degrees Fahrenheit. The units of heat are then, respectively, the *pound-calorie* and the *British thermal unit* (Btu). Where the Btu is too small for a given application, the *therm* ($=10^5$ Btu) is normally used.

Thus the following definitions of heat quantities apply:

System	Mass Unit	Temperature Scale (Degrees)	Unit of Heat
cgs	gram	Celsius	calorie
mks	kilogram	Celsius	kilocalorie
fps	pound	Celsius	pound calorie or Centigrade heat unit (CHU)
fps	pound	Fahrenheit	British thermal unit (Btu) 1 CHU = 1.8 Btu

In all of these systems, by definition, the specific heat capacity of water is unity. It may be noted that, by comparing the definitions used in the SI and the mks systems, the kilocalorie is equivalent to 4186.8 J/kg K. This quantity has often been referred to as the *mechanical equivalent of heat J*.

1.2.8 Molar Units

When working with ideal gases and systems in which a chemical reaction is taking place, it is usual to work in terms of *molar* units rather than mass. The *mole* (*mol*) is defined in the SI system as the quantity of material that contains as many entities (atoms, molecules or formula units) as there are in 12 g of carbon 12. It is more convenient, however, to work in terms of the *kilomole* (*kmol*) which relates to 12 kg of carbon 12, and the kilomole is used exclusively in this book. The number of molar units is denoted by dimensional symbol **N**. The number of kilomoles of a substance **A** is obtained by dividing its mass in kilograms (**M**) by its *molecular weight*, M_A. M_A thus has the dimensions $\mathbf{MN}^{-1}$. The Royal Society recommends the use of the term *relative molecular mass* instead of *molecular weight*, but *molecular weight* is normally used here because of its general adoption in the literature and in the processing industries.

1.2.9 Electrical Units

Electrical current (I) has been chosen as the basic SI unit in terms of which all other electrical quantities are defined. Unit current, the *ampere* (A, or amp), is defined as the force exerted between two parallel conductors in which a current of 1 amp is flowing. Since the unit of power, the Watt, is the product of current and potential difference, the *volt* (V) is defined as watts per amp and therefore has dimensions of $\mathbf{M\,L^2T^{-3}I^{-1}}$. From Ohm's law, the unit of resistance, the *ohm*, is given by the ratio volts/amps and therefore has dimensions of $\mathbf{M\,L^2T^{-3}I^{-2}}$. A similar procedure may be followed for the evaluation of the dimensions of other electrical units.

1.3 Conversion of Units

Conversion of units from one system to another is simply carried out if the quantities are expressed in terms of the fundamental units of mass, length, time, and temperature. Typical conversion factors for the British and metric systems are:

Mass	$1\,\text{lb} = \left(\frac{1}{32.2}\right)\,\text{slug} = 453.6\,\text{g} = 0.4536\,\text{kg}$
Length	$1\,\text{ft} = 30.48\,\text{cm} = 0.3048\,\text{m}$
Time	$1\,\text{s} = \left(\frac{1}{3600}\right)\,\text{h}$
Temperature difference	$1\,°\text{F} = \left(\frac{1}{1.8}\right)\,°\text{C} = \left(\frac{1}{1.8}\right)\,\text{K (or degK)}$
Force	$1\,\text{pound force} = 32.2\,\text{poundal} = 4.44 \times 10^5\,\text{dyne} = 4.44\,\text{N}$

Other conversions are now illustrated.

Example 1.1

Convert 1 poise to British Engineering units and SI units.

Solution

$$
\begin{aligned}
1\,\text{Poise} = 1\,\text{g/cm s} &= \frac{1\,\text{g}}{1\,\text{cm} \times 1\,\text{s}} \\
&= \frac{(1/453.6)\,\text{lb}}{(1/30.48)\,\text{ft} \times 1\,\text{s}} \\
&= 0.0672\,\text{lb/ft s} \\
&= \underline{\underline{242\,\text{lb/ft h}}}
\end{aligned}
$$

$$
\begin{aligned}
1\,\text{Poise} = 1\,\text{g/cm s} &= \frac{1\,\text{g}}{1\,\text{cm} \times 1\,\text{s}} \\
&= \frac{(1/1000)\,\text{kg}}{(1/100)\,\text{m} \times 1\,\text{s}} \\
&= 0.1\,\text{kg/m s} \\
&= \underline{\underline{0.1\,\text{N s/m}^2}}\left[(\text{kg m/s}^2)\,\text{s/m}^2\right] \text{ or } 0.1\,\text{Pa s}
\end{aligned}
$$

Example 1.2

Convert 1 kW to h.p.

Solution

$$1\,\text{kW} = 10^3\,\text{W} = 10^3\,\text{J/s}$$

$$= 10^3 \times \left(\frac{1\,\text{kg} \times 1\,\text{m}^2}{1\,\text{s}^3}\right)$$

$$= \frac{10^3 \times (1/0.4536)\,\text{lb} \times (1/0.3048)^2\,\text{ft}^2}{1\,\text{s}^3}$$

$$= 23{,}730\,\text{lb ft}^2/\text{s}^3$$

$$= \left(\frac{23{,}730}{32.2}\right) = 737\,\text{slug ft}^2/\text{s}^3$$

$$= 737\,\text{lb}_f\,\text{ft/s}$$

$$= \left(\frac{737}{550}\right) = \underline{\underline{1.34\,\text{h.p.}}}$$

or:

1 h.p. = 0.746 kW.

Conversion factors to SI units from other units are given in Table 1.2 which is based on a publication by Mullin.[3]

In closing this subsection, it is worthwhile to emphasise here the fact that while an engineering problem may be described in terms of mixed systems of units, it is perhaps better to convert all known quantities into a consistent system of units to facilitate obtaining the solution to the problem in an unambiguous manner. The familiar Hazen-Williams equation relating the pressure drop-flow rate of water in a pipe is one such example. It is written as:

$$\frac{\Delta p}{l} = 4.52\left(\frac{Q^{1.85}}{C}\right)d^{-4.87} \tag{1.6}$$

In Eq. (1.6): $(\Delta p/l)$ is in Psi/100 ft of pipe length; Q is the volumetric flow rate in gallons (US) per minute; C is a dimensionless roughness factor; and d is pipe internal diameter in inches! The numerical constant of 4.52 is thus not unitless and its value will change from one system of units to another. Therefore, it is perhaps best to rewrite Eq. (1.6) in a consistent set of units.

Table 1.2 Conversion factors for some common SI units[4]

Length	*1 in.	: 25.4 mm
	*1 ft	: 0.304,8 m
	*1 yd	: 0.914,4 m
	1 mile	: 1.609,3 km
	* 1 Å (angstrom)	: 10^{-10} m
Time	*1 min	: 60 s
	*1 h	: 3.6 ks
	*1 day	: 86.4 ks
	1 year	: 31.5 Ms
Area	*1 in.^{2}	: 645.16 mm^2
	1 ft^2	: 0.092,903 m^2
	1 yd^2	: 0.836,13 m^2
	1 acre	: 4046.9 m^2
	1 mile^2	: 2.590 km^2
Volume	1 in.^{3}	: 16.387 cm^3
	1 ft^3	: 0.028,32 m^3
	1 yd^3	: 0.764,53 m^3
	1 UK gal	: 4546.1 cm^3
	1 US gal	: 3785.4 cm^3
Mass	1 oz	: 28.352 g
	*1 lb	: 0.453,592,37 kg
	1 cwt	: 50.802,3 kg
	1 ton	: 1016.06 kg
Force	1 pdl	: 0.138,26 N
	1 lbf	: 4.448,2 N
	1 kgf	: 9.806,7 N
	1 tonf	: 9.964,0 kN
	*1 dyn	: 10^{-5} N
Temperature difference	*1 deg F (deg R)	: $\frac{5}{9}$deg C (deg K)
Energy (work, heat)	1 ft lbf	: 1.355,8 J
	1 ft pdl	: 0.042,14 J
	* 1 cal (international table)	: 4.186,8 J
	1 erg	: 10^{-7} J
	1 Btu	: 1.055,06 kJ
	1 hp h	: 2.684,5 MJ
	*1 kW h	: 3.6 MJ
	1 therm	: 105.51 MJ
	1 thermie	: 4.185,5 MJ
Calorific value (volumetric)	1 Btu/ft^3	: 37.259 kJ/m^3
Velocity	1 ft/s	: 0.304,8 m/s
	1 mile/h	: 0.447,04 m/s
Volumetric flow	1 ft^3/s	: 0.028,316 m^3/s
	1 ft^3/h	: 7.865,8 cm^3/s
	1 UK gal/h	: 1.262,8 cm^3/s
	1 US gal/h	: 1.051,5 cm^3/s
Mass flow	1 lb/h	: 0.126,00 g/s
	1 ton/h	: 0.282,24 kg/s
Mass per unit area	1 lb/in.^{2}	: 703.07 kg/m^2
	1 lb/ft^2	: 4.882,4 kg/m^2
	1 ton/sq mile	: 392.30 kg/km^2

Table 1.2 Conversion factors for some common SI units—Cont'd

Density	1 lb/in^3	: 27.680 g/cm^3
	1 lb/ft^3	: 16.019 kg/m^3
	1 lb/UK gal	: 99.776 kg/m^3
	1 lb/US gal	: 119.83 kg/m^3
Pressure	1 lbf/in.^{2}	: 6.894,8 kN/m^2
	1 tonf/in.^{2}	: 15.444 MN/m^2
	1 lbf/ft^2	: 47.880 N/m^2
	* 1 standard atm	: 101.325 kN/m^2
	*1 atm (1 kgf/cm^2)	: 98.066,5 kN/m^2
	*1 bar	: 10^5 N/m^2
	1 ft water	: 2.989,1 kN/m^2
	1 in. water	: 249.09 N/m^2
	1 in. Hg	: 3.386,4 kN/m^2
	1 mm Hg (1 torr)	: 133.32 N/m^2
Power (heat flow)	1 hp (British)	: 745.70 W
	1 hp (metric)	: 735.50 W
	1 erg/s	: 10^{-7} W
	1 ft lbf/s	: 1.355,8 W
	1 Btu/h	: 0.293,07 W
	1 ton of refrigeration	: 3516.9 W
Moment of inertia	1 lb ft^2	: 0.042,140 kg m^2
Momentum	1 lb ft/s	: 0.138,26 kg m/s
Angular momentum	1 lb ft^2/s	: 0.042,140 kg m^2/s
Viscosity, dynamic	*1 P (Poise)	: 0.1 N s/m^2
	1 lb/ft h	: 0.413,38 mN s/m^2
	1 lb/ft s	: 1.488,2 Ns/m^2
Viscosity, kinematic	*1 S (Stokes)	: 10^{-4} m^2/s
	1 ft^2/h	: 0.258,06 cm^2/s
Surface energy	1 erg/cm^2	: 10^{-3} J/m^2
Surface tension	1 dyn/cm	: 10^{-3} N/m
Mass flux density	1 lb/h ft^2	: 1.356,2 g/s m^2
Heat flux density	1 Btu/h ft^2	: 3.154,6 W/m^2
	*1 kcal/h m^2	: 1.163 W/m^2
Heat transfer coefficient	1 Btu/h ft^2°F	: 5.678,3 W/m^2 K
Specific enthalpy (latent heat, etc.)	*1 Btu/lb	: 2.326 kJ/kg
Specific heat capacity	*1 Btu/lb °F	: 4.186,8 kJ/kg K
Thermal conductivity	1 Btu/h ft °F	: 1.730,7 W/m K
	1 kcal/h m °C	: 1.163 W/m K

An asterisk * denotes an exact relationship.

1.4 Dimensional Analysis

Dimensional analysis depends upon the fundamental principle that any equation or relation between variables must be *dimensionally consistent*; that is, each term in the relationship must have the same dimensions. Thus, in the simple application of the principle, an equation may consist of a number of terms, each representing, and therefore having, the dimensions of length. It is not permissible to add, say, lengths and velocities in an algebraic equation because they are quantities of different attributes. The corollary of this principle is that if the whole equation is divided through by any one of the terms, each remaining term in the equation must be dimensionless. The use of these *dimensionless groups*, or *dimensionless numbers* as they are called, is of considerable value in developing relationships in chemical engineering, which are applicable in any consistent system of units.

The requirement of dimensional consistency places a number of constraints on the form of the functional relation between variables in a problem and forms the basis of the technique of *dimensional analysis*, which enables the variables in a problem to be grouped into the form of dimensionless groups. Since the dimensions of the physical quantities may be expressed in terms of a number of fundamentals, usually mass, length, time, and sometimes temperature and thermal energy, the requirement of dimensional consistency must be satisfied with respect to each of the fundamentals. Dimensional analysis gives no information about the form of the functions, nor does it provide any means of evaluating numerical proportionality constants.

The study of problems in fluid dynamics and in heat transfer is made difficult by the many parameters, which appear to affect them. In most instances further study shows that the variables may be grouped together in dimensionless groups, thus reducing the effective number of variables. It is rarely possible, and certainly time consuming, to try to vary these many variables separately, and the method of dimensional analysis in providing a smaller number of independent groups is most helpful to the investigated. The application of the principles of dimensional analysis may best be understood by considering an example.

It is found, as a result of experiments, that the pressure difference (ΔP) between two ends of a pipe in which a fluid is flowing is a function of the pipe diameter d, the pipe length l, the fluid velocity u, the fluid density ρ, and the fluid viscosity μ.

The relationship between these variables may be written as:

$$\Delta P == f_1(d, l, u, \rho, \mu) \tag{1.7}$$

The form of the function is unknown, though since any function can be expanded as a power series, the function may be regarded as the sum of a number of terms each consisting of products of powers of the variables. The simplest form of relation will be where the function consists simply of a single term, or:

$$\Delta P = \text{const } d^{n_1} l^{n_2} u^{n_3} \rho^{n_4} \mu^{n_5} \tag{1.8}$$

The requirement of dimensional consistency is that the combined term on the right-hand side will have the same dimensions as that on the left; that is, it must have the dimensions of pressure.

Each of the variables in Eq. (1.8) may be expressed in terms of mass, length, and time. Thus, dimensionally:

$$\Delta P \equiv \mathbf{ML^{-1}T^{-2}} \qquad u \equiv \mathbf{LT^{-1}}$$
$$d \equiv \mathbf{L} \qquad \rho \equiv \mathbf{ML^{-3}}$$
$$l \equiv \mathbf{L} \qquad \mu \equiv \mathbf{ML^{-1}T^{-1}}$$

and: $\mathbf{ML^{-1}T^{-2}} = \mathbf{L}^{n_1}\mathbf{L}^{n_2}(\mathbf{LT^{-1}})^{n_3}(\mathbf{ML^{-3}})^{n_4}(\mathbf{ML^{-1}T^{-1}})^{n_5}$

The conditions of dimensional consistency must be met for each of the fundamentals of **M**, **L**, and **T** and the indices of each of these variables may be equated. Thus:

$$\begin{aligned} \mathbf{M}&: \quad 1 = n_4 + n_5 \\ \mathbf{L}&: \quad -1 = n_1 + n_2 + n_3 - 3n_4 - n_5 \\ \mathbf{T}&: \quad -2 = -n_3 - n_5 \end{aligned}$$

Three equations and five unknowns result and the equations may be solved in terms of any two unknowns. Solving in terms of ft n_2 and n_5:

$$n_4 = 1 - n_5 \text{ (from the equation in } \mathbf{M})$$
$$n_3 = 2 - n_5 \text{ (from the equation in } \mathbf{T})$$

Substituting in the equation for **L**:

$$-1 = n_1 + n_2 + (2 - n_5) - 3(1 - n_5) - n_5$$

or:

$$0 = n_1 + n_2 + n_5$$

and:

$$n_1 = -n_2 - n_5$$

Thus, substituting into Eq. (1.8):

$$\Delta P = \text{const } d^{-n_2-n_5} l^{n_2} u^{2-n_5} \rho^{1-n_5} \mu^{n_5}$$

Slight arrangement yields:

$$\frac{\Delta P}{\rho u^2} = \text{const} \left(\frac{l}{d}\right)^{n_2} \left(\frac{\mu}{du\rho}\right)^{n_5} \tag{1.9}$$

Since n_2 and n_5 are arbitrary constants, this equation can only be satisfied if each of the terms $\Delta P/\rho u^2$, l/d, and $\mu/du\rho$ is dimensionless. Evaluating the dimensions of each of these groups shows that this is, in fact, the case.

The group $ud\rho/\mu$, known as the *Reynolds number*, is one which frequently arises in the study of fluid flow and affords a criterion by which the type of flow in a given geometry may be characterised. Eq. (1.9) involves the reciprocal of the Reynolds number, although this may be rewritten as:

$$\frac{\Delta P}{\rho u^2} = \text{const}\left(\frac{l}{d}\right)^{n_2}\left(\frac{ud\rho}{\mu}\right)^{-n_5} \tag{1.10}$$

The right-hand side of Eq. (1.10) is a typical term in the function for $\Delta P/\rho u^2$. More generally:

$$\frac{\Delta P}{\rho u^2} = f_2\left(\frac{1}{d}, \frac{ud\rho}{\mu}\right) \tag{1.11}$$

Comparing Eqs (1.7) and (1.11), it is seen that a relationship between six variables has been reduced to a relationship between three dimensionless groups. In subsequent sections of this chapter, this statement will be generalised to show that the number of dimensionless groups is *normally* the number of variables less the number of fundamentals (but see the note in Section 1.5).

A number of important points emerge from a consideration of the preceding example:

1. If the index of a particular variable is found to be zero, this indicates that this variable is of no significance in the problem.
2. If two of the fundamentals always appear in the same combination, such as **L** and **T** always occurring as powers of $\mathbf{LT}^{-1}$, for example, then the same equation for the indices will be obtained for both **L** and **T** and the number of effective fundamentals is thus reduced by one.
3. The form of the final solution will depend upon the method of solution of the simultaneous equations. If the equations had been solved, say, in terms of n_3 and n_4 instead of n_2 and n_5, the resulting dimensionless groups would have been different, although these new groups would simply have been products of powers of the original groups. Any number of fresh groups can be formed in this way.

Clearly, the maximum degree of simplification of the problem is achieved by using the greatest possible number of fundamentals since each yields a simultaneous equation of its own. In certain problems, force may be used as a fundamental in addition to mass, length, and time, provided that at no stage in the problem is force defined in terms of mass and acceleration. In heat transfer problems, temperature is usually an additional fundamental, and heat can also be used as a fundamental provided it is not defined in terms of mass and temperature and provided that the equivalence of mechanical and thermal energy is not utilised. Considerable experience is needed in the proper use of dimensional analysis, and its application in a number of areas of fluid flow and heat transfer is seen in the relevant chapters of this volume.

The choice of physical variables to be included in the dimensional analysis must be based on an understanding of the nature of the phenomenon being studied although, on occasions there may be some doubts as to whether a particular quantity is relevant or not.

If a variable is included which does not exert a significant influence on the problem, the value of the dimensionless group in which it appears will have little effect on the final numerical solution of the problem, and therefore the exponent of that group must approach zero. This presupposes that the dimensionless groups are so constituted that the variable in question appears in only one of them. On the other hand if an important variable is omitted, it may be found that there is no unique relationship between the dimensionless groups.

Chemical engineering analysis requires the formulation of relationships, which will apply over a wide range of size of the individual items of a plant. This problem of scale up is vital and is much helped by dimensional analysis.

Since linear size is included among the variables, the influence of scale, which may be regarded as the influence of linear size without change of shape or other variables, has been introduced. Thus in viscous flow past an object, a change in linear dimension **L** will alter the Reynolds number and therefore the flow pattern around the solid, however if the change in scale is accompanied by a change in any other variable in such a way that the Reynolds number remains unchanged, then the flow pattern around the solid will not be altered. This ability to change scale and still maintain a design relationship is one of the many attractions of dimensional analysis.

It should be noted here that it is permissible to take a function only of a dimensionless quantity. It is easy to appreciate this argument when the fact is considered that any function may be expanded as a power series, each term of which must have the same dimensions, and the requirement of dimensional consistency can be met only if these terms and the function are dimensionless. Where this principle appears to have been invalidated, it is generally because the equation includes a further term, such as an integration constant, which will restore the requirements of dimensional consistency. For example, $\int_{x_0}^{x} \frac{dx}{x} = \ln x - \ln x_0$, and if x is not dimensionless, it appears at first sight that the principle has been infringed. Combining the two logarithmic terms, however, yields $\ln\left(\frac{x}{x_0}\right)$, and $\frac{x}{x_0}$ is clearly dimensionless. In the case of the indefinite integral, $\ln x_o$ would, in effect, have been the integration constant.

1.5 Buckingham's Π Theorem

The need for dimensional consistency imposes a restraint with respect to each of the fundamentals involved in the dimensions of the variables. This is apparent from the previous discussion in which a series of simultaneous equations was solved, one equation for each of the fundamentals. A generalisation of this statement is provided in Buckingham's Π theorem[4] which states that the number of dimensionless groups is equal to the number of variables minus the number of fundamental dimensions. In mathematical terms, this can be expressed as follows:

If there are n variables, Q_1, Q_2, ..., Q_n, the functional relationship between them may be written as:

$$f_3(Q_1, Q_2, \ldots, Q_n) = 0 \quad (1.12)$$

If there are m fundamental dimensions, there will be $(n-m)$ dimensionless groups (Π_1, Π_2, ..., Π_{n-m}) and the functional relationship between them may be written as:

$$f_4(\Pi_1, \Pi_2, \ldots, \Pi_{n-m}) = 0 \quad (1.13)$$

The groups Π_1, Π_2, and so on must be independent of one another, and no one group should be capable of being formed by multiplying together powers of the other groups.

By making use of this theorem, it is possible to obtain the dimensionless groups more easily than by solving the simultaneous equations for the indices. Furthermore, the functional relationship can often be obtained in a form, which is of more immediate use.

The method involves choosing m of the original variables to form what is called a *recurring set.* Any set m of the variables may be chosen with the following two provisions:

(1) Each of the fundamentals must appear in at least one of the m variables.
(2) It must not be possible to form a dimensionless group from some or all of the variables within the recurring set. If it were so possible, this dimensionless group would, of course, be one of the Π terms. Thus, the number of dimensionless groups is increased by one for each of the independent groups that can be so formed.

The procedure is then to take each of the remaining $(n-m)$ variables on its own and to form it into a dimensionless group by combining it with one or more members of the recurring set. In this way the $(n-m)$ Π groups are formed, the only variables appearing in more than one group being those that constitute the recurring set. Thus, if it is desired to obtain an explicit functional relation for one particular variable, that variable should not be included in the recurring set.

In some cases, the number of dimensionless groups will be greater than predicted by the Π theorem. For instance, if two of the fundamentals always occur in the same combination, length and time always as $\mathbf{LT}^{-1}$, for example, they will constitute a single fundamental instead of two fundamentals. By referring back to the method of equating indices, it is seen that each of the two fundamentals gives the same equation, and therefore only a single constraint is placed on the relationship by considering the two variables. Thus, although m is normally the number of fundamentals, it is more strictly defined as the *maximum number of variables from which a dimensionless group cannot be formed.*

The procedure is more readily understood by consideration of the illustration given previously. The relationship between the variables affecting the pressure drop for flow of fluid in a pipe may be written as:

$$f_5(\Delta P, d, l, \rho, \mu, u) = 0 \quad (1.14)$$

Eq. (1.14) includes six variables, and three fundamental quantities (mass, length, and time). Thus:

$$\text{Number of dimensionless groups} = (6-3) = 3$$

The recurring set must contain three variables that cannot themselves be formed into a dimensionless group and must include all fundamental units. This imposes the following two restrictions:

(1) Both l and d cannot be chosen as they can be formed into the dimensionless group l/d.
(2) ΔP, ρ, and u cannot be used since $\Delta P/\rho u^2$ is dimensionless.

Outside these constraints, any three variables can be chosen. It should be remembered, however, that the variables forming the recurring set will appear in all the dimensionless groups. As this problem deals with the effect of conditions on the pressure difference ΔP, it is convenient if ΔP appears in only one group, and therefore it is preferable not to include it in the recurring set.

If the variables d, u, and ρ are chosen as the recurring set, this fulfils all the above conditions. Dimensionally:

$$d \equiv \mathbf{L}$$
$$u \equiv \mathbf{LT}^{-1}$$
$$\rho \equiv \mathbf{ML}^{-3}$$

Each of the dimensions **M**, **L**, and **T** may then be obtained explicitly in terms of the variables d, u, and ρ, to give:

$$\mathbf{L} \equiv d$$
$$\mathbf{M} \equiv \rho d^3$$
$$\mathbf{T} \equiv du^{-1}$$

The three dimensionless groups are thus obtained by taking each of the remaining variables ΔP, l, and μ in turn.

ΔP has dimensions $\mathbf{ML}^{-1}\mathbf{T}^{-2}$, and $\Delta P\mathbf{M}^{-1}\mathbf{LT}^{2}$ is therefore dimensionless. Now substituting for **M**, **L**, and **T** in terms of d, u, and s:

Group Π_1 is therefore,

$$\Delta P\left(\rho d^3\right)^{-1}(d)\left(du^{-1}\right)^2 = \frac{\Delta P}{\rho u^2}$$

l has dimensions **L**, and $l\mathbf{L}^{-1}$ is therefore dimensionless.

Group Π_2 is therefore:

$$l\left(d^{-1}\right) = \frac{l}{d}$$

μ has dimensions $\mathbf{ML}^{-1}\mathbf{T}^{-1}$, and $\mu\mathbf{M}^{-1}\mathbf{LT}$ is therefore dimensionless.

Group Π_3 is, therefore:

$$\mu\left(\rho d^3\right)^{-1}(d)\left(du^{-1}\right)=\frac{\mu}{du\rho}$$

Thus:

$$f_6\left(\frac{\Delta P}{\rho u^2},\frac{l}{d},\frac{\mu}{ud\rho}\right)=0 \quad \text{or} \quad \frac{\Delta P}{\rho u^2}=f_7\left(\frac{l}{d},\frac{ud\rho}{\mu}\right)$$

$\mu/ud\rho$ is arbitrarily inverted because the Reynolds number is usually expressed in the form $ud\rho/\mu$.

Some of the important dimensionless groups used in Chemical Engineering applications are listed in Table 1.3 along with the relevant applications.

Table 1.3 Some important dimensionless groups

Symbol	Name of Group	In Terms of Other Groups	Definition	Application
Ar	Archimedes	Ga	$\frac{\rho(\rho_s-\rho)gd^3}{\mu^2}$	Gravitational settling of particles in fluids
Bn	Bingham number		$\frac{R_Y d}{\mu u}$	Flow of yield stress fluids
Db	Deborah		$\frac{t_P}{t_F}$	Flow of viscoelastic fluids
Eu	Euler		$\frac{P}{\rho u^2}$	Pressure and momentum in fluid
Fo	Fourier		$\frac{D_H t}{l^2}, \frac{Dt}{l^2}$	Unsteady state heat/mass transfer
Fr	Froude		$\frac{u^2}{gl}$	Fluid flow with free surface
Ga	Galileo	Ar	$\frac{\rho(\rho_s-\rho)gd^3}{\mu^2}$	Gravitational settling of particles in fluids
Gr	Grashof		$\frac{l^3\rho^2\beta g\Delta T}{\mu^2}$	Heat transfer by natural convection
Gz	Graetz		$\frac{GC_p}{kl}$	Heat transfer to fluid in tube
He	Hedström	$Re\cdot Bn$	$\frac{R_Y\rho d^2}{\mu_p^2}$	Flow of fluid exhibiting yield stress
Le	Lewis	$Sc\cdot Pr^{-1}$	$\frac{k}{C_p\rho D}=\frac{D_H}{D}$	Simultaneous heat and mass transfer
Ma	Mach		$\frac{u}{u_w}$	Gas flow at high velocity (compressible flow)
Nu	Nusselt		$\frac{hl}{k}$	Heat transfer in fluid
Pe	Peclet	$Re\cdot Pr$	$\frac{ul}{D_H}$	Fluid flow and heat transfer
		$Re\cdot Sc$	$\frac{ul}{D}$	Fluid flow and mass transfer
Pr	Prandtl		$\frac{C_p\mu}{k}$	Heat transfer in flowing fluid
Re	Reynolds		$\frac{ul\rho}{\mu}$	Fluid flow involving viscous and inertial forces
Sc	Schmidt		$\frac{\mu}{\rho D}$	Mass transfer in flowing fluid
Sh	Sherwood		$\frac{h_D l}{D}$	Mass transfer in fluid
St	Stanton	$Nu\cdot Pr^{-1}\cdot Re^{-1}$	$\frac{h}{C_p\rho u}$	Heat transfer in flowing fluid
We	Weber		$\frac{\rho u^2 l}{\sigma}$	Fluid flow with interfacial forces
ϕ	Friction factor		$\frac{R}{\rho u^2}$	Fluid drag at surface
N_p	Power number		$\frac{\mathbf{P}}{\rho N^3 d^5}$	Power consumption for mixers

Example 1.3

A glass particle settles under the action of gravity in a liquid. Obtain a dimensionless grouping of the variables involved. The falling velocity is found to be proportional to the square of the particle diameter when the other variables are constant. What would be the effect on the falling velocity of doubling the viscosity of the liquid?

Solution

It may be expected that the variables likely to influence the terminal velocity of a glass particle settling in a liquid, u_0, are:

particle diameter, d; particle density, ρ_s; liquid density, ρ; liquid viscosity, μ and the acceleration due to gravity, g.

Particle density ρ_s is important because it determines the gravitational (accelerating) force on the particle, i.e. weight of the particle itself. However when immersed in a liquid the particle receives an upthrust which is proportional to the liquid density ρ. The effective density of the particles $(\rho_s - \rho)$ is therefore used in this analysis. However, only two out of these three ρ, ρ_s, and $(\rho_s - \rho)$ are independent variables. Then:

$$u_0 = f(d, (\rho_s - \rho), \rho, \mu, g)$$

The dimensions of each variable are:

$$u_0 = \mathbf{LT}^{-1},\ d = \mathbf{L},\ \rho_s - \rho = \mathbf{ML}^{-3},\ \rho = \mathbf{ML}^{-3},$$
$$\mu = \mathbf{ML}^{-1}\mathbf{T}^{-1} \text{ and } g = \mathbf{LT}^{-2}.$$

With six variables and three fundamental dimensions, $(6-3)=3$ dimensionless groups are expected. Choosing d, ρ, and μ as the recurring set:

$$\begin{array}{ll} d \equiv \mathbf{L} & \mathbf{L} = d \\ \rho \equiv \mathbf{ML}^{-3} & \mathbf{M} = \rho\mathbf{L}^3 = \rho d^3 \\ \mu \equiv \mathbf{ML}^{-1}\mathbf{T}^{-1} & \mathbf{T} = \mathbf{M}/\mu L = \rho d^3/(\mu d) = \rho d^2/\mu \end{array}$$

Thus:

$$\begin{array}{ll} \text{dimensionless group 1:} & u_0\mathbf{TL}^{-1} = u_0\rho d^2/(\mu d) = u_0\rho d/\mu \\ \text{dimensionless group 2:} & (\rho_s - \rho)\mathbf{L}^3\mathbf{M}^{-1} = \rho_s d^3/(\rho d^3) = (\rho_s - \rho)/\rho \\ \text{dimensionless group 3:} & g\mathbf{T}^2\mathbf{L}^{-1} = g\rho^2 d^4/(\mu^2 d) = g\rho^2 d^3/\mu^2 \end{array}$$

and:

$$(u_0\rho d/\mu) \propto ((\rho_s - \rho)/\rho)(g\rho^2 d^3/\mu^2)$$

or:

$$(u_0\rho d/\mu) = \kappa((\rho_s - \rho)/\rho)^{n_1}(g\rho^2 d^3/\mu^2)^{n_2}$$

when $u_0 \propto d^2$, when $(3n_2 - 1) = 2$ and $n_2 = 1$.

Thus:

$$(u_0\rho d/\mu) = \kappa((\rho_s - \rho)/\rho)^{n_1}(g\rho^2 d^3/\mu^2)$$

or:

$$u_0 = \kappa((\rho_s - \rho)/\rho)^{n_1}(d^2\rho g/\mu)$$

and:

$$u_0 \propto (1/\mu)$$

In this case, doubling the viscosity of the liquid will *halve the terminal velocity of the particle*, suggesting that the flow is in the Stokes' law regime. Note that since we are interested in u_0 as the dependent variable here, it is not prudent to include it in the recurring set.

Example 1.4

A drop of liquid spreads over a horizontal surface. Obtain dimensionless groups of the variables that will influence the rate at which the liquid spreads.

Solution

The rate at which a drop spreads, say u_R m/s, will be influenced by:

$$\begin{aligned}
&\text{viscosity of the liquid, } \mu \text{ — dimensions} = \mathbf{ML^{-1}T^{-1}}\\
&\text{volume of the drop, } V \text{ — dimensions} = \mathbf{L^3}\\
&\text{density of the liquid, } \rho \text{ — dimensions} = \mathbf{ML^{-3}}\\
&\text{acceleration due to gravity, } g \text{ — dimensions} = \mathbf{LT^{-2}}.
\end{aligned}$$

and possibly, surface tension of the liquid, σ–dimensions $= \mathbf{MT^{-2}}$.

Noting the dimensions of u_R as $\mathbf{LT^{-1}}$, there are six variables and hence $(6-3)=3$ dimensionless groups. Taking V, ρ, and g as the recurring set:

$$\begin{aligned}
&V \equiv \mathbf{L}^3 \text{ and } \mathbf{L} = V^{1/3}\\
&\rho \equiv \mathbf{ML}^{-3} \text{ and } \mathbf{M} = \rho\mathbf{L}^3 = \rho V\\
&g \equiv \mathbf{LT}^{-2} \text{ and } \mathbf{T}^2 = \mathbf{L}/g \text{ or } \mathbf{T} = V^{1/6}/g^{1/2}
\end{aligned}$$

Thus:

$$\begin{aligned}
&\text{dimensionless group 1:} \quad u_R\mathbf{TL}^{-1} = u_R V^{1/6}/(V^{1/3}g^{1/2}) = u_R/(V^{1/3}g)^{1/2}\\
&\text{dimensionless group 2:} \quad \mu\mathbf{LTM}^{-1} = \mu V^{1/3}(V^{1/6}/g^{1/2})(\rho V)^{-1} = \mu/(\rho g^{1/2}V^{1/2})\\
&\text{dimensionless group 3:} \quad \sigma\mathbf{T}^2\mathbf{M}^{-1} = \sigma(V^{1/3}/g)/(\rho V) = \sigma/(g\rho V^{2/3})
\end{aligned}$$

and:

$$u_R/\left(V^{1/3}g\right)^{1/2} \propto \left(\mu/\left(g^{1/2}\rho V^{1/2}\right)\right)/\left(\sigma/\left(g\rho V^{2/3}\right)\right)$$

or:

$$\underline{\underline{u_R^2/V^{1/3}g = \kappa(\mu^2/g\rho^2 V)^{n_1}\left(\sigma/g\rho V^{2/3}\right)^{n_2}}}$$

The values of k, n, and n can only be determined either by experiments or by other insights developed using theoretical analysis.

1.6 Scale Up

In addition to the application of dimensional analysis in reducing the number of variables and/ or help in planning the experiments, analysis and representation of data in a cogent manner, it also provides the so-called scaling laws that allow us to use data from small-scale laboratory tests to design large prototype systems. The history of scaled-model testing in fluid mechanics dates back to the nineteenth century when Froude used a water basin to design ship hulls for the British admiralty. Similarly, at the beginning of the 20th century, the Wright brothers exploited the systematic use of wind tunnels to design the *kitty hawk* and the *flyers*. These and other similar studies firmly established the principles of scale model testing. The basic ideas are presented here whereas more detailed treatments are available in many excellent books (see further reading list)

Most engineering systems are not convenient for their study at the prototype (real) level because of their size (too small/big), too fast/slow, or due to the other practical difficulties. The scaled-models are the model replicas proportioned after the original system in such a manner that the quantitative behaviour of the prototype can be inferred from that of a suitably designed scaled model. This process hinges on the concept of similarity. There are generally three types of similarities in fluid mechanics applications: The *geometric* similarity implies that the ratios of all corresponding linear dimensions in the model and prototype are equal, the so-called scale factor. The *kinematic* similarity exists in two geometrically similar systems when the velocities, acceleration, at the corresponding points in the model and prototype are in a constant ratio. Also, the paths of fluid motion (flow patterns) must be similar. The two geometrically and kinematically similar systems are said to be *dynamically* similar if all corresponding forces (pressure, inertial, viscous, etc.) at homologous points in the two systems bear a constant ratio. Indeed, the dimensionless numbers like Froude, Weber, Euler, Reynolds (listed in Table 1.3) can be interpreted to be the ratio of forces. Thus, for instance, the Reynolds number can easily be seen to be the ratio of the inertial to viscous forces; The Froude number being the ratio of the inertial to gravity forces, etc. It thus stands to reason that if the overall behaviour of a system can be described in terms of the pertinent dimensionless groups, this relationship is expected to be independent of the equipment scale as long as the two systems are similar. Conversely, the equality of the relevant dimensionless groups in the scaled model and prototype device ensures that the results of lab tests can be scaled up with confidence to design the large-scale equipment. This principle is illustrated in Example 1.5.

Example 1.5

It is desired to estimate the force on the propeller blade of a small airplane by constructing a 1:10 scale model for laboratory-studies. If the laboratory tests are to be conducted in air, i.e. the working fluid is the same for both the scaled model and prototype, calculate the ratio of the forces on the propeller blade and that of the velocities of the airplane in the model and the prototype. How will your answer change if the model studies are performed in water instead of air?

Solution

First, we need to list the variables likely to influence the force, F, exerted on the blade. These are:

$$\begin{aligned}&\text{forward velocity of propeller (speed of plane)}:\ u(\mathbf{LT}^{-1})\\&\text{diameter of propeller}:\ d(\mathbf{L})\\&\text{density of air}:\ \rho(\mathbf{ML}^{-3})\\&\text{viscosity of air}:\ \mu(\mathbf{ML}^{-1}\mathbf{T}^{-1})\\&\text{rotational speed of propeller}:\ N(\mathbf{T}^{-1})\end{aligned}$$

Hence, this functional relationship can be written as:

$$F=f(u,d,\rho,\mu,N)$$

The Buckingham π-theorem yields the number of dimensionless groups as $6-3=3$. Choosing ρ, N, and d as the recurring set, we obtain $\mathbf{L}\equiv d$, $\mathbf{M}\equiv\rho d^3$ and $\mathbf{T}\equiv N^{-1}$.

Thus,

$$\text{dimensionless group 1}:\ F\mathbf{M}^{-1}\mathbf{L}^{-1}\mathbf{T}^2=F(\rho d^3)^{-1}(d)^{-1}(N^{-1})^2=\frac{F}{\rho d^4N^2}$$

$$\text{dimensionless group 2}:\ u\mathbf{L}^{-1}\mathbf{T}=ud^{-1}N^{-1}=\frac{u}{Nd}$$

$$\text{dimensionless group 3}:\ \mu\mathbf{M}^{-1}\mathbf{LT}=\mu(\rho d^3)^{-1}(d)(N^{-1})=\frac{\mu}{\rho d^2N}$$

$$\therefore\quad \frac{F}{\rho d^4N^2}=\phi\left(\frac{u}{Nd},\frac{\rho d^2N}{\mu}\right)$$

Therefore, the similarity between the two systems demands that the values of $\frac{u}{Nd}$ and $\frac{\rho d^2N}{\mu}$ be identical, i.e.

$$\left.\frac{u}{Nd}\right|_p=\left.\frac{u}{Nd}\right|_m\quad\text{and}\quad\left.\frac{\rho d^2N}{\mu}\right|_p=\left.\frac{\rho d^2N}{\mu}\right|_m$$

Once this requirement is satisfied, the values of the third dimensionless group in the two facilities must be equal, i.e.

$$\left.\frac{F}{\rho d^4N^2}\right|_p=\left.\frac{F}{\rho d^4N^2}\right|_m$$

Since the working fluid is air in both situations, i.e. $\rho_p=\rho_m$ and $\mu_p=\mu_m$, the ratio of the forces on the propeller is given by:

$$\frac{F_p}{F_m}=\left(\frac{d_p}{d_m}\right)^4\left(\frac{N_p}{N_m}\right)^2$$

Here, $\frac{d_m}{d_p}=\frac{1}{10}$ and rearranging the equality of Reynolds numbers as:

$$(d^2N)_p=(d^2N)_m$$

Now equating the other dimensionless group:

$$\left.\frac{u}{Nd}\right|_p = \left.\frac{u}{Nd}\right|_m \quad \text{or} \quad \frac{u_p}{u_m} = \frac{Nd|_p}{Nd|_m} = \frac{d_m}{d_p} = \frac{1}{10}$$

Now calculating the ratio of forces to get:

$$\frac{F_p}{F_m} = \left(\frac{d_p^2 N_p}{d_m^2 N_m}\right)^2 = 1, \quad \text{i.e.,} \; F_p = F_m$$

This result is surprising to say the least! However, the prototype device will have velocity, which is only one tenth of the lab-scale study!

Lab Tests in Water

$$\rho_m = 1000\,\text{kg/m}^3; \;\; \mu_m = 10^{-3}\,\text{Pas}$$
$$\rho_p = 1.25\,\text{kg/m}^3; \;\; \mu_p = 1.8 \times 10^{-5}\,\text{Pas}$$

Equating the two Reynolds numbers and forward (dimensionless) velocity yields:

$$\left.\frac{\rho d^2 N}{\mu}\right|_p = \left.\frac{\rho d^2 N}{\mu}\right|_m \quad \text{and} \quad \left.\frac{u}{Nd}\right|_m = \left.\frac{u}{Nd}\right|_p$$

Substituting values and solving:

$$\frac{N_p}{N_m} = 0.144 \quad \text{and} \quad \frac{u_p}{u_m} = 1.44$$

Now calculating the ratio of the forces:

$$\frac{F_p}{F_m} = \left(\frac{\rho_p}{\rho_m}\right)\left(\frac{d_p}{d_m}\right)^4\left(\frac{N_p}{N_m}\right)^2$$

Substituting values:

$$\frac{F_p}{F_m} = \left(\frac{1.25}{1000}\right)(10)^4(0.144)^2 = 0.26$$

In this case, the force in the prototype will be one fourth of that in the model while the velocity in the prototype will be about 70% of that in the model study.

In actual practice, it is often not possible to achieve complete similarity (due to limitations on material properties, operating conditions, etc.) and one must therefore match the important dimensionless groups in order to use the so-called partial or distorted similarity. Additional difficulties may arise while dealing with heat transfer, chemical reactions, and structured materials in terms of obtaining the so-called '*chemical or material*' similarity.[5]

1.7 Redefinition of the Length and Mass Dimensions

1.7.1 Vector and Scalar Quantities

It is important to recognise the differences between *scalar* quantities, which have a magnitude but no direction, and vector quantities, which have both magnitude and direction. Most length terms are vectors in the Cartesian system and may have components in the *X*, *Y*, and *Z* directions which may be expressed as $\mathbf{L_X}$, $\mathbf{L_Y}$, and $\mathbf{L_Z}$. There must be dimensional consistency in all equations and relationships between physical quantities, and therefore there is the possibility of using all three length dimensions as fundamentals in dimensional analysis. This means that the number of dimensionless groups that are formed will be less.

Combinations of length dimensions in areas, such as $\mathbf{L_X L_Y}$, and velocities, accelerations, and forces are all *vector* quantities. On the other hand, mass, volume, and heat are all scalar quantities with no directional significance. The power of dimensional analysis is thus increased as a result of the larger number of fundamentals which are available for use. Furthermore, by expressing the length dimension as a vector quantity, it is possible to obviate the difficulty of two quite different quantities having the same dimensions. For example, the units of *work* or *energy* may be obtained by multiplying a force in the *X*-direction (say) by a distance also in the *X*-direction. The dimensions of energy are therefore:

$$\left(\mathbf{M L_X T^{-2}}\right)\left(\mathbf{L_X}\right) = \mathbf{M L_X^2 T^{-2}}$$

It should be noted in this respect that a *torque* is obtained as a product of a force in the *X*-direction and an arm of length $\mathbf{L_Y}$, say, in a direction at right-angle to the *Y*-direction. Thus, the dimensions of torque are $\mathbf{M L_X L_Y T^{-2}}$, which distinguish it from energy.

Another benefit arising from the use of vector lengths is the ability to differentiate between the dimensions of frequency and angular velocity, both of which are $\mathbf{T^{-1}}$ if length is treated as a scalar quantity. Although an angle is dimensionless in the sense that it can be defined by the ratio of two lengths, its dimensions become $\mathbf{L_X/L_Y}$ if these two lengths are treated as vectors. Thus angular velocity then has the dimensions $\mathbf{L_X L_Y^{-1} T^{-1}}$ compared with $\mathbf{T^{-1}}$ for frequency.

Of particular interest in fluid flow is the distinction between shear stress and pressure (or pressure difference), both of which are defined as force per unit area. For steady-state flow of a fluid in a pipe, the forces attributable to the pressure difference and the shear stress must balance. The pressure difference acts in the axial *X*-direction, say, and the area *A* on which it acts lies in the *Y*–*Z* plane and its dimensions can therefore be expressed as $\mathbf{L_Y L_Z}$. On the other hand, the shear stress *R* which is also exerted in the *X*-direction acts on the curved surface of the walls whose area *S* has the dimensions $\mathbf{L_X L_R}$ where $\mathbf{L_R}$ is the length in the radial direction. Because there is axial symmetry, $\mathbf{L_R}$ can be expressed as $\mathbf{L_Y^{1/2} L_Z^{1/2}}$ and the dimensions of *S* are then $\mathbf{L_X L_Y^{1/2} L_Z^{1/2}}$.

The force F acting on the fluid in the X (axial)-direction has dimensions $\mathbf{ML_XT^{-2}}$, and hence:

$$\Delta P = F/A \text{ has dimensions } \mathbf{ML_XT^{-2}/L_YL_Z = ML_XL_Y^{-1}L_Z^{-1}T^{-2}}$$

and

$$R = F/S \text{ has dimensions } \mathbf{ML_XT^{-2}/L_XL_Y^{1/2}L_Z^{1/2} = ML_Y^{-1/2}L_Z^{-1/2}T^{-2}}$$

giving dimensions of $\Delta P/R$ as $\mathbf{L_XL_Y^{-1/2}L_Z^{-1/2}}$ or $\mathbf{L_XL_R^{-1}}$ (which would have been dimensionless had lengths not been treated as vectors).

For a pipe of radius r and length l, the dimensions of r/l are $\mathbf{L_X^{-1}L_R}$ and hence $(\Delta P/R)\ (r/l)$ is a dimensionless quantity. The role of the ratio r/l would not have been established had the lengths not been treated as vectors. It is seen in Chapter 3 that this conclusion is consistent with the results obtained there by taking a force balance on the fluid.

1.7.2 Quantity Mass and Inertia Mass

The term 'mass **M**' is used to denote two distinct and different properties:

1 The quantity of matter $\mathbf{M}_\mu$, and
2 The inertial property of the matter $\mathbf{M}_i$.

These two quantities are proportional to one another and may be numerically equal, although they are essentially different in their physical nature and are therefore not identical. The distinction is particularly useful when considering the energy of a body or of a fluid.

Because inertial mass is involved in mechanical energy, the dimensions of all energy terms are $\mathbf{M_iL^2T^{-2}}$. Inertial mass, however, is not involved in thermal energy (heat) and therefore specific heat capacity C_p has the dimensions $\mathbf{M_iL^2T^2/M_\mu}\theta = \mathbf{M_iM_\mu^{-1}L^2T^{-2}}\theta^{-1}$ or $\mathbf{HM_\mu^{-1}}\theta^{-1}$ according to whether energy is expressed in, joules or kilocalories, for example.

In practical terms, this can lead to the possibility of using both mass dimensions as fundamentals, thereby achieving similar advantages to those arising from taking length as a vector quantity. This subject is discussed in more detail by Huntley.[6]

Warning

Dimensional analysis is a very powerful tool in the analysis of problems involving a large number of variables. However, there are many pitfalls for the unwary, and the technique should never be used without a thorough understanding of the underlying basic principles of the physical problem that is being analysed.

1.8 Nomenclature

		Units in SI System	Dimensions in M, N, L, T, θ, H, I
C_P	Specific heat capacity at constant pressure	J/kg K	$\mathbf{L^2T^{-2}\theta^{-1}}$ $\mathbf{(HM^{-1}\theta^{-1})}$
D	Diffusion coefficient, molecular diffusivity	m^2/s	$\mathbf{L^2T^{-1}}$
D_H	Thermal diffusivity $k/C_p\rho$	m^2/s	$\mathbf{L^2T^{-1}}$
d	Diameter	m	**L**
f	A function	–	–
G	Mass rate of flow	kg/s	$\mathbf{MT^{-1}}$
g	Acceleration due to gravity	m/s^2	$\mathbf{LT^{-2}}$
g_c	Numerical constant equal to standard value of 'g'	–	–
h	Heat transfer coefficient	W/m^2 K	$\mathbf{MT^{-3}\theta^{-1}}$
h_D	Mass transfer coefficient	m/s	$\mathbf{LT^{-1}}$
I	Electric current	A	**I**
J	Mechanical equivalent of heat	–	–
k	Thermal conductivity	W/m K	$\mathbf{MLT^{-3}\theta^{-1}}$
l	Characteristic length or length of pipe	m	**L**
M_A	Molecular weight (relative molecular mass) of A	kg/kmol	$\mathbf{MN^{-1}}$
m	Number of fundamental dimensions	–	–
N	Rotational speed	s^{-1}	$\mathbf{T^{-1}}$
n	Number of variables	–	–
P	Power	W	$\mathbf{ML^2T^{-3}}$
P	Pressure	N/m^2	$\mathbf{ML^{-1}T^{-2}}$
$\triangle P$	Pressure difference	N/m^2	$\mathbf{ML^{-1}T^{-2}}$
Q	Physical quantity	–	–
R	Shear stress	N/m^2	$\mathbf{ML^{-1}T^{-2}}$
R_y	Yield stress	N/m^2	$\mathbf{ML^{-1}T^{-2}}$
R	Universal gas constant	8314 J/kmol K	$\mathbf{MN^{-1}L^2T^{-2}\theta^{-1}}$ $\mathbf{(HN^{-1}\theta^{-1})}$
r	Pipe radius	m	**L**
$\triangle T$	Temperature difference	K	$\boldsymbol{\theta}$
t	Time	s	**T**
t_F	Characteristic time for fluid	s	**T**

t_P	Characteristic time for process	s	$\mathbf{T}$
u	Velocity	m/s	$\mathbf{LT^{-1}}$
u_w	Velocity of a pressure wave	m/s	$\mathbf{LT^{-1}}$
V	Potential difference	V	$\mathbf{ML^2T^{-3}I^{-1}}$
β	Coefficient of cubical expansion of fluid	K^{-1}	$\boldsymbol{\theta}^{-1}$
Π	A dimensionless group	–	–
μ	Viscosity	N s/m^2	$\mathbf{ML^{-1}T^{-1}}$
μ_p	Plastic viscosity	N s/m^2	$\mathbf{ML^{-1}T^{-1}}$
ρ	Density of fluid	kg/m^3	$\mathbf{ML^{-3}}$
ρ_s	Density of solid	kg/m^3	$\mathbf{ML^{-3}}$
σ	Surface or interfacial tension	N/m	$\mathbf{MT^{-2}}$
Ω	Electrical resistance	Ohm	$\mathbf{ML^2T^{-3}I^{-2}}$

Dimensions

H	Heat		
I	Electric current	Amp	I
L	Length		
$\mathbf{L_X}$ $\mathbf{L_Y}$ $\mathbf{L_Z}$	Length vectors in *X–Y–Z* directions		
M	Mass		
$\mathbf{M_i}$	Inertial mass		
$\mathbf{M}_\mu$	Quantity mass		
N	Moles		
T	Time		
$\boldsymbol{\theta}$	Temperature		

References

1 British Standards Institution Publication PD 5686. *The use of SI units*; 1967.
2 *SI. The International System of Units*. HMSO; 1970.
3 Mullin JW. SI units in chemical engineering. *Chem Eng (Lond)* 1967;**211**:176.
4 Buckingham E. On physically similar systems: illustrations of the use of dimensional equations. *Phys Rev Ser* 1914;**2**(4):345.
5 Delaplace G, Loubiere K, Ducept F, Jeantet R. *Dimensional analysis of food processes*. London: Elsevier; 2015.
6 Huntley HE. *Dimensional analysis*. London: Macdonald and Co (Publishers) Ltd.; 1952

Further reading

1 Astarita G. Dimensional analysis, scaling and orders of magnitude. *Chem Eng Sci* 1997;**52**:4681.
2 Blackman DR. *SI units in engineering*. Melbourne: Macmillan; 1969.
3 Bridgman PW. *Dimensional analysis*. New Haven: Yale University Press; 1931.
4 Focken CM. *Dimensional methods and their applications*. London: Edward Arnold; 1953.

5 Gibbings JC. *Dimensional analysis.* New York: Springer; 2011.
6 Hewitt GF. *Proceedings of the 11th international heat transfer conference, Kyongju, Korea*; 1998. p. 1.
7 Ipsen DC. *Units, dimensions, and dimensionless numbers.* New York: McGraw-Hill; 1960.
8 Johnstone RE, Thring MW. *Pilot plants, models, and scale-up in chemical engineering.* New York: McGraw-Hill; 1957.
9 Klinkenberg A, Mooy HH. Dimensionless groups in fluid friction, heat and material transfer. *Chem Eng Prog* 1948;**44**:17.
10 Massey BS. *Units, dimensional analysis and physical similarity.* London: van Nostrand Reinhold; 1971.
11 Mullin JW. Recent developments in the change-over to the International System of Units (SI). *Chem Eng (Lond)* 1971;**254**:352.
12 Palmer AC. *Dimensional analysis and intelligent experimentation.* Singapore: World Scientific; 2008.
13 Szirtes T. *Applied dimensional analysis and modelling.* 2nd ed. Oxford: Butterworth-Heinemann; 2006.
14 Zohuri B. *Dimensional analysis beyond the Pi theorem.* New York: Springer; 2017.
15 Quantities, units and symbols. The Symbols Committee of the Royal Society; 1971.

CHAPTER 2

Flow of Fluids—Energy and Momentum Relationships

2.1 Introduction

Chemical and process engineers are interested in several aspects of problems associated with the flow of fluids. In the first place, as is common with many other engineers, they are concerned with the transport of fluids from one location to another through pipes or open ducts, which requires the determination of the pressure drops in the system, and hence of the power required for pumping, selection of the most suitable type of pump, and measurement of the flow rates. In many cases, the fluid contains solid particles in suspension and it is necessary to determine the effect of these particles on the flow characteristics of the fluid or, alternatively, the drag force exerted by the fluid on the particles. In some cases, such as filtration, the particles are in the form of a fairly stable bed and the fluid has to pass through the tortuous channels formed by the pore spaces. In other cases, the shape of the boundary surfaces must be so arranged that a particular flow pattern is obtained: for example, when solids are maintained in suspension in a liquid by means of agitation, the desired effect can be obtained with the minimum expenditure of energy as the most suitable flow pattern is produced in the fluid. Further, in those processes where heat transfer or mass transfer to a flowing fluid occurs, the nature of the flow may have a profound effect on the heat/mass transfer coefficient for the process.

It is necessary to be able to calculate the energy and momentum of a fluid at various positions in a flow system. It will be seen that energy occurs in a number of forms and that some of these are influenced by the motion of the fluid. In the first part of this chapter, the thermodynamic properties of fluids will be discussed. It will then be seen how the thermodynamic relations are modified if the fluid is in motion. In later chapters, the effects of frictional forces will be considered, and the principal methods of measuring flow will be described.

2.2 Internal Energy

Generally, when a fluid flows from one location to another, energy will be converted from one form to another. The energy, which is attributable to the physical state of the fluid is known as internal energy; it is arbitrarily taken as zero at some reference state, such as at the absolute zero of temperature or the melting point of ice at atmospheric pressure. A change in

Coulson and Richardson's Chemical Engineering. https://doi.org/10.1016/B978-0-08-101099-0.00002-1

the physical state of a fluid will, in general, cause an alteration in the internal energy. An elementary reversible change results from an infinitesimal change in one of the intensive factors acting on the system; the change proceeds at an infinitesimal rate and a small change in the intensive factor in the opposite direction would have caused the process to take place in the reverse direction. Truly reversible changes denote idealisations which never occur in practice but they provide a useful standard with which actual processes can be compared. In an irreversible process, changes are caused by a finite difference in the intensive factor and take place at a finite rate. In general, the process will be accompanied by the conversion of electrical or mechanical energy into heat, or by the reduction of the temperature difference between different parts of the system.

For a stationary material, the change in the internal energy is equal to the difference between the net amount of heat added to the system and the net amount of work done by the system on its surroundings. For an infinitesimal change:

$$dU = \delta q - \delta W \tag{2.1}$$

where dU is the small change in the internal energy, δq is the small amount of heat added, and δW is the net amount of work done on the surroundings.

In this expression, consistent units must be used. In the SI system, each of the terms in Eq. (2.1) is expressed in Joules per kilogram (J/kg). In other systems, either heat units (e.g. cal/g) or mechanical energy units (e.g. erg/g) may be used, dU is a small change in the internal energy, which is a property of the system; it is therefore a perfect differential. On the other hand, δq and δW are small quantities of heat and work; they are not properties of the system and their values depend on the manner in which the change is effected; they are, therefore, path functions and not perfect differentials. For a reversible process, however, both δq and δW can be expressed in terms of the properties of the system. For convenience, reference will be made to systems of unit mass and the effects on the surroundings will be disregarded.

A property called entropy is defined by the relation:

$$dS = \frac{\delta q}{T} \tag{2.2}$$

where dS is the small change in entropy resulting from the addition of a small quantity of heat δq, at a temperature T, under reversible conditions. From the definition of the thermodynamic scale of temperature, $\oint \delta q/T = 0$ for a reversible cyclic process, the net change in the entropy is also zero. Thus, for a particular condition of the system, the entropy has a definite value and must be a property of the system; dS is, therefore, a perfect differential.

For an irreversible process:

$$\frac{\delta q}{T} < dS = \frac{\delta q}{T} + \frac{\delta F}{T} \quad \text{(say)} \tag{2.3}$$

δF is then a measure of the degree of irreversibility of the process. It represents the amount of mechanical energy converted into heat or the conversion of heat energy at one temperature to heat energy at another temperature. For a finite process:

$$\int_{S_1}^{S_2} T\,\mathrm{d}S = \sum \delta q + \sum \delta F = q + F \quad \text{(say)} \tag{2.4}$$

When a process is isentropic, $q = -F$; a reversible process is isentropic when $q = 0$, that is, a reversible adiabatic process is isentropic.

The increase in the entropy of an irreversible process may be illustrated in the following manner. Considering the spontaneous transfer of a quantity of heat δq from one part of a system at a temperature T_1 to another part of the system at a temperature T_2, the net change in the entropy of the system as a whole is given as:

$$\mathrm{d}S = \frac{\delta q}{T_2} - \frac{\delta q}{T_1}$$

T_1 must be greater than T_2 and $\mathrm{d}S$ is therefore positive. If the process had been carried out reversibly, there would have been an infinitesimal difference between T_1 and T_2 and the change in entropy would have been zero.

The change in the internal energy may be expressed in terms of the properties of the system itself. For a reversible process:

$$\delta q = T\,\mathrm{d}S \quad \text{(from Eq. 2.2)} \quad \text{and} \quad \delta W = P\,\mathrm{d}v$$

if the only work done is that resulting from a change in volume, dv.

Thus, from Eq. (2.1):

$$\mathrm{d}U = T\,\mathrm{d}S - P\,\mathrm{d}v \tag{2.5}$$

Since this relation is in terms of the intrinsic properties of the system, it must also apply to a system in motion and to irreversible changes where the only work done is the result of change of volume.

Thus, in an irreversible process, for a stationary system:

From Eqs (2.1), (2.2):

$$\mathrm{d}U = \delta q - \delta W = T\,\mathrm{d}S - P\,\mathrm{d}v$$

and from Eq. (2.3):

$$\begin{aligned} \delta q + \delta F &= T\,\mathrm{d}S \\ \delta W &= P\,\mathrm{d}v - \delta F \end{aligned} \tag{2.6}$$

that is, the useful work performed by the system is less than $P\,dv$ by an amount δF, which represents the amount of mechanical energy converted irreversibly into heat energy.

The relation between the internal energy and the temperature of a fluid will now be considered. In a system consisting of unit mass of material and where the only work done is that resulting from volume change, the change in internal energy after a reversible change is given by:

$$dU = \delta q - P\,dv \text{ (from Eq. 2.1)}$$

If there is no volume change:

$$dU = \delta q = C_v dT \tag{2.7}$$

where C_v is the specific heat at constant volume.

As this relation is in terms of the properties of the system, it must be applicable to all changes at constant volume.

In an irreversible process:

$$\begin{aligned} dU &= \delta q - (P\,dv - \delta F) \quad \text{(from Eqs 2.1, 2.6)} \\ &= \delta q + \delta F \quad \text{(under conditions of constant volume)} \end{aligned} \tag{2.8}$$

This quantity δF thus represents the mechanical energy which has been converted into heat and which is therefore available for increasing the temperature.

$$\delta q + \delta F = C_v dT = dU \tag{2.9}$$

For changes that take place under conditions of constant pressure, it is more satisfactory to consider variations in the enthalpy H. The enthalpy is defined by the relation:

$$H = U + Pv \tag{2.10}$$

Thus:

$$\begin{aligned} dH &= dU + P\,dv + v\,dP \\ &= \delta q - P\,dv + \delta F + P\,dv + v\,dP \quad \text{(from Eq. 2.8)} \end{aligned}$$

for an irreversible process: (For a reversible process $\delta F = 0$)

$$\begin{aligned} \therefore\ dH &= \delta q + \delta F + v\,dP \\ &= \delta q + \delta F \quad \text{(at constant pressure)} \end{aligned} \tag{2.11}$$

$$= C_p dT \tag{2.12}$$

where C_p is the specific heat of the fluid at a constant pressure.

No assumptions have been made concerning the properties of the system and, therefore, the following relations apply to all fluids.

From Eq. (2.7):

$$\left(\frac{\partial U}{\partial T}\right)_v = C_v \tag{2.13}$$

From Eq. (2.12):

$$\left(\frac{\partial H}{\partial T}\right)_P = C_p \tag{2.14}$$

For an ideal gas, the two specific heats are related by the simple relation $C_p - C_v = \mathbf{R}$ where $\mathbf{R}$ is the universal gas constant, as shown in Eq. (2.27) in a later section.

2.3 Types of Fluid

Fluids may be classified in two different ways; either according to their behaviour under the action of externally applied pressure, or according to the effects produced by the action of a shear stress.

If the volume of an element of fluid is independent of its pressure and temperature, the fluid is said to be incompressible; if its volume changes it is said to be compressible. No real fluid is completely incompressible though liquids may generally be regarded as such when their flow is considered. Gases have a very much higher compressibility than liquids, and appreciable changes in volume may occur if the pressure or temperature is altered. However, if the percentage change in the pressure or in the absolute temperature is small, for practical purposes a gas may also be regarded as incompressible. Thus, in practice, volume changes are likely to be important only when the pressure or temperature of a gas changes by a large proportion. The relation between pressure, temperature, and volume of a real gas is generally complex though, except at very high pressures the behaviour of gases approximates to that of the ideal gas for which the volume of a given mass is inversely proportional to the pressure and directly proportional to the absolute temperature. However, at high pressures or when pressure changes are large, there may be appreciable deviations from this law and an approximate equation of state must then be used.

The behaviour of a fluid under the action of a shear stress is important in that it determines the way in which it will flow. The most important physical property affecting the stress distribution within the fluid is its viscosity. For a gas, the viscosity is low and even at high rates of shear, the viscous stresses are small. Under such conditions, the gas approximates in its behaviour to an *inviscid* fluid. In many problems involving the flow of a gas or a liquid, the viscous stresses (given by the product of viscosity and shear rate) are important and give rise to appreciable velocity gradients within the fluid, and dissipation of energy occurs as a result of the frictional forces set up. In gases and in most pure liquids, the ratio of the shear stress to the rate of shear is constant and equal to the viscosity of the fluid. These fluids are said to be *Newtonian* in their behaviour. Thus, their flow behaviour is characterised by its viscosity. However, in some liquids, particularly of high molecular weight and/or those containing a second phase in suspension, the ratio of the shear stress to the shear rate is not constant and the apparent viscosity of the fluid is a function of the rate of shear or of the shear stress. The fluid is then said to be *non-Newtonian* and to exhibit rheological properties. The importance of the viscosity of the fluid in determining velocity profiles and friction losses is discussed in Chapter 3.

The effect of pressure on the properties of an incompressible fluid, an ideal gas, and a nonideal gas is now considered.

2.3.1 The Incompressible Fluid (Liquid)

By definition, v is independent of P, such that $(\partial v/\partial P)_T = 0$. The internal energy will be a function of temperature but not a function of pressure. Most liquids under normal applications conform to this condition.

2.3.2 The Ideal Gas

An *ideal gas* is defined as a gas whose properties obey the law:

$$PV = n\mathbf{R}T \tag{2.15}$$

where V is the volume occupied by n molar units of the gas, $\mathbf{R}$ the universal gas constant, and T the absolute temperature. Here n is expressed in kmol when using the SI system of units.

This law is closely obeyed by real gases under conditions where the actual volume of the molecules is small compared with the total volume, and where the molecules exert only a very small attractive force on one another. These conditions are met at very low pressures when the distance between the individual molecules is large. The value of $\mathbf{R}$ is then the same for all gases and in SI units has the value of 8314 J/kmol K.

When the only external force on a gas is the fluid pressure, the equation of state is:

$$f(P, V, T, n) = 0$$

Any property may be expressed in terms of any three other properties. Considering the dependence of the internal energy on temperature and volume, then:

$$U = f(T, V, n)$$

For unit mass of gas:

$$U = f(T, v)$$

and:

$$Pv = \frac{\mathbf{R}T}{M} \tag{2.16}$$

where M is the molecular weight of the gas and v is the volume per unit mass.

Thus:

$$dU = \left(\frac{\partial U}{\partial T}\right)_v dT + \left(\frac{\partial U}{\partial v}\right)_T dv \tag{2.17}$$

and:

$$T\,dS = dU + P\,dv \text{ (from Eq. 2.5)}$$

$$\therefore \quad T\,dS = \left(\frac{\partial U}{\partial T}\right)_v dT + \left[P + \left(\frac{\partial U}{\partial v}\right)_T\right] dv$$

and:

$$dS = \left(\frac{\partial U}{\partial T}\right)_v \frac{dT}{T} + \frac{1}{T}\left[P + \left(\frac{\partial U}{\partial v}\right)_T\right] dv \tag{2.18}$$

Thus:

$$\left(\frac{\partial S}{\partial T}\right)_v = \frac{1}{T}\left(\frac{\partial U}{\partial T}\right)_v \tag{2.19}$$

and:

$$\left(\frac{\partial S}{\partial v}\right)_T = \frac{1}{T}\left[P + \left(\frac{\partial U}{\partial v}\right)_T\right] \tag{2.20}$$

Then differentiating Eq. (2.19) with respect to v and Eq. (2.20) with respect to T and equating:

$$\frac{1}{T}\frac{\partial^2 U}{\partial T \partial v} = \frac{1}{T}\left[\left(\frac{\partial P}{\partial T}\right)_v + \frac{\partial^2 U}{\partial v \partial T}\right] - \frac{1}{T^2}\left[P + \left(\frac{\partial U}{\partial v}\right)_T\right]$$

or

$$\left(\frac{\partial U}{\partial v}\right)_T = T\left(\frac{\partial P}{\partial T}\right)_v - P \tag{2.21}$$

This relation applies to any fluid. In the case of an ideal gas, since $Pv = \mathbf{R}T/M$ (Eq. 2.16):

$$T\left(\frac{\partial P}{\partial T}\right)_v = T\frac{\mathbf{R}}{Mv} = P$$

so that:

$$\left(\frac{\partial U}{\partial v}\right)_T = 0 \tag{2.22}$$

and:

$$\left(\frac{\partial U}{\partial P}\right)_T = \left(\frac{\partial U}{\partial v}\right)_T \left(\frac{\partial v}{\partial P}\right)_T = 0 \tag{2.23}$$

Thus the internal energy of an ideal gas is a function of temperature only. The variation of internal energy and enthalpy with temperature will now be calculated as:

$$\begin{aligned} \mathrm{d}U &= \left(\frac{\partial U}{\partial T}\right)_v \mathrm{d}T + \left(\frac{\partial U}{\partial v}\right)_T \mathrm{d}v \quad (\text{Eq. } 2.17) \\ &= C_v \mathrm{d}T \quad (\text{from Eqs } 2.13,\ 2.22) \end{aligned} \tag{2.24}$$

Thus for an ideal gas under all conditions:

$$\frac{\mathrm{d}U}{\mathrm{d}T} = C_v \tag{2.25}$$

In general, this relation applies only to changes at constant volume. In the case of an ideal gas, however, it applies under all circumstances.

Again, since $H = f(T, P)$:

$$\begin{aligned} \mathrm{d}H &= \left(\frac{\partial H}{\partial T}\right)_P \mathrm{d}T + \left(\frac{\partial H}{\partial P}\right)_T \mathrm{d}P \\ &= C_p \mathrm{d}T + \left(\frac{\partial U}{\partial P}\right)_T \mathrm{d}P + \left(\frac{\partial (Pv)}{\partial P}\right)_T \mathrm{d}P \quad (\text{from Eqs } 2.12,\ 2.10) \\ &= C_p \mathrm{d}T \end{aligned}$$

since $(\partial U/\partial P)_T = 0$ and $[\partial(Pv)/\partial P]_T = 0$ for an ideal gas.

Thus under all conditions for an ideal gas:

$$\frac{\mathrm{d}H}{\mathrm{d}T} = C_p \tag{2.26}$$

$$\therefore \quad C_p - C_v = \frac{\mathrm{d}H}{\mathrm{d}T} - \frac{\mathrm{d}U}{\mathrm{d}T} = \frac{\mathrm{d}(Pv)}{\mathrm{d}T} = \frac{\mathbf{R}}{M} \tag{2.27}$$

Isothermal processes

In fluid flow, it is important to know how the volume of a gas will vary as the pressure changes. Two important idealised conditions, which are rarely obtained in practice are changes at constant temperature and changes at constant entropy. Although not actually reached, these conditions are approached in many flow problems.

For an *isothermal* change in an ideal gas, the product of pressure and volume is a constant. For unit mass of gas:

$$Pv = \frac{\mathbf{R}T}{M} = \text{constant} \quad (\text{Eq. 2.16})$$

Isentropic processes

For an *isentropic* process, the enthalpy may be expressed as a function of the pressure and volume:

$$\begin{aligned} H &= f(P, v) \\ \mathrm{d}H &= \left(\frac{\partial H}{\partial P}\right)_v \mathrm{d}P + \left(\frac{\partial H}{\partial v}\right)_P \mathrm{d}v \\ &= \left(\frac{\partial H}{\partial T}\right)_v \left(\frac{\partial T}{\partial P}\right)_v \mathrm{d}P + \left(\frac{\partial H}{\partial T}\right)_P \left(\frac{\partial T}{\partial v}\right)_P \mathrm{d}v \end{aligned}$$

Since:

$$\left(\frac{\partial H}{\partial T}\right)_P = C_p \quad (\text{Eq. 2.14})$$

and:

$$\begin{aligned} \left(\frac{\partial H}{\partial T}\right)_v &= \left(\frac{\partial U}{\partial T}\right)_v + \left(\frac{\partial (Pv)}{\partial T}\right)_v \\ &= C_v + v\left(\frac{\partial P}{\partial T}\right)_v \quad (\text{from Eq. 2.13}) \end{aligned}$$

Further:

$$\begin{aligned} \mathrm{d}H &= \mathrm{d}U + P\mathrm{d}v + v\mathrm{d}P \quad (\text{from Eq. 2.10}) \\ &= T\mathrm{d}S - P\mathrm{d}v + P\mathrm{d}v + v\mathrm{d}P \quad (\text{from Eq. 2.5}) \qquad (2.28) \\ &= T\mathrm{d}S + v\mathrm{d}P \\ &= v\mathrm{d}P \quad (\text{for isentropic process}) \qquad (2.29) \end{aligned}$$

Thus, for an isentropic process:

$$v\mathrm{d}P = \left[C_v + v\left(\frac{\partial P}{\partial T}\right)_v\right]\left(\frac{\partial T}{\partial P}\right)_v \mathrm{d}P + C_p\left(\frac{\partial T}{\partial v}\right)_P \mathrm{d}v$$

or:

$$\left(\frac{\partial T}{\partial P}\right)_v \mathrm{d}P + \frac{C_p}{C_v}\left(\frac{\partial T}{\partial v}\right)_P \mathrm{d}v = 0$$

From the equation of state for an ideal gas (Eq. 2.15):

$$\left(\frac{\partial T}{\partial P}\right)_v = \frac{T}{P} \quad \text{and} \quad \left(\frac{\partial T}{\partial v}\right)_P = \frac{T}{v}$$

$$\therefore \quad \left(\frac{\mathrm{d}P}{P}\right) + \gamma\left(\frac{\mathrm{d}v}{v}\right) = 0$$

where $\gamma = C_p/C_v$.

Integration gives:

$$\ln P + \gamma \ln v = \text{constant}$$

or:

$$Pv^{\gamma} = \text{constant} \tag{2.30}$$

This relation holds only approximately, even for an ideal gas, since γ has been taken as a constant in the integration. It does, however, vary somewhat with pressure.

2.3.3 The Nonideal Gas

For a nonideal gas, Eq. (2.15) is modified by including a compressibility factor Z which is a function of both temperature and pressure:

$$PV = Zn\mathbf{R}T \tag{2.31}$$

At very low pressures, deviations from the ideal gas law are caused mainly by the attractive forces between the molecules, and the compressibility factor has a value less than unity. At higher pressures, deviations are caused mainly by the fact that the volume of the molecules themselves, which can be regarded as incompressible, becomes significant compared with the total volume of the gas.

Many equations have been developed to denote the approximate relation between the properties of a nonideal gas. Of these the simplest, and probably the most commonly used, is the van der Waals' equation:

$$\left(P + a\frac{n^2}{V^2}\right)(V - nb) = n\mathbf{R}T \tag{2.32}$$

where b is a quantity which is a function of the incompressible volume of the molecules themselves, and a/V^2 is a function of the attractive forces between the molecules. Values of a

and b can be expressed in terms of the critical pressure P_c and the critical temperature T_c as $a=\frac{27\mathbf{R}^2T_c^2}{64P_c}$ and $b=\frac{\mathbf{R}T_c}{8P_c}$. It is seen that as P approaches zero and V approaches infinity, this equation reduces to the equation of state for the ideal gas.

A chart that correlates experimental $P-V-T$ data for all gases is included as Fig. 2.1 and this is known as the generalised compressibility-factor chart.[1] Use is made of *the reduced* coordinates where the *reduced temperature* T_R, the *reduced pressure* P_R, and the *reduced volume* V_R are defined as the ratio of the actual temperature, pressure, and volume of the gas to the corresponding values of these properties at the critical state. It is found that, at a given value of T_R and P_R, nearly all gases have the same molar volume, compressibility factor, and other thermodynamic properties. This empirical relationship applies to within about 2% for most gases; the most important exception to the rule is ammonia.

The generalised compressibility-factor chart is not to be regarded as a substitute for experimental $P-V-T$ data. If accurate data are available, as they are for some of the more common gases, they should be used.

It will be noted from Fig. 2.1 that Z approaches unity for all temperatures as the pressure approaches zero. This serves to confirm the statement made previously that all gases approach

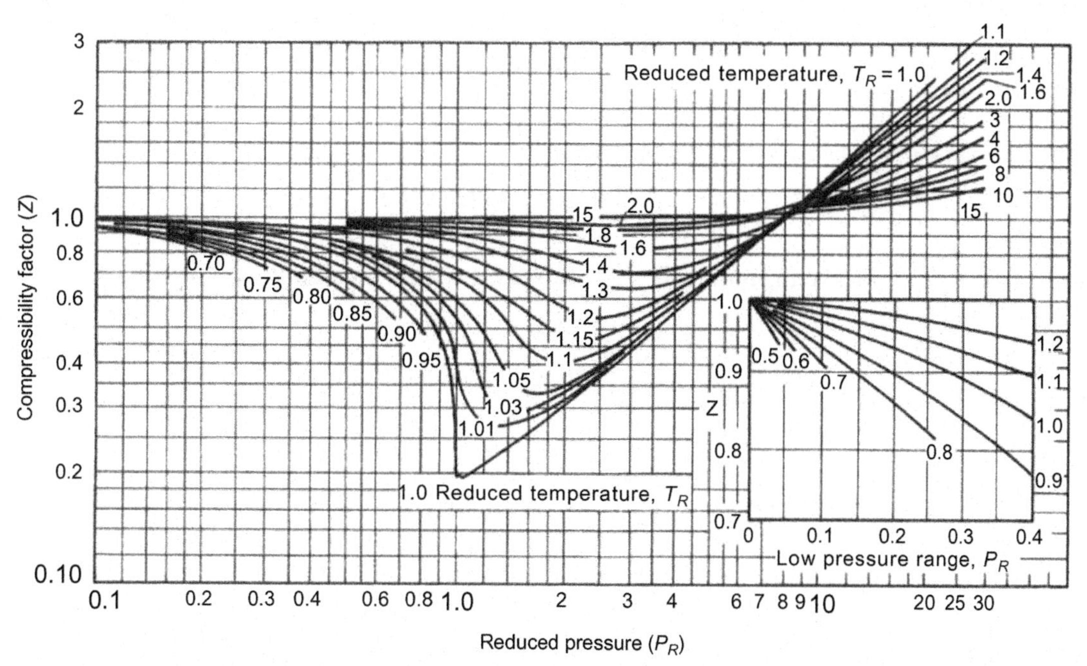

Fig. 2.1
Compressibility factors of gases and vapours.

ideality as the pressure is reduced to zero. For most gases, the critical pressure is 3 MN/m^2 or greater. Thus at atmospheric pressure (101.3 kN/m^2), P_R is 0.033 or less. At this pressure, for any temperature above the critical temperature ($T_R=1$), it will be seen that Z deviates from unity by no more than 1%. Thus at atmospheric pressure for temperatures greater than the critical temperature, the assumption that the ideal gas law is valid usually leads to errors of less than 1%. It should also be noted that for reduced temperatures between 3 and 10, the compressibility factor is nearly unity for reduced pressures up to a value of 6. For very high temperatures, the isotherms approach a horizontal line at $Z=1$ for all pressures. Thus all gases tend towards ideality as the temperature approaches infinity.

Example 2.1

It is required to store 1 kmol of methane at 320 K and 60 MN/m^2. Using the following methods, estimate the volume of the vessel that must be provided:

(a) ideal gas law;
(b) van der Waals' equation;
(c) generalised compressibility-factor chart;

Solution

For 1 kmol of methane, i.e. $n=1$ kmol,

(a) $PV=1\times\mathbf{R}T$, where $\mathbf{R}=8314$ J/kmol K.
In this case:

$$P=60\times10^6\,\text{N/m}^2;\ T=320\,\text{K}$$

$$V=8314\times\frac{320}{(60\times10^6)}=\underline{\underline{0.0443\ \text{m}^3}}$$

(b) In van der Waals' equation (2.32), the constants may be taken as:

$$a=\frac{27\mathbf{R}^2T_c^2}{64P_c};\ b=\frac{\mathbf{R}T_c}{8P_c}$$

where the critical temperature $T_c=191$ K and the critical pressure $P_c=4.64\times10^6$ N/m^2 for methane:

$$\therefore\ a=\frac{27\times8314^2\times191^2}{(64\times4.64\times10^6)}=229{,}300\ (\text{N/m}^2)(\text{m}^3)^2/(\text{kmol})^2$$

and

$$b=8314\times\frac{191}{(8\times4.64\times10^6)}=0.0427\,\text{m}^3/\text{kmol}$$

Thus in Eq. (2.32):

$$\left(60\times10^6+229{,}300\times\frac{1}{V^2}\right)(V-(1\times0.0427))=1\times8314\times320$$

or:

$$V^3-0.0427V^2+0.000382=0.0445$$

Solving by trial and error:

$$V = \underline{\underline{0.066\,\mathrm{m}^3}}$$

(c) $T_r = \frac{T}{T_c} = \frac{320}{191} = 1.68$

$$P_r = \frac{P}{P_c} = \frac{60 \times 10^6}{4.64 \times 10^6} = 12.93$$

Thus from Fig. 2.1, $Z = 1.33$ and:

$$V = \frac{Zn\mathbf{R}T}{P} \quad \text{(from Eq. 2.31)}$$

$$= \frac{1.33 \times 1.0 \times 8314 \times 320}{(60 \times 10^6)} = \underline{\underline{0.0589\,\mathrm{m}^3}}$$

While the use of the van de Waals equation and of the generalised compressibility chart yields comparable values, these are about 50% greater than that obtained using the ideal gas law.

Example 2.2

Obtain expressions for the variation of:

(a) internal energy with change of volume,
(b) internal energy with change of pressure, and
(c) enthalpy with change of pressure,

all at constant temperature, for a gas whose equation of state is given by van der Waals' law.

Solution

van der Waals' equation (2.32) may be written for n kmol of gas as:

$$\left[P + \left(an^2/V^2\right)\right](V - nb) = n\mathbf{R}T \quad \text{(Eq. 2.32)}$$

or:

$$P = \left[n\mathbf{R}T/(V - nb)\right] - \left(an^2V^2\right)$$

(a) Internal energy and temperature are related by:

$$\left(\frac{\partial U}{\partial V}\right)_T = T\left(\frac{\partial P}{\partial T}\right)_V - P \quad \text{(Eq. 2.21)}$$

From van der Waals' equation:

$$\left(\frac{\partial P}{\partial T}\right)_V = \frac{n\mathbf{R}}{(V - nb)} \quad \text{and} \quad T\left(\frac{\partial P}{\partial T}\right)_V = n\mathbf{R}T/(V - nb)$$

Hence:

$$\underline{\underline{\left(\frac{\partial U}{\partial V}\right)_T = \frac{n\mathbf{R}T}{(V - nb)} - P = \frac{an^2}{V^2}}}$$

(For an ideal gas: $b = 0$ and $(\partial U/\partial V)_T = (n\mathbf{R}T/V) - P = 0$)

(b) $\left(\frac{\partial U}{\partial P}\right)_T = \left(\frac{\partial U}{\partial V}\right)_T \left(\frac{\partial V}{\partial P}\right)_T$ and $(\partial V/\partial P)_T = 1/(\partial P/\partial V)_T$

Hence using Eq. (2.32):

$$\left(\frac{\partial P}{\partial V}\right)_T = \frac{-n\mathbf{R}T}{(V-nb)^2} + \frac{2an^2}{V^3} = \frac{2an^2(V-nb)^2 - n\mathbf{R}TV^3}{V^3(V-nb)^2}$$

and thus:

$$\left(\frac{\partial U}{\partial V}\right)_T = \frac{n\mathbf{R}T}{(V-nb)} - P = \frac{an^2}{(V^2}$$

$$\left(\frac{\partial U}{\partial P}\right)_T = \left(\frac{n\mathbf{R}T}{(V-b)} - P\right)\left(\frac{V^3(V-b)}{2a(V-b)^2 - n\mathbf{R}TV^3}\right)$$

$$= \frac{[n\mathbf{R}T - P(V-b)][V^3(V-b)]}{[2a(V-b)^2 - n\mathbf{R}TV^3]}$$

(For an ideal gas, $a=b=0$ and $(\partial U/\partial P)_T = 0$)

(c) Since H is a function of T and P:

$$dH = \left(\frac{\partial H}{\partial T}\right)_P dT + \left(\frac{\partial H}{\partial P}\right)_T dP$$

$$= C_p dT + \left(\frac{\partial U}{\partial P}\right)_T dP + \left(\frac{\partial (PV)}{\partial P}\right)_T dP$$

For a constant temperature process:

$$C_p dT = 0 \quad \text{and} \quad \left(\frac{\partial H}{\partial P}\right)_T = \left(\frac{\partial U}{\partial P}\right)_T + \left(\frac{\partial (PV)}{\partial P}\right)_T$$

One can write the following expression for PV: $PV = P(V - nb + nb) = P(V - nb) + nPb$

Substituting for $P(V-nb)$:

$$PV = \frac{nRT}{\left(1 + \frac{an^2}{PV^2}\right)} + nPb$$

Differentiating both sides with respect to P at constant temperature:

$$\left.\frac{\partial (PV)}{\partial P}\right|_T = \frac{an^3RT}{V^3\left(P + \frac{an^2}{V^2}\right)^2}\left[V + 2P\left.\frac{\partial V}{\partial P}\right|_T\right] + nb$$

This can now be combined with the expression for $(\partial U/\partial p)_T$ obtained in part (b) to get:

$$\left(\frac{\partial H}{\partial P}\right)_T = \left\{\frac{2an^2(V-nb)^3}{2an^2(V-nb)^2 - nRTV^3} + nb\right\}$$

For an ideal gas: $a=0$, $b=0$ and therefore as expected $\left({}^{\partial H}/_{\partial P}\right)_T = 0$.

Joule–Thomson effect

It has already been shown that the change of internal energy of unit mass of fluid with volume at constant temperature is given by the relation:

$$\left(\frac{\partial U}{\partial v}\right)_T = T\left(\frac{\partial P}{\partial T}\right)_v - P \quad (Eq. 2.21)$$

For a nonideal gas:

$$T\left(\frac{\partial P}{\partial T}\right)_v \neq P$$

and therefore $(\partial U/\partial v)_T$ and $(\partial U/\partial P)_T$ are not equal to zero.

Thus the internal energy of the nonideal gas is a function of pressure as well as temperature. As the gas is expanded, the molecules are separated from each other against the action of the attractive forces between them. Energy is therefore stored in the gas; this is released when the gas is compressed and the molecules are allowed to approach one another again.

A characteristic feature of the nonideal gas is that it has a finite *Joule–Thomson effect.* This relates to the amount of heat that must be added during an expansion of a gas from a pressure P_1 to a pressure P_2 in order to maintain isothermal conditions. Imagine a gas flowing from a cylinder, fitted with a piston at a pressure P_1 to a second cylinder at a pressure P_2 (Fig. 2.2).

The net work done by unit mass of the gas on the surroundings in expanding from P_1 to P_2 is given by:

$$W = P_2 v_2 - P_1 v_1 \tag{2.33}$$

A quantity of heat (say q) is added during the expansion so as to maintain isothermal conditions. The change in the internal energy is therefore given by:

$$\Delta U = q - W \quad (\text{from Eq. 2.1})$$

$$\therefore \quad q = \Delta U - P_1 v_1 + P_2 v_2 \tag{2.34}$$

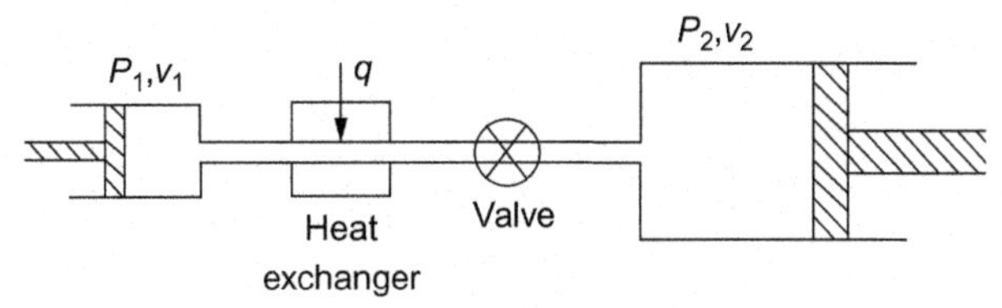

Fig. 2.2
Joule–Thomson effect.

For an ideal gas, under isothermal conditions, $\Delta U = 0$ and $P_2 v_2 = P_1 v_1$. Thus $q = 0$ and the ideal gas is said to have a zero Joule–Thomson effect. A nonideal gas has a Joule–Thomson effect which may be either positive or negative.

2.4 The Fluid in Motion

When a fluid flows through a duct or over a surface, the velocity over a plane at right angles to the fluid stream is not normally uniform. The variation of velocity can be shown by the use of streamlines, which are lines so drawn that the velocity vector is always tangential to them. In other words, there is no flow across a streamline. The flowrate between any two streamlines is therefore always the same. Constant velocity over a cross-section is shown by equidistant streamlines and an increase in velocity by closer spacing of the streamlines. There are two principal types of flow, which are discussed in detail later, namely streamline and turbulent flow. In streamline flow, movement across streamlines occurs solely as the result of diffusion on a molecular scale and the flowrate is steady. In turbulent flow, the presence of circulating current results in transference of fluid on a larger scale, and random fluctuations occur in the flowrate, though the time-average flowrate remains constant.

A group of streamlines can be taken together to form a streamtube, and thus the whole area for flow can be regarded as being composed of bundles of streamtubes.

Figs 2.3–2.5 show the flow patterns in a straight tube, through a constriction and past an immersed object. In the first case, the streamlines are all parallel to one another, whereas in the other two cases the streamlines approach one another as the passage becomes constricted, indicating that the velocity is increasing, i.e. the fluid is accelerating.

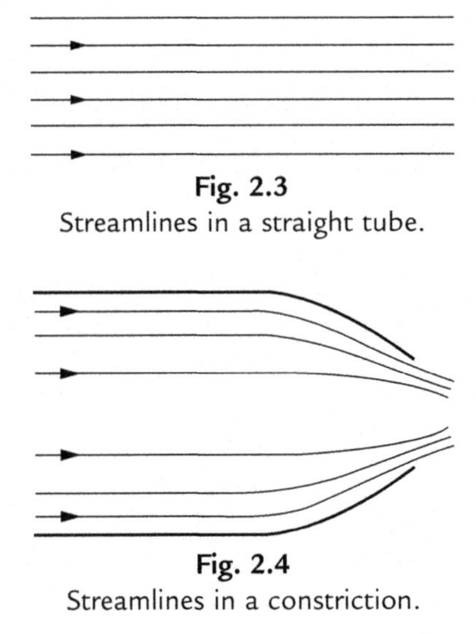

Fig. 2.3
Streamlines in a straight tube.

Fig. 2.4
Streamlines in a constriction.

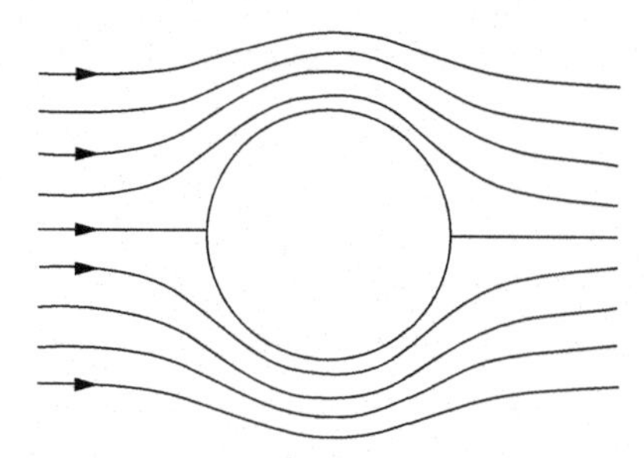

Fig. 2.5
Streamlines for flow past an immersed object.

2.4.1 Continuity

Considering the flow of a fluid through a streamtube, as shown in Fig. 2.6, then equating the mass rates of flow at sections 1 and 2:

$$dG = \rho_1 \dot{u}_1 dA_1 = \rho_2 \dot{u}_2 dA_2 \tag{2.35}$$

where ρ_1, ρ_2 are the densities; $\dot{u}_1, \dot{u}_2$ the velocities in the streamtube; and dA_1, dA_2 the flow areas at sections 1 and 2 respectively.

On integration:

$$G = \int \rho_1 \dot{u}_1 dA_1 = \int \rho_2 \dot{u}_2 dA_2 = \rho_1 u_1 A_1 = \rho_2 u_2 A_2 \tag{2.36}$$

where u_1, u_2 are the average velocities (defined by the previous equations) at the two sections. In many problems, the mass flowrate per unit area, G', is an important quantity.

$$G' = \frac{G}{A} = \rho u \tag{2.37}$$

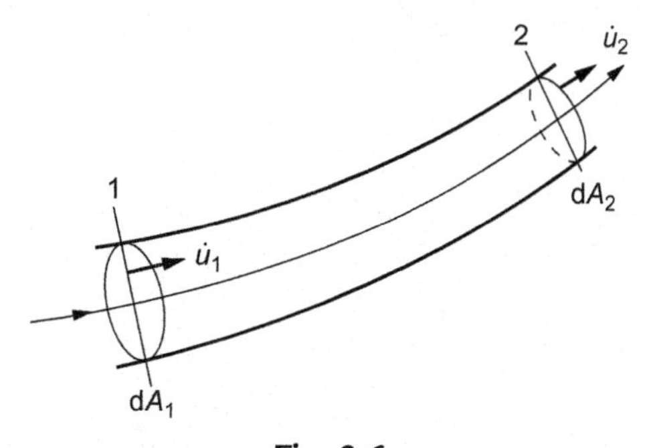

Fig. 2.6
Flow through a streamtube.

For an incompressible fluid, such as a liquid or a gas where the pressure changes are small (thus $\rho_1 = \rho_2$):

$$u_1 A_1 = u_2 A_2 \tag{2.38}$$

It is seen that it is important to be able to determine the velocity profile so that the flowrate can be calculated, and this is done in Chapter 3. For streamline flow in a pipe, the mean velocity is 0.5 times the maximum stream velocity which occurs at the axis of the pipe. For turbulent flow, the velocity profile is flatter and the ratio of the mean velocity to the maximum velocity is about 0.82.

2.4.2 Momentum Changes in a Fluid

As a fluid flows through a duct, its momentum and pressure may change. The magnitude of the changes can be considered by applying the momentum equation (force equals the rate of change of momentum) to the fluid in a streamtube and then integrating over the cross-section of the duct. The effect of the frictional forces will be neglected at first and the relations thus obtained will strictly apply only to an inviscid (frictionless) fluid. Considering an element of length dl of a streamtube of cross-sectional area dA, increasing to $dA + (d(dA)/dl)dl$, as shown in Fig. 2.7, then the upstream pressure $= P$ and the force attributable to the upstream pressure $= P\,dA$.

The downstream pressure $= P + \left(\frac{dP}{dl}\right) dl$

This pressure acts over an area $dA + (d(dA)/dl)dl$ which gives rise to a total force of $-\{P + (dP/dl)dl\}\{dA + [d(dA)/dl]dl\}$.

In addition, the mean pressure of $P + \frac{1}{2}(dP/dl)dl$ acting on the sides of the streamtube will give rise to a force having a component $[P + \frac{1}{2}(dP/dl)dl][d(dA)/dl]dl$ along the streamtube. Thus, the net force along the streamtube due to the pressure gradient is:

$$P dA - \left(P + \frac{dP}{dl}dl\right)\left(dA + \frac{d(dA)}{dl}dl\right) + \left(P + \frac{1}{2}\frac{dP}{dl}dl\right)\frac{d(dA)}{dl}dl \asymp -\frac{dP}{dl}dl\,dA$$

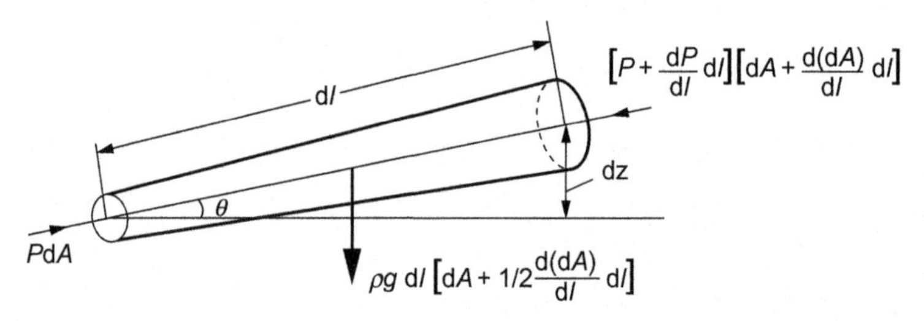

Fig. 2.7
Forces on fluid in a streamtube.

The other relevant force acting on the fluid is the weight of the fluid itself

$$=\rho g \mathrm{d}l\left(\mathrm{d}A+\frac{1}{2}\frac{\mathrm{d}(\mathrm{d}A)}{\mathrm{d}l}\mathrm{d}l\right)$$

The component of this force along the streamtube is

$$-\rho g \mathrm{d}l\left(\mathrm{d}A+\frac{1}{2}\frac{\mathrm{d}(\mathrm{d}A)}{\mathrm{d}l}\mathrm{d}l\right)\sin\theta$$

Neglecting second order terms and noting that $\sin\theta = \mathrm{d}z/\mathrm{d}l$:

$$\text{Total force exerted on the element of fluid} = -\frac{\mathrm{d}P}{\mathrm{d}l}\mathrm{d}l\mathrm{d}A - \frac{\mathrm{d}z}{\mathrm{d}l}\rho g \mathrm{d}l\mathrm{d}A \tag{2.39}$$

The rate of change of momentum of the fluid along the streamtube

$$\begin{aligned}&=(\rho\dot{u}\,\mathrm{d}A)\left[\left(\dot{u}+\frac{\mathrm{d}\dot{u}}{\mathrm{d}l}\mathrm{d}l\right)-\dot{u}\right]\\&=\rho\dot{u}\frac{\mathrm{d}\dot{u}}{\mathrm{d}l}\mathrm{d}l\mathrm{d}A\end{aligned} \tag{2.40}$$

Equating Eqs (2.39), (2.40):

$$\rho\dot{u}\,\mathrm{d}l\mathrm{d}A\frac{\mathrm{d}\dot{u}}{\mathrm{d}l} = -\mathrm{d}l\mathrm{d}A\frac{\mathrm{d}P}{\mathrm{d}l} - \rho g\mathrm{d}l\mathrm{d}A\frac{\mathrm{d}z}{\mathrm{d}l}$$

$$\therefore\quad \dot{u}\,\mathrm{d}\dot{u} + \frac{\mathrm{d}P}{\rho} + g\mathrm{d}z = 0 \tag{2.41}$$

On integration:

$$\frac{\dot{u}^2}{2} + \int\frac{\mathrm{d}P}{\rho} + gz = \text{constant} \tag{2.42}$$

For the simple case of an incompressible fluid, ρ is independent of pressure, and:

$$\frac{\dot{u}^2}{2} + \frac{P}{\rho} + gz = \text{constant} \tag{2.43}$$

Eq. (2.43) is known as the Bernoulli's equation, which relates the pressure at a point in the fluid to its position and velocity for the frictionless flow. Each term in Eq. (2.43) represents energy per unit mass of fluid. Thus, if all the fluid is moving with a velocity u, the total energy per unit mass ψ is given by:

$$\psi = \frac{u^2}{2} + \frac{P}{\rho} + gz \tag{2.44}$$

Dividing Eq. (2.44) by g:

$$\frac{u^2}{2g}+\frac{P}{\rho g}+z=\text{constant} \tag{2.45}$$

In Eq. (2.45), each term represents energy per unit weight of fluid and has the dimensions of length and can be regarded as representing a contribution to the total fluid head.

Thus:

$\frac{u^2}{2g}$ is the velocity head
$\frac{P}{\rho g}$ is the pressure head
and: z is the potential head

Eq. (2.42) can also be obtained by considering the energy changes in the fluid.

Example 2.3

Water leaves the 25 mm diameter nozzle of a fire hose at a velocity of 25 m/s. What will be the reaction force at the nozzle that the fireman will need to counterbalance?

Solution

$$\text{Mass rate of discharge of water, } G=\rho uA$$
$$=1000\times 25\times\frac{\pi}{4}(0.025)^2$$
$$=12.27\,\text{kg/s}$$

$$\text{Momentum of fluid per second}=Gu$$
$$=12.27\times 25$$
$$=307\,\text{N}$$

$$\text{Reaction force}=\text{Rate of change of momentum}=\underline{\underline{307\,\text{N}}}$$

This result neglects the force arising from the changes in pressure and the linear momentum of the water stream upstream of the nozzle.

Example 2.4

Water is flowing at 5 m/s in a 50 mm diameter pipe which incorporates a 90° bend, as shown in Fig. 2.8. What is the additional force to which a retaining bracket will be subjected to, as a result of the momentum changes in the liquid, if it is arranged symmetrically in the pipe bend?

Solution

Momentum per second of approaching liquid in Y-direction

$$=\rho u^2A$$
$$=1000\times 25\times\frac{\pi}{4}(0.050)^2$$
$$=49.1\,\text{N}$$

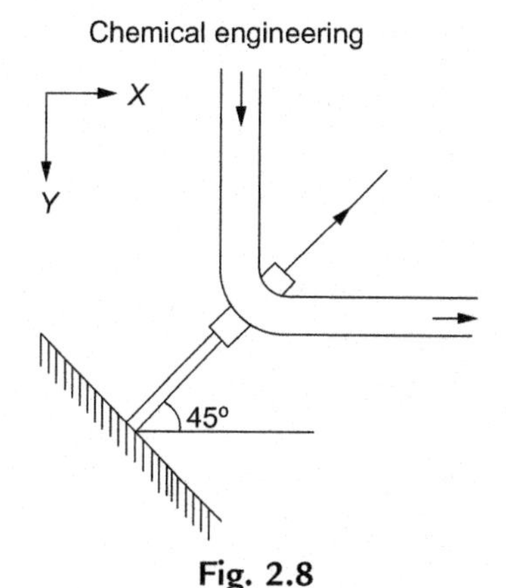

Fig. 2.8
Force on support for a pipe bend.

The pipe bracket must therefore exert a reaction force of −49.1 N in the *Y*-direction, the direction in which the fluid is accelerating. Similarly, the force in the *X*-direction = 49.1 N.

The resultant force in the direction of arm of bracket = 49.1 cos 45° + 49.1 sin 45°

$$= 49.1\left(\frac{1}{\sqrt{2}} + \frac{1}{\sqrt{2}}\right)$$

$$= \underline{\underline{69.4\,\text{N}}}$$

Water hammer

If the flowrate of a liquid in a pipeline is suddenly reduced, such as by rapid closure of a valve, or a pump failure for example, its rate of change of momentum can be sufficiently high for very large forces to be set up which may cause damage to the installation. In a pipeline carrying water, the resulting pressure wave may have a velocity as high as 1200 m/s. The behaviour of the pipe network will be influenced by a large number of factors, including the density and the bulk modulus of elasticity of the liquid, Young's modulus for the material of the pipe, and the design and layout of the installation. The phenomenon is complex[2] and reference should be made to one of the specialised texts, such as those by Parmakian,[3] Sharp,[4] Wylie and Streeter[5] and Douglas et al.[6] for a detailed analysis of this and other similar effects. The situation can arise with the flow of any liquid, but it is usually referred to as *water hammer* on account of the characteristic sound arising from water distribution systems.

2.4.3 Energy of a Fluid in Motion

The total energy of a fluid in motion is made up of a number of components. For unit mass of fluid and neglecting changes in the magnetic and electrical energy, the magnitudes of the various forms of energy are as follows.

Internal energy U

This has already been discussed in Section 2.2.

Pressure energy

This represents the work that must be done in order to introduce the fluid, without change in volume, into the system. It is therefore given by the product Pv, where P is the pressure of the system and v is the volume per unit mass of fluid.

Potential energy

The potential energy of the fluid, due to its position in the earth's gravitational field, is equal to the work which must be done on it in order to raise it to that position from some arbitrarily chosen datum level at which the potential energy is taken as zero. Thus, if the fluid is situated at a height z above the datum level, the potential energy is zg, where g is the acceleration due to gravity which is taken as constant unless otherwise stated.

Kinetic energy

The fluid possesses kinetic energy by virtue of its motion with reference to some arbitrarily fixed body, normally taken as the earth. If the fluid is moving with a velocity $\dot{u}$, the kinetic energy is $\dot{u}^2/2$.

The total energy of unit mass of fluid is, therefore:

$$U + Pv + gz + \frac{\dot{u}^2}{2} \tag{2.46}$$

If the fluid flows from section 1 to section 2 (where the values of the various quantities are denoted by suffixes 1 and 2 respectively) and q is the net heat absorbed from the surroundings and W_s is the net work done by the fluid on the surroundings, other than that done by the fluid in entering or leaving the section under consideration, then using the first law of thermodynamics:

$$U_2 + P_2v_2 + gz_2 + \frac{\dot{u}_2^2}{2} = U_1 + P_1v_1 + gz_1 + \frac{\dot{u}_1^2}{2} + q - W_s \tag{2.47}$$

$$\Delta U + \Delta(Pv) + g\Delta z + \Delta\frac{u^2}{2} = q - W_s \tag{2.48}$$

where Δ denotes a finite change in the quantities.

Thus:

$$\Delta H + g\Delta z + \frac{\Delta\dot{u}^2}{2} = q - W_s \tag{2.49}$$

It should be noted that the shaft work W_s is related to the total work W by the relation:

$$W = W_s + \Delta(Pv) \tag{2.50}$$

For a small change in the system:

$$dH + g\,dz + \dot{u}\,d\dot{u} = \delta q - \delta W_s \tag{2.51}$$

For many purposes, it is convenient to eliminate H by using Eq. (2.11):

$$dH = \delta q + \delta F + v\,dP \quad \text{(Eq. 2.11)}$$

Here δF represents the amount of mechanical energy irreversibly converted into heat.

Thus:

$$\dot{u}\,d\dot{u} + g\,dz + v\,dP + \delta W_s + \delta F = 0 \tag{2.52}$$

When no work is done by the fluid on the surroundings and when friction can be neglected, it can be noted that Eq. (2.52) is identical to Eq. (2.41) derived from consideration of a momentum balance, since:

$$v = \frac{1}{\rho}$$

Integrating this equation for flow from section 1 to section 2 and summing the terms δW_S and δF:

$$\Delta\frac{\dot{u}^2}{2} + g\Delta z + \int_{P_1}^{P_2} v\,dP + W_s + F = 0 \tag{2.53}$$

Eqs (2.41)–(2.53) are quite general and apply therefore to any type of fluid.

With incompressible fluids, the energy F is either lost to the surroundings or causes a very small rise in temperature. If the fluid is compressible, however, the rise in temperature may result in an increase in the pressure energy and part of it may be available for doing useful work.

If the fluid is flowing through a channel or pipe, a frictional drag arises in the region of the boundaries and gives rise to a velocity distribution across any section perpendicular to the direction of flow. For the unidirectional flow of fluid, the mean velocity of flow has been defined by Eq. (2.36) as the ratio of the volumetric flowrate to the cross-sectional area of the channel. When Eq. (2.52) is applied over the whole cross-section, therefore, allowance must be made for the fact that the mean square velocity is not equal to the square of the mean velocity, and a correction factor α must therefore be introduced into the kinetic energy term. Thus, considering the fluid over the whole cross-section, for small changes:

$$\frac{u\,du}{\alpha} + g\,dz + v\,dP + \delta W_s + \delta F = 0 \tag{2.54}$$

and for finite changes:

$$\Delta\left(\frac{u^2}{2\alpha}\right)+g\Delta z+\int_{P_1}^{P_2} v\mathrm{d}P+W_s+F=0 \tag{2.55}$$

Before Eq. (2.55) may be applied to any particular flow problem, the term $\int_{P_1}^{P_2} v\, dP$ must be evaluated.

Eq. (2.49) becomes:

$$\Delta\left(\frac{u^2}{2\alpha}\right)+g\Delta z+\Delta H=q-W_s \tag{2.56}$$

For flow in a pipe of circular cross-section, α will be shown to be exactly 0.5 for streamline flow and approximate to unity for turbulent flow.

For turbulent flow, and where no external work is done, Eq. (2.54) becomes:

$$u\mathrm{d}u+g\mathrm{d}z+v\mathrm{d}P=0 \tag{2.57}$$

if frictional effects can be neglected.

For horizontal flow, or where the effects of change of elevation may be neglected, as is normally the case with gases, Eq. (2.57) simplifies to:

$$u\mathrm{d}u+v\mathrm{d}P=0 \tag{2.58}$$

2.4.4 Pressure and Fluid Head

In Eq. (2.54), each term represents energy per unit mass of fluid. If the equation is multiplied throughout by density ρ, each term has the dimensions of pressure and represents energy per unit volume of fluid:

$$\rho\frac{u\mathrm{d}u}{\alpha}+\rho g\mathrm{d}z+\mathrm{d}P+\rho\delta W_s+\rho\delta F=0 \tag{2.59}$$

If Eq. (2.54) is divided throughout by g, each term has the dimensions of length, and, as already noted, may be regarded as a component of the total head of the fluid and represents energy per unit weight:

$$\frac{1}{g}\frac{u\mathrm{d}u}{\alpha}+\mathrm{d}z+v\frac{\mathrm{d}P}{g}+\frac{\delta W_s}{g}+\frac{\delta F}{g}=0 \tag{2.60}$$

For an incompressible fluid flowing in a horizontal pipe of constant cross-section, in the absence of work being done by the fluid on the surroundings, the pressure change due to frictional effects is given by:

$$v\frac{\mathrm{d}P_f}{g}+\frac{\delta F}{g}=0$$

or:

$$-\mathrm{d}P_f = \frac{\delta F}{v} = \rho g \mathrm{d}h_f \tag{2.61}$$

where $\mathrm{d}h_f$ is the loss in head corresponding to a change in pressure due to friction of $\mathrm{d}P_f$.

2.4.5 Constant Flow Per Unit Area

When the flow rate of the fluid per unit area G' is constant, Eq. (2.37) can be written as:

$$\frac{G}{A} = G' = \frac{u_1}{v_1} = \frac{u_2}{v_2} = \frac{u}{v} \tag{2.62}$$

or:

$$G' = u_1\rho_1 = u_2\rho_2 = u\rho \tag{2.63}$$

Eq. (2.58) is the momentum balance for horizontal turbulent flow:

$$u\mathrm{d}u + v\mathrm{d}P = 0 \quad (\text{Eq. } 2.58)$$

or:

$$u\frac{\mathrm{d}u}{v} + \mathrm{d}P = 0$$

Because u/v is constant, on integration this gives:

$$\frac{u_1(u_2 - u_1)}{v_1} + P_2 - P_1 = 0$$

or:

$$\frac{u_1^2}{v_1} + P_1 = \frac{u_2^2}{v_2} + P_2 \tag{2.64}$$

2.4.6 Separation

It may be noted that the energy and mass balance equations assume that the fluid is continuous. This is so in the case of a liquid, provided that the pressure does not fall to such a low value that boiling, or the evolution of dissolved gases, takes place. For water at normal temperatures, the pressure should not be allowed to fall below the equivalent of a head of 1.2 m of liquid. With gases, there is no lower limit to the pressures at which the fluid remains continuous, but the various equations that are derived need modifications if the pressures are so low that the linear dimensions of the channels become comparable to the mean free path of the molecules, that is when the so-called *molecular flow* sets in.

2.5 Pressure-Volume Relationships

2.5.1 Incompressible Fluids

For incompressible fluids, v is independent of pressure so that

$$\int_{P_1}^{P_2} v\mathrm{d}P = (P_2 - P_1)v \tag{2.65}$$

Therefore Eq. (2.55) becomes:

$$\frac{u_1^2}{2\alpha_1} + gz_1 + P_1 v = \frac{u_2^2}{2\alpha_2} + gz_2 + P_2 v + W_s + F \tag{2.66}$$

or:

$$\Delta\frac{u^2}{2\alpha} + g\Delta z + v\Delta P + W_s + F = 0 \tag{2.67}$$

In a frictionless system in which the fluid does not work on the surroundings and α_1 and α_2 are taken as unity (turbulent flow), then:

$$\frac{u_1^2}{2} + gz_1 + P_1 v = \frac{u_2^2}{2} + gz_2 + P_2 v \tag{2.68}$$

Example 2.5

Water flows from a tap at a pressure of 250 $\mathrm{kN/m^2}$ above atmospheric pressure. What is the velocity of the jet if frictional effects are neglected?

Solution

From Eq. (2.68):

$$0.5\left(u_2^2 - u_1^2\right) = g(z_1 - z_2) + \frac{(P_1 - P_2)}{\rho}$$

Using suffix 1 to denote conditions in the pipe and suffix 2 to denote conditions in the jet and neglecting the velocity of approach in the pipe:

$$0.5\left(u_2^2 - 0\right) = 9.81 \times 0 + \frac{250 \times 10^3}{1000}$$

$$\underline{\underline{u_2 = 22.4\,\mathrm{m/s}}}$$

2.5.2 Compressible Fluids

For a gas, the appropriate relation between specific volume and pressure must be used although for small changes in pressure or temperature, little error is introduced by using a mean value of the specific volume.

The term $\int_{P_1}^{P_2} v\,dP$ will now be evaluated for the ideal gas under various conditions. In most cases, the results so obtained may be applied to the nonideal gas without introducing an error greater than that involved in estimating the other factors concerned in the process. The only common exception to this occurs in the flow of gases at very high pressures and for the flow of steam, when it is necessary to employ one of the approximate equations for the state of a nonideal gas, in place of the equation for the ideal gas. Alternatively, Eq. (2.56) may be used and work expressed in terms of changes in enthalpy. For a gas, the potential energy term is usually small compared with the other energy terms.

The relation between the pressure and the volume of an ideal gas depends on the rate of transfer of heat to the surroundings and the degree of irreversibility of the process. The following conditions will be considered.

(a) an isothermal process;
(b) an isentropic process;
(c) a reversible process which is neither isothermal nor adiabatic;
(d) an irreversible process which is not isothermal.

Isothermal process

For an isothermal process, $Pv = \mathbf{R}T/M = P_1v_1$, where the subscript 1 denotes the initial values and M is the molecular weight.

Thus

$$\int_{P_1}^{P_2} v\,dP = P_1v_1 \int_{P_1}^{P_2} \frac{1}{P}\,dP = P_1v_1 \ln\frac{P_2}{P_1} \tag{2.69}$$

Isentropic process

From Eq. (2.30), for an isentropic process:

$$\begin{aligned}
Pv^{\gamma} &= P_1v_1^{\gamma} = \text{constant} \\
\int_{P_1}^{P_2} v\,dP &= \int_{P_1}^{P_2} \left(\frac{P_1v_1^{\gamma}}{P}\right)^{1/\gamma} dP \\
&= P_1^{1/\gamma}v_1 \int_{P_1}^{P_2} P^{-1/\gamma}\,dP \\
&= P_1^{1/\gamma}v_1 \frac{1}{1-(1/\gamma)}\left(P_2^{1-(1/\gamma)} - P_1^{1-(1/\gamma)}\right) \\
&= \frac{\gamma}{\gamma-1}P_1v_1\left[\left(\frac{P_2}{P_1}\right)^{(\gamma-1)/\gamma} - 1\right]
\end{aligned} \tag{2.70}$$

$$= \frac{\gamma}{\gamma - 1}\left[P_1\left(\frac{P_2}{P_1}\right)^{(\gamma-1)/\gamma}\left(\frac{P_2}{P_1}\right)^{1/\gamma} v_2 - P_1 v_1\right] \qquad (2.71)$$

$$= \frac{\gamma}{\gamma - 1}(P_2 v_2 - P_1 v_1)$$

Further, from Eqs (2.29), (2.26), taking C_p as constant:

$$\int_{P_1}^{P_2} v\,\mathrm{d}P = \int_{H_1}^{H_2} \mathrm{d}H = C_p \Delta T \qquad (2.72)$$

The above relations apply for an ideal gas to a reversible adiabatic process, which as already shown, is isentropic.

Reversible process—Neither isothermal nor adiabatic

In general the conditions under which a change in state of a gas takes place are neither isothermal nor adiabatic and the relation between pressure and volume is approximately of the form Pv^k = constant for a reversible process, where k is a numerical quantity whose value depends on the heat transfer between the gas and its surroundings. k usually lies between 1 and γ though it may, under certain circumstances, lie outside these limits; it will have the same value for a reversible compression as for a reversible expansion under similar conditions. Under these conditions therefore, Eq. (2.70) becomes:

$$\int_{P_1}^{P_2} v\,\mathrm{d}P = \frac{k}{k-1} P_1 v_1 \left[\left(\frac{P_2}{P_1}\right)^{(k-1)/k} - 1\right] \qquad (2.73)$$

Irreversible process

For an irreversible process, it may not be possible to express the relation between pressure and volume as a continuous mathematical function though, by choosing a suitable value for the constant k, an equation of the form Pv^k = constant may be used over a limited range of conditions. Eq. (2.73) may then be used for the evaluation of $\int_{P_1}^{P_2} v\,\mathrm{d}P$. It may be noted that, for an irreversible process, k will have different values for compression and expansion under otherwise similar conditions. Thus, for the irreversible adiabatic compression of a gas, k will be greater than γ, and for the corresponding expansion k will be less than γ. This means that more energy has to be put into an irreversible compression than will be recovered when the gas expands to its original condition.

In many instances, it is preferable to use the actual pressure–density data when neither of the foregoing idealisations is justified. This method is illustrated by Darby and Chhabra.[7]

2.6 Rotational or Vortex Motion in a Fluid

In many chemical engineering applications, a liquid undergoes rotational motion, such as for example, in a centrifugal pump, in a stirred vessel, in the basket of a centrifuge, or in a cyclone-type separator. In the first instance, the effects of friction may be disregarded and consideration will be given to how the forces acting on the liquid determine the pressure distribution. If the liquid may be considered to be rotating about a vertical axis, it will then be subjected to vertical forces due to gravity and centrifugal forces in a horizontal plane. The total force on the liquid and the pressure distribution is then obtained by summing the two components. The vertical pressure gradient attributed to the force of gravity is given by:

$$\frac{\partial P}{\partial z} = -\rho g = -\frac{g}{v} \tag{2.74}$$

where z is measured vertically upward.

The centrifugal force acts in a horizontal plane and the resulting pressure gradient may be obtained by taking a force balance on a small element of liquid as shown in Fig. 2.9.

At radius r, the pressure is P.

At radius $r + \mathrm{d}r$, the pressure is $P + (\partial P/\partial r)\mathrm{d}r$.

For small values of $\mathrm{d}r$, the pressure on the 'cut' faces may be taken as $P + \frac{1}{2}(\partial P/\partial r)\mathrm{d}r$.

Then, a force balance in the radial direction on an element of inner radius r, outer radius $r + \mathrm{d}r$, depth $\mathrm{d}z$ and subtending a small angle $\mathrm{d}\theta$ at the centre gives:

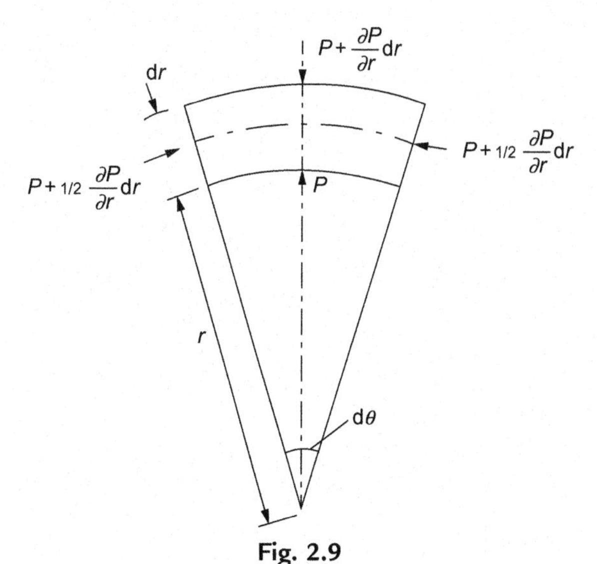

Fig. 2.9
Forces acting on an element of fluid in a vortex.

$$\left(P+\frac{\partial P}{\partial r}\mathrm{d}r\right)(r+\mathrm{d}r)\mathrm{d}\theta\mathrm{d}z - Pr\,\mathrm{d}\theta\mathrm{d}z - 2\left(P+\frac{1}{2}\frac{\partial P}{\partial r}\mathrm{d}r\right)\mathrm{d}r\,\mathrm{d}z\sin\left(\frac{\mathrm{d}\theta}{2}\right)$$
$$-r\,\mathrm{d}\theta\mathrm{d}r\,\mathrm{d}z\rho r\omega^2 = 0$$

Simplifying and neglecting small quantities of second order and putting $\sin(\mathrm{d}\theta/2)$ equal to $\mathrm{d}\theta/2$ for a small angle:

$$\frac{\partial P}{\partial r} = \rho\omega^2 r = \frac{\rho u_t^2}{r} \tag{2.75}$$

where u_t is the tangential component of the liquid velocity at radius r.

Now:

$$\mathrm{d}P = \frac{\partial P}{\partial z}\mathrm{d}z + \frac{\partial P}{\partial r}\mathrm{d}r \tag{2.75a}$$

Substituting for $\partial P/\partial z$ and $\partial P/\partial r$ from Eqs (2.74), (2.75):

$$\mathrm{d}P = (-\rho g)\mathrm{d}z + \left(r\rho\omega^2\right)\mathrm{d}r \tag{2.76}$$

Eq. (2.76) may be integrated provided that the relation between ω and r is specified. Two important cases are considered here:

(a) The *forced vortex* in which ω is constant and is independent of r, and
(b) The *free vortex* in which the energy per unit mass of the liquid is constant.

2.6.1 The Forced Vortex

In a forced vortex, the angular velocity of the liquid is maintained constant by mechanical means, such as by an agitator rotating in the liquid or by rotation in the basket of a centrifuge, and:

$$\frac{u_t}{r} = \omega = \text{constant} \tag{2.77}$$

Thus, on integration of Eq. (2.76), for a constant value of ω:

$$P = -\rho g z + \frac{\rho\omega^2 r^2}{2} + \text{constant}$$

If the z-coordinate is z_a at the point on the axis of rotation, which coincides with the free surface of the liquid (or the extension of the free surface), then the corresponding pressure P_0 must be that of the atmosphere in contact with the liquid.

That is, when $r=0$, $z=z_a$ and $P=P_0$, as shown in Fig. 2.10.

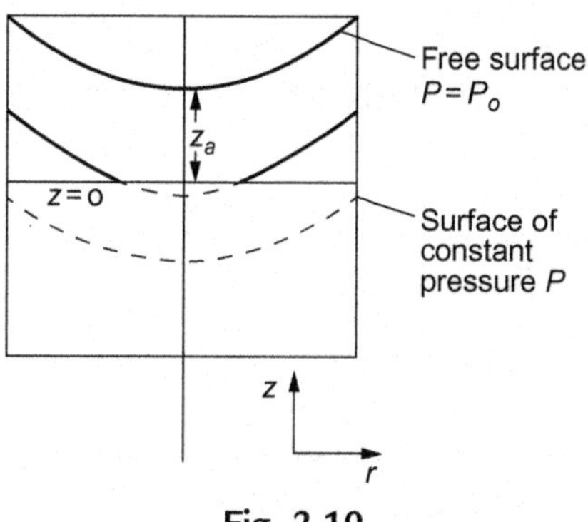

Fig. 2.10
Forced vortex.

Then, on evaluation of the constant:

$$P-P_0=\frac{\rho\omega^2r^2}{2}-\rho g(z-z_a) \tag{2.78}$$

For any constant pressure P, Eq. (2.78) is the equation of a parabola, and therefore all surfaces of constant pressure are paraboloids of revolution. The free surface of the liquid everywhere is at the pressure P_0 of the surrounding atmosphere and therefore is itself a paraboloid of revolution. Putting $P=P_0$ in Eq. (2.78) for the free surface ($r=r_0$, $z=z_0$):

$$(z_0-z_a)=\frac{w^2}{2g}r_0^2 \tag{2.79}$$

Differentiating Eq. (2.79):

$$\frac{dz_0}{dr_0}=\frac{r_0\omega^2}{g} \tag{2.80}$$

Thus the greater the speed of rotation ω, the steeper is the slope. If $r_0\omega^2 \gg g$, $dz_0/dr_0 \to \infty$ and the surface is nearly vertical, and if $r_0\omega^2 \ll g$, $dz_0/dr_0 \to 0$ and the surface is almost horizontal.

The total energy ψ per unit mass of fluid is given by Eq. (2.44):

$$\psi=\frac{u_t^2}{2}+\frac{P}{\rho}+gz \quad \text{(Eq. 2.44)}$$

where u_t denotes the tangential velocity of the liquid.

Substituting $u_t=\omega r$ and for P/ρ from Eq. (2.78):

$$\begin{aligned}\psi&=\frac{\omega^2r^2}{2}+\left(\frac{P_0}{\rho}+\frac{\omega^2r^2}{2}-g(z-z_a)\right)+gz\\&=\omega^2r^2+\frac{P_0}{\rho}+gz_a\end{aligned} \tag{2.81}$$

Thus, the energy per unit mass increases with radius r and is independent of depth z.
In the absence of an agitator or mechanical means of rotation, energy transfer will take place to equalise ψ between all elements of fluid. Thus the *forced vortex* tends to decay into a *free vortex* (where energy per unit mass is independent of radius).

Application of the forced vortex—The centrifuge

Some of the important cases of forced vortexes are:

(a) The movement of liquid within the impeller of a centrifugal pump when there is no flow as, for example, when the outlet valve is closed.
(b) The rotation of liquid within the confines of a stirrer in an agitated tank.
(c) The rotation of liquid in the basket of a centrifuge. This application will now be considered. The operation of centrifuges is considered in detail in Chapter 1 of Vol. 1B.

If the liquid is contained in a cylindrical basket, which is rotated about a vertical axis, the surfaces of constant pressure, including the free surface are paraboloids of revolution.

Thus, in general, the pressure at the walls of the basket is not constant, but varies with height. However, at normal operating speeds the centrifugal force will greatly exceed the gravitational force, and the inner surface of the liquid will be approximately vertical and the wall pressure will be nearly constant. At high operating speeds, where the gravitational force is relatively small, the functioning of the centrifuge is independent of the orientation of the axis of rotation. If mixtures of liquids or suspensions are to be separated in a centrifuge it is necessary to calculate the pressure at the walls arising from the rotation of the basket.

From Eq. (2.75):

$$\frac{\partial P}{\partial r} = \rho \omega^2 r \quad \text{(Eq. 2.75)}$$

If it is assumed that there is no slip between the liquid and the basket, ω is constant, and a forced vortex is created.

For a basket of radius R and with the radius of the inner surface of the liquid equal to r_0, the pressure P_R at the walls of the centrifuge is given by integration of Eq. (2.75) for a given value of z:

$$P_R - P_0 = \frac{\rho \omega^2}{2}\left(R^2 - r_0^2\right) \tag{2.82}$$

that is the pressure difference across the liquid at any horizontal level is

$$\frac{\rho \omega^2}{2}\left(R^2 - r_0^2\right) \tag{2.82a}$$

Example 2.6

Water is contained in the basket of a centrifuge of 0.5 m internal diameter, rotating at 50 revolutions per second. If the inner radius of the liquid is 0.15 m, what is the pressure at the walls of the basket?

Solution

Angular speed of rotation $=(2\pi \times 50)=314$ rad/s

The wall pressure is given by Eq. (2.82) as:

$$\frac{(1000 \times 314^2)}{2}(0.25^2 - 0.15^2)$$
$$= \underline{\underline{1.97 \times 10^6 \text{ N/m}^2}}$$

2.6.2 The Free Vortex

In a free vortex, the energy per unit mass of fluid is constant, i.e. constant ψ, and thus a free vortex is inherently stable. The variation of pressure with radius is obtained by differentiating Eq. (2.44) with respect to radius at constant depth z to give:

$$u_t \frac{\partial u_t}{\partial r} + \frac{1}{\rho}\frac{\partial P}{\partial r} = 0 \tag{2.83}$$

But:

$$\frac{\partial P}{\partial r} = \frac{\rho u_t^2}{r} \quad \text{(Eq. 2.75)}$$

$$\therefore \quad \frac{\partial u_t}{\partial r} + \frac{u_t}{r} = 0$$

and by integration:

$$u_t r = \text{constant} = \kappa \tag{2.84}$$

Hence the angular momentum of the liquid is everywhere constant.

Thus:

$$\frac{\partial P}{\partial r} = \frac{\rho \kappa^2}{r^3} \tag{2.85}$$

Substituting from Eqs (2.74), (2.85) into Eq. (2.75a) and integrating:

$$P - P_\infty = (z_\infty - z)\rho g - \frac{\rho \kappa^2}{2r^2} \tag{2.86}$$

where P_∞ and z_∞ are the values of P and z at $r = \infty$.

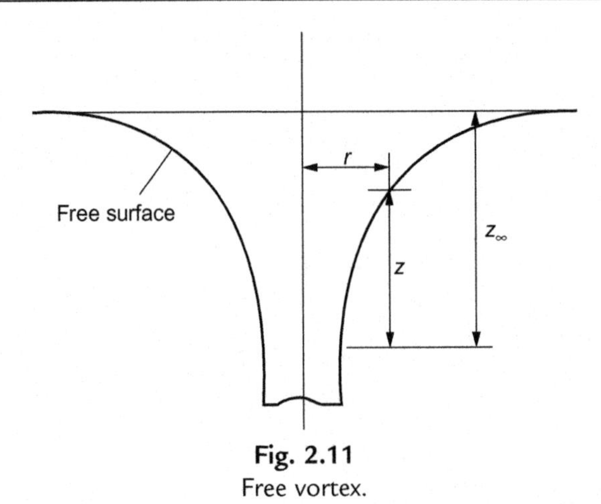

Fig. 2.11
Free vortex.

Putting $P = P_\infty$ = atmospheric pressure:

$$z = z_\infty - \frac{\kappa^2}{2r^2 g} \tag{2.87}$$

Substituting into Eq. (2.44) gives:

$$\psi = \frac{P_\infty}{\rho} + g z_\infty \tag{2.88}$$

ψ is constant by definition and equal to the value at $r = \infty$ where $u = 0$.

A free vortex (Fig. 2.11) exists:

(a) outside the impeller of a centrifugal pump;
(b) outside the region of the agitator in a stirred tank;
(c) in a cyclone separator or hydrocyclone;
(d) in the flow of liquid into a drain, as in a sink or bath;
(e) in liquid flowing round a bend in a pipe.

In all of these cases, the free vortex may be modified by the frictional effect exerted by the external walls, which has been neglected here.

2.7 Nomenclature

		Units in SI System	Dimensions in M, N, L, T, θ
A	Area perpendicular to direction of flow	m^2	$\mathbf{L^2}$
a	Coefficient in van der Waals' equation	$(kN/m^2)(m^3)^2/(kmol)^2$	$\mathbf{M\,N^{-2}\,L^5\,T^{-2}}$
b	Coefficient in van der Waals' equation	$m^3/kmol$	$\mathbf{N^{-1}\,L^3}$

C_P	Specific heat at constant pressure per unit mass	J/kg K	$\mathbf{L}^2\,\mathbf{T}^{-2}\boldsymbol{\theta}^{-1}$
C_v	Specific heat at constant volume per unit mass	J/kg K	$\mathbf{L}^2\,\mathbf{T}^{-2}\boldsymbol{\theta}^{-1}$
F	Energy per unit mass degraded because of irreversibility of process	J/kg	$\mathbf{L}^2\,\mathbf{T}^{-2}$
G	Mass rate of flow	kg/s	$\mathbf{M}\,\mathbf{T}^{-1}$
G'	Mass rate of flow per unit area	kg/m^2 s	$\mathbf{M}\,\mathbf{L}^{-2}\,\mathbf{T}^{-1}$
g	Acceleration due to gravity	9.81 m/s^2	$\mathbf{L}\,\mathbf{T}^{-2}$
H	Enthalpy per unit mass	J/kg	$\mathbf{L}^2\,\mathbf{T}^{-2}$
h_f	Head lost due to friction	m	$\mathbf{L}$
k	Numerical constant used as index for compression	–	–
l	Length of streamtube	m	$\mathbf{L}$
M	Molecular weight	kg/kmol	$\mathbf{M}\,\mathbf{N}^{-1}$
n	Number of molar units of fluid	kmol	$\mathbf{N}$
P	Pressure	N/m^2	$\mathbf{M}\,\mathbf{L}^{-1}\,\mathbf{T}^{-2}$
P_B	Pressure at wall of centrifuge basket	N/m^2	$\mathbf{M}\,\mathbf{L}^{-1}\,\mathbf{T}^{-2}$
P_R	Reduced pressure	–	–
q	Net heat flow into system	J/kg	$\mathbf{L}^2\,\mathbf{T}^{-2}$
R	Radius of centrifuge basket	m	$\mathbf{L}$
R	Universal gas constant	(8314) J/kmol K	$\mathbf{M}\,\mathbf{N}^{-1}\,\mathbf{L}^2\,\mathbf{T}^{-2}\boldsymbol{\theta}^{-1}$
r	Radius	m	$\mathbf{L}$
S	Entropy per unit mass	J/kg K	$\mathbf{L}^2\,\mathbf{T}^{-2}\boldsymbol{\theta}^{-1}$
T	Absolute temperature	K	$\boldsymbol{\theta}$
T_R	Reduced temperature	–	–
t	Time	s	$\mathbf{T}$
U	Internal energy per unit mass	J/kg	$\mathbf{L}^2\,\mathbf{T}^{-2}$
u	Mean velocity	m/s	$\mathbf{L}\,\mathbf{T}^{-1}$
u_t	Tangential velocity	m/s	$\mathbf{L}\,\mathbf{T}^{-1}$
$\dot{u}$	Velocity in streamtube	m/s	$\mathbf{L}\,\mathbf{T}^{-1}$
V	Volume of fluid	m^3	$\mathbf{L}^3$
V_R	Reduced volume	–	–
v	Volume per unit mass of fluid	m^3/kg	$\mathbf{M}^{-1}\,\mathbf{L}^3$
W	Net work per unit mass done by system on surroundings	J/kg	$\mathbf{L}^2\,\mathbf{T}^{-2}$
W_s	Shaft work per unit mass	J/kg	$\mathbf{L}^2\,\mathbf{T}^{-2}$
Z	Compressibility factor for nonideal gas	–	–
z	Distance in vertical direction	m	$\mathbf{L}$

z_a	Value of z_0 at $r_0=0$	m	**L**
α	Constant in expression for kinetic energy of fluid	–	–
γ	Ratio of specific heats C_p/C_v	–	–
θ	Angle	–	–
ρ	Density of fluid	kg/m^3	$\mathbf{M\,L^{-3}}$
ψ	Total mechanical energy per unit mass of fluid	J/kg	$\mathbf{L^2\,T^{-2}}$
ω	Angular velocity of rotation	rad/s	$\mathbf{T^{-1}}$

Suffix

c	Value at critical condition
0	Value at free surface
∞	Value at $r=\infty$

References

1. Hougen OA, Watson KM. *Chemical process principles.* New York: Wiley; 1964.
2. Clarke D. *The chemical engineer (London)*, vol. 452; 1988. p. 34. Waterhammer: 1.
3. Parmakian J. *Waterhammer analysis.* New York: Dover; 1963.
4. Sharp BB. *Water hammer: problems and solutions.* London: Edward Arnold; 1981.
5. Wylie EB, Streeter VL. *Fluid transients in systems.* Englewood Cliffs, NJ: Prentice Hall; 1993.
6. Douglas JF, Gasiorek JM, Swaffield JA. *Fluid mechanics.* 4th ed. London: Pearson; 2001.
7. Darby R, Chhabra RP. *Chemical engineering fluid mechanics.* 3rd ed. Boca Raton, FL: CRC Press; 2017.

Further Reading

1. Elliot JR, Lira CT. *Introductory chemical engineering thermodynamics.* 2nd ed. Englewood Cliffs, NJ: Prentice Hall; 2012.
2. Massey BS. *Mechanics of fluids.* 6th ed. London: Chapman and Hall; 1989.
3. Milne-Thomson LM. *Theoretical hydromechanics.* London: Macmillan; 1968.
4. Panton RL. *Incompressible flow.* 3rd ed. New York: Wiley; 2005.
5. Schlichting H. *Boundary layer theory.* 5th ed. New York: McGraw-Hill; 1968.
6. Smith JM, van Ness HC. *Introduction to chemical engineering thermodynamics.* 7th ed. New York: McGraw-Hill; 2005.

CHAPTER 3

Flow of liquids in Pipes and Open Channels

3.1 Introduction

In the processing and allied industries, it is often necessary to pump fluids over long distances, and there may be a substantial drop in pressure in both the pipeline and in individual units. Intermediate products are often pumped from one factory site to another, and raw materials such as natural gas and petroleum products may be pumped across very long distances to domestic or industrial consumers. It is necessary, therefore, to consider the problems associated with calculating the power requirements for pumping, problems with designing the most suitable flow system, with estimating the most economical sizes of pipes, with measuring the rate of flow, and frequently with controlling this flow at a steady rate. Fluid flow may take place at high pressures, when process streams are fed to a reactor, for instance; or at low pressures when, for example, vapour leaves the top of a vacuum distillation column or an evaporator.

Fluids may be conveniently categorised in a number of different ways. First, the response of the fluid to change of pressure, needs to be considered. In general, liquids may be regarded as incompressible in the sense that their densities are substantially independent of the pressure to which they are subjected, and volume changes are insufficient to affect their flow behaviour in most applications of practical interest. On the other hand, gases are highly compressible, and their isothermal densities are approximately directly proportional to the pressure-exactly so when their behaviour follows the *ideal gas law*. Again, compressibility is of only minor importance if the pressure changes by only a small proportion of the total pressure; it is then satisfactory to regard the gas as an incompressible fluid whose properties may be taken as those at the mean pressure. When pressure ratios differ markedly from unity, the effects of compressibility may give rise to fundamental changes in the flow behaviour.

All gases and most liquids of simple molecular structure (low molecular weight) exhibit what is termed *Newtonian* behaviour, and their viscosities are independent of the way in which they flow. Temperature and pressure may however, exert a strong influence on viscosity which, for highly viscous liquids, will show a rapid decrease as the temperature is increased. Gases, show the reverse tendency, however, with viscosity rising with increasing temperature, and also with increase of pressure.

Coulson and Richardson's Chemical Engineering. https://doi.org/10.1016/B978-0-08-101099-0.00003-3

Liquids of complex structure, such as polymer solutions and melts, pseudo-homogeneous suspensions of fine particles, foams, emulsions and worm-like micellar fluids, will generally exhibit *non-Newtonian* behaviour, with their apparent viscosities depending on the rate at which they are sheared, and the time for which they have been subjected to shear. They may also exhibit significant elastic properties – similar to those normally associated with solids. The flow behaviour of such fluids is therefore very much more complicated than that of Newtonian fluids.[1–3]

The fluids discussed so far consist essentially of a single phase and the composition does not vary from one point to another within the flow field. Many of the fluids encountered in processing operations consist of more than one phase however, and the flow behaviour depends on how the phases are distributed. Important cases, which will be considered include the flow of gas–liquid mixtures (such as the flow of oil–natural gas, and oil–water mixtures), where the flow pattern will be influenced by the properties of the two phases, their relative proportions and the flow velocity, and by the geometry of the flow passages. Liquids, both Newtonian and non-Newtonian, are frequently used for the transport of particulate solids both in pipelines and in open channels, and it is important to be able to design such systems effectively so that they will operate both reliably and economically. Gases are also used for the transportation of suspended solids in pipelines and, in this case, there is the added complication that the transporting fluid is compressible and the flow velocity will increase along the length of the pipeline. Similarly, there are many instances where two immiscible liquids like oil and water flow together in a pipe.

The treatment of fluid flow in this Volume is structured as follows:

Chapter 3 Flow of Newtonian and non-Newtonian Liquids
Chapter 4 Flow of Compressible Fluids (Gases)
Chapter 5 Flow of Multiphase Systems (Gas–Liquid, Liquid–Solids, Gas–Solids, Liquid–Liquid, Gas–Liquid–Solids)
Chapter 6 Flow Measurement
Chapter 7 Mixing of Liquids
Chapter 8 Pumping of Liquids and Gases

3.2 The Nature of Fluid Flow

When a fluid flows through a tube or over a surface, the pattern of the flow varies with the velocity, the physical properties of the fluid, and the geometry of the surface. This problem was first examined by Reynolds[4] in 1883 using an apparatus shown in Fig. 3.1. A glass tube with a flared entrance was immersed in a glass tank fed with water and by means of the valve, the rate of flow from the tank through the glass tube was controlled. By introducing a fine filament of coloured water from a small reservoir centrally into the flared entrance of the glass tube, the nature of the flow was observed. At low rates of flow, the coloured filament

Coulson and Richardson's Chemical Engineering. https://doi.org/10.1016/B978-0-08-101099-0.00003-3

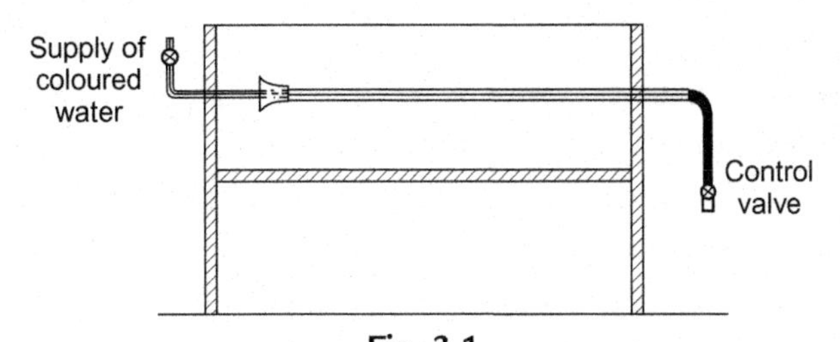

Fig. 3.1
Reynolds' method for tracing flow patterns.

remained at the axis of the tube indicating that the flow was in the form of parallel streams which did not interact with each other in the lateral direction. Such flow is called *laminar* or *streamline* and is characterised by the absence of bulk movement at right angles to the main stream direction, though a small amount of radial dispersion will occur as a result of molecular diffusion. As the flowrate was increased, oscillations appeared in the coloured filament which broke up into eddies causing dispersion across the tube section. This type of flow, known as *turbulent flow*, is characterised by the rapid movement of fluid as eddies randomly in all three directions across the tube. The general pattern is as shown in Fig. 3.2. These experiments clearly showed the nature of the transition from streamline to turbulent flow. Below the critical velocity, oscillations in the flow were unstable and any disturbance quickly disappeared by the action of the stabilising viscous forces. At higher velocities, however, the oscillations were stable and increased in amplitude, causing a high degree of radial mixing. It was found, however, that even when the main flow was turbulent there was a region near the wall (the *laminar sublayer)* in which streamline flow persisted.

In the present discussion, only the problem of steady flow will be considered in which the time average velocity in the main stream direction X is constant and equal to u_x. In laminar flow, the instantaneous velocity at any point then has a steady value of u_x and does not fluctuate. In turbulent flow, the instantaneous velocity at a point will vary about the mean value of u_x. It is convenient to consider the components of the eddy velocities in two directions – one along the main stream direction X and the other at right angles to the stream flow Y. Since the net flow in the X-direction is steady, the instantaneous velocity u_i may be imagined as being made up of a steady velocity component u_x and a fluctuating component u_{Ex}, so that:

$$u_i = u_x + u_{Ex} \tag{3.1}$$

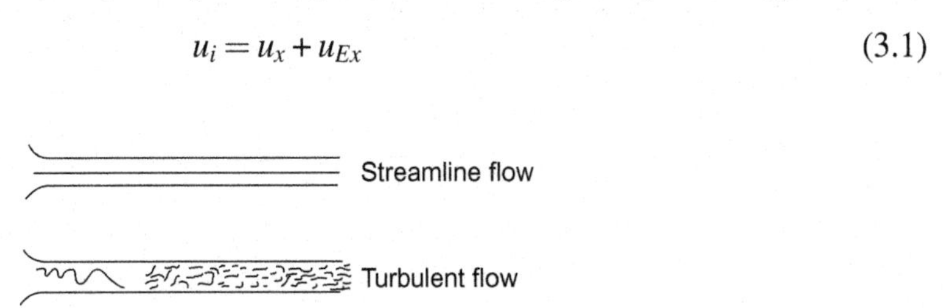

Fig. 3.2
Break-up of laminar thread in Reynolds' experiment.

Since the average value of the main stream velocity is u_x, the average value of u_{Ex}, is zero, although the fluctuating component may at any instant amount to a significant proportion of the stream velocity. The fluctuating velocity in the Y-direction also varies but again, this must have an average value of zero since there is no net flow at right angles to the stream flow. Turbulent flow is of great importance in fluids processing because it causes rapid mixing of the fluid elements and is therefore responsible for promoting high rates of heat and mass transfer and chemical reactions.

3.2.1 Flow Over a Surface

When a fluid flows over a surface, the fluid elements in contact with the surface will be brought to rest (due to the familiar no-slip or wall adherence condition) and the adjacent layers retarded by the viscous drag of the fluid. Thus the velocity in the neighbourhood of the surface will change with distance at right angles to the stream flow. It is important to realise that this change in velocity originates at the walls or a solid surface. If a fluid flowing with uniform velocity approaches a plane surface, as shown in Fig. 3.3, a velocity gradient is set up at right angles to the surface because of the viscous forces acting within the fluid. The fluid in contact with the surface must be brought to rest as otherwise there would be an infinite velocity gradient at the wall, and a corresponding infinite stress. If u_x is the velocity in the X-direction at distance y from the surface, u_x will increase from being zero at the surface ($y=0$) and will gradually approach the freestream velocity u_s at some distance from the surface. Thus, if the values of u_x are measured, the velocity profile will be as shown in Fig. 3.3. The velocity distributions are shown for three different distances downstream from the leading edge of the plane surface, and it is seen that in each case there is a rapid change in velocity near the wall and that the thickness of the layer in which the fluid is retarded becomes greater with distance in the direction of flow. The line AB divides the stream into two sections; in the lower part (below line AB) the velocity increases with distance from the surface, whilst in the upper portion the velocity is approximately equal to u_s. This line indicates the limits of the zone of retarded fluid which was termed the *boundary layer* by Prandtl.[5] As shown in Chapter 3 of Vol. 1B, the main stream velocity is approached asymptotically, and therefore the boundary layer strictly has no precise outer limit. However, it is convenient to define the boundary layer thickness such that the velocity at its outer edge equals 99% or 99.9% of the stream velocity. Other definitions are given later. Thus, by making certain assumptions

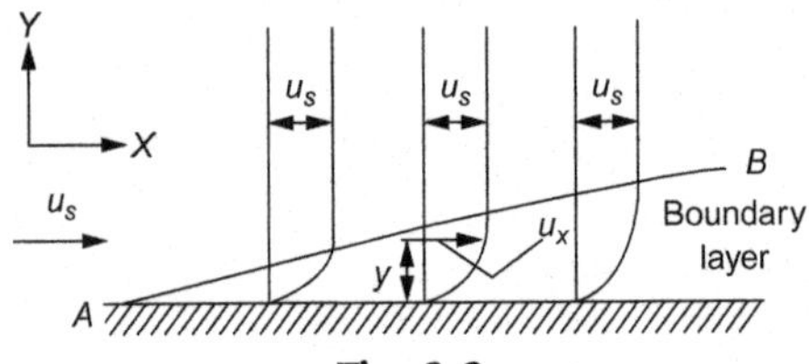

Fig. 3.3
Development of the momentum boundary layer on a plane surface.

concerning the velocity profile, it is shown in Chapter 3 of Vol. 1B that the boundary layer thickness δ at a distance x from the leading edge of a surface is dependent on the local value (based on X) of the Reynolds number.

Near the leading edge of the surface, the flow in the boundary layer is laminar, and then at a critical distance, eddies start to form giving rise to a turbulent boundary layer. In the turbulent layer, there is a thin region near the surface where the flow remains laminar, and this is known as the *laminar sublayer*. The change from laminar to turbulent flow in the boundary layer occurs at different distances downstream depending on the roughness of the surface and the physical properties of the fluid. This is discussed at length in Chapter 3 of Vol. 1B.

3.2.2 Flow in a Pipe

When a fluid flowing at a uniform velocity enters a pipe (such as from a large storage tank), the layers of fluid adjacent to the walls are slowed down as they are on a plane surface and a boundary layer forms at the entrance. This builds up in thickness as the fluid passes into the pipe. At some distance downstream from the entrance, the boundary layer thickness equals the pipe radius, after which conditions remain constant and *fully developed flow* exists. If the flow in the boundary layers is streamline where they meet, laminar flow exists in the pipe. If the transition has already taken place before they meet, turbulent flow will persist in the region of fully developed flow. The region before the boundary layers join is known as the *entry length* or the developing flow and this is discussed in greater detail in Chapter 3 of Vol. 1B.

3.3 Newtonian Fluids

3.3.1 Shearing Characteristics of a Newtonian Fluid

As a fluid is deformed because of flow and applied external forces, frictional effects are exhibited by the motion of molecules relative to each other. The effects are encountered in all fluids and are due to their *viscosities*. Considering a thin layer of fluid between two parallel planes, distance y apart as shown in Fig. 3.4 with the lower plane fixed and a shearing

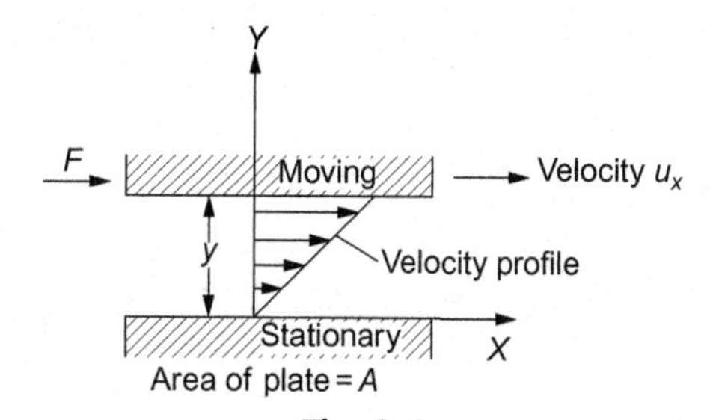

Fig. 3.4
Shear stress and velocity gradient in a fluid.

force F applied to the other, and since fluids deform continuously under shear, the upper plane moves at a steady velocity u_x relative to the fixed lower plane. When conditions are steady, the externally applied force F is balanced by an internal force in the fluid due to its viscosity and the shear force per unit area is proportional to the velocity gradient in the fluid, or:

$$\frac{F}{A} = R_y \alpha \frac{u_x}{y} \propto \frac{du_x}{dy} \tag{3.2}$$

R is the shear stress in the fluid and du_x/dy is the velocity gradient or the rate of shear. It may be noted that R corresponds to τ used by many authors to denote shear stress; similarly, shear rate may be denoted by either du_x/dy or $\dot{\gamma}$. The proportionality sign may be replaced by the introduction of the proportionality factor μ, which is the coefficient of viscosity, to give:

$$R_y = \pm \mu \frac{du_x}{dy} \tag{3.3}$$

A *Newtonian* fluid is one in which, provided that the temperature and pressure remain constant, the shear rate increases linearly with shear stress over a wide range of shear rates. As the shear stress tends to retard the fluid near the centre of the pipe and accelerate the slow moving fluid towards the walls, at any radius within the pipe it is acting simultaneously in a negative direction on the fast moving fluid and in the positive direction on the slow moving fluid. In strict terms Eq. (3.3) should be written with the incorporation of modulus signs to give:

$$\mu = \frac{|R_y|}{|du_x/dy|} \tag{3.4}$$

The viscosity strongly influences the shear stresses and hence the pressure drop for the flow. Viscosities for liquids are generally two orders of magnitude greater than that for gases at atmospheric pressure. For example, at 294 K, $\mu_{water} = 1.0 \times 10^{-3}$ N s/m^2 and $\mu_{air} = 1.8 \times 10^{-5}$ N s/m^2. Thus for a given shear rate, the shear stresses are considerably greater for

Table 3.1 Typical values of viscosity for common substances at room temperature

Substance	Viscosity (mPas)
Air	10^{-2}
Benzene	0.65
Water	1
Ethyl alcohol	1.20
Mercury (293 K)	1.55
Ethylene glycol	20
Olive oil	100
Castor oil	600
Pure glycerine (293 K)	1500
Honey	10^4
Corn syrup	10^5
Bitumen	10^{11}
Molten glass	10^{15}

liquids. It may be noted that with increase in temperature, the viscosity of a liquid decreases and that of a gas increases. At high pressures, especially near the critical point, the viscosity of a gas increases with increase in pressure. Table 3.1 lists typical viscosity values for a range of substances.

3.3.2 Pressure Drop for Flow of Newtonian Liquids Through a Pipe

Experimental work by Reynolds,[4] Nikuradse,[6] Stanton and Pannell,[7] Moody,[8] and others[9] on the drop in pressure for flow through a pipe is most conveniently represented by plotting the head loss per unit length against the average velocity through the pipe. In this way, the curve shown in Fig. 3.5 is obtained in which $i=h_f/l$ is known as the hydraulic gradient. At low velocities, the plot is linear showing that i is directly proportional to the velocity, but at higher velocities the pressure drop increases more rapidly. If logarithmic axes are used, as in Fig. 3.6, the results fall into three sections. Over the region of low velocity, the line PB has a slope of unity although beyond this region, over section BC, there is instability with poorly defined data. At higher velocities, the line CQ has a slope of about 1.8–2.0. If QC is produced, it cuts PB at the point A, corresponding in Reynolds' earlier experiments to the change from laminar to turbulent flow and representing the critical velocity. Thus for streamline flow the pressure gradient is directly proportional to the velocity, and for turbulent flow the pressure gradient is proportional to the velocity raised to the power of approximately 1.8–2.0.

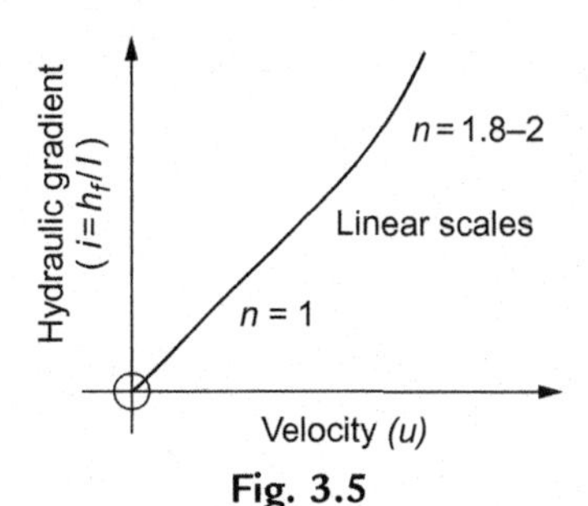

Fig. 3.5
Hydraulic gradient versus velocity.

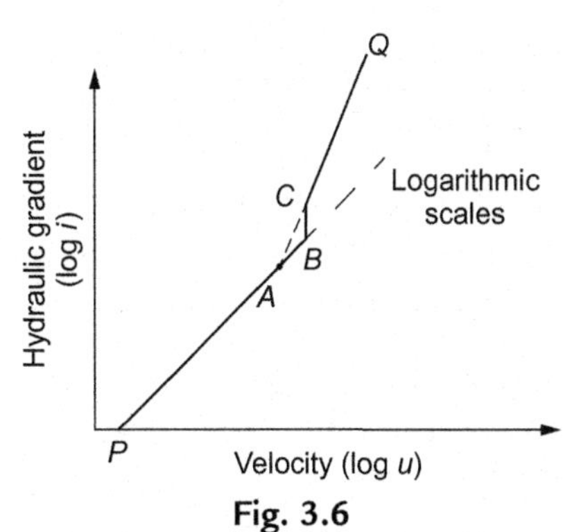

Fig. 3.6
Hydraulic gradient versus velocity showing laminar-turbulent transition.

The velocity corresponding to point A is taken as the lower critical velocity and that corresponding to B as the higher critical velocity. Experiments with pipes of various sizes showed that the critical velocity was inversely proportional to the diameter, and that it was less at higher temperatures where the viscosity was lower. This led Reynolds to develop a criterion based on the velocity of the fluid, the diameter of the tube, and the viscosity and density of the fluid. The dimensionless group $du\rho/\mu$ is termed the *Reynolds number* (Re) and this is of vital importance in the study of fluid flow. It has been found that for values of Re less than about 2000, the flow is usually laminar and for values above 4000 the flow is usually turbulent. The precise velocity at which the transition occurs depends on the geometry and on the pipe roughness. It is important to realise that there is no such thing as stable transitional flow.

If a turbulent fluid passes into a pipe so that the Reynolds number there is less than 2000, the flow pattern will change and the fluid will become streamline at some distance from the point of entry. On the other hand, if the flow is initially streamline ($Re < 2000$), the diameter of the pipe can be gradually increased so that the Reynolds number exceeds 2000 and yet streamline flow will persist in the absence of any disturbance. Unstable streamline flow has been obtained in this manner at Reynolds numbers as high as 40,000. The initiation of turbulence requires a small force at right angles to the flow to promote the formation of eddies.

The property of the fluid which appears in the Reynolds number is the kinematic viscosity μ/ρ. The kinematic viscosity of water at 294 K and atmospheric pressure is 10^{-6} m^2/s compared with 15.5×10^{-6} m^2/s for air. Thus, gases typically have higher kinematic viscosities than liquids at atmospheric pressure.

Shear stress in fluid

It is now convenient to relate the pressure drop due to fluid friction $-\Delta P_f$ to the shear stress R_0, at the walls of a pipe. If R_y is the shear stress at a distance y from the wall of the pipe, the corresponding value at the wall R_0 is given by:

$$R_0 = -\mu \left(\frac{du_x}{dy}\right)_{y=0} \quad \text{(from Eq. 3.4)}$$

In this equation, the negative sign is introduced in order to maintain a consistency of sign convention when shear stress is related to momentum transfer as in Chapter 3 of Vol. 1B. Since $(du_x/dy)_{y=0}$ must be positive (velocity increases towards the pipe centre), R_0 is negative. It is therefore more convenient to work in terms of R, the shear stress exerted by the fluid on the surface ($=-R_0$) when calculating friction data.

If a fluid is flowing through a pipe of length l and of radius r (diameter d), in which the change in pressure due to friction is ΔP_f, then a force balance on the fluid in the direction of flow in the pipe gives:

$$-\Delta P_f \pi r^2 = 2\pi r l(-R_0) = 2\pi r l R$$

or:

$$R = -R_0 = -\Delta P_f \frac{r}{2l} \tag{3.5}$$

or:

$$-R_0 = R = -\Delta P_f \frac{r}{2l} = -\Delta P_f \frac{d}{4l} \tag{3.6}$$

If a force balance is taken over the central core of fluid of radius s:

$$-\Delta P_f \pi s^2 = 2\pi s l(-R_y)$$

or:

$$-R_y = -\Delta P_f \frac{s}{2l} \tag{3.7}$$

Thus from Eqs. (3.5), (3.7):

$$\frac{R_y}{-R_0} = \frac{-R_y}{-R_0} = \frac{s}{r} = 1 - \frac{y}{r} \tag{3.8}$$

where y is the distance measured from the wall.

Thus the shear stress increases linearly from zero at the centre of the pipe to a maximum at the walls, and:

$$\frac{|R_y|}{R} = 1 - \frac{y}{r} \tag{3.9}$$

It may be noted that at the pipe walls, the shear stress acting on the walls R (positive) is equal and opposite to the shear stress acting on the fluid in contact with the walls R_0 (negative)

Thus:

$$R = -R_0 \tag{3.10}$$

Resistance to flow in pipes

Stanton and Pannell[7] measured the drop in pressure due to friction for a number of fluids flowing in pipes of various diameters and surface roughness. The results were expressed using the concept of a friction factor, defined as the dimensionless group $R/\rho u^2$, which is plotted as a function of the Reynolds number, where $R(=-R_0)$ represents the resistance to flow per unit area of pipe surface. For a given pipe surface, a single curve was found to express the results for all fluids, pipe diameters, and velocities. As with the results of Reynolds, the curve was in three parts, as shown in Fig. 3.7. At low values of Reynolds number ($Re < 2000$), $R/\rho u^2$ was independent of the surface roughness, but at high values ($Re > 2500$), $R/\rho u^2$ varied (in fact

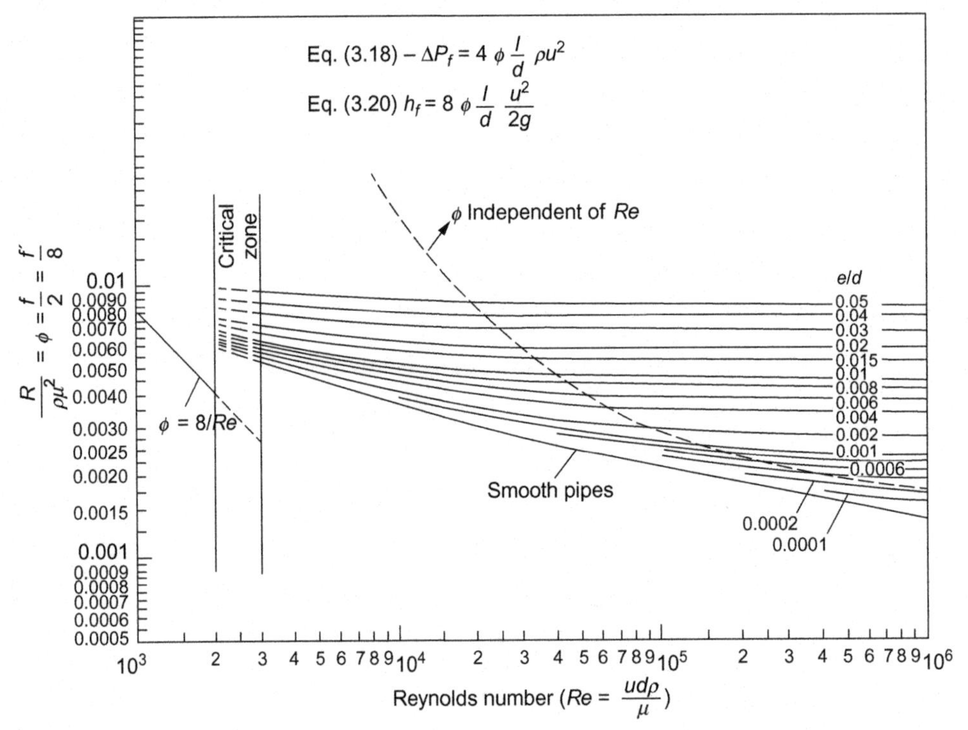

Fig. 3.7
Pipe friction chart ϕ versus Re.

increased) with the surface roughness. At very high Reynolds numbers, the friction factor became independent of Re and a function of the surface roughness only. Over the transition region of Re, from 2000 to 2500, $R/\rho u^2$ increased very rapidly, showing the great increase in friction as soon as turbulent motion commenced. This general relationship is one of the most widely used in all problems associated with fluid motion, heat transfer, and mass transfer. Moody[8] worked in terms of a friction factor (here denoted by f') equal to $8R/\rho u^2$ and expressed this factor as a function of the two dimensionless terms Re and e/d where e is a length representing the magnitude of the surface roughness. This relationship may be obtained from dimensional analysis.

Thus if R is a function of u, d, ρ, μ, and e, the dimensional analysis gives:

$$\frac{R}{\rho u^2} = \text{function of } \frac{ud\rho}{\mu} \text{ and } \frac{e}{d}$$

Thus a single curve will correlate the friction factor with the Reynolds group for all pipes with the same degree of roughness of e/d. This curve is of very great importance since it not only determines the pressure loss in the flow but can often be related to heat transfer or mass transfer, as shown in Chapter 4 of Vol. 1B. Such a series of curves for varying values of e/d is shown in Fig. 3.7 where the values of ϕ and the Fanning friction f and of the Moody (or Darcy) friction factor f' are related to the Reynolds number. Four separate regions may be distinguished:

Region 1 ($Re < 2000$) corresponds to streamline motion and a single curve represents all the data, irrespective of the roughness of the pipe surface. The equation of the curve is $R/\rho u^2 = 8/Re$.
Region 2 ($2000 < Re < 3000$) is a transition region between streamline and turbulent flow conditions. Reproducible values of pressure drop cannot be obtained in this region, but the value of $R/\rho u^2$ is considerably higher than that in the streamline region. If an unstable form of streamline flow does persist at a value of Re greater than 2000, the frictional force will correspond to that given by the curve $R/\rho u^2 = 8/Re$, extrapolated to values of Re greater than 2000.
Region 3 ($Re > 3000$) corresponds to turbulent motion of the fluid and $R/\rho u^2$ is a function of both Re and e/d, with rough pipes giving high values of $R/\rho u^2$. For smooth pipes, there is a lower limit below which $R/\rho u^2$ does not fall for any particular value of Re.
Region 4 corresponds to rough pipes at high values of Re. In this region the friction factor becomes independent of Re and depends only on (e/d) as follows:

$$\frac{e}{d} = 0.05, \quad Re > 1 \times 10^5: \quad \phi = \frac{R}{\rho u^2} = 0.0087$$
$$\frac{e}{d} = 0.0075, \quad Re > 1 \times 10^5: \quad \phi = \frac{R}{\rho u^2} = 0.0042$$
$$\frac{e}{d} = 0.001, \quad Re > 1 \times 10^6: \quad \phi = \frac{R}{\rho u^2} = 0.0024$$

A number of expressions have been proposed for calculating $R/\rho u^2 (= \phi)$ in terms of the Reynolds number including the following:

$$\text{Smooth pipes}: \quad 2.5 \times 10^3 < Re < 10^5: \quad \phi = 0.0396 Re^{-0.25} \tag{3.11}$$

$$\text{Smooth pipes}: \quad 2.5 \times 10^3 < Re < 10^7:$$
$$\phi^{-0.5} = 2.5 \ \ln\left(Re\phi^{0.5}\right) + 0.3 \quad (\text{see Eq. 4.77, volume 1B}) \tag{3.12}$$

$$\text{Rough pipes}: \quad \phi^{-0.5} = -2.5 \ \ln\left(0.27\frac{e}{d} + 0.885 Re^{-1}\phi^{-0.5}\right) \tag{3.13}$$

$$\text{Rough pipes}: \quad \frac{e}{d}Re\phi^{0.5} \gg 3.3: \quad \phi^{-0.5} = 3.2 - 2.5 \ \ln\frac{e}{d} \tag{3.14}$$

Eq. (3.11) is due to Blasius[10] and the others are derived from considerations of velocity profile. In addition to the Moody or Darcy friction factor $f' = 8R/\rho u^2$, the Fanning friction factor $f = 2R/\rho u^2$ is often used. It is extremely important therefore to be clear about the exact definition of the friction factor when using this term in calculating head losses due to friction.

Calculation of pressure drop for liquid flowing in a pipe

For the flow of a fluid in a pipe of length l and diameter d, the total frictional force at the walls is the product of the shear stress R and the surface area of the pipe ($R\pi dl$). This frictional force results in a change in pressure ΔP_f so that for a horizontal pipe:

$$R\pi dl = -\Delta P_f \pi \frac{d^2}{4} \tag{3.15}$$

or:

$$-\Delta P_f = 4R\frac{l}{d} = 4\frac{R}{\rho u^2}\frac{l}{d}\rho u^2 = 4\phi\frac{l}{d}\rho u^2 \tag{3.16}$$

and:

$$\phi = \frac{R}{\rho u^2} = \frac{-\Delta P d}{4l\rho u^2} \tag{3.17}$$

The head lost due to friction is then:

$$h_f = \frac{-\Delta P_f}{\rho g} = 4\frac{R}{\rho u^2}\frac{l}{d}\frac{u^2}{g} \tag{3.18}$$

The energy dissipated per unit mass F is then given by Eq. (3.19):

$$F = \frac{-\Delta P_f}{\rho} = 4\frac{R}{\rho u^2}\frac{l}{d}u^2 = 4\phi\frac{l}{d}u^2 \tag{3.19}$$

To calculate $-\Delta P_f$ it is therefore necessary to evaluate e/d and obtain the corresponding value of $\phi = R/\rho u^2$ from a knowledge of the value of *Re using Fig.* 3.7 *or an expression like Eq.* (3.13) *or* (3.14), *etc.* This value of ϕ is then used in Eq. (3.16) to give —ΔP_f or the head loss due to friction h_f as:

$$h_f = \frac{-\Delta P_f}{\rho g} = 4\phi\frac{l}{d}\frac{u^2}{g} = 8\phi\frac{l}{d}\frac{u^2}{2g} \tag{3.20}$$

With the friction factors used by Moody and Fanning, f' and f respectively, the head loss due to friction is obtained from the following equations:

$$\text{Moody}: \quad h_f = f'\frac{l}{d}\frac{u^2}{2g} \tag{3.21}$$

$$\text{Fanning}: \quad h_f = 4f\frac{l}{d}\frac{u^2}{2g} \tag{3.22}$$

The energy dissipated per unit mass due to the irreversibility of the process is given by $F = -\Delta P_f/\rho = 4\phi(l/d)u^2$ (Eq. 3.19).

If it is necessary to calculate the flow in a pipe where the pressure drop is specified, the velocity u is required but the Reynolds number is unknown, and this approach cannot be used to give $R/\rho u^2$ directly. One alternative here is to estimate the value of $R/\rho u^2$ and calculate the velocity and hence the corresponding value of Re. The value of $R/\rho u^2$ is then determined and, if different from the assumed value, a further trial becomes necessary.

An alternative approach to this problem is to use a friction group formed by combining ϕ and Re as follows:

$$\phi Re^2 = \frac{R}{\rho u^2}\left(\frac{\rho u d}{\mu}\right)^2 = \frac{Rd^2\rho}{\mu^2} = \frac{-\Delta P_f d^3 \rho}{4l\mu^2} \tag{3.23}$$

If Re is plotted as a function of ϕRe^2 (independent of fluid velocity) and e/d as shown in Fig. 3.8, the group $(-\Delta P_f d^3 \rho / 4l\mu^2) = \phi Re^2$ may be evaluated directly as it is independent of velocity. Hence Re may be found from the graph and the required velocity $\left(u = \frac{Re\mu}{\rho d}\right)$ obtained. Similarly, if the diameter of pipe is required to transport fluid at a mass rate of flow G with a given fall in pressure, the following group, which is independent of d, may be used to obviate the need for a trial and error solution:

$$\left(\frac{R}{\rho u^2}\right)\left(\frac{ud\rho}{\mu}\right)^{-1} = \phi Re^{-1} = \frac{-\Delta P_f \mu}{4\rho^2 u^3 l} \tag{3.24}$$

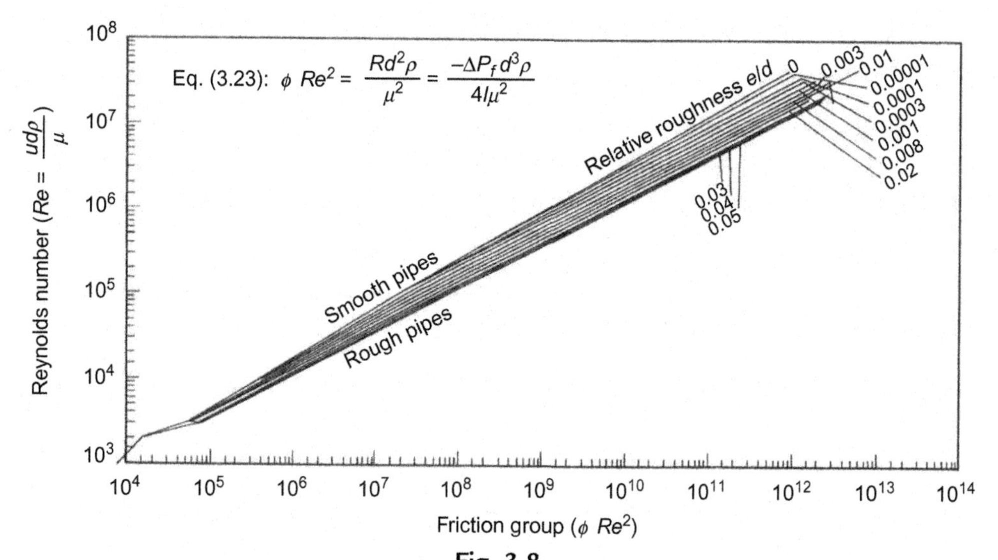

Fig. 3.8
Pipe friction chart in terms of ϕRe^2 versus Re for various values of e/d.

Table 3.2 Values of absolute roughness *e*

	ft	mm
Drawn tubing	0.000005	0.0015
Commercial steel and wrought-iron	0.00015	0.046
Asphalted cast-iron	0.0004	0.12
Galvanised iron	0.0005	0.15
Cast-iron	0.00085	0.26
Wood stave	0.0006–0.003	0.18–0.9
Concrete	0.001–0.01	0.3–3.0
Riveted steel	0.003–0.03	0.9–9.0

Effect of roughness of pipe surfaces

The estimation of the roughness of the surface of the pipe often presents considerable difficulty. The use of an incorrect value is not usually serious, however, even for turbulent flow at low Reynolds numbers because the pressure drop is not critically dependent on the roughness in this region. However, at high values of the Reynolds number, the effect of pipe roughness is considerable, as may be seen from the plot of $R/\rho u^2$ against the Reynolds number shown in Fig. 3.7. The values of the absolute roughness have been measured for a number of materials and typical data are given in Table 3.2. Where the value for the pipe surface in question is not given, it is necessary to estimate an approximate value based on the available data. Where pipes have become corroded, the value of the roughness is commonly increased, up to tenfold.

Values of roughness applicable to materials used in the construction of open channels are also included in Table 3.2.

With practical installations it must be remembered that the frictional losses cannot be estimated with very great accuracy because the roughness will change with use and the pumping unit must therefore always have ample excess capacity to overcome the extra frictional loss on this account.

Example 3.1

Ninety-eight per cent sulphuric acid is pumped at 4.5 tonne/h (1.25 kg/s) through a 25 mm diameter pipe, 30 m long, to a reservoir 12 m higher than the feed point. Calculate the pressure drop in the pipeline.

$$\text{Viscosity of acid} = 25 \text{ mN s/m}^2 \quad \text{or} \quad 25 \times 10^{-3} \text{ N s/m}^2$$

$$\text{Density of acid} = 1840 \text{ kg/m}^3$$

Solution
Reynolds number:

$$Re = \frac{ud\rho}{\mu} = \frac{4G}{\pi \mu d}$$

$$= \frac{4 \times 1.25}{\pi \times 25 \times 10^{-3} \times 25 \times 10^{-3}}$$

$$= 2545$$

For a mild steel pipe, suitable for conveying the acid, the roughness e will be between 0.05 and 0.5 mm (0.00005 and 0.0005 m).

The relative roughness is thus:

$$\frac{e}{d} = 0.002 - 0.02$$

From Fig. 3.7:

$$\frac{R}{\rho u^2} = 0.006 \text{ over this range of } \frac{e}{d}$$

and the velocity is:

$$u = \frac{G}{\rho A} = \frac{1.25}{1840 \times (\pi/4)(0.025)^2}$$

$$= 1.38 \text{ m/s}$$

The kinetic energy attributable to this velocity will be dissipated when the liquid enters the reservoir. The pressure drop may now be calculated from the energy balance equation and Eq. (3.19). For turbulent flow ($x = 1$) of an incompressible fluid:

$$\Delta \frac{u^2}{2} + g\Delta z + v(P_2 - P_1) + 4\frac{R}{\rho u^2}\frac{l}{d}u^2 = 0 \quad (\text{from Eq. 2.67 with } W_s = 0)$$

$$\therefore \quad -\Delta P = (P_1 - P_2) = \rho\left[0.5 + 4\frac{R}{\rho u^2}\frac{l}{d}\right]u^2 + \rho g \Delta z$$

$$= 1840\left\{\left[0.5 + 4 \times 0.006\frac{30}{0.025}\right]1.38^2 + 9.81 \times 12\right\}$$

$$= 3.19 \times 10^5 \text{ N/m}^2$$

or:

$$\underline{\underline{-\Delta P = 320 \text{ kN/m}^2}}$$

Example 3.2

Water flows in a 50 mm pipe, 100 m long, whose roughness e is equal to 0.013 mm. If the pressure drop across this length of pipe is not to exceed 50 kN/m^2, what is the maximum allowable water velocity? The density and viscosity of water may be taken as 1000 kg/m^3 and 1.0 mN s/m^2, respectively.

Solution

From Eq. (3.23):

$$\phi Re^2 = \frac{R}{\rho u^2} Re^2 = -\frac{\Delta P_f d^3 \rho}{4 l \mu^2}$$

$$-\frac{\Delta P_f d^3 \rho}{4 l u^2} = \frac{50,000(0.05)^3 1000}{4 \times 100(1 \times 10^{-3})^2}$$

$$= 1.56 \times 10^7$$

and:

$$\frac{e}{d} = \frac{0.013}{50} = 0.00026$$

From Fig. 3.8, for $\phi Re^2 = 1.56 \times 10^7$ and $(e/d) = 0.00026$, then:

$$Re = \frac{\rho u d}{\mu} = 7.9 \times 10^4$$

Hence:

$$u = \frac{7.9 \times 10^4 (1 \times 10^{-3})}{1000 \times 0.05}$$

$$= \underline{\underline{1.6 \text{ m/s}}}$$

Example 3.3

A cylindrical tank, 5 m in diameter, discharges water through a horizontal mild steel pipe, 100 m long and 225 mm in diameter, connected to the base. What is the time taken for the water level in the tank to drop from 3 to 0.3 m above the bottom? The viscosity of water may be taken as 1 mN s/m^2.

Solution

If at time t, the liquid level is D m above the bottom of the tank, then designating point 1 as the liquid level and point 2 as the pipe outlet, and applying the energy balance Eq. (2.67) for turbulent flow, then:

$$\Delta \frac{u^2}{2} + g\Delta z + v(P_2 - P_1) + F = 0$$

$$P_2 = P_1 = 101.3 \text{ kN/m}^2$$

$$\frac{u_1}{u_2} = \left(\frac{0.225}{5}\right)^2 = 0.0020, \text{ and hence } u_1 \text{ may be neglected}$$

and:

$$\Delta z = -D$$

Thus:

$$\frac{u_2^2}{2} - Dg + 4\frac{R}{\rho u^2}\frac{l}{d}u_2^2 = 0$$

or:

$$u_2^2 - 19.62D + 8\frac{R}{\rho u^2}\left(\frac{100}{0.225}\right)u_2^2 = 0$$

from which:

$$u_2 = \frac{4.43\sqrt{D}}{\sqrt{1 + 3552(R/\rho u^2)}}$$

As the level of liquid in the tank changes from D to $(D + \mathrm{d}D)$ in time interval dt, one can write the mass balance as: $\frac{\pi}{4}(5)^2\left(-\frac{\mathrm{d}D}{\mathrm{d}t}\right) = \frac{\pi}{4}(0.225)^2 U_2$ i.e. $-\frac{\mathrm{d}D}{\mathrm{d}t} = 2.025x10^3 U_2$

Now substituting for u_2 from above, the time taken for the level to change by an amount dD is given by:

$$\mathrm{d}t = \frac{-19.63\ \mathrm{d}D}{(\pi/4)0.225^2 \times 4.43\sqrt{D}/\sqrt{[1 + 3552(R/\rho u^2)]}}$$

$$= -111.5\sqrt{\left[1 + 3552\frac{R}{\rho u^2}\right]}D^{-0.5}\mathrm{d}D$$

and the total time:

$$t = -\int_3^{0.3} 111.5\sqrt{\left[1 + 3552\frac{R}{\rho u^2}\right]}\ D^{-0.5}\,\mathrm{d}D$$

Assuming that $R/\rho u^2$ is constant over the range of flow rates considered, then:

$$t = 264\sqrt{\left[1 + 3552\frac{R}{\rho u^2}\right]}\ \mathrm{s}$$

If it is assumed that the kinetic energy of the liquid is small compared with the frictional losses, then an approximate value of $R/\rho u^2$ may be calculated.

Pressure drop along the pipe $= D\rho g = \frac{4Rl}{d}$:

$$\frac{R}{\rho u^2}Re^2 = \frac{Rd^3\rho}{\mu^2} = \frac{Dg\rho^2 d^3}{4l\mu^2}$$

$$= \frac{(D \times 9.81 \times 1000^2 \times 0.225^3)}{(4 \times 100 \times 0.001^2)}$$

$$= 2.79 \times 10^8 D$$

As D varies from 3 to 0.3 m, $(R/\rho u^2)Re^2$ varies from 8.38×10^8 to 0.838×10^8. It is of interest to consider whether the difference in roughness of a new or old pipe has a significant effect at this stage.

For a mild steel pipe:

$$\text{new:} \quad e = 0.00005 \text{ m}, \quad \frac{e}{d} = 0.00022, \quad Re = 7.0 - 2.2 \times 10^5 \quad \text{(from Fig. 3.7)}$$

$$\text{old:} \quad e = 0.0005 \text{ m}, \quad \frac{e}{d} = 0.0022, \quad Re = 6.0 - 2.2 \times 10^5 \quad \text{(from Fig. 3.7)}$$

For a new pipe, $R/\rho u^2$ therefore varies from 0.0019 to 0.0020 and for an old pipe, $R/\rho u^2 = 0.0029$ (from Fig. 3.7).

Taking a value of 0.002 for a new pipe, and assuming that this is constant, then:

$$t = 264\sqrt{(1 + 3552 \times 0.002)} = 264\sqrt{8.1} = 750 \text{ s}$$

The pressure drop due to friction is approximately (7.1/8.1) = 0.88% or 88% of the total pressure drop, that is the drop due to friction plus the change in kinetic energy.

Thus:

$$\frac{R}{\rho u^2} Re^2 \text{ varies from about } 7.4 \times 10^8 \text{ to } 0.74 \times 10^8$$

$$Re \text{ varies from about } 6.2 \times 10^5 \text{ to } 1.9 \times 10^5$$

and:

$$\frac{R}{\rho u^2} \text{ varies from about } 0.0019 \text{ to } 0.0020$$

which is sufficiently close to the assumed value of $R/\rho u^2 = 0.002$.

The time taken for the level to fall is therefore about 750 s or 12.5 min.

Example 3.4

Two storage tanks, A and B, containing a petroleum product, discharge through pipes each 0.3 m in diameter and 1.5 km long to a junction at D, as shown in Fig. 3.9. From D the liquid is passed through a 0.5 m diameter pipe to a third storage tank C, 0.75 km away. The surface of the liquid in A is initially 10 m above that in C and the liquid level in B is 6 m higher than that in A. Calculate the initial rate of discharge of liquid into tank C assuming the pipes are of mild steel. The density and viscosity of the liquid are 870 kg/m^3 and 0.7 mN s/m^2, respectively.

Solution

Because the pipes are long, the kinetic energy of the fluid and minor losses at the entry to the pipes may be neglected.

It may be assumed, as a first approximation, that $R/\rho u^2$ is the same in each pipe and that the velocities in pipes AD, BD, and DC are u_1, u_2, and u_3 respectively, if the pressure at D is taken as P_D and point D is z_d m above the datum for the calculation of potential energy, the liquid level in C.

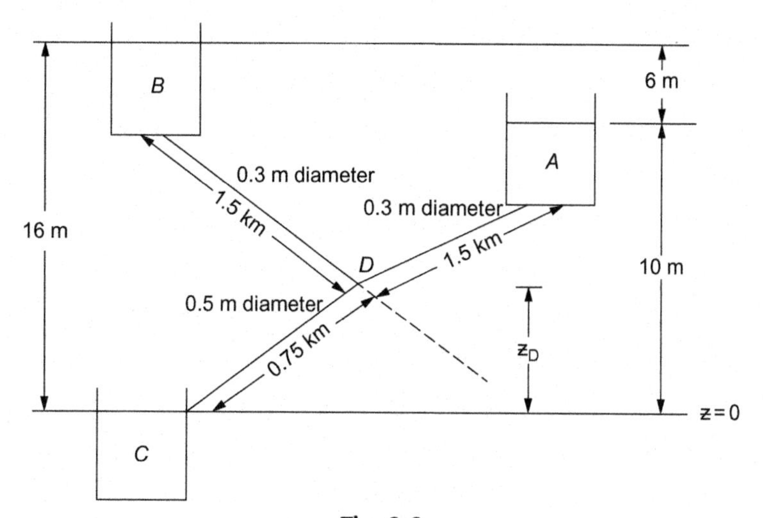

Fig. 3.9
Tank layout for Example 3.4.

Then applying the energy balance equation between *D* and the liquid level (free surface) in each of the tanks gives:

$$A-D: \quad (z_d-10)g + vP_D + 4\frac{R}{\rho u^2}\left(\frac{1500}{0.3}\right)u_1^2 = 0 \tag{1}$$

$$B-D: \quad (z_d-16)g + vP_D + 4\frac{R}{\rho u^2}\left(\frac{1500}{0.3}\right)u_2^2 = 0 \tag{2}$$

$$D-C: \quad -z_d g - vP_D + 4\frac{R}{\rho u^2}\left(\frac{750}{0.5}\right)u_3^2 = 0 \tag{3}$$

From Eqs. (1), (2):

$$6g + 20{,}000\frac{R}{\rho u^2}\left(u_1^2 - u_2^2\right) = 0 \tag{4}$$

and from Eqs. (2), (3):

$$-16g + 20{,}000\frac{R}{\rho u^2}\left(u_2^2 + 0.30u_3^2\right) = 0 \tag{5}$$

Taking the roughness of mild steel pipe *e* as 0.00005 m, *e*/*d* varies from 0.0001 to 0.00017.

As a first approximation, $R/\rho u^2$ may be taken as 0.002 in each pipe, and substituting this value in Eqs. (4), (5) then:

$$58.9 + 40\left(u_1^2 - u_2^2\right) = 0 \tag{6}$$

$$-156.0 + 40\left(u_2^2 + 0.30u_3^2\right) = 0 \tag{7}$$

The flowrate in *DC* is equal to the sum of the flowrates in *AD* and *BD*.

$$\therefore \quad \frac{\pi}{4}0.3^2u_1 + \frac{\pi}{4}0.3^2u_2 = \frac{\pi}{4}0.5^2u_3$$

or:

$$u_1 + u_2 = 2.78u_3 \tag{8}$$

From Eq. (6):

$$u_1^2 = u_2^2 - 1.47 \tag{9}$$

From Eqs. (7)–(9):

$$-156.0 + 40\left\{u_2^2 + 0.3 \times \left(\frac{1}{2.78}\right)^2 \left[u_2^2 + u_2^2 - 1.47 + 2u_2\sqrt{(u_2^2 - 1.47)}\right]\right\} = 0$$

$$\therefore \quad u_2\sqrt{(u_2^2 - 1.47)} = 50.7 - 13.8u_2^2$$

or:

$$u_2^4 - 7.38u_2^2 + 13.57 = 0$$

$$\therefore \quad u_2^2 = 0.5\left[7.38 \pm \sqrt{(54.46 - 54.28)}\right] = 3.90 \text{ or } 3.48$$

and:

$$u_2 = 1.975 \text{ or } 1.87 \text{ m/s}$$

Substituting in Eq. (9):

$$u_1 = 1.56 \text{ or } 1.42 \text{ m/s}$$

Substituting in Eq. (8):

$$u_3 = 1.30 \text{ or } 1.18 \text{ m/s}$$

When these values of u_1, u_2, and u_3 are substituted in Eq. (7), the lower set of values satisfies the equation and the higher set, introduced as false roots during squaring, does not.

Thus:

$$u_1 = 1.42 \text{ m/s}$$
$$u_2 = 1.87 \text{ m/s}$$

and:

$$u_3 = 1.18 \text{ m/s}$$

The assumed value of 0.002 for $R/\rho u^2$ must now be checked:

$$\text{For pipe } AD: \quad Re = \frac{0.3 \times 1.42 \times 870}{0.7 \times 10^{-3}} = 5.3 \times 10^5$$

$$\text{For pipe } BD: \quad Re = \frac{0.3 \times 1.87 \times 870}{0.7 \times 10^{-3}} = 6.9 \times 10^5$$

$$\text{For pipe } DC: \quad Re = \frac{0.5 \times 1.18 \times 870}{0.7 \times 10^{-3}} = 7.3 \times 10^5$$

For $e/d = 0.0001$–0.00017 and this range of *Re*, $R/\rho u^2$ varies from 0.0019 to 0.0017. The assumed value of 0.002 is therefore sufficiently close.

Thus, the volumetric flowrate is:

$$\frac{\pi}{4}\left(0.5^2 \times 1.18\right) = \underline{\underline{0.23\ \mathrm{m^3/s}}}$$

Example 3.5

A chemical solvent ($\rho = 900$ kg/m^3; $\mu = 5$ mPa s) stored in an open top tank flows into a vessel through a 20 m long pipe. The top of the liquid level in the storage tank is 4.5 m above the discharge point. (i) Determine the pipe diameter to achieve the mass flow rate of 2100 kg/h. (ii) How will your answer change if the flow rate is increased by 50%?

Solution

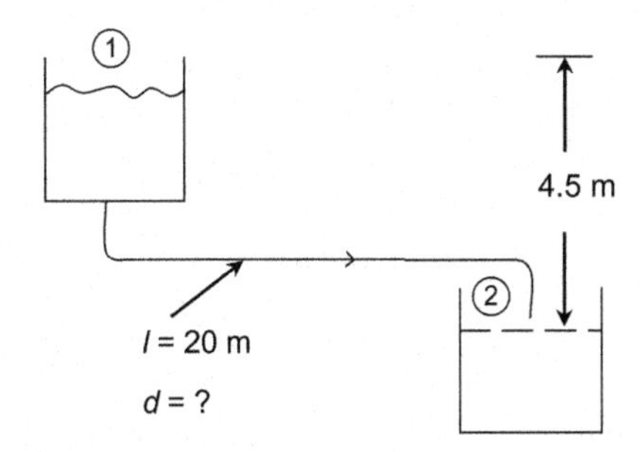

(i) Applying the energy balance between point 1 and 2 on the free surface in two tanks:

$$\frac{P_1}{\rho g} + \frac{u_1^2}{2\alpha_1 g} + z_1 + h_s = \frac{P_2}{\rho g} + \frac{u_2^2}{2\alpha_2 g} + z_2 + h_f$$

Here $P_1 = P_2 = 1$ atm; $u_1 \asymp 0$; $u_2 \asymp 0$; $h_s = 0$

$$\therefore\ \ h_f = (z_1 - z_2) = 4.5\ \mathrm{m}$$

The head loss due to friction is given by Eq. (3.18):

$$4\phi \frac{l}{d}\frac{u^2}{g} = 4.5$$

where $l = 20$ m; u is the velocity in the pipe, and ϕ is the friction factor which itself is a function of the Reynolds number, $Re = \rho u d/\mu$. Because d is unknown, u cannot be estimated. Therefore, a trial and error solution is needed in general. On the other hand, if the effect of roughness (e/d) is neglected and if it is assumed that $Re < 10^5$, one can rearrange Eq. (3.11) as follows:

$$\phi = 0.0396\,Re^{-0.25}\ \ (\mathrm{Eq.\,3.11})$$

Dividing this relation by *Re*:

$$\frac{\phi}{Re} = 0.0396\,Re^{-1.25}$$

Using Eq. (3.24) to rewrite this equation as

$$\frac{-\Delta P_f \cdot \mu}{4\rho^2 u^3 l} = 0.0396\,Re^{-1.25}$$

Now using $u = \frac{4G}{\pi d^2 \rho}$:

$$\frac{-\Delta P_f \cdot \mu}{4\rho^2 \left(\frac{4G}{\pi d^2 \rho}\right)^3 l} = 0.0396\left\{\frac{\rho\left(\frac{4G}{\pi d^2 \rho}\right)\cdot d}{\mu}\right\}^{-1.25}$$

Noting that $-\Delta P_f = \rho g h_f$

$$\frac{\rho g h_f \mu . \pi^3 d^6 \rho^3}{4\rho^2 l (4G)^3} = 0.0396\left(\frac{4G}{\pi d \mu}\right)^{-1.25}$$

This simplifies as:

$$\left(\frac{\pi^3}{256}\right)\frac{\rho^2 h_f \mu g d^6}{lG^3} = 0.0396\left(\frac{4G}{\pi d \mu}\right)^{-1.25}$$

Now substituting values and solving for *d*:

$$\left(\frac{3.14^3}{256}\right)\frac{900^2 \times 4.5 \times 5 \times 10^{-3} \times 9.81 \times d^{4.75}}{20 \times (2100/3600)^3} = 0.0396\left(\frac{4 \times \frac{2100}{3600}}{3.14 \times 5 \times 10^{-3}}\right)^{-1.25}$$

$d = 0.0222$ m or $d = 22.2$ mm.

Now checking the value of the Reynolds number:

$$Re = \frac{4G}{\pi d \mu} = \frac{4 \times (2100/3600)}{3.14 \times 5 \times 10^{-3} \times 0.0222} = 6690$$

This value is well within the range of validity of Eq. (3.11).

(ii) In this case, one can still use the following equation obtained in part (i):

$$\left(\frac{\pi^3}{256}\right)\frac{\rho^2 h_f \mu g d^6}{lG^3} = 0.0396\left(\frac{4G}{\pi d \mu}\right)^{-1.25}$$

Now substituting values and using the new value of G of $2100 \times 1.5 = 3150$ kg/h, it gives for

$$d^{4.75} = 0.0396\left(\frac{4 \times (3150/3600)}{3.14 \times 0.005}\right)^{-1.25} \times \frac{20 \times (3150/3600)^3 \times 256}{\pi^3 \times 900^2 \times 4.5 \times 0.005 \times 9.81}$$

$$d = 0.0258 \text{ m or } 25.8 \text{ mm}$$

The new value of Reynolds number, $Re = 8642$ which is still within the range of validity of Eq. (3.11).

In principle, one can use either of the Eqs. (3.11)–(3.14) depending upon the expected flow regime and/or whether or not the pipe roughness is important, algebra may become more tedious in some cases.

3.3.3 Reynolds Number and Shear Stress

For a fluid flowing through a pipe the momentum per unit cross-sectional area is given by ρu^2. This quantity, which is proportional to the inertial force per unit area, is the force required to counterbalance the momentum flux.

The ratio u/d represents the velocity gradient in the fluid, and thus the group $(\mu u/d)$ is proportional to the shear stress in the fluid, so that $(\rho u^2)/(\mu u/d) = (du\rho/\mu) = Re$ is proportional to the ratio of the inertial forces to the viscous forces. This is an important physical interpretation of the Reynolds number. In turbulent flow with high values of Re, the inertial forces become predominant and the viscous shear stress becomes correspondingly less important.

In steady streamline flow the direction and velocity of flow at any point remain constant and the shear stress R_y at a point where the velocity gradient at right angles to the direction of flow is $\mathrm{d}u_x/\mathrm{d}y$ and is given, for a Newtonian fluid, by the relation:

$$R_y = -\mu\frac{\mathrm{d}u_x}{\mathrm{d}y} = -\frac{\mu}{\rho}\frac{\mathrm{d}(\rho u_x)}{\mathrm{d}y} \tag{3.25}$$

which gives the relation between shear stress and momentum per unit volume (ρu_x) (Eq. 3.3). The negative sign in Eq. (3.25) indicates that the shear stress on the fluid exerts a retarding force on the faster-moving fluid.

In turbulent motion, the presence of circulating or eddy currents brings about a much-increased exchange of momentum in all three directions of the stream flow, and these eddies are responsible for the random fluctuations in velocity u_E. The high rate of transfer in turbulent flow is accompanied by a much higher shear stress for a given velocity gradient.

Thus:

$$R_y = -\left(\frac{\mu}{\rho} + E\right)\frac{\mathrm{d}(\rho u_x)}{\mathrm{d}y} \tag{3.26}$$

where E is known as the *eddy kinematic viscosity* of the fluid, which will depend upon the degree of turbulence in the fluid, is not a physical property of the fluid and varies with position.

In streamline flow, E is very small and approaches zero, so that μ/ρ determines the shear stress. In turbulent flow, E is negligible at the wall and increases very rapidly with distance from the wall. Laufer,[11] using very small hot-wire anemometers, measured the velocity fluctuations and gave a valuable account of the structure of turbulent flow. In the operations of mass, heat, and momentum transfer, the transfer has to be effected through the laminar layer near the wall, and it is here that the greatest resistance to transfer lies.

The Reynolds group will often be used to characterise the nature of flow where a moving fluid is concerned. Thus the drag produced as a fluid flows past a particle is related to the Reynolds number in which the diameter of the particle is used in place of the diameter of the pipe. Under these conditions the transition from streamline to turbulent flow occurs at a very much lower value. Again for the flow of fluid through a bed composed of granular particles, a mean dimension of the particles is often used, and the velocity is usually calculated by dividing the flowrate by the total area of the bed. In this case, there is no sharp transition from streamline to turbulent flow because the sizes of the individual flow passages vary.

If the surface over which the fluid is flowing contains a series of relatively large projections, turbulence may arise at a very low Reynolds number. Under these conditions, the frictional force will be increased but so will the coefficients for heat transfer and mass transfer, and therefore turbulence is often purposely induced by this method.

3.3.4 Velocity Distributions and Volumetric Flowrates for Streamline Flow

The velocity across the cross-section of a fluid flowing in a pipe is not uniform. Whilst this distribution in velocity over a diameter can be calculated for streamline flow, this is not possible in the same basic manner for turbulent flow.

The pressure drop due to friction and the velocity distribution resulting from the shear stresses within a fluid in streamline Newtonian flow are considered for three cases: (a) the flow through a pipe of circular cross-section, (b) the flow between two parallel plates, and (c) the flow through an annulus. The velocity at any distance from the boundary surfaces is calculated, and this and the mean velocity of the fluid are related to the pressure gradient in the system. For flow through a circular pipe, the kinetic energy of the fluid may be calculated in terms of the mean velocity of flow thereby obtaining an expression for α appearing in the energy equation.

Pipe of circular cross-section

A horizontal pipe with a concentric element marked $ABCD$ is shown in Fig. 3.10. Since the flow is steady, the net force on this element must be zero. The forces acting are the normal pressures over the ends and shear forces over the curved sides.

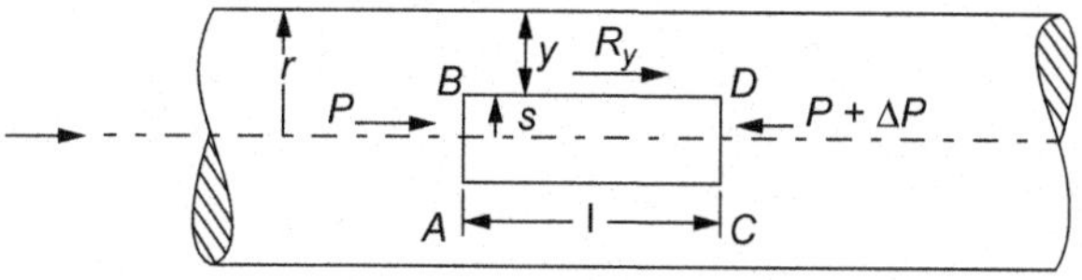

Fig. 3.10
Flow through pipe.

The force over $AB = P\pi s^2$

The force over $CD = -(P + \Delta P)\pi s^2$

and the force over curved surface $= 2\pi s l R_y$

$$\text{where the shear } R_y = \mu \frac{du_x}{ds} \left(= -\mu \frac{du_x}{dy} \right) \tag{3.27}$$

Taking a force balance on the fluid element ABDC:

$$P\pi s^2 - (P + \Delta P)\pi s^2 + 2\pi s l \, \mu \frac{du_x}{ds} = 0$$

or:

$$\left(\frac{-\Delta P}{l} \right) s + 2\mu \frac{du_x}{ds} = 0 \tag{3.28}$$

From Eq. (3.27):

$$\frac{du_x}{dy} = -\frac{du_x}{ds}$$

and hence in Eq. (3.28):

$$\frac{du_x}{dy} = \left(\frac{-\Delta P}{l} \right) \frac{s}{2\mu}$$

and the shear rate at the wall is given by:

$$\left(\frac{du_x}{dy} \right)_{y=0} = \left(\frac{-\Delta P}{l} \right) \frac{r}{2\mu} = \left(\frac{-\Delta P}{l} \right) \frac{d}{4\mu} \tag{3.29}$$

The velocity at any distance s from the axis of the pipe may now be found by integrating Eq. (3.28) to give:

$$u_x = \frac{1}{2\mu} \left(\frac{\Delta P}{l} \right) \frac{s^2}{2} + \text{constant}$$

At the walls of the pipe, that is where $s = r$, the velocity u_x must be zero in order to satisfy the condition of zero wall slip. Substituting the value $u_x = 0$, when $s = r$, then:

$$\text{constant} = \frac{1}{2\mu}\left(\frac{-\Delta P}{l}\right)\frac{r^2}{2}$$

and:

$$u_x = \frac{1}{4\mu}\left(\frac{-\Delta P}{l}\right)(r^2 - s^2) \tag{3.30}$$

Thus the velocity over the cross-section varies in a parabolic manner with the distance from the axis of the pipe. The velocity of flow is seen to be a maximum when $s=0$, that is at the pipe axis.

Thus the maximum velocity, at the pipe axis, is given by u_{CL} where:

$$u_{max} = u_{CL} = \frac{1}{4\mu}\left(\frac{-\Delta P}{l}\right)r^2 = \left(\frac{-\Delta P}{l}\right)\frac{d^2}{16\mu} \tag{3.31}$$

Hence:

$$\frac{u_x}{u_{CL}} = 1 - \frac{s^2}{r^2} \tag{3.32}$$

or:

$$= 1 - \frac{4s^2}{d^2} \tag{3.33}$$

The velocity is thus seen to vary in a parabolic manner over the cross-section of the pipe, and this agrees well with experimental measurements.

Volumetric rate of flow and average velocity

If the velocity is taken as constant over an annulus of radii s and $(s+\mathrm{d}s)$, the volumetric rate of flow $\mathrm{d}Q$ through the annulus is given by:

$$\begin{aligned} \mathrm{d}Q &= 2\pi s\,\mathrm{d}s\,u_x \\ &= 2\pi u_{CL} s\left(1 - \frac{s^2}{r^2}\right)\mathrm{d}s \end{aligned} \tag{3.34}$$

The total flow over the cross-section is then given by integrating Eq. (3.34):

$$\begin{aligned} Q &= 2\pi u_{CL}\int_0^r s\left(1 - \frac{s^2}{r^2}\right)\mathrm{d}s \\ &= 2\pi u_{CL}\left[\frac{s^2}{2} - \frac{s^4}{4r^2}\right]_0^r \\ &= \frac{\pi}{2}r^2 u_{CL} = \frac{\pi}{8}d^2 u_{CL} \end{aligned} \tag{3.35}$$

Thus the average velocity u is given by:

$$u=\frac{Q}{(\pi d^2/4)}$$

On substituting from Eq. (3.35) into Eq. (3.31):

$$u=\left(\frac{-\Delta P}{l}\right)\frac{r^2}{8\mu}=\left(\frac{-\Delta P}{l}\right)\frac{d^2}{32\mu}=\frac{u_{CL}}{2}=\frac{u_{\max}}{2} \tag{3.36}$$

This relation was derived by Hagen[12] in 1839 and independently by Poiseuille[13] in 1840.

From Eqs. (3.16), (3.36):

$$32\mu u\frac{l}{d^2}=-\Delta P=4\frac{R}{\rho u^2}\frac{l}{d}(\rho u^2)$$

and:

$$\frac{R}{\rho u^2}=\frac{8\mu}{ud\rho}=8Re^{-1} \tag{3.37}$$

as shown in Fig. 3.7.

From Eqs. (3.32), (3.35):

$$\frac{u_x}{u}=\frac{2(d^2-4s^2)}{d^2}=2\left(1-\frac{s^2}{r^2}\right) \tag{3.38}$$

Eq. (3.38) is plotted in Fig. 3.11 which shows the shape of the velocity profile for streamline flow.

The velocity gradient at any point is obtained by differentiating Eq. (3.38) with respect to s, or:

$$\frac{\mathrm{d}u_x}{\mathrm{d}y}=-\frac{\mathrm{d}u_x}{\mathrm{d}s}=\frac{4su}{r^2}$$

$$\text{At the wall, } s=r \text{ and } \left(\frac{\mathrm{d}u_x}{\mathrm{d}y}\right)_{y=0}=\frac{4u}{r}=\frac{8u}{d} \tag{3.39}$$

$$\text{Thus, the shear stress at the wall } R=\frac{8\mu u}{d} \tag{3.40}$$

Kinetic energy of fluid

In order to obtain the kinetic energy term for use in the energy balance equation, it is necessary to obtain the average kinetic energy per unit mass in terms of the mean velocity.

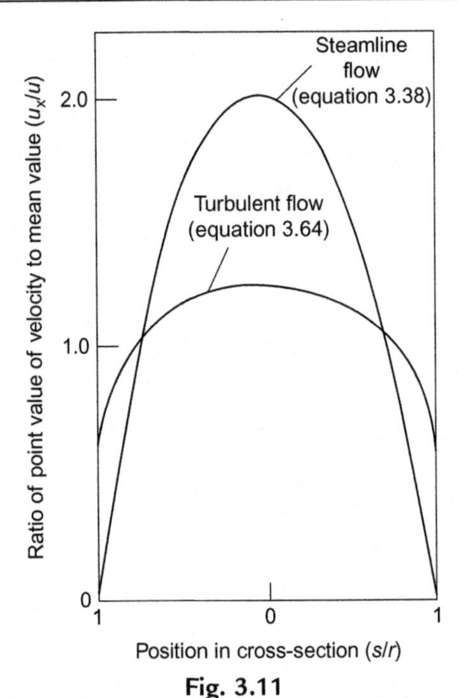

Fig. 3.11
Shape of velocity profiles for streamline and turbulent flow.

The kinetic energy of the fluid flowing per unit time through the annulus between s and $(s+\mathrm{d}s)$ is given by:

$$(\rho 2\pi s\ \mathrm{d}s\ u_x)\frac{u_x^2}{2} = \pi\rho s u_x^3\ \mathrm{d}s$$

$$= 8\pi\rho u^3 s\left[1-\frac{s^2}{r^2}\right]^3 \mathrm{d}s \quad \text{(from Eq. 3.38)}$$

The total kinetic energy per unit time of the fluid flowing in the pipe is then:

$$= -4\pi\rho u^2 r^2 \int_{s=0}^{s=r}\left(1-\frac{s^2}{r^2}\right)^3 d\left(1-\frac{s^2}{r^2}\right)$$

$$\text{Integrating gives:} \quad = -\frac{4\pi\rho u^3 r^2}{4}\left[\left(1-\frac{s^2}{r^2}\right)^4\right]_{s=0}^{s=r}$$

$$= \rho\pi r^2 u^3$$

Hence the kinetic energy per unit mass is given by:

$$\frac{\rho\pi r^2 u^3}{\rho\pi r^2 u} = u^2 \tag{3.41}$$

Since in the energy balance equation, the kinetic energy per unit mass is expressed as $u^2/2\alpha$, hence $\alpha=0.5$ for the streamline flow of a fluid in a round pipe.

Flow between two parallel plates

Considering the steady, fully developed flow of an incompressible fluid between two plates of unit width, at a distance d_a apart, as shown in Fig. 3.12, then for the equilibrium of an element *ABCD*, a force balance may be set up in a similar manner to that used for flow through pipes to give:

$$P2s - (P + \Delta P)2s + 2l\mu \frac{du_x}{ds} = 0$$

or:

$$\frac{-\Delta P}{l} s + \mu \frac{du_x}{ds} = 0 \tag{3.42}$$

and:

$$u_x = -\frac{1}{\mu}\left(\frac{-\Delta P}{l}\right)\frac{s^2}{2} + \text{constant}$$

When $s=d_a/2$, $u_x=0$, and:

$$\text{constant} = \frac{d_a^2}{8}\frac{1}{\mu}\left(\frac{-\Delta P}{l}\right)$$

or:

$$u_x = \frac{1}{2\mu}\left(\frac{-\Delta P}{l}\right)\left(\frac{d_a^2}{4} - s^2\right) \tag{3.43}$$

The total rate of flow of fluid between the plates is obtained by calculating the flow through two laminae of thickness ds and situated at a distance s from the centre plane and then integrating it. The flow through the laminae is given by:

$$dQ' = \frac{1}{2\mu}\left(\frac{-\Delta P}{l}\right)\left(\frac{d_a^2}{4} - s^2\right)2ds$$

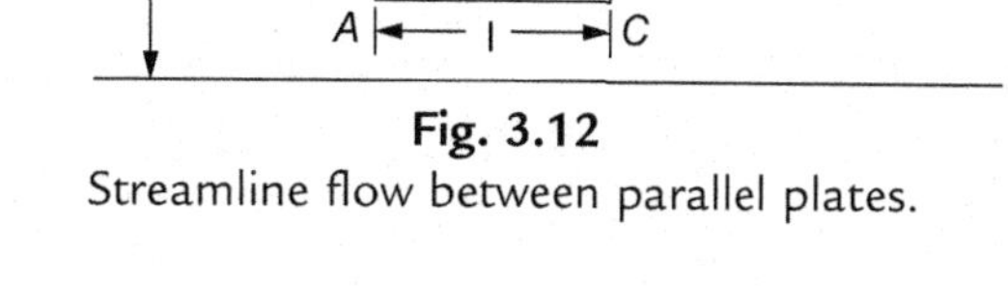

Fig. 3.12
Streamline flow between parallel plates.

On integrating between the limits of s from 0 to $d_a/2$,

the total rate of flow:

$$\begin{aligned} Q' &= \frac{1}{\mu}\left(\frac{-\Delta P}{l}\right)\left(\frac{d_a^3}{8}-\frac{d_a^3}{24}\right) \\ &= \left(\frac{-\Delta P}{l}\right)\frac{d_a^3}{12\mu} \end{aligned} \tag{3.44}$$

The average velocity of the fluid is:

$$\begin{aligned} u &= \frac{Q'}{d_a} \\ &= \left(\frac{-\Delta P}{l}\right)\frac{d_a^2}{12\mu} \end{aligned} \tag{3.45}$$

The maximum velocity occurs at the centre plane, and this is obtained by putting $s=0$ in Eq. (3.43) to give:

$$\text{Maximum velocity} = u_{\max} = \left(\frac{-\Delta P}{l}\right)\frac{d_a^2}{8\mu} = 1.5u \tag{3.46}$$

It has been assumed that the width of the plates is large compared with the distance between them so that the flow may be considered as unidirectional.

Flow through an annulus

The velocity distribution and the mean velocity of a fluid flowing through an annulus of outer radius r and inner radius r_i is more complex. If, as shown in Fig. 3.13, the pressure changes by an amount ΔP as a result of friction over a length l of annulus, the resulting force may be equated to the shearing force acting on the fluid. For the flow of the fluid situated at a distance

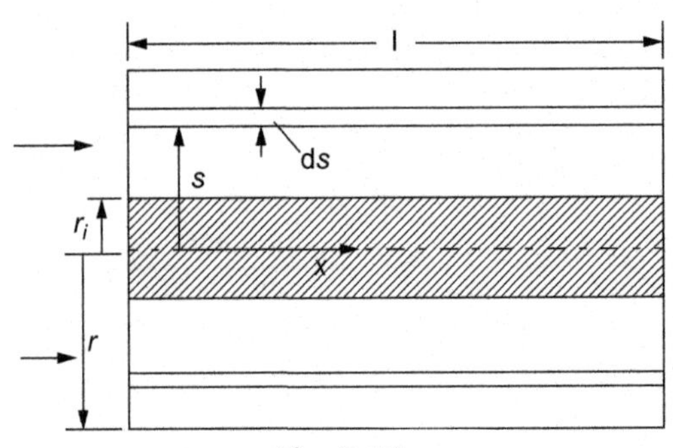

Fig. 3.13
Streamline flow through a concentric annulus.

not greater than s from the centre line of the annuli, the shear force acting on this fluid consists of two parts; one is the drag on its outer surface which may be expressed in terms of the viscosity of the fluid and the velocity gradient at that radius, and the other is the drag occurring at the inner boundary of the annulus, which cannot be estimated at present and will be denoted by the symbol λ for unit length of pipe.

Then:

$$\left(\frac{-\Delta P}{l}\right) l\pi\left(s^2-r_i^2\right)=\mu 2\pi s\, l\left(\frac{\mathrm{d}u_x}{\mathrm{d}s}\right)+\lambda l \tag{3.47}$$

where u_x is the velocity of the fluid at radius s.

Thus:

$$\mathrm{d}u_x=\frac{s^2-r_i^2}{2\mu s}\frac{\Delta P}{l}\mathrm{d}s-\lambda\frac{\mathrm{d}s}{2\pi\mu s} \tag{3.48}$$

On integration:

$$u_x=\frac{1}{2\mu}\frac{\Delta P}{l}\left(\frac{s^2}{2}-r_i^2\ \ln s\right)-\frac{\lambda}{2\pi\mu}\ \ln s+u_c \tag{3.49}$$

where u_c is an integration constant with dimensions of velocity.

Substituting the boundary conditions $s=r_i$. $u_x=0$, and $s=r$, $u_x=0$ in Eq. (3.49) and solving for λ and u_c gives:

$$\lambda=\pi\frac{\Delta P}{l}\left(\frac{r^2-r_i^2}{2\ \ln(r/r_i)}-r_i^2\right) \tag{3.50}$$

and:

$$u_c=\frac{1}{2\mu}\left(\frac{-\Delta P}{l}\right)\left(\frac{r^2}{2}-\frac{r^2-r_i^2}{2\ \ln(r/r_i)}\ \ln r\right) \tag{3.51}$$

Substituting these values of λ and u_c into Eq. (3.49), and simplifying:

$$u_x=\frac{1}{4\mu}\left(\frac{-\Delta P}{l}\right)\left(r^2-s^2+\frac{r^2-r_i^2}{\ln(r/r_i)}\ \ln\frac{s}{r}\right) \tag{3.52}$$

The rate of flow of fluid through a small annulus of inner radius s and outer radius $(s+\mathrm{d}s)$, is given by:

$$\begin{aligned}\mathrm{d}Q&=2\pi s\ \mathrm{d}s u_x\\&=\frac{\pi}{2\mu}\left(\frac{-\Delta P}{l}\right)\left(r^2 s-s^3+\frac{r^2-r_i^2}{\ln(r/r_i)}s\ \ln\frac{s}{r}\right)\ \mathrm{d}s\end{aligned} \tag{3.53}$$

Integrating between the limits $s=r_i$ and $s=r$:

$$Q=\frac{-\Delta P\pi}{8\mu l}\left(r^2+r_i^2-\frac{r^2-r_i^2}{\ln(r/r_i)}\right)\left(r^2-r_i^2\right) \tag{3.54}$$

The average velocity is given by:

$$u=\frac{Q}{\pi\left(r^2-r_i^2\right)}$$

$$=\frac{-\Delta P}{8\mu l}\left(r^2+r_i^2-\frac{r^2-r_i^2}{\ln(r/r_i)}\right) \tag{3.55}$$

3.3.5 The Transition From Laminar to Turbulent Flow in a Pipe

Laminar flow ceases to be stable when a small perturbation or disturbance in the flow tends to increase in magnitude rather than decay. For flow in a pipe of circular cross-section, the critical condition occurs at a Reynolds number of about 2100. Thus although laminar flow can take place at much higher values of Reynolds number, that flow is no longer stable and a small disturbance to the flow will lead to the growth of the disturbance and the onset of turbulence. Similarly, if turbulence is artificially promoted at a Reynolds number of less than 2100, the flow will ultimately revert to a laminar condition in the absence of any further disturbance.

In connection with the transition, Ryan and Johnson[14] have proposed a *stability parameter Z*. If the critical value Z_c of that parameter is exceeded at any point on the cross-section of the pipe, then turbulence will ensue. Based on a concept of a balance between energy supply to a perturbation and energy dissipation, it was proposed that Z could be defined as:

$$Z=\frac{r\rho u_x}{R}\frac{\mathrm{d}\mu_x}{\mathrm{d}y} \tag{3.56}$$

Z will be zero at the pipe wall ($u_x=0$) and at the axis ($\mathrm{d}u_x/\mathrm{d}y=0$), and it will reach a maximum at some intermediate position in the cross-section. From Eq. (3.6), the wall shear stress R for laminar flow may be expressed in terms of the pressure gradient along the pipe ($-\Delta P/l$):

$$R=\frac{-\Delta Pd}{4l}=\frac{-\Delta Pr}{2l} \quad \text{(Eq. 3.6)}$$

Thus:

$$Z=\frac{2l\rho u_x}{-\Delta P}\frac{du_x}{dy}$$

The velocity distribution over the pipe cross-section is given by:

$$u_x = \frac{-\Delta P}{4\mu l}\left(r^2 - s^2\right)$$

and:

$$\frac{\mathrm{d}u_x}{\mathrm{d}y} = -\frac{\mathrm{d}u_x}{\mathrm{d}s} = \frac{-\Delta Ps}{2\mu l}$$

Thus:

$$Z = \frac{2l\rho}{-\Delta P}\left[\frac{-\Delta P}{4\mu l}\left(r^2 - s^2\right)\right]\frac{-\Delta Ps}{2\mu l}$$

$$= \frac{-\Delta P\rho}{4\mu^2 l}\left(r^2 s - s^3\right)$$

The value of s at which Z is a maximum is obtained by differentiating Z with respect to s and equating the derivative to zero.

$$\frac{\mathrm{d}Z}{\mathrm{d}s} = \frac{-\Delta P\rho}{4\mu^2 l}\left(r^2 - 3s^2\right)$$

and, when $\frac{\mathrm{d}Z}{\mathrm{d}s} = 0$: $s = \frac{r}{\sqrt{3}}$

Thus, the position in the cross-section where laminar flow will first break down is where $s/r = 1/\sqrt{3}$and:

$$Z_{\max} = \frac{-\Delta P\rho r^3}{6\sqrt{3}\,\mu^2 l} \tag{3.57}$$

Substituting for $(-\Delta P/l)$ from Eq. (3.36):

$$u = \frac{-\Delta P r^2}{8\mu l} \quad \text{(Eq. 3.36)}$$

$$Z_{\max} = \frac{4}{3\sqrt{3}}\frac{u\rho r}{\mu}$$

$$= \frac{2}{3\sqrt{3}}\frac{ud\rho}{\mu} = 0.384\, Re_{crit} \tag{3.58}$$

Taking $Re_{crit} = 2100$, then:

$$Z_{\max} = 808$$

3.3.6 Velocity Distributions and Volumetric Flowrates for Turbulent Flow

No exact mathematical analysis of the conditions within a turbulent fluid has yet been developed, though a number of semitheoretical expressions for the shear stress at the walls of a pipe of circular cross-section have been suggested, including that proposed by Blasius.[10]

The shear stresses within the fluid are responsible for the frictional force at the walls and the velocity distribution over the cross-section. A given assumption for the shear stress at the walls therefore implies some particular velocity distribution. It will be shown in Chapter 3 of Vol. 1B that the velocity at any point in the cross-section will be proportional to the one-seventh power of the distance from the walls if the shear stress is given by the Blasius equation (Eq. 3.11). This may be expressed as:

$$\frac{u_x}{u_{CL}} = \left(\frac{y}{r}\right)^{1/7} \tag{3.59}$$

where u_x is the velocity at a distance y from the walls, u_{CL} the velocity at the axis of the pipe, and r the radius of the pipe.

This equation is sometimes referred to as the *Prandtl one-seventh power law*. This equation approximates the experimental data satisfactorily over the range $\sim 0.1 \leq y/r \leq \sim 1$ in the interval $\sim 3000 \leq Re \leq 10^5$. It neither works well close to the wall nor does it satisfy the condition of $dx/dr = 0$ at the axis of the tube. Detailed discussions can be found in the literature, e.g., see Lumley and Tennekes[15] and Davidson.[16]

Pipe of circular cross-section

Mean velocity

In a thin annulus of inner radius s and outer radius $s+ds$, the velocity u_x may be taken as constant (see Fig. 3.10) and:

$$\therefore \; dQ = 2\pi s \, ds \, u_x \tag{3.60}$$

$$= -2\pi(r-y)dy \, u_x \; (\text{since } s+y=r)$$

$$= -2\pi(r-y)dy \, u_{CL}\left(\frac{y}{r}\right)^{1/7} \tag{3.61}$$

Thus:

$$Q = \int_{y=r}^{y=0} 2\pi r^2 u_{CL}\left(1-\frac{y}{r}\right)\left(\frac{y}{r}\right)^{1/7} d\left(\frac{y}{r}\right)$$

$$= 2\pi r^2 u_{CL}\left[\frac{7}{8}\left(\frac{y}{r}\right)^{8/7} - \frac{7}{15}\left(\frac{y}{r}\right)^{15/7}\right]_{y=r}^{y=0}$$

$$= 2\pi r^2 u_{CL}\left[\frac{7}{8} - \frac{7}{15}\right]$$

$$=\frac{49}{60}\pi r^2 u_{CL}=0.817\pi r^2 u_{CL} \tag{3.62}$$

The mean velocity of flow is then:

$$u=\frac{Q}{\pi r^2}=\frac{49}{60}u_{CL}=0.817u_{CL} \tag{3.63}$$

This relation holds provided that the one-seventh power law may be assumed to apply over the whole of the cross-section of the pipe. This is strictly the case only at high Reynolds numbers when the thickness of the laminar sublayer is small. By combining Eqs. (3.59), (3.63), the velocity profile is given by:

$$\begin{aligned}\frac{u_x}{u}&=\frac{1}{0.817}\left(\frac{y}{r}\right)^{1/7}\\&=1.22\left(1-\frac{s}{r}\right)^{1/7}\end{aligned} \tag{3.64}$$

Eq. (3.64) is plotted in Fig. 3.11, from which it may be noted that the velocity profile is very much flatter than that for streamline flow.

The variation of (u/u_{max}) with Reynolds number is shown in Fig. 3.14, from which the sharp change at a Reynolds number between 2000 and 3000 may be noted.

Kinetic energy

Since:

$$dQ=-2\pi(r-y)dy\ u_x \quad \text{(Eq. 3.60)}$$

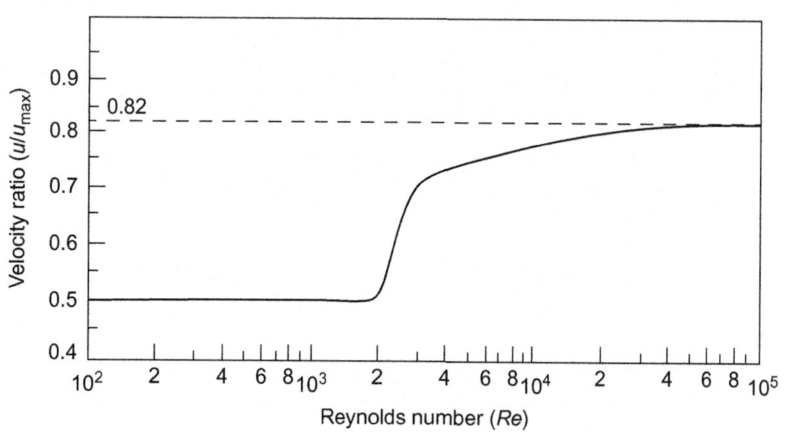

Fig. 3.14
Variation of (u/u_{max}) with Reynolds number in a pipe.

the kinetic energy per unit time of the fluid passing through the annulus is given by:

$$= -2\pi(r-y)\rho \mathrm{d}y\, u_x \frac{u_x^2}{2}$$

$$= -\pi u_{CL}^3 \rho r^2 \left(1-\frac{y}{r}\right)\left(\frac{y}{r}\right)^{3/7} \mathrm{d}\left(\frac{y}{r}\right) \tag{3.65}$$

Integration of Eq. (3.65) gives the total kinetic energy per unit time as:

$$= \pi u_{CL}^3 \rho r^2 \left[\frac{7}{10} - \frac{7}{17}\right]$$

$$= \frac{49}{170} \pi r^2 \rho u_{CL}^3 \tag{3.66}$$

The mean kinetic energy per unit mass of fluid from Eq. (3.63) is:

$$\begin{aligned}
&= \frac{\left(\frac{49}{170}\pi r^2 \rho u_{CL}^3\right)}{\left(\frac{49}{60}\pi r^2 \rho u_{CL}\right)} \\
&= \frac{6}{17} u_{CL}^2 \\
&= \frac{6}{17}\left(\frac{60}{49}u\right)^2 \text{ (from Eq. 3.60)} \qquad (3.67) \\
&= 0.53 u^2 \approx \frac{u^2}{2} \\
&= \frac{u^2}{2\alpha} \text{ (from the definition of } \alpha)
\end{aligned}$$

Thus for turbulent flow at high Reynolds numbers, where the thickness of the laminar sublayer may be neglected, $\alpha \asymp 1$.

When the thickness of the laminar sublayer may not be neglected, α will be slightly less than 1.

Example 3.6

On the assumption that the velocity profile in a fluid in turbulent flow is given by the Prandtl one-seventh power law, calculate the radius at which the flow between it and the centre is equal to that between it and the wall, for a pipe 100 mm in diameter.

Solution

The Prandtl one-seventh power law gives the velocity at a distance y from the wall, u_x, as:

$$u_x = u_{CL}(y/r)^{1/7} \text{ (Eq. 3.59)}$$

where u_{CL} is the velocity at the centre line of the pipe, and r is the radius of the pipe.

The total flow is then given by Eq. (3.62):

$$Q = 2\pi r^2 u_{CL} \int_0^1 \left[\left(\frac{y}{r}\right)^{1/7} - \left(\frac{y}{r}\right)^{8/7}\right] d\left(\frac{y}{r}\right)$$

$$= 2\pi r^2 u_{CL} \left[\frac{7}{8}\left(\frac{y}{r}\right)^{8/7} - \frac{7}{15}\left(\frac{y}{r}\right)^{15/7}\right]_0^1$$

$$= \frac{49}{60}\pi r^2 u_{CL} \quad \text{(Eq. 3.62)}$$

When the flow in the central core is equal to the flow in the surrounding annulus, then taking $a = y/r$, the flow in the central core is:

$$Q_c = 2\pi r^2 u_{CL} \left[\frac{7}{8}a^{8/7} - \frac{7}{15}a^{15/7}\right]_0^a$$

$$= \frac{\pi r^2}{60} u_{CL}\left(105a^{8/7} - 56a^{15/7}\right)$$

Since:

$$\text{flow in the core} = 0.5(\text{flow in the whole pipe})$$

then:

$$0.5 \times \left(\frac{49}{60}\right) \pi r^2 u_{CL} = \left(\frac{\pi r^2}{60}\right) u_{CL}\left(105a^{8/7} - 56a^{15/7}\right)$$

or:

$$105a^{8/7} - 56a^{15/7} = 24.5$$

Solving by trial and error:

$$a = y/r = 0.33$$

and:

$$y = (0.33 \times 50) = \underline{\underline{16.5\ \text{mm}}}$$

Noncircular ducts

For turbulent flow in a duct of noncircular cross-section, the hydraulic mean diameter may be used in place of the pipe diameter and the formulae for circular pipes may then be applied without introducing a large error. This approach is entirely empirical.

The *hydraulic mean diameter* d_m is defined as four times the cross-sectional area available for flow divided by the wetted perimeter. For a circular pipe, for example, the hydraulic mean diameter is:

$$d_m = \frac{4(\pi/4)d^2}{\pi d} = d \tag{3.68}$$

For an annulus of outer radius r and inner radius r_i:

$$d_m = \frac{4\pi\left(r^2 - r_i^2\right)}{2\pi(r + r_i)} = 2(r - r_i) = d - d_i \tag{3.69}$$

and for a duct of rectangular cross-section d_a by d_b:

$$\begin{aligned} d_m &= \frac{4d_a d_b}{2(d_a + d_b)} \\ &= \frac{2d_a d_b}{d_a + d_b} \end{aligned} \tag{3.70}$$

The method is not entirely satisfactory for streamline flow, and exact expressions relating the pressure drop to the velocity may be obtained only for ducts of certain shapes. The currently available analytical and numerical results for a range of shapes have been summarised by Shah and London[17] in the streamline flow in the form of the product $(\phi \cdot Re)$ based on the use of the hydraulic diameter instead of the tube diameter. Surprisingly, the results for a range of shapes including a triangle, rectangle, elliptic, etc., deviate only by 10%–15% from the corresponding value of $\phi \cdot Re = 8$ for a pipe.

3.3.7 Flow Through Curved Pipes

If a pipe is not straight, the velocity distribution over the section is altered and the direction of flow of the fluid is continuously changing. The frictional losses are therefore somewhat greater than that for a straight pipe of the same length. If the radius of the pipe divided by the radius of the bend is less than about 0.002, the effects of the curvature are negligible, however.

White[18] found that stable streamline flow persists at higher values of the Reynolds number in coiled pipes. Thus, for example, when the ratio of the diameter of the pipe to the diameter of the coil is between 1 to 15, the transition occurs at a Reynolds number of about 8000 as opposed to at $Re = 2000$–2100 in a straight circular tube.

3.3.8 Miscellaneous Friction Losses

Friction losses occurring as a result of a sudden enlargement or contraction in the cross-section of the pipe, and the resistance of various standard pipe fittings, are now considered.

Sudden enlargement

If the diameter of the pipe suddenly increases, as shown in Fig. 3.15, the effective area available for flow gradually increases from that of the smaller pipe to that of the larger one, and the velocity of flow progressively decreases. Thus fluid with a relatively high velocity will be injected into relatively slow moving fluid; turbulence will be set up and much of the excess kinetic energy will be converted into heat and therefore wasted. If the change of cross-section is gradual, the kinetic energy may be recovered as pressure energy.

For the fluid flowing as shown in Fig. 3.15, from section 1 (the pressure just inside the enlargement is found to be equal to that at the end of the smaller pipe) to section 2, the net force = the rate of change of momentum, or:

$$P_1A_2 - P_2A_2 = \rho_2 A_2 u_2 (u_2 - u_1)$$

and:

$$(P_1 - P_2)v_2 = u_2^2 - u_1 u_2 \tag{3.71}$$

For an incompressible fluid:

$$(P_1 - P_2)v_2 = -\int_{P_1}^{P_2} v \, \mathrm{d}P \tag{3.72}$$

Applying the energy equation between the two sections:

$$\frac{u_1^2}{2\alpha_1} = \frac{u_2^2}{2\alpha_2} + \int_1^2 v \, \mathrm{d}P + F \tag{3.73}$$

or:

$$F = \frac{u_1^2}{2\alpha_1} - \frac{u_2^2}{2\alpha_2} + u_2^2 - u_1 u_2 \tag{3.74}$$

For fully turbulent flow:

$$\alpha_1 = \alpha_2 = 1$$

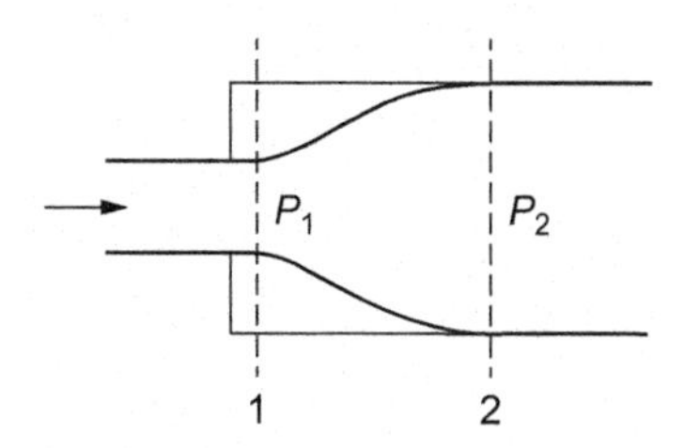

Fig. 3.15
Sudden enlargement in a pipe.

and:

$$F = \frac{(u_1 - u_2)^2}{2} \tag{3.75}$$

The change in pressure $-\Delta P_f$ is therefore given by:

$$-\Delta P_f = \rho \frac{(u_1 - u_2)^2}{2} \tag{3.76}$$

The loss of head h_f is given by:

$$h_f = \frac{(u_1 - u_2)^2}{2g} \tag{3.77}$$

Substituting from the continuity relation $u_1 A_1 = u_2 A_2$ into Eq. (3.76):

$$-\Delta P_f = \frac{\rho u_1^2}{2}\left[1 - \left(\frac{A_1}{A_2}\right)\right]^2 \tag{3.78}$$

The loss can be substantially reduced if a tapering enlarging section is used. For a circular pipe, the optimum angle of taper is about 7 degree and for a rectangular duct it is about 11 degree.

If A_2 is very large compared with A_1, as, for instance, at a pipe exit, then:

$$-\Delta P_f = \frac{\rho u_1^2}{2} \tag{3.79}$$

The use of a small angle enlarging section is a feature of the venturi meter, as discussed in Chapter 6.

Example 3.7

Water flows at 7.2 m^3/h through a sudden enlargement from a 40 to a 50 mm diameter pipe. What is the loss in head?

Solution

$$\text{Velocity in 50 mm pipe} = \frac{(7.2/3600)}{(\pi/4)(50 \times 10^{-3})^2} = 1.02 \text{ m/s}$$

$$\text{Velocity in 40 mm pipe} = \frac{(7.2/3600)}{(\pi/4)(40 \times 10^{-3})^2} = 1.59 \text{ m/s}$$

The head lost is given by Eq. (3.77) as:

$$h_f = \frac{(u_1 - u_2)^2}{2g}$$

$$= \frac{(1.59 - 1.02)^2}{(2 \times 9.81)} = 0.0165 \text{ m water}$$

or:

16.5 mm water

Since the Reynolds number here is of the order of 40,000 to 60,000, the assumption of $\alpha_1 = \alpha_2 = 1$ is justified.

Sudden contraction

As shown in Fig. 3.16, the effective area for flow gradually decreases as a sudden contraction is approached and then continues to decrease, for a short distance, to what is known as the *vena contracta*. After the vena contracta, the flow area gradually approaches that of the smaller pipe. As the fluid moves towards the vena contracta, it is accelerated and pressure energy is converted into kinetic energy; this process does not give rise to eddy formation and losses are very small. Beyond the vena contracta, however, the velocity falls as the flow area increases and conditions are equivalent to those for a sudden enlargement. The expression for the loss at a sudden enlargement can therefore be applied for the fluid flowing from the vena contracta to some section a small distance downstream, where the whole of the cross-section of the pipe is available for flow.

Applying Eq. (3.75) between sections C and 2, as shown in Fig. 3.16, the frictional loss per unit mass of fluid is then given by:

$$F = \frac{(u_c - u_2)^2}{2} = \frac{u_2^2}{2}\left[\frac{u_c}{u_2} - 1\right]^2 \tag{3.80}$$

Denoting the ratio of the area at section C to that at section 2 by a coefficient of contraction C_c:

$$F = \frac{u_2^2}{2}\left[\frac{1}{C_c} - 1\right]^2 \tag{3.81}$$

Thus the change in pressure ΔP_f is $-(\rho u_2^2/2)[(1/C_c)-1]^2$ and the head lost is:

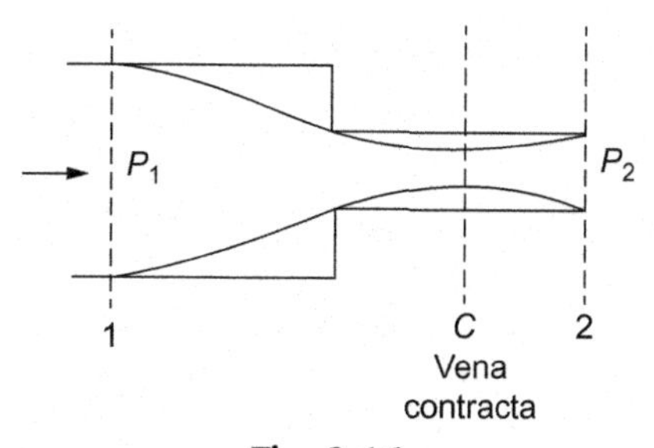

Fig. 3.16
Sudden contraction in a pipe.

$$\frac{u_2^2}{2g}\left[\frac{1}{C_c}-1\right]^2$$

C_c varies from about 0.6 to 1.0 as the ratio of the pipe diameters varies from 0 to 1. For a common value of C_c of 0.67:

$$F=\frac{u_2^2}{8} \tag{3.82}$$

It may be noted that the maximum possible frictional loss which can occur at a change in cross-section is equal to the entire kinetic energy of the fluid.

Pipe fittings

Most pipes are fabricated from steel with or without small alloying ingredients, and they are welded or drawn to give a seamless pipe. Tubes with diameters of 6–50 mm are frequently made from nonferrous metals such as copper, brass, or aluminium, and these are very widely used in heat exchangers. For special purposes, there is a very wide range of materials including industrial glass, many varieties of plastics, rubber, stoneware, and ceramic materials. The normal metal piping is supplied in standard lengths of about 6 m and these are joined to give longer lengths as required. Such jointing is by screw flanging or welding, and small diameter copper or brass tubes are often brazed or soldered or jointed by compression fittings.

A very large range of pipe fittings is available to enable branching and changes in size to be incorporated into industrial pipe layouts. For the control of the flow of a fluid, valves of various designs are used, the most important being gate, globe, and needle valves. In addition, check valves are supplied for relieving the pressure in pipelines, and reducing valves are available for controlling the pressure on the downstream side of the valve. Gate and globe valves are supplied in all sizes and may be controlled by motor units operated from an automatic control system. Hand wheels are usually fitted for emergency use.

In general, gate valves give coarse control, globe valves give finer and needle valves give the finest control of the rate of flow. Diaphragm valves are also widely used for the handling of corrosive fluids since the diaphragm may be made of corrosion resistant materials.

Some representative figures are given in Table 3.3 for the friction losses in various pipe fittings for turbulent flow of fluid, and are expressed in terms of the equivalent length of straight pipe with the same resistance, and as the number of velocity heads ($u^2/2g$) lost. Considerable variation occurs according to the exact construction of the fittings.

Typical pipe-fittings are shown in Figs. 3.17 and 3.18 and details of other valve types are given in Volume 6. More details concerning the estimation of fitting losses are available in a recent book by Darby and Chhabra.[19]

Table 3.3 Friction losses in pipe fittings

	Number of Pipe Diameters	Number of Velocity Heads ($u^2/2g$)
45 degree elbows (A)[a]	15	0.3
90 degree elbows (standard radius) (B)	30–40	0.6–0.8
90 degree square elbows (C)	60	1.2
Entry from leg of T-piece (D)	60	1.2
Entry into leg of T-piece (D)	90	1.8
Unions and couplings (E)	Very small	Very small
Globe valves fully open	60–300	1.2–6.0
Gate valves fully open	7	0.15
$\frac{3}{4}$ open	40	1
$\frac{1}{2}$ open	200	4
$\frac{1}{4}$ open	800	16

[a]See Fig. 3.17.

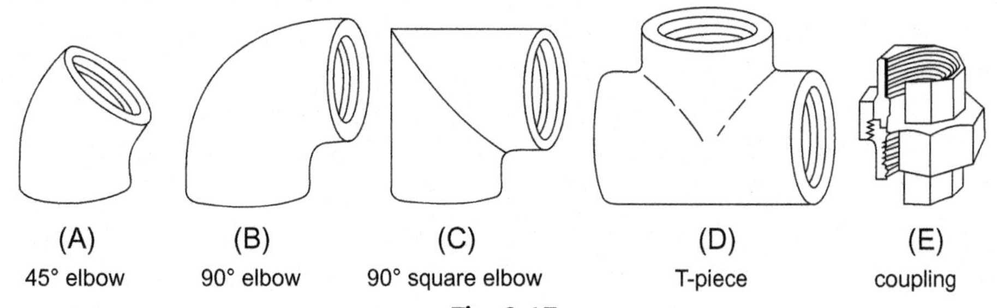

Fig. 3.17
Standard pipe fittings.

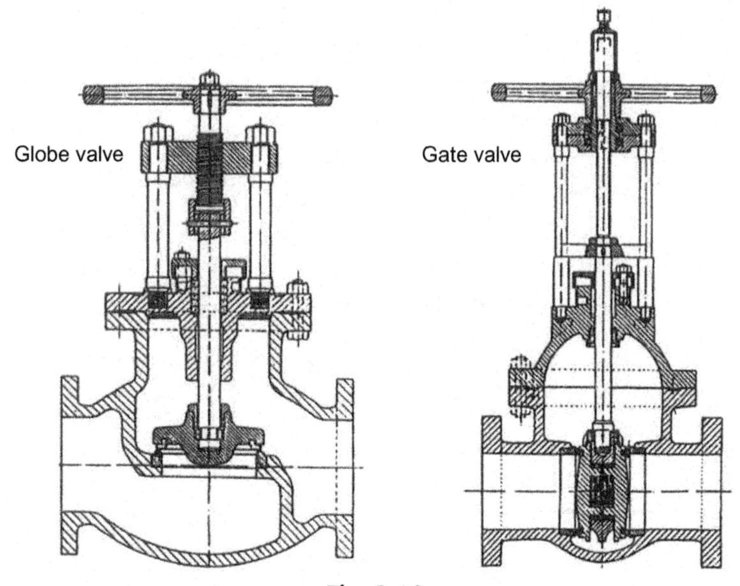

Fig. 3.18
Standard valves.

Example 3.8

2.27 m^3/h water at 320 K is pumped in a 40 mm i.d. pipe through a distance of 150 m in a horizontal direction and then up through a vertical height of 10 m. In the pipe there is a control valve for which the friction loss may be taken as equivalent to 200 pipe diameters and also other pipe fittings equivalent to 60 pipe diameters. Also in the line is a heat exchanger across which there is a loss in head of 1.5 m of water. If the main pipe has a roughness of 0.2 mm, what power must be supplied to the pump if it is 60% efficient?

Solution

$$\text{Relative roughness of the main pipe: } \frac{e}{d} = \left(\frac{0.2}{40}\right) = 0.005$$

Viscosity of water at 320 K: $\mu = 0.65\ \text{mN s/m}^2$ or $0.65 \times 10^{-3}\ \text{N s/m}^2$
Flowrate $= 2.27\ \text{m}^3/\text{h} = 6.3 \times 10^{-4}\ \text{m}^3/\text{s}$
Area for flow $= \frac{\pi}{4}(40 \times 10^{-3})^2 = 1.26 \times 10^{-3}\ \text{m}^2$

Thus:

$$\text{Velocity} = \frac{6.3 \times 10^{-4}}{1.26 \times 10^{-3}} = 0.50\ \text{m/s}$$

and:

$$Re = \frac{40 \times 10^{-3} \times 0.50 \times 1000}{0.65 \times 10^{-3}} = 30{,}770$$

giving:

$$\frac{R}{\rho u^2} = 0.004\ \text{(from Fig. 3.7)}$$

$$\text{Equivalent length of pipe} = 150 + 10 + \left(260 \times 40 \times 10^{-3}\right) = 170.4\ \text{m}$$

$$h_f = 4\frac{R}{\rho u^2}\frac{1}{d}\frac{u^2}{g}$$

$$= 4 \times 0.004\left(\frac{170.4}{40 \times 10^{-3}}\right)\left(\frac{0.5^2}{9.81}\right)$$

$$= 1.74\ \text{m}$$

$$\text{Total head to be developed} = (1.74 + 1.5 + 10) = 13.24\ \text{m}$$

$$\text{Mass throughput} = \left(6.3 \times 10^{-4} \times 1000\right) = 0.63\ \text{kg/s}$$

$$\therefore\ \text{Power required} = (0.63 \times 13.24 \times 9.81) = 81.8\ \text{W}$$

Since the pump efficiency is 60%, the power required $= \left(\frac{81.8}{0.60}\right) = 136.4$ W or 0.136 kW.

The kinetic energy head, $u^2/2g$ amounts to $0.5^2/(2 \times 9.81) = 0.013$ m, and this may be neglected.

Example 3.9

Water in a tank flows through an outlet 25 m below the water level into a 0.15 m diameter horizontal pipe 30 m long, with a 90 degree elbow at the end leading to a vertical pipe of the same diameter 15 m long. This is connected to a second 90 degree elbow which leads to a horizontal pipe of the same diameter, 60 m long, containing a fully open globe valve and discharging to atmosphere 10 m below the level of the water in the tank. Taking $e/d=0.01$ and the viscosity of water as 1 mN s/m^2, what is the initial rate of discharge?

Solution

From Eq. (3.20), the head lost due to friction is given by:

$$h_f = 4\phi \frac{1}{d}\frac{u^2}{g} \text{ m water}$$

The total head loss is:

$$h = \frac{u^2}{2g} + h_f + \text{losses in fittings}$$

From Table 3.2, the losses in the fittings are:

$$= \frac{2\times 0.8u^2}{2g} \text{ (for the elbows)} + \frac{5.0u^2}{2g} \text{ (for the valve)}$$

$$= \frac{6.6u^2}{2g} \text{ m water}$$

Taking ϕ as 0.0045, then:

$$10 = \frac{(6.6+1)u^2}{2g} + 4\times 0.0045\left[\frac{30+15+60}{0.15}\right]\frac{u^2}{g}$$

$$= \frac{(3.8+6.3)u^2}{g}$$

from which

$$u^2 = 9.71 \text{ m}^2/\text{s}^2$$

and:

$$u = 3.12 \text{ m/s}$$

The assumed value of ϕ may now be checked.

$$Re = \frac{du\rho}{\mu} = \frac{(0.15\times 3.12\times 1000)}{(1\times 10^{-3})} = 4.68\times 10^5$$

For $Re = 4.68\times 10^5$ and $e/d=0.01$, $\phi=0.0045$ (from Fig. 3.7) which agrees with the assumed value. Thus the initial rate of discharge $= 3.12\times(\pi/4)0.15^2 = 0.0\ 55\text{m}^3/\text{s}$ or $(0.0\ 55\times 1000) = 55\text{kg/s}$.

3.3.9 Flow Over Banks of Tubes

The frictional loss for a fluid flowing parallel to the axes of the tubes may be calculated in the normal manner by considering the hydraulic mean diameter of the system, although this applies strictly to turbulent flow only.

For flow at right angles to the axes of the tubes, the cross-sectional area is continually changing, and the problem may be treated as one involving a series of sudden enlargements and sudden contractions. Thus the friction loss would be expected to be directly proportional to the number of banks of pipes j in the direction of flow and to the kinetic energy of the fluid. The pressure drop $-\Delta P_f$ may be written as:

$$-\Delta P_f = \frac{C_f j \rho u_t^2}{6} \tag{3.83}$$

where C_f is a coefficient dependent on the arrangement of the tubes and the Reynolds number. The values of C_f given in Chapter 1 of Vol. 1B (Tables 1.3 and 1.4 of Vol. 1B) are based on the velocity u_t of flow at the narrowest cross-section.

3.3.10 Flow With a Free Surface

If a liquid is flowing with a free surface exposed to the surroundings, the pressure at the liquid surface will everywhere be constant and equal to atmospheric pressure. Flow will take place therefore only as a result of the action of the gravitational force, and the surface level will necessarily fall in the direction of flow.

Two cases are considered. The first, the laminar flow of a thin film down an inclined surface, is important in the heat transfer from a condensing vapour where the main resistance to transfer lies in the condensate film, as discussed in Chapter 1 of Vol. 1B (Section 1.6.1 of vol. 1B). The second is the flow in open channels which are frequently used for transporting liquids down a slope on an industrial site.

Laminar flow down an inclined surface

In any liquid flowing down a surface, a velocity profile is established with the velocity increasing from zero at the surface itself to a maximum where it is in contact with the surrounding atmosphere. The velocity distribution may be obtained in a manner similar to that used in connection with pipe flow, but noting that the driving force is that due to gravity rather than a pressure gradient.

For the flow of a liquid of depth s down a plane surface of width w inclined at an angle θ to the horizontal, as shown in Fig. 3.19, a force balance in the X-direction (parallel to the surface) may be written. In an element of length dx, the gravitational force acting on that part of the liquid which is at a distance greater than y from the surface is given by:

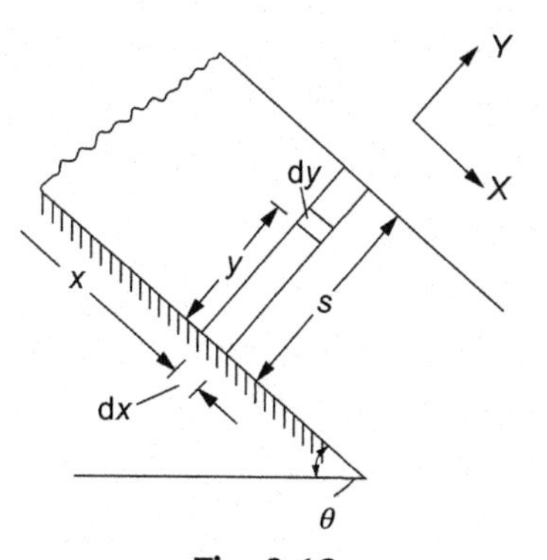

Fig. 3.19
Flow of liquid over a surface.

$$(s-y)w\,\mathrm{d}x\rho g\sin\theta$$

If the drag force of the atmosphere is negligible, the retarding force for laminar flow is attributable to the viscous drag in the liquid at the distance y from the surface given by:

$$\mu\frac{\mathrm{d}u_x}{\mathrm{d}y}w\,\mathrm{d}x$$

where u_x is the velocity of the fluid at that position.

Thus, at equilibrium:

$$(s-y)w\,\mathrm{d}x\,\rho g\,\sin\theta=\mu\frac{\mathrm{d}u_x}{\mathrm{d}y}w\,\mathrm{d}x \tag{3.84}$$

Since there will normally be no slip between the liquid and the surface, then $u_x=0$ when $y=0$, and:

$$\int_0^{u_x}\mathrm{d}u_x=\frac{\rho g\,\sin\theta}{\mu}\int_0^y(s-y)\mathrm{d}y$$

and:

$$u_x=\frac{\rho g\,\sin\theta}{\mu}\left(sy-\frac{1}{2}y^2\right) \tag{3.85}$$

The mass rate of flow G of liquid down the surface is now calculated.

$$\begin{aligned}G&=\int_0^s\left(\frac{\rho g\,\sin\theta}{\mu}w\left(sy-\tfrac{1}{2}y^2\right)\right)\rho\,\mathrm{d}y\\&=\frac{\rho^2 g\sin\theta}{\mu}w\left(\frac{s^3}{2}-\frac{s^3}{6}\right)\\&=\frac{\rho^2 g\,\sin\theta w s^3}{3\mu}\end{aligned} \tag{3.86}$$

The mean velocity of the fluid is then:

$$u=\frac{G}{\rho ws}=\frac{\rho gsin\theta\ s^2}{3\mu} \tag{3.87}$$

For a vertical surface:

$$\sin\theta=1 \text{ and } u=\frac{\rho g s^2}{3\mu}$$

The maximum velocity, which occurs at the free surface is given by:

$$u_s=\frac{\rho g \sin\theta\ s^2}{2\mu} \tag{3.88}$$

and this is 1.5 times the mean velocity of the liquid (Eq. 3.87).

Flow in open channels

For flow in an open channel, only turbulent flow is considered because streamline flow occurs in practice only when the liquid is flowing as a thin layer, as discussed in the previous section. The transition from streamline to turbulent flow occurs over the range of Reynolds numbers, $u\rho d_m/\mu=4000-11{,}000$, where d_m is the hydraulic mean diameter discussed earlier under *Flow in noncircular ducts.*

Three different types of turbulent flow may be obtained in open channels. They are *tranquil flow*, *rapid flow*, and *critical flow*. In tranquil flow, the velocity is less than that at which some disturbance, such as a surge wave, will be transmitted, and the flow is influenced by conditions at both the upstream and the downstream end of the channel. In rapid flow, the velocity of the fluid is greater than the velocity of a surge wave and the conditions at the downstream end do not influence the flow. Critical flow occurs when the velocity is exactly equal to the velocity of a surge wave.

Uniform flow

For a channel of constant cross-section and slope, the flow is said to be uniform when the depth of the liquid D is constant throughout the length of the channel. For these conditions, as shown in Fig. 3.20, for a length l of channel:

$$\text{The accelerating force acting on liquid}=lA\rho g\sin\theta$$
$$\text{The force resisting the motion}=R_m pl$$

where R_m is the mean value of the shear stress at the solid surface of the channel, p the wetted perimeter, and A the cross-sectional area of the flowing liquid.

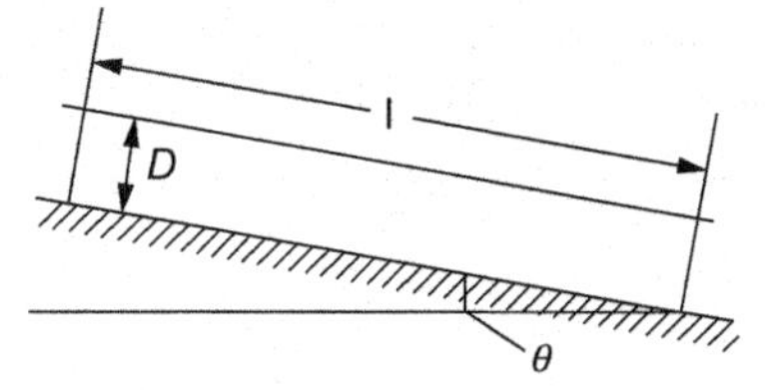

Fig. 3.20
Uniform flow in open channel.

For uniform motion:

$$lA\rho g\sin\theta = R_m pl$$

$$R_m = \frac{A}{p}\rho g\sin\theta \quad (3.89)$$

$$= \frac{1}{4}d_m\rho g\sin\theta$$

where d_m is the hydraulic mean diameter $=4\ A/p$ (see Eqs. 3.68–3.70).

Dividing both sides of the equation by ρu^2, where u is the mean velocity in the channel gives:

$$\frac{R_m}{\rho u^2} = \frac{d_m}{4}\frac{g\sin\theta}{u^2} \quad (3.90)$$

or:

$$u^2 = \frac{d_m}{4(R_m/\rho u^2)}g\sin\theta \quad (3.91)$$

For turbulent flow, $R_m/\rho u^2$ is almost independent of velocity, although it is a function of the surface roughness of the channel. Thus the resistance force is proportional to the square of the velocity. $R_m/\rho u^2$ is found experimentally to be proportional to the one-third power of the relative roughness of the channel surface and may be conveniently written as:

$$\frac{R_m}{\rho u^2} = \frac{1}{16}\left(\frac{e}{d_m}\right)^{1/3} \quad (3.92)$$

for the values of the roughness e given in Table 3.4.

Thus:

$$u^2 = 4d_m\left(\frac{d_m}{e}\right)^{1/3} g\ \sin\theta \quad (3.93)$$

$$= 4d_m^{4/3}e^{-1/3}g\ \sin\theta \quad (3.94)$$

Table 3.4 Values of the roughness *e*, for use in Eq. (3.92)

	ft	mm
Planed wood or finished concrete	0.00015	0.046
Unplaned wood	0.00024	0.073
Unfinished concrete	0.00037	0.11
Cast iron	0.00056	0.17
Brick	0.00082	0.25
Riveted steel	0.0017	0.51
Corrugated metal	0.0055	1.68
Rubble	0.012	3.66

The volumetric rate of flow Q is then given by:

$$\begin{aligned} Q &= uA \\ &= 2Ad_m^{2/3}e^{-1/6}\sqrt{g\sin\theta} \end{aligned} \tag{3.95}$$

The loss of energy due to friction for unit mass of fluid flowing isothermally through a length l of channel is equal to its loss of potential energy because the other forms of energy remain unchanged.

Thus:

$$F = gl\,\sin\theta = 4\frac{R_m}{\rho u^2}\frac{1}{d_m}u^2 \tag{3.96}$$

An empirical equation for the calculation of the velocity of flow in an open channel is the *Chezy equation*, which may be expressed as:

$$u = C\sqrt{\frac{1}{4}d_m \sin\theta} \tag{3.97}$$

where the value of the coefficient C is a function of the units of the other quantities in the equation. This expression takes no account of the effect of surface roughness on the velocity of flow.

The velocity of the liquid varies over the cross-section and is usually a maximum at a depth of between 0.05 D and 0.25 D below the surface, at the centre line of the channel. The velocity distribution may be measured by means of a pitot tube as described in Chapter 6.

The shape and the proportions of the channel may be chosen for any given flow rate so that the perimeter, and hence the cost of the channel, is a minimum.

From Eq. (3.95):

$$Q = 2Ad_m^{2/3}e^{-1/6}\sqrt{g\sin\theta}$$

Assuming that slope and the roughness of the channel are fixed, then for a given flowrate Q, from Eq. (3.95):

$$Ad_m^{2/3} = \text{constant}$$

or:

$$A^{5/3}p^{-2/3} = \text{constant} \tag{3.98}$$

The perimeter is therefore a minimum when the cross-section for flow is a minimum.

For a rectangular channel of depth D and width B (using $p = B + 2D$):

$$A = DB = D(p - 2D)$$
$$\therefore \quad D^{5/3}(p - 2D)^{5/3}p^{-2/3} = \text{constant}$$

p is a minimum when $\mathrm{d}p/\mathrm{d}D = 0$. Differentiating with respect to D and putting $\mathrm{d}p/\mathrm{d}D$ equal to 0:

$$p^{3/5} - p^{-2/5}4D = 0$$
$$\therefore \quad 4D = p = 2D + B$$

or:

$$B = 2D \tag{3.99}$$

Thus the most economical rectangular section is one where the width is equal to twice the depth of liquid flowing. The most economical proportions for other shapes may be determined in a similar manner.

Specific energy of liquid

For a liquid flowing in a channel inclined at an angle θ to the horizontal as shown in Fig. 3.21, the various energies associated with unit mass of fluid at a depth h below the liquid surface (measured at right angles to the bottom of the channel) are:

$$\begin{aligned} &\text{Internal energy:} && = U \\ &\text{Pressure energy:} && = (P_a + h\rho g \sec\theta)v = P_a v + hg \sec\theta \end{aligned}$$

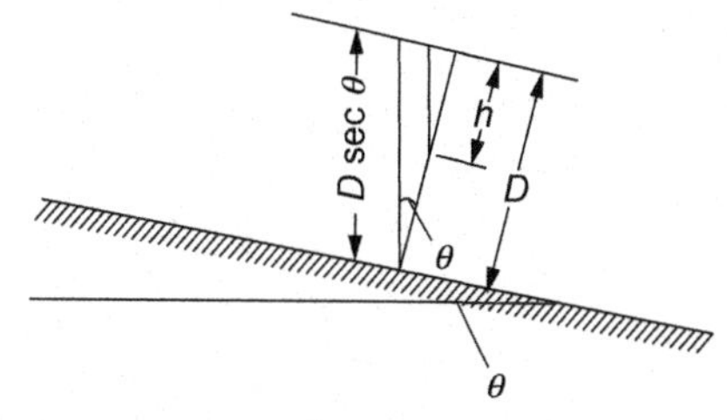

Fig. 3.21
Energy of fluid in open channel.

where P_a is atmospheric pressure.

$$\text{Potential energy:} \quad = zg + (D-h)g\sec\theta$$

where z is the height of the bottom of the channel above the datum level at which the potential energy is reckoned as zero.

$$\text{Kinetic energy} = \frac{u^2}{2}$$

The total energy per unit mass is, therefore:

$$U + P_a v + zg + Dg\sec\theta + \frac{u^2}{2} \tag{3.100}$$

It may be seen that at any cross-section, the total energy is independent of the depth h below the liquid surface. As the depth is increased, the pressure energy increases at the same rate as the potential energy decreases. If the fluid flows through a length dl of channel, the net change in energy per unit mass is given by:

$$\delta q - \delta W_s = \mathrm{d}U + g\mathrm{d}z + g\sec\theta\ \mathrm{d}D + u\ \mathrm{d}u \tag{3.101}$$

For an irreversible process:

$$\begin{aligned} \mathrm{d}U &= T\mathrm{d}S - P\mathrm{d}v \quad \text{(Eq. 2.5)} \\ &= \delta q + \delta F \quad \text{(Eq. 2.9)} \end{aligned}$$

assuming the fluid is incompressible, and dv=0.

If no work is done on the surroundings, $\delta W_s = 0$ and:

$$g\ \mathrm{d}z + g\sec\theta\ \mathrm{d}D + u\ \mathrm{d}u + \delta F = 0 \tag{3.102}$$

For a fluid at a constant temperature, the first three terms in Eq. (3.100) are independent of the flow conditions within the channel. On the other hand, the last two terms are functions of the velocity and the depth of liquid. The *specific energy* of the fluid is defined by:

$$J = Dg\sec\theta + \frac{u^2}{2} \tag{3.103}$$

For a horizontal channel, rectangular in section:

$$J = Dg + \frac{u^2}{2} = Dg + \frac{Q^2}{2B^2D^2}$$

The specific energy will vary with the velocity of the liquid and will be a minimum for some critical value of D; for a given rate of flow Q, the minimum energy will occur when $\mathrm{d}J/\mathrm{d}D = 0$.

Thus:

$$g + \frac{(-2Q^2)}{2B^2D^3} = 0$$

$$\frac{u^2}{D} = g$$

and:

$$u = \sqrt{gD} \tag{3.104}$$

This value of u is known as the *critical velocity*.

The corresponding values of D the *critical depth* and J are given by:

$$D = \left(\frac{Q^2}{B^2 g}\right)^{1/3} \tag{3.105}$$

and:

$$J = Dg + \frac{Dg}{2} = \frac{3Dg}{2} \tag{3.106}$$

Similarly it may be shown that, at the critical conditions, the flowrate is a maximum for a given value of the specific energy J. At the critical velocity, (u^2/gD) is equal to unity. This dimensionless group is known as the *Froude number* Fr. For velocities greater than the critical velocity, Fr is greater than unity, and vice versa. It may be shown that the velocity with which a small disturbance is transmitted through a liquid in an open channel is equal to the critical velocity, and hence the Froude number is the criterion by which the type of flow, tranquil or rapid, is determined. Tranquil flow occurs when Fr is less than unity and rapid flow when Fr is greater than unity.

Velocity of transmission of a wave

For a liquid which is flowing with a velocity u in a rectangular channel of width B, the depth of liquid is initially D_1. As a result of a change in conditions at the downstream end of the channel, the level there suddenly increases to some value D_2. A wave therefore tends to move upstream against the motion of the oncoming fluid. For two sections, 1 and 2, one on each side of the wave at any instant, as shown in Fig. 3.22, the rate of accumulation of fluid between the two sections is given by:

$$u_1 D_1 B - U_2 D_2 B$$

This accumulation of fluid results from the propagation of the wave and is therefore equal to $u_w B(D_2 – D_1)$.

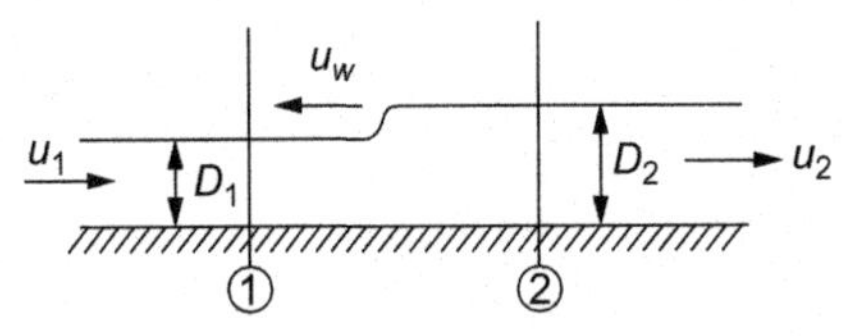

Fig. 3.22
Transmission of a wave.

Thus:

$$B(u_1D_1 - u_2D_2) = u_w(D_2 - D_1)B$$

or:

$$u_2 = \frac{u_w(D_1 - D_2) + u_1D_1}{D_2} \tag{3.107}$$

The velocity of the fluid is changed from u_1 to u_2 by the passage of the wave. The rate of travel of the wave relative to the upstream liquid is $(u_1 + u_w)$ and therefore the mass of fluid whose velocity is changed in unit time is:

$$(u_1 + u_w)BD_1\rho \tag{3.108}$$

The force acting on the fluid at any section where the liquid depth is D is:

$$\int_0^D (h\rho g)B \, dh = \frac{1}{2}B\rho g D^2 \tag{3.109}$$

where h is any depth below the liquid surface.

The net force acting on the fluid between sections 1 and 2 in Fig. 3.22 is:

$$\frac{1}{2}B\rho g\left(D_1^2 - D_2^2\right) \text{ in the direction of flow}$$

Thus, neglecting the frictional drag of the walls of the channel between sections 1 and 2, the net force can be equated to the rate of increase of momentum and thus:

$$\frac{1}{2}B\rho g\left(D_1^2 - D_2^2\right) = (u_1 + u_w)BD_1\rho(u_2 - u_1)$$

$$\therefore \quad (u_1 + u_w)D_1\left\{\frac{1}{D_2}[u_w(D_1 - D_2) + u_1D_1] - u_1\right\} = \frac{1}{2}g\left(D_1^2 - D_2^2\right)$$

and:

$$\begin{aligned}(u_1 + u_w)^2 &= \frac{D_2}{D_1}(D_1 + D_2)\frac{g}{2} \\ &= \frac{gD_2}{2}\left(1 + \frac{D_2}{D_1}\right)\end{aligned} \tag{3.110}$$

where $u_1 + u_w$ is the velocity of the wave relative to the oncoming fluid. For a very small wave, $D_1 \rightarrow D_2$ and:

$$u_1 + u_w = \sqrt{(gD_2)} \tag{3.111}$$

It is thus seen that the velocity of an elementary wave is equal to the critical velocity, at which the specific energy of the fluid is a minimum for a given flowrate. The criterion for critical conditions is therefore that the Froude number, (u^2/gD), be equal to unity. Detailed treatment of open channel flow is given, amongst others by Chanson.[20]

Hydraulic jump

If a liquid enters a channel under a gate, it will flow at a high velocity through and just beyond the gate and the depth will be correspondingly low. This is an unstable condition, and at some point the depth of the liquid may suddenly increase and the velocity fall. This change is known as the *hydraulic jump*, and it is accompanied by a reduction of the specific energy of the liquid as the flow changes from rapid to tranquil, any excess energy being dissipated as a result of turbulence.

If a liquid is flowing in a rectangular channel in which a hydraulic jump occurs between sections 1 and 2, as shown in Fig. 3.23, then the conditions after the jump can be determined by equating the net force acting on the liquid between the sections to the rate of change of momentum, if the frictional forces at the walls of the channel may be neglected.

The net force acting on the fluid in the flow direction is given by:

$$\frac{1}{2}B\rho g\left(D_1^2 - D_2^2\right)$$

The rate of change of momentum of fluid is given by:

$$= u_1 B D_1 \rho (u_2 - u_1)$$

or:

$$\frac{1}{2}g\left(D_1^2 - D_2^2\right) = u_1 D_1 (u_2 - u_1)$$

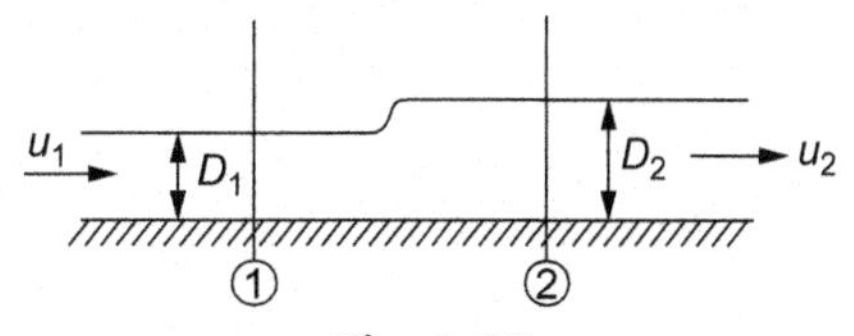

Fig. 3.23
Hydraulic jump.

The volumetric rate of flow of the fluid is the same at sections 1 and 2.

Thus:

$$Q = u_1 B D_1 = u_2 B D_2$$

or:

$$\frac{1}{2}g\left(D_1^2 - D_2^2\right) = u_1^2 D_1 \left[\left(\frac{D_1}{D_2}\right) - 1\right]$$

If $D_1 \neq D_2$, then:

$$\frac{1}{2}g(D_1 + D_2) = \frac{u_1^2 D_1}{D_2}$$

and:

$$D_2^2 + D_1 D_2 - \frac{2u_1^2 D_1}{g} = 0 \tag{3.112}$$

or:

$$D_2 = \frac{1}{2}\left(-D_1 \pm \sqrt{D_1^2 + \frac{8u_1^2 D_1}{g}}\right) \tag{3.113}$$

For obvious reasons, only the positive sign in Eq. (3.113) is admissible.

This expression gives D_2 as a function of the conditions at the upstream side of the hydraulic jump. The corresponding velocity u_2 is obtained by substituting in the equation:

$$u_1 D_1 = u_2 D_2$$

Corresponding values of D_1 and D_2 are referred to as *conjugate depths.*

The minimum depth at which a hydraulic jump can occur is found by putting $D_1 = D_2 = D$ in Eq. (3.112) giving:

$$2D^2 = 2D\frac{u_1^2}{g}$$

or:

$$D = \frac{u_1^2}{g} \tag{3.114}$$

This value of D corresponds to the *critical depth* for flow in a channel.

Thus a hydraulic jump can occur provided that the depth of the liquid is less than the critical depth. After the jump, the depth will be greater than the critical depth, the flow having changed from rapid to tranquil.

The energy dissipated in the hydraulic jump is now calculated. For a small change in the flow of a fluid in an open channel:

$$u\,\mathrm{d}u + g\,\mathrm{d}z + g\sec\theta\,\mathrm{d}D + \delta F = 0 \quad \text{(Eq. 3.102)}$$

Then, for a horizontal channel:

$$F = g(D_1 - D_2) + \frac{(u_1^2 - u_2^2)}{2}$$

From Eq. (3.112):

$$u_1^2 = \frac{gD_2(D_1 + D_2)}{2D_1}$$

Similarly:

$$u_2^2 = \frac{gD_1(D_1 + D_2)}{2D_2}$$

Thus:

$$\begin{aligned} F &= g(D_1 - D_2) + \frac{1}{4}g(D_1 + D_2)\left(\frac{D_2}{D_1} - \frac{D_1}{D_2}\right) \\ &= \tfrac{1}{4}g(D_1 - D_2)\left[4 - (D_1 + D_2)\frac{(D_1 + D_2)}{D_1 D_2}\right] \\ &= \frac{(D_2 - D_1)^3 g}{4D_1 D_2} \end{aligned} \quad (3.115)$$

The hydraulic jump may be compared with the shock wave for the flow of a compressible fluid, discussed in Chapter 4.

Example 3.10

A hydraulic jump occurs during the flow of a liquid discharging from a tank into an open channel under a gate so that the liquid is initially travelling at a velocity of 1.5 m/s with a depth of 75 mm. Calculate the corresponding velocity and the liquid depth after the jump.

Solution

The depth of fluid in the channel after the jump is given by:

$$D_2 = 0.5\left\{-D_1 + \sqrt{\left[D_1^2 + \left(8u_1^2 D_1/g\right)\right]}\right\} \quad \text{(Eq. 3.113)}$$

where D_1 and u_1 are the depth and velocity of the fluid before the jump.

Thus:

$$D_1 = 0.075 \text{ m} \quad \text{and} \quad u_1 = 1.5 \text{ m/s}$$

and hence:

$$\begin{aligned} D_2 &= 0.5\left\{-0.075 + \sqrt{\left[0.075^2 + \left(8 \times 1.5^2 \times 0.075/9.81\right)\right]}\right\} \\ &= 0.152 \text{ m} = \underline{\underline{152 \text{ mm}}} \end{aligned}$$

If the channel is of uniform cross-sectional area, then:

$$u_1 D_1 = u_2 D_2$$

and:

$$\begin{aligned} u_2 &= u_1 D_1 / D_2 \\ &= (1.5 \times 0.075)/0.152) = \underline{\underline{0.74 \text{ m/s}}} \end{aligned}$$

3.4 Non-Newtonian Fluids

In the previous sections of this chapter, the calculation of frictional losses associated with the flow of simple *Newtonian* fluids has been discussed. A Newtonian fluid at a given temperature and pressure has a constant viscosity μ which does not depend on the shear rate or shear stress, and for streamline (laminar) flow, is equal to the ratio of the shear stress (R_y) to the shear rate (du_x/dy) as shown in Eq. (3.4), or:

$$\mu = \frac{|R_y|}{|du_x/dy|} \tag{3.116}$$

The modulus sign is used because shear stresses within a fluid act in both the positive and negative senses. Gases and simple low molecular weight liquids are all Newtonian, and viscosity may be treated as constant in any flow problem unless there are significant variations of temperature or pressure.

Many fluids, including some that are encountered very widely both industrially and domestically, exhibit non-Newtonian behaviour and their apparent viscosities may depend on the rate at which they are sheared and on their previous shear history. At any position and time in the fluid, the apparent viscosity μ_a which is defined as the ratio of the shear stress to the shear rate at that point is given by:

$$\mu_a = \frac{|R_y|}{|du_x/dy|} \tag{3.117}$$

When the apparent viscosity is a function of the shear rate alone, the behaviour is said to be *shear-dependent*; when it is a function of the duration of shearing at a particular rate, it is referred to as *time-dependent*. Any shear-dependent fluid must to some extent be time-dependent because, if the shear rate is suddenly changed, the apparent viscosity does not alter instantaneously, but gradually moves towards its new equilibrium value. In many cases, however, the time-scale for the flow process may be sufficiently long for the effects of time-dependence to be negligible.

The apparent viscosity of a fluid may either decrease or increase as the shear rate is raised. The more common effect is for the apparent viscosity to fall as the shear rate is raised; such behaviour is referred to as *shear-thinning*. Most paints are shear-thinning: in the can and when loaded on to the brush, the paint is subject only to low rates of shear and has a high apparent viscosity. When applied to the surface, the paint is sheared by the brush, its apparent viscosity becomes less, and it flows readily to give an even film. However, when the brushing ceases, it recovers its high apparent viscosity and does not drain from the surface under its own weight. This non-Newtonian behaviour is an important characteristic of a good paint. It is frequently necessary to build-in non-Newtonian characteristics to give a product the desired properties. Paints will usually exhibit appreciable time-dependent behaviour. Thus, as the paint is stirred, its apparent viscosity will decrease progressively until it reaches an asymptotic equilibrium value characteristic of that particular rate of shear.

Some materials have the characteristics of both solids and liquids. For instance, tooth paste behaves as a solid in the tube, but when the tube is squeezed the paste flows as a plug. The essential characteristic of such a material is that it will not flow until a certain critical shear stress, known as the *yield stress* is exceeded. Thus, it behaves as a solid at low shear stresses and as a fluid at high shear stress. It is a further example of a shear-thinning fluid, with an infinite apparent viscosity at stress values below the yield value, and a falling finite value as the stress is progressively increased beyond this point.

A further important property which may be shown by a non-Newtonian fluid is *elasticity* – which causes the fluid to try to regain its former condition as soon as the stress is removed, i.e., it exhibits memory effects. Again, the material is showing some of the characteristics of both a solid and a liquid. An ideal (Newtonian) liquid is one in which the stress is proportional to the *rate* of shear (or rate of strain). On the other hand, for an ideal solid (obeying Hooke's Law) the stress is proportional to the strain. A fluid showing elastic behaviour is termed *viscoelastic* or *elasticoviscous*.

The branch of science which is concerned with the flow of both simple (Newtonian) and complex (non-Newtonian) fluids is known as *rheology*. The flow characteristics are represented by a *rheogram*, which is a plot of shear stress against rate of shear, and normally consists of a collection of experimentally determined points through which a curve may be drawn. If an equation can be fitted to the curve, it facilitates calculation of the behaviour of the fluid. It must be borne in mind, however, that such equations are approximations to the actual behaviour of the fluid and should not be used outside the range of conditions (particularly shear rates) for which they were determined.

An understanding of non-Newtonian behaviour is important to the chemical engineer from two points of view. Frequently, non-Newtonian properties are desirable in that they can confer desirable properties on the material which are essential if it is to fulfil the purpose for which it is required. The example of paint has already been given. Toothpaste should not flow out of the tube until it is squeezed and should stay in place on the brush until it is applied to the teeth. The texture of foodstuffs is largely attributable to rheology.

Second, it is necessary to take account of non-Newtonian behaviour in the design of process plant and pipelines. Heat and mass transfer coefficients are considerably affected by the behaviour of the fluid, and special attention must be devoted to the selection of appropriate flow metering devices, mixing equipment, and pumps.

In this section, some of the important aspects of non-Newtonian behaviour will be quantified, and some of the simpler approximate equations of state will be discussed. An attempt has been made to standardise nomenclature in the British Standard, BS 5168.[21]

Shear stress is denoted by R in order to be consistent with other parts of the book; τ or σ is frequently used elsewhere to denote shear stress. R without suffix denotes the shear stress acting on a surface in the direction of flow and $R_0(=-R)$ denotes the shear stress exerted by the surface on the fluid. R_s denotes the positive value in the fluid at a radius s and R_y the positive value at a distance y from a surface. Strain is defined as the ratio dx/dy, where dx is the shear displacement of two elements of fluid a distance dy apart and is often denoted by γ. The rate of strain or rate of shear is $(dx/dt)/dy$ or du_x/dy and is denoted by $\dot{\gamma}$.

Thus Eq. (3.117) may be written:

$$\mu_a = \frac{\tau}{\dot{\gamma}} \tag{3.118}$$

3.4.1 Steady-State Shear-Dependent Behaviour

In this section, consideration will be given to the equilibrium relationships between shear stress and shear rate for fluids exhibiting non-Newtonian behaviour. Whenever the shear stress or the shear rate is altered, the fluid will gradually move towards its new equilibrium state and for the present, the period of 'adjustment' between the two equilibrium states will be ignored.

The fluid may be either shear-thinning or, less often, shear-thickening, and in either case the shear stress and the *apparent viscosity* μ_a are functions of shear rate, or:

$$|R_y| = f_1\left(\left|\frac{du_x}{dy}\right|\right) \tag{3.119}$$

and:

$$\mu_a = |R_y| \Big/ \left|\frac{du_x}{dy}\right| = f_2\left(\left|\frac{du_x}{dy}\right|\right) \tag{3.120}$$

Here y is the distance measured from a boundary surface.

Typical forms of the curve of shear stress versus shear rate are shown in Fig. 3.24 for a shear-thinning (or *pseudoplastic*) fluid, a *Bingham-plastic*, a Newtonian fluid, and a shear-thickening (or *dilatant*) fluid. For the particular case chosen, the apparent viscosity is the same for all four at the shear rate where they intersect. At lower shear rates, the shear-thinning fluid is more viscous than the Newtonian fluid, and at higher shear rates it is less viscous. For the shear-thickening fluid, the situation is reversed. The corresponding curves showing the variation of apparent viscosity are given in Fig. 3.25. Because the rates of shear of interest can cover several orders of magnitude, it is convenient to replot the curves using log-log coordinates, as shown in Figs. 3.26 and 3.27.

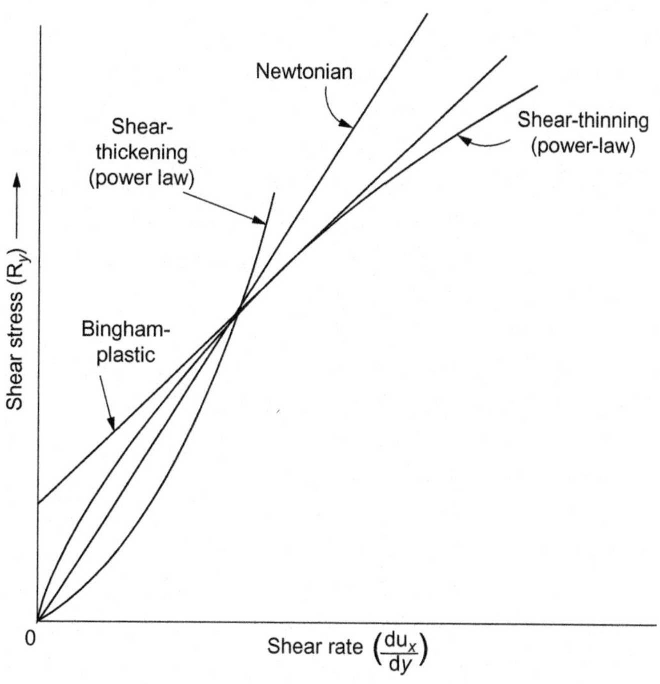

Fig. 3.24
Shear stress–shear rate behaviour of Newtonian and non-Newtonian fluids plotted using linear coordinates.

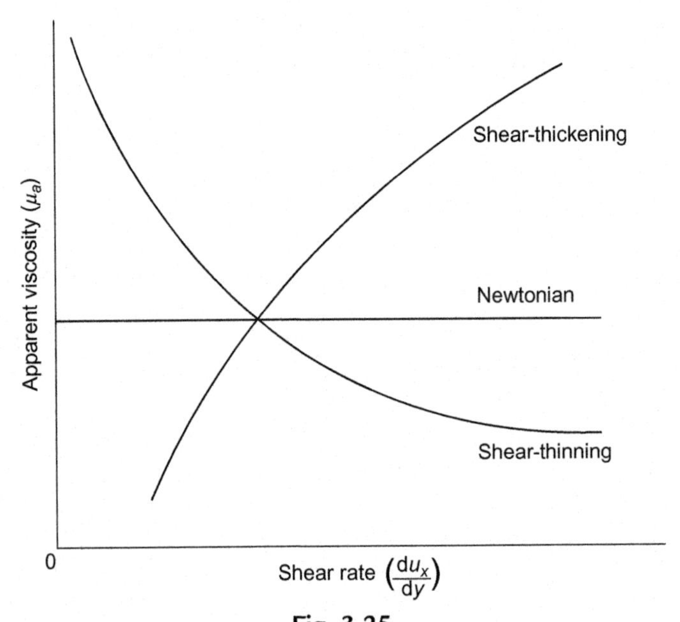

Fig. 3.25
Shear rate dependence of apparent viscosity for Newtonian and non-Newtonian fluids plotted on linear coordinates.

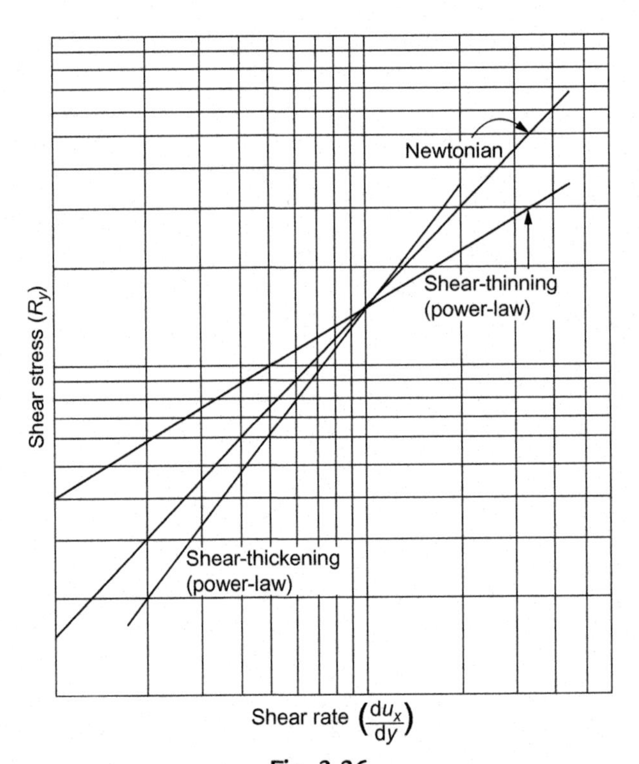

Fig. 3.26
The relation between shear-stress and shear-rate using logarithmic axes.

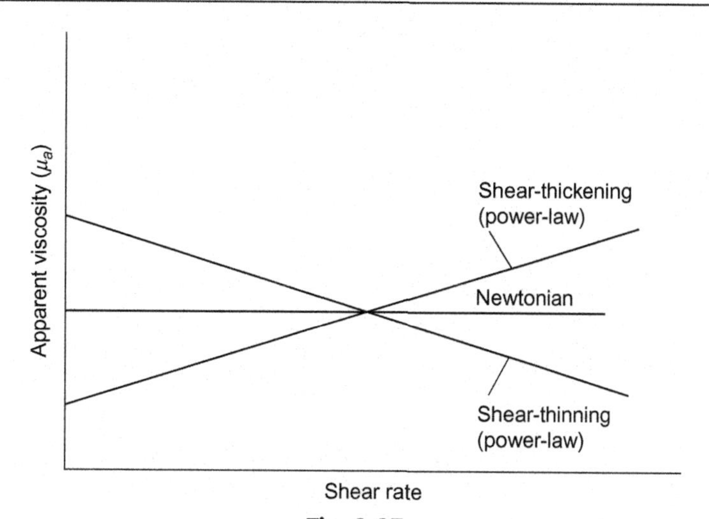

Fig. 3.27
Plot of apparent viscosity–shear rate relation using logarithmic axes.

The relation between shear stress and shear rate for the Newtonian fluid is defined by a single parameter μ, the viscosity of the fluid. No single parameter model will describe non-Newtonian behaviour and models involving two or even more parameters only approximate to the characteristics of real fluids, and can be used only over a limited range of shear rates.

A useful two-parameter model is the *power-law* model, or Ostwald–de Waele law to identify its first proponents. The relation between shear stress and shear rate is given by:

$$|R_y| = k\left(\left|\frac{du_x}{dy}\right|\right)^n \tag{3.121}$$

where n is known as the *power-law index* and k is known as the *consistency* coefficient. This equation may be written as:

$$|R_y| = k\left(\left|\frac{du_x}{dy}\right|\right)^{n-1}\left|\frac{du_x}{dy}\right| \tag{3.122}$$

It is therefore seen that the apparent viscosity μ_a is given by:

$$\mu_a = k\left(\left|\frac{du_x}{dy}\right|\right)^{n-1} \tag{3.123}$$

From Eq. (3.123):

when $n > 1$, μ_a increases with increase of shear rate, and shear-thickening behaviour is described;

when $n = 1$, μ_a is constant and equal to the Newtonian viscosity μ of the fluid;

when $n < 1$, μ_a decreases with increase of shear rate, and the behaviour is that of a shear-thinning fluid.

Thus, by selecting an appropriate value of n, both shear-thinning and shear-thickening behaviour can be represented, with $n=1$ representing Newtonian behaviour which essentially marks the transition from shear-thinning to shear-thickening characteristics.

It will be noted that the dimensions of k are $\mathbf{ML}^{-1}\ \mathbf{T}^{n-2}$, that is they are dependent on the value of n. Values of k for fluids with different n values cannot therefore be compared. *Numerically*, k is the value of the apparent viscosity (or shear stress) at unit shear rate and this numerical value will depend on the units used; for example the value of k at a shear rate of 1 s^{-1} will be different from that at a shear rate of 1 h^{-1}. Therefore, caution is necessary when comparing the power-law constants for two fluids with different values of the power-law index.

In Fig. 3.28, the shear stress is shown as a function of shear rate for a typical shear-thinning fluid, using logarithmic coordinates. Over the shear rate range (*ca* 10^0 to 10^3 s^{-1}), the fluid behaviour is described by the power-law equation with an index n of 0.6, that is the line CD has a slope of 0.6. If the power-law was followed at all shear rates, the extrapolated line $C'\ CDD'$ would be applicable. Fig. 3.29 shows the corresponding values of apparent viscosity and the line $C'\ CDD'$ has a slope of $n-1=-0.4$. It is seen that it extrapolates to $\mu_a=\infty$ at zero shear rate and to $\mu_a=0$ at infinite shear rate.

The viscosities of most real shear-thinning fluids (polymer solutions and melts, at least) approach constant values both at *very low shear rates* and at *very high shear rates*; that is, they tend to show Newtonian properties at the extremes of shear rates. The limiting viscosity at low shear rates μ_0 is referred to as the lower-Newtonian (or zero-shear) viscosity (see lines AB in

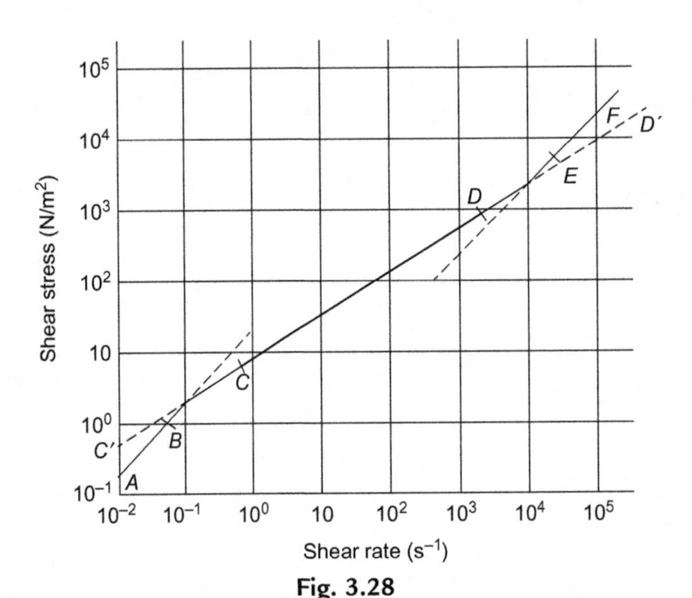

Fig. 3.28

Behaviour of a typical shear-thinning fluid plotted logarithmically for several orders of shear rate.

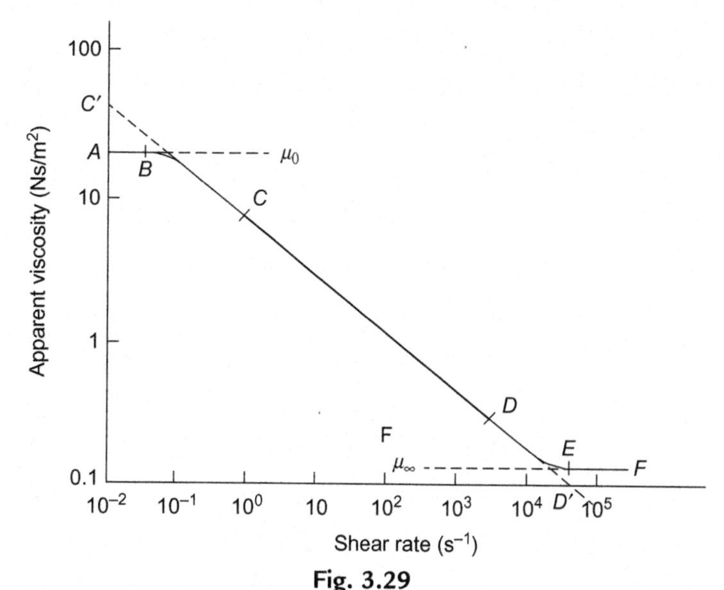

Fig. 3.29
Apparent viscosity corresponding to the data of Fig. 3.28.

Figs. 3.28 and 3.29), and that at high shear rates μ_∞ is the upper-Newtonian (or infinite-shear) viscosity (see lines *EF* in Figs. 3.28 and 3.29).

By reference to Fig. 3.29, it is seen that the power law model can, at best therefore, represent the behaviour of a real fluid over only a limited range of shear rates. Effectively, it represents what is happening in the region where apparent viscosity is changing most rapidly with shear rate. However, over any limited range of shear rates, it is possible to approximate the curve in Fig. 3.29 by a straight line, whose slope $(n-1)$ will determine the 'best' value of the power-law index n which can be used over that range.

In practice, shear rates ranging from 10^{-6} to $10^{6}\ \text{s}^{-1}$, or 12 orders of magnitudes are encountered and over that wide range the behaviour depicted in Fig. 3.29 is of greater importance. Frequently, however, a much lower range of shear rates occurs in a particular application and rheological measurements can then be restricted to this narrower shear-rate range and a relatively simple model, such as the power-law, may be used. It is always convenient to use the simplest model which adequately describes the rheology over the range of interest. As an example, the shear-rate ranges which apply in the production and subsequent use of foodstuffs are given in Table 3.5.

A convenient form of 3-parameter equation which extrapolates to a constant limiting apparent viscosity (μ_0 or μ_∞) as the shear rate approaches both zero and infinity has been proposed by Cross[22]:

$$\frac{\mu_0 - \mu_\infty}{\mu_a - \mu_\infty} = 1 + \beta\dot{\gamma}^{2/3} \tag{3.124}$$

Table 3.5 Typical shear-rate ranges for food products and processes

Situation	Shear Rate Range (s^{-1})	Application
Sedimentation of particles in a suspending fluid	10^{-6}–10^{-3}	Spices in salad dressing
Levelling due to surface tension	0.01–0.1	Frosting
Draining under gravity	0.1–10	Vats, small food containers
Extrusion	1–1000	Snack foods, cereals, pasta
Calendering	10–100	Dough sheeting
Chewing and swallowing	10–100	All foods
Dip coating	10–100	Confectionery
Mixing and stirring	10–1000	Numerous
Pumping/pipeflow	1–1000	Numerous
Spraying and brushing	10^3–10^4	Spray drying

Courtesy Neil Alderman, AEA Technology. See also Barnes HA, Hutton JF, Walters K. An introduction to rheology. *Amsterdam: Elsevier; 1989.*

where $\dot{\gamma}$ is the shear rate and μ_0, μ_∞, and β must be determined from experimental data for each fluid. For some materials, a slight modification of the exponent of $\dot{\gamma}^{2/3}$ leads to improved agreement between predicted and experimental results.

This equation is based on the assumption that pseudoplastic (shear-thinning) behaviour is associated with the formation and rupture of structural linkages. It is based on an experimental study of a wide range of fluids – including aqueous suspensions of flocculated inorganic particles, aqueous polymer solutions and nonaqueous suspensions and solutions – over a wide range of shear rates ($\dot{\gamma} \sim 10$–$10^4\ s^{-1}$).

For a shear-thickening fluid, the same arguments can be applied, with the apparent viscosity rising from zero at zero shear rate to infinity at infinite shear rate, on application of the power law model. However, shear-thickening is generally observed over very much narrower ranges of shear rate and it is difficult to generalise on the type of curve which will be obtained in practice. It is not uncommon for a substance to exhibit shear-thinning behaviour at low shear rates and shear-thickening at high shear rates.[23]

In order to overcome the shortcomings of the power-law model, several alternative forms of equation between shear rate and shear stress have been proposed. These are all more complex involving three or more parameters. Reference should be made to specialist works on non-Newtonian flow[2,24–26] for details of these *Constitutive Equations.*

Some fluids exhibit a *yield stress*. When subjected to stresses below the yield stress they do not flow and effectively can be regarded as fluids of infinite viscosities, or alternatively as solids. When the yield stress is exceeded they flow as fluids. Such behaviour cannot be described by a power-law model.

The simplest type of behaviour for a fluid exhibiting a yield stress is known as *Bingham-plastic*. The shear rate is directly proportional to the amount by which the stress exceeds the yield stress.

Thus:

$$|R_y| - R_Y = \mu_p \left|\frac{du_x}{dy}\right| \quad (|R_y| > R_Y) \tag{3.125}$$

$$\frac{du_x}{dy} = 0 \quad (|R_y| R_Y)$$

μ_p is known as the *plastic viscosity*.

The apparent viscosity μ_a is given, by definition, as:

$$\begin{aligned} \mu_a &= |R_y| / \left|\frac{du_x}{dy}\right| \quad (\text{Eq. } 3.117) \\ &= \mu_p + R_Y / \left|\frac{du_x}{dy}\right| \quad (|R_y| R_Y) \end{aligned} \tag{3.126}$$

Thus, the apparent viscosity falls from infinity at zero shear rate ($|R_y \overline{<} R_Y$) to μ_p at infinite shear rate, i.e. the fluid shows shear-thinning characteristics.

Because it is very difficult to measure the flow characteristics of a material at very low shear rates, behaviour at zero shear rate can often only be assessed by extrapolation of experimental data obtained over a limited range of shear rates. This extrapolation can be difficult, if not impossible.[27] From Example 3.10 in Section **3.4.7**, it can be seen that it is sometimes possible to approximate the behaviour of a fluid *over the range of shear rates for which experimental results are available*, either by a power-law or by a Bingham-plastic equation.

Some materials show more complex behaviour and the plot of shear stress against shear rate approximates to a curve, rather than to a straight line with an intercept R_Y on the shear stress axis. An equation of the following form may then be used:

$$|R_y| - R_Y = \mu'_p \left(\left|\frac{du_x}{dy}\right|\right)^m \tag{3.127}$$

Thus, Eq. (3.127), which includes three parameters, is effectively a combination of Eqs. (3.121), (3.125). It is sometimes called the *generalised Bingham equation* or *Herschel–Bulkley equation*, and the fluids are sometimes referred to as having *false body*. Figs. 3.30 and 3.31 show shear stress and apparent viscosity, respectively, for Bingham plastic and false body fluids, using linear coordinates.

In many flow geometries, the shear stress will vary over the cross-section and there will then be regions where the shear stress exceeds the yield stress and the fluid will then be sheared. In other regions, the shear stress will be less than the yield value and the fluid will flow there as an

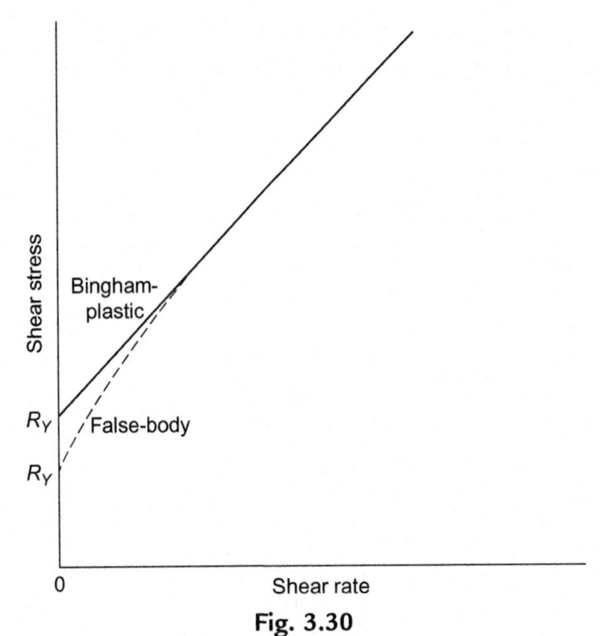

Fig. 3.30
Shear stress-shear rate data for Bingham-plastic and false-body fluids using linear scale axes.

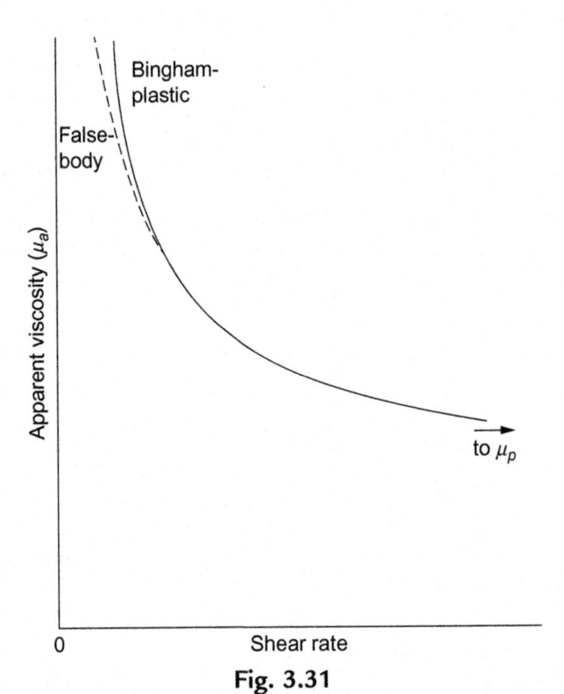

Fig. 3.31
Apparent viscosity for Bingham-plastic and false-body fluids using linear axes.

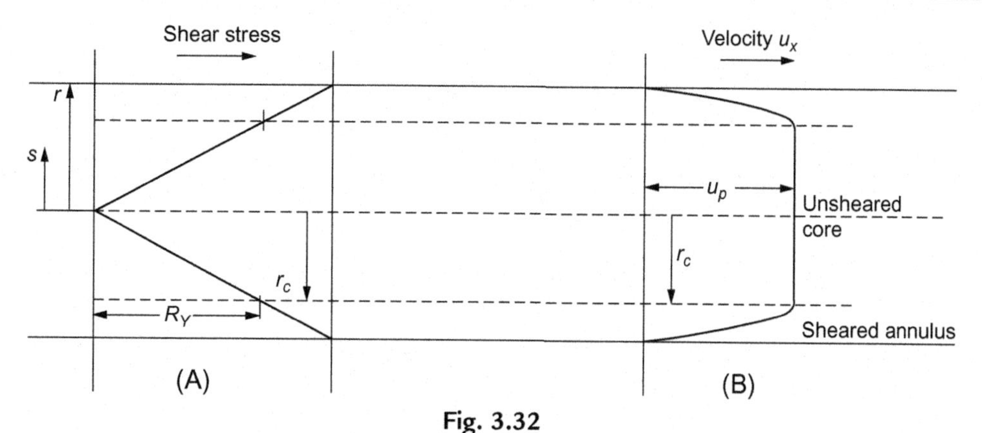

Fig. 3.32
(A) Shear stress distribution in pipe. (B) Velocity profile for Bingham plastic fluid in pipe.

unsheared plug. Care must be taken therefore in the application of Eqs. (3.125), (3.127). Thus, for the case of flow in a circular tube, which will be considered later, the shear stress varies linearly from a maximum value at the wall, to zero at the centre-line (see Eq. 3.9). If the shear stress at the wall is less than the yield stress, no flow will occur. If it is greater than the yield stress, shear will take place in the region between the wall and the point where the shear stress equals the yield stress (Fig. 3.32). Inside this region the fluid will flow as an unsheared plug. As the pressure difference over the tube is increased, the wall shear stress will increase in direct proportion and the unsheared plug will be of smaller radius. For any fluid subjected to a finite pressure gradient, there must always be some region, however small, near the pipe axis in which shear does not take place. Naturally, the homogenisation and mixing, heating or cooling of such fluids is much more challenging than their Newtonian counterparts.[28]

3.4.2 Time-Dependent Behaviour

In a *time-dependent fluid*, the shear rate depends upon the time for which it has been subjected to a given shear stress. Conversely, if the shear rate is kept constant, the shear stress will change with time. However, with all *time-dependent* fluids an equilibrium condition is reached if the imposed condition (e.g., shear rate or shear stress) is maintained constant. Some fluids respond so quickly to changes that the effect of time dependence can be neglected. Others may have a much longer constant, and in changing flow situations will never be in the equilibrium state.

In general, for shear-thinning pseudoplastic fluids, the apparent viscosity will gradually decrease with time if there is a step increase in its rate of shear. This phenomenon is known as *thixotropy*. Similarly, with a shear-thickening fluid the apparent viscosity increases under these circumstances and the fluid exhibits *rheopexy* or *negative-thixotropy*.

The effect of increasing and then decreasing the rate of shear of a thixotropic fluid is shown in Fig. 3.33 in which the shear stress is plotted against shear rate in an experiment; the

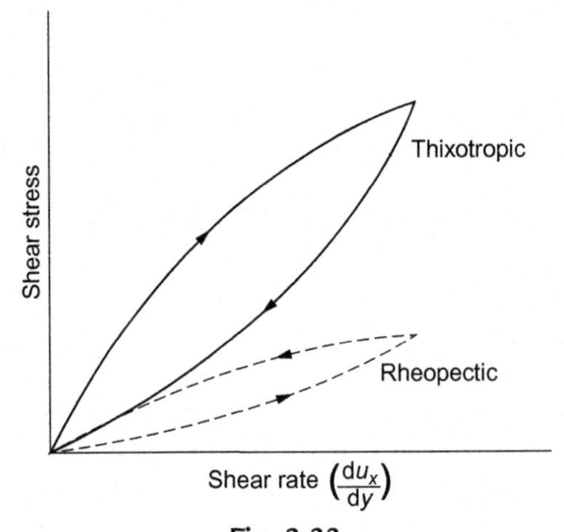

Fig. 3.33
Thixotropic and rheopectic behaviour.

shear rate is steadily increased from zero and subsequently decreased again. It is seen that a hysteresis loop is formed, with the shear stress always lagging behind its equilibrium value. If the shear rate is changed rapidly the hysteresis loop will have a large area. At low rates of change, the area will be small and will eventually become zero as the two curves coincide when sufficient time is allowed for equilibrium to be reached at each point on the curve. In Fig. 3.34, the effect on the apparent viscosity is seen of a step increase in the shear rate; it gradually decreases from the initial to the final equilibrium value. This is a picture of what happens to a material which does not suffer any irreversible changes as a result of shearing. Some materials, particular gels, suffer structural breakdown when subjected to a shear field, but this effect is not considered here.

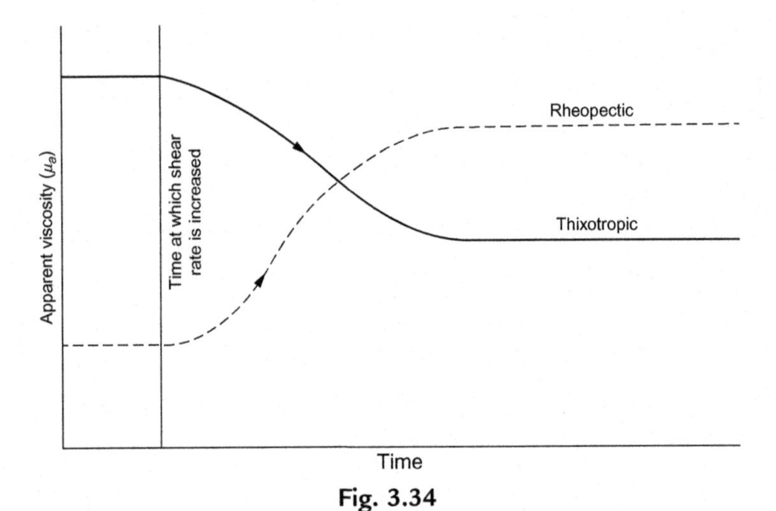

Fig. 3.34
Effect of sudden change of shear rate on apparent viscosity of time-dependent fluid.

The behaviour of a *rheopectic* fluid is the reverse of that of a thixotropic fluid and is illustrated by the broken lines in Figs. 3.33 and 3.34.

3.4.3 Viscoelastic Behaviour

A true fluid flows when it is subjected to a shear field and motion ceases as soon as the stress is removed. In contrast, an ideal solid which has been subjected to a stress recovers its original state as soon as the stress is removed. The two extremes of behaviour are therefore represented by:

(a) The ideal Newtonian fluid in which:

$$|R_y| = \mu \left|\frac{du_x}{dy}\right| \quad \text{(Eq. 3.4)}$$

$$(\text{or } \tau = \mu\dot{\gamma})$$

where $du_x/dy(=\dot{\gamma})$ is the rate of shear.

(b) The ideal elastic solid which obeys Hooke's law in which the relation between shear distortion and stress is:

$$|R_y| = \mathbf{G}\left|\frac{dx}{dy}\right| \tag{3.128}$$

$$(\text{or } \tau = \mathbf{G}\gamma)$$

where **G** is Young's modulus, and the shear $dx/dy(=\gamma)$ is the ratio of the shear displacement of two elements to their distance apart.

Many materials of practical interest (such as polymer solutions and melts, foodstuffs, and biological fluids) exhibit *viscoelastic* characteristics; they have some ability to store and recover shear energy and therefore show some of the properties of both a solid and a liquid. Thus a solid may be subject to *creep* and a fluid may exhibit *elastic* properties. Several phenomena ascribed to fluid elasticity including normal stress effects, die swell, rod climbing (Weissenberg effect), the tubeless siphon, bouncing of a sphere, turbulent drag reductions, and the development of secondary flow patterns at low Reynolds numbers, have recently been illustrated in an excellent photographic study.[29] Two common and easily observable examples of viscoelastic behaviour in a liquid are:

(a) The liquid in a cylindrical vessel is given a swirling motion by means of a stirrer. When the stirring is stopped, the fluid gradually comes to rest and, if viscoelastic, may then start to rotate in the opposite direction, that is to unwind a little bit.

(b) A viscoelastic fluid, on emerging from a tube or from a die, may form a jet which is of larger diameter than the aperture. The phenomenon, referred to above as 'die-swell', results from the sudden removal of a constraining force on the fluid.

Viscoelastic fluids are thus capable of exerting *normal stresses*. Because most materials, under appropriate circumstances, show simultaneously *solid-like* and *fluid-like* behaviours in varying proportions, the notion of an ideal elastic solid or of a purely viscous fluid represents the commonly encountered limiting condition. For instance, the viscosity of ice and the elasticity of water may both pass unnoticed! The response of a material may also depend upon the type of deformation to which it is subjected. A material may behave like a highly elastic solid in one flow situation, and like a viscous fluid in another.

Generally, viscoelastic effects are of particular importance in unsteady state flow and where there are rapid changes in the pressure to which the fluid is subjected. They can give rise to very complex behaviour and mechanical analogues have sometimes been found useful for calculating their flow behaviour, at least qualitatively. In this approach, the fluid properties are represented by a 'dashpot', a piston in a cylinder with a small outlet in which the flowrate is linearly related to the pressure difference across it. The elastic properties are represented by a spring. By combination of such elements in a variety of ways it is possible to simulate the behaviour of very complex materials. A single dashpot and spring in series is known as the Maxwell model; and the dashpot and spring in parallel is the Voigt model, as shown in Fig. 3.35.

Because of the assumption that linear relations exist between shear stress and shear rate (Eq. 3.4) and between distortion and stress (Eq. 3.128), both of these models, namely the Maxwell and Voigt models, and all other such models involving combinations of springs and dashpots, are restricted to small strains and small strain rates. Accordingly, the equations describing these models are known as linear viscoelastic equations. Several theoretical and semitheoretical approaches are available to account for nonlinear viscoelastic effects, and reference should be made to specialist works[24–26,30] for further details.

For process design calculations, there are two basic matters of concern:

(1) to characterise the viscoelastic behaviour of a substance, and
(2) to ascertain whether viscoelastic effects are significant in a given flow situation.

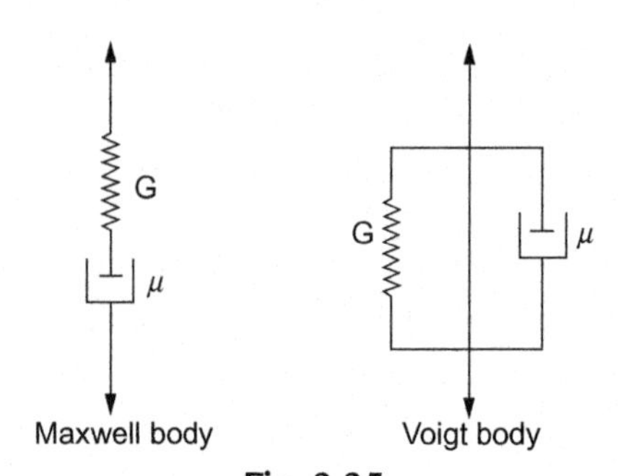

Fig. 3.35
Mechanical analogues for viscoelastic fluids.

Depending upon the particular application and the type of deformation to which an element is likely to be subjected, there are several ways of characterising viscoelastic behaviour of a material, not all of which are mutually exclusive; in fact, in some cases, the information deduced using different experimental techniques may be interconnected. In general, viscoelastic effects are of greater significance in flow domains remote from boundary surfaces. If an infinitesimal element of viscoelastic fluid flowing in the x-direction is subjected to a steady-state shear stress which arises from the velocity gradient in the y-direction it will, unlike an inelastic fluid, be subject to normal stresses I_x, I_y, and I_z acting on the faces, A, C, and B, respectively, as shown in Fig. 3.36. In this case, the fluid has zero velocity components in the y- and z-directions and there is no velocity gradient in either the x- or the z-direction. Of great importance in characterising viscoelastic behaviour are the so-called normal shear-stress differences $N_1(=I_x-I_y)$ and $N_2(=I_y-I_z)$ which are more easily measured than the normal shear stresses themselves. Generally, N_1 is considerably greater than N_2, and the ratio N_1/R_c, where R_c is the shear stress acting on face C, is an indication of the degree of viscoelasticity of the fluid. The higher the value of N_1/R_c, the greater is the elasticity, the ratio being zero for an inelastic fluid.

The other commonly encountered type of deformation is stretching, or extensional flow. This type of flow occurs in the film-blowing process, and in flow through dense porous media where it arises from the successive convergent-divergent nature of the flow passages. Viscoelastic fluids in extensional flow behave in a significantly different manner from purely viscous fluids. It is well established that the so-called *Trouton ratio*, defined as the ratio of the extensional viscocity (tension/rate of stretch) to the shear viscosity is 3 for a Newtonian fluid and it is independent of shear/stretch rates. For many viscoelastic fluids, the Trouton ratio is again 3 at low deformation (shear/stretch) rates, but values as high as 1000 have sometimes been

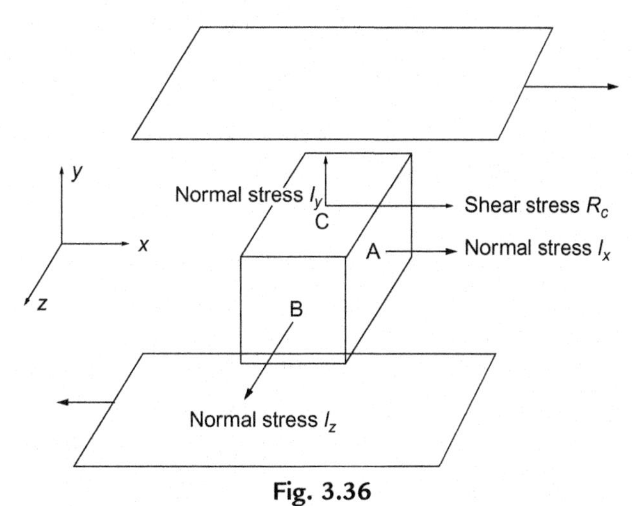

Fig. 3.36
Shear stress and normal stresses on element of fluid.

observed. Again, the higher the value of the Trouton ratio, the more elastic is the fluid behaviour. In general, extensional viscosities are high when the aspect ratios of the molecules are large. Thus, it is quite common for a fluid to exhibit shear-thinning in simple shear but strain-hardening in extensional flow.

Many other techniques of measuring viscoelastic parameters, such as transient shear, creep, and sinusoidally-varying shear, are available. A good description, together with the merits and demerits of each of these techniques, is available in Whorlow[31] and Macosko.[32]

3.4.4 Characterisation of Non-Newtonian Fluids

The characterisation of non-Newtonian fluids is a major area of science in itself and the equipment required for measurement of all the relevant properties, including the normal stress differences and Trouton ratio of viscoelastic fluids, is extremely complex. The range of shear rates of interest is very great covering many orders of magnitude, and it is frequently better to use a number of instruments each covering a relatively narrow range, rather than to try to carry out all the measurements with a single instrument. Some instruments are designed to operate at a series of constant shear rates and the resulting stresses are measured. Others, constant stress instruments, are more suitable for measurements at low shear rate conditions, particularly with materials that have yield stresses.

Even the measurement of the steady-state characteristics of shear-dependent fluids is more complex than the determination of viscosities for Newtonian fluids. In simple geometries, such as capillary tubes, the shear stress and shear rate vary over the cross-section and consequently, at a given operating condition, the apparent viscosity will vary with location. Rheological measurements are therefore usually made with instruments in which the sample to be sheared is subjected to the same rate of shear throughout its whole mass. This condition is achieved in concentric cylinder geometry (Fig. 3.37) where the fluid is sheared in the narrow annular space between a fixed and a rotating cylinder; if the gap is small compared with the diameters of the cylinders, the shear rate is approximately constant. Alternatively, a cone and plate geometry (Fig. 3.38) gives a constant shear rate, provided the angle θ between the cone and the plate is sufficiently small (~1–2 degree) so that $\sin\theta \approx \theta$.

In making rheological measurements, and in applying the results to a particular flow geometry, it should be noted that many non-Newtonian fluids exhibit the phenomenon of *wall slip* as a result of which the fluid layers in contact with a surface have a finite velocity relative to that surface. This may occur when the fluid consists of a solution of macromolecules or of a melt, of a polymer of high molecular weight, or of a suspension of very fine particles. If the size of the individual molecules or particles is large compared with the roughness dimensions of the surface, the particles or macromolecules may *ride over* the tops of the small protuberance. The difficulty can usually be overcome by artificial roughening of the surface.

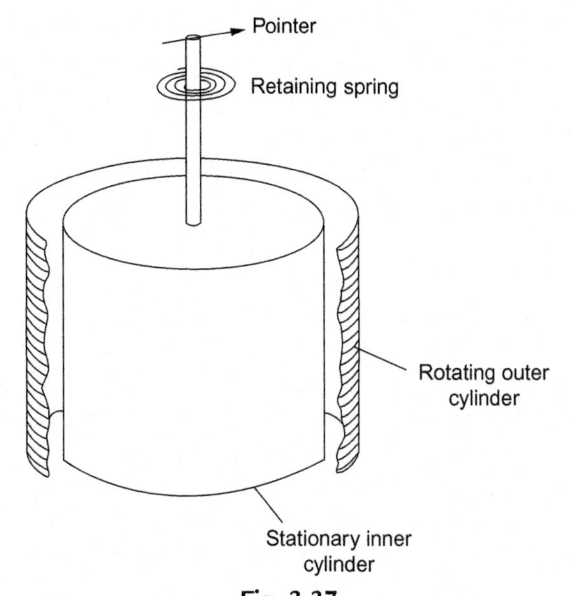

Fig. 3.37
Partial section of a concentric-cylinder viscometer.

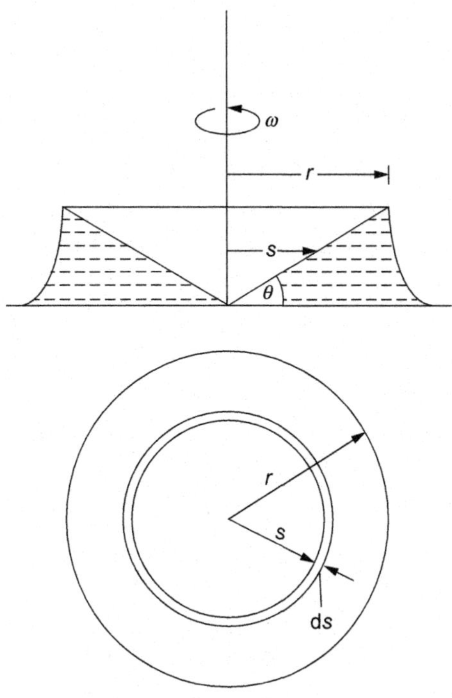

Fig. 3.38
Cone and plate viscometer.

If it is known that a particular form of relation, such as the *power-law* model, is applicable, it is not necessary to maintain a constant shear rate. Thus, for instance, a capillary tube viscometer can be used for determination of the values of the two parameters in the model. In this case, it is usually possible to allow for the effects of wall-slip by making measurements with tubes covering a range of bores and extrapolating the results to a tube of infinite diameter. Details of the method are given by Farooqi and Richardson[33] and Chhabra and Richardson.[2]

3.4.5 Dimensionless Characterisation of Viscoelastic Flows

The enormous practical value obtained from the consideration of viscous flows in terms of Reynolds numbers has led to the consideration of the possibility of using analogous dimensionless groups to characterise elastic behaviour. The Reynolds group represents the ratio of inertial to viscous forces, and it might be expected that a ratio involving elastic and inertial forces would be relevant. Attempts to achieve useful correlations have not been very successful, perhaps most frequently because of the complexity of natural situations and of real materials. One simple parameter that proves to be of value is the ratio of a *characteristic time of deformation* to a *natural time constant of the fluid.* The precise definition for these times is somewhat arbitrary, but it is evident that, for processes involving very slow deformation of the fluid elements, it is possible for the elastic forces to be released by the normal processes of relaxation as fast as they build up. Examples of the flow of rigid (apparently infinitely viscous) material over long periods of time include the thickening of the lower parts of medieval glass windows, and the plastic flow and deformation that lead to the folded strata of geological structures. In operations that are carried out rapidly, the extent of viscous flow will be minimal and the deformation will be followed by rapid recovery when the stress is removed. To obtain some idea of the possible regions in which such an analysis can provide guidance, consideration may be given to the flow of a 1% solution of polyacrylamide in water for which the relaxation time (in Maxwell model terms) is of the order of 10^{-2} s. If the fluid were flowing through a packed bed, it would be subject to alternating acceleration and deceleration as it flowed through the interstices of the bed. With a particle size of the order of 25 mm diameter and a superficial flowrate of 0.25 m/s, the elastic properties would not be expected to influence the flow significantly. However, in a free jet discharge with a velocity of the order of 30 m/s through a 3 mm diameter nozzle, some evidence of elastic behaviour would be expected to be evident near to the point of discharge.

The *Deborah number Db* has been defined by Metzner et al.[35] as:

$$Db = \frac{\text{Characteristic fluid time}}{\text{Characteristic process time}}$$

In the above example of the packed bed, Db would be ~ 0.1; for the jet and nozzle case, Db would be ~ 100. The larger the value of Db, the more likely is the fluid elasticity to be of practical significance. Thus, $Db \rightarrow o$ corresponds to the viscous fluid behaviour whereas $Db \rightarrow oo$ denotes the perfectly elastic deformation.

Unfortunately, this group Db depends on the assignment of a single characteristic time to the fluid (perhaps a *relaxation time*). While this has led to some success, it appears to be inadequate for many viscoelastic materials which show different relaxation behaviour under differing conditions.

For further information on viscoelastic behaviour, reference should again be made to specialist sources.[24–26,34]

3.4.6 Relation Between Rheology and Structure of Material

An understanding of the contribution of the relevant physical and chemical properties of the system to rheological behaviour is an area which has made little progress until recent years, but a clearer picture is now emerging of how the various types of rheological behaviour arise.

Most non-Newtonian fluids are either two-phase systems, or single phase systems in which large molecules are in solution in a liquid which itself may be Newtonian or non-Newtonian, or are in the form of a melt.

Flocculated suspensions of fine particles (coal, china clay, pigments, etc.) are usually shear-thinning. The flocs are weak mechanically and tend to break up in a shear field and the individual particles are capable of regrouping to give structures offering a lower resistance to shear. When the shear field is removed, the original structure is regained. The particles are generally of the order of 1 μm in size and the flocs (loose agglomerates occluding liquid) may be 10–100 μm in size. The specific surface of the particles is very high, and surface chemistry plays an important role in determining the structure of the flocs. The rheology of many of these suspensions can be approximated over a reasonable range of flowrates by a power-law equation ($0.1 < n < 0.6$). There is evidence that some have a finite yield stress and conform to the Bingham-plastic model. Frequently however, the rheology can equally well be described by a power-law or Bingham-plastic form of equation, as seen in Example 3.11.

Non-flocculated suspensions can exist at very much higher concentrations and, at all but the highest volumetric concentrations, are often Newtonian. When such suspensions are sheared, some *dilation* occurs as a result of particles trying to 'climb over each other'. If the amount of liquid present is then insufficient fully to fill the void spaces, particle-particle solid friction can come into play and the resistance to shear increases. This is just one way in which shear-thickening can occur.

Many polymers form shear-thinning solutions in water. The molecules are generally long and tend to be aligned and to straighten out in a shear field, and thus to offer less resistance to flow. Such solutions are sometimes viscoelastic and this effect may be attributable to a tendency of the molecules to recover their previous configuration once the stress is removed. Molten polymers are usually viscoelastic.

3.4.7 Streamline Flow in Pipes and Channels of Regular Geometry

As in the case of Newtonian fluids, one of the most important practical problems involving non-Newtonian fluids is the calculation of the pressure drop for flow in pipelines. The flow is much more likely to be streamline, or laminar, because non-Newtonian fluids usually have very much higher apparent viscosities than most simple Newtonian fluids. Furthermore, the difference in behaviour is much greater for laminar flow where viscosity plays such an important role than for turbulent flow. Attention will initially be focused on laminar-flow, with particular reference to the flow of power-law and Bingham-plastic fluids.

In order to predict the transition point from stable streamline to stable turbulent flow, it is necessary to define a modified Reynolds number, though it is not clear that the same sharp transition in flow regime always occurs. Particular attention will be paid to the steady flow in pipes of circular cross-section, but the methods are applicable to other geometries (annuli, between flat plates, and so on) as in the case of Newtonian fluids, and the methods described earlier for flow between plates, through an annulus or down a surface can be adapted to take account of non-Newtonian characteristics of the fluid.

Power-law fluids

The distribution of shear stress over the cross-section of a pipe is determined by a force balance and is independent of the nature of the fluid or the type of flow.

From Eq. (3.8) and Fig. 3.32A it is seen that the shear stress $|R_s|$ at a radius s in a pipe of radius r is given by:

$$\frac{|R_s|}{R_0} = \frac{s}{r} \tag{3.129}$$

that is, the shear stress varies linearly from the centre of the pipe ($s=0$) to the wall ($s=r$).

When the fluid behaviour can be described by a power-law, the apparent viscosity for a shear-thinning fluid will be a minimum at the wall where the shear stress is a maximum, and will rise to a theoretical value of infinity at the pipe axis where the shear stress and shear rate both are zero. On the other hand, for a shear-thickening fluid, the apparent viscosity will fall to zero at the pipe axis. It is apparent therefore, that there will be some error in applying the power-law near the pipe axis since all real fluids have a limiting viscosity μ_0 at zero shear stress. The procedure is exactly analogous to that used for the Newtonian fluid, except that the power-law relation is used to relate shear stress to shear rate, as opposed to the simple Newtonian equation.

For a power-law fluid, Eq. (3.28) becomes:

$$\left(\frac{-\Delta P}{l}\right)s - 2k\left(-\frac{du_x}{ds}\right)^n = 0$$

$$\therefore \ du_x = -\left(\frac{-\Delta P}{2kl}\right)^{1/n} s^{1/n}\, ds$$

and:

$$u_x = -\left(\frac{-\Delta P}{2kl}\right)^{1/n} \frac{n}{n+1} s^{(n+1)/n} + \text{constant}$$

At the pipe wall, $s=r$ and for the no-slip condition, $u_x=0$.

So:

$$\text{constant} = \left(\frac{-\Delta P}{2kl}\right)^{1/n} \frac{n}{n+1} r^{(n+1)/n}$$

$$\therefore \ u_x = \left(\frac{-\Delta P}{2kl}\right)^{1/n} \frac{n}{n+1}\left(r^{(n+1)/n} - s^{(n+1)/n}\right) \tag{3.130}$$

The velocity at the centre line, u_{CL}, is then obtained by putting $s=0$:

$$u_{CL} = \left(\frac{-\Delta P}{2kl}\right)^{1/n} \frac{n}{n+1} r^{(n+1)/n} \tag{3.131}$$

Dividing Eq. (3.130) by Eq. (3.131) gives:

$$\frac{u_x}{u_{CL}} = 1 - \left(\frac{s}{r}\right)^{(n+1)/n} \tag{3.132}$$

The mean velocity of flow u is given by:

$$u = \frac{1}{\pi r^2}\int_0^r u_x 2\pi s\, ds$$

$$= \frac{2}{r^2} u_{CL} r^2 \int_0^1 \left(1 - \frac{s}{r}\right)^{(n+1)/n} \frac{s}{r} d\left(\frac{s}{r}\right)$$

$$= 2u_{CL}\left[\frac{1}{2}\left(\frac{s}{r}\right)^2 - \frac{n}{3n+1}\left(\frac{s}{r}\right)^{(3n+1)/n}\right]_0^1$$

$$= 2u_{CL}\left(\frac{1}{2} - \frac{n}{3n+1}\right)$$

or:

$$\frac{u}{u_{CL}}=\frac{n+1}{3n+1} \tag{3.133}$$

Combining Eqs. (3.132), (3.133), gives the velocity profile in terms of the mean velocity u in place of the centre-line velocity u_{CL} as:

$$\frac{u_x}{u}=\frac{3n+1}{n+1}\left[1-\left(\frac{s}{r}\right)^{(n+1)/n}\right] \tag{3.134}$$

Substituting into Eq. (3.131):

$$u=\left(\frac{-\Delta P}{2kl}\right)^{1/n}\frac{n}{3n+1}r^{(n+1)/n} \tag{3.135}$$

Working in terms of pipe diameter:

$$u=\left(\frac{-\Delta P}{4kl}\right)^{1/n}\frac{n}{6n+2}d^{(n+1)/n} \tag{3.136}$$

For a Newtonian fluid $n=1$, $k=\mu$ and:

$$u=\frac{-\Delta P d^2}{32kl}$$

This is identical to Eq. (3.36), bearing in mind that $k=\mu$ for a Newtonian fluid.

The shear rate (velocity gradient) at the tube wall is obtained by differentiating Eq. (3.134) with respect to s, and then putting $s=r$.

$$\frac{1}{u}\frac{du_x}{ds}=\frac{3n+1}{n+1}\left[-\frac{n+1}{n}\left(\frac{s}{r}\right)^{(n+1)/n}\frac{1}{r}\right]$$

If y is distance from the wall, $y+s=r$ and:

$$\left(-\frac{du_x}{ds}\right)_{s=r}=\left(\frac{du_x}{dy}\right)_{y=0}=\left(\frac{3n+1}{n}\right)\frac{u}{r}=\left(\frac{6n+2}{n}\right)\frac{u}{d} \tag{3.137}$$

For a Newtonian fluid, Eq. (3.137) gives a wall shear rate of $8u/d$ (corresponding to Eq. 3.39) and a shear stress of $8\mu u/d$ (corresponding to Eq. 3.40).

For a Newtonian fluid, the data for pressure drop may be represented on a pipe friction chart as a friction factor $\phi=(R/\rho u^2)$ expressed as a function of Reynolds number $Re=(udp/\mu)$. The friction factor is independent of the rheological properties of the fluid, but the Reynolds number involves the viscosity which, for a non-Newtonian fluid, is dependent on shear rate. Metzner and Reed[35] defined a Reynolds number Re_{MR} for a power-law fluid in such a way that it is related to the friction factor for streamline flow in exactly the same way as for a Newtonian fluid.

Thus, from Eq. (3.39):

$$\phi = \frac{R}{\rho u^2} = 8Re_{MR}^{-1} \tag{3.138}$$

For the flow of a power-law fluid in a pipe of length l, the pressure drop $-\Delta P$ is obtained from Eq. (3.136) as:

$$-\Delta P = \left(\frac{6n+2}{n}\right)^n 4klu^n d^{-(n+1)}$$

Thus:

$$\begin{aligned} \phi &= \frac{R}{\rho u^2} = \frac{-\Delta P d}{4l\rho u^2} \quad \text{(from Eq. 3.17)} \\ &= \left(\frac{6n+2}{n}\right)^n \frac{ku^{n-2}d^{-n}}{\rho} \end{aligned} \tag{3.139}$$

Substituting into Eq. (3.138):

$$Re_{MR} = 8\left(\frac{n}{6n+2}\right)^n \frac{\rho u^{2-n} d^n}{k} \tag{3.140}$$

As indicated in Section 3.7.9, this definition of Re_{MR} may be used to determine the limit of stable streamline flow. The transition value $(Re_{MR})_c$ is approximately the same as for a Newtonian fluid, but there is some evidence that, for moderately shear-thinning fluids, streamline flow may persist to somewhat higher values of Re_{MR}. Putting $n=1$ in Eq. (3.140) leads to the standard definition of the Reynolds number.

The effect of power-law index on the velocity profile is seen by plotting Eq. (3.134) for various values of n, as shown in Fig. 3.39.

Compared with the parabolic profile for a Newtonian fluid ($n=1$), the profile is flatter for a shear-thinning fluid ($n<1$) and sharper for a shear-thickening fluid ($n>1$). The ratio of the centre line (u_{CL}) to mean (u) velocity, calculated from Eq. (3.133), is:

n	2.0	1.5	1.0	0.8	0.6	0.4	0.2	0.1
u_{CL}/u	2.33	2.2	2	1.89	1.75	1.57	1.33	1.18

Bingham-plastic fluids

For the flow of a Bingham-plastic fluid, the cross-section may be considered in two parts, as shown in Fig. 3.32:

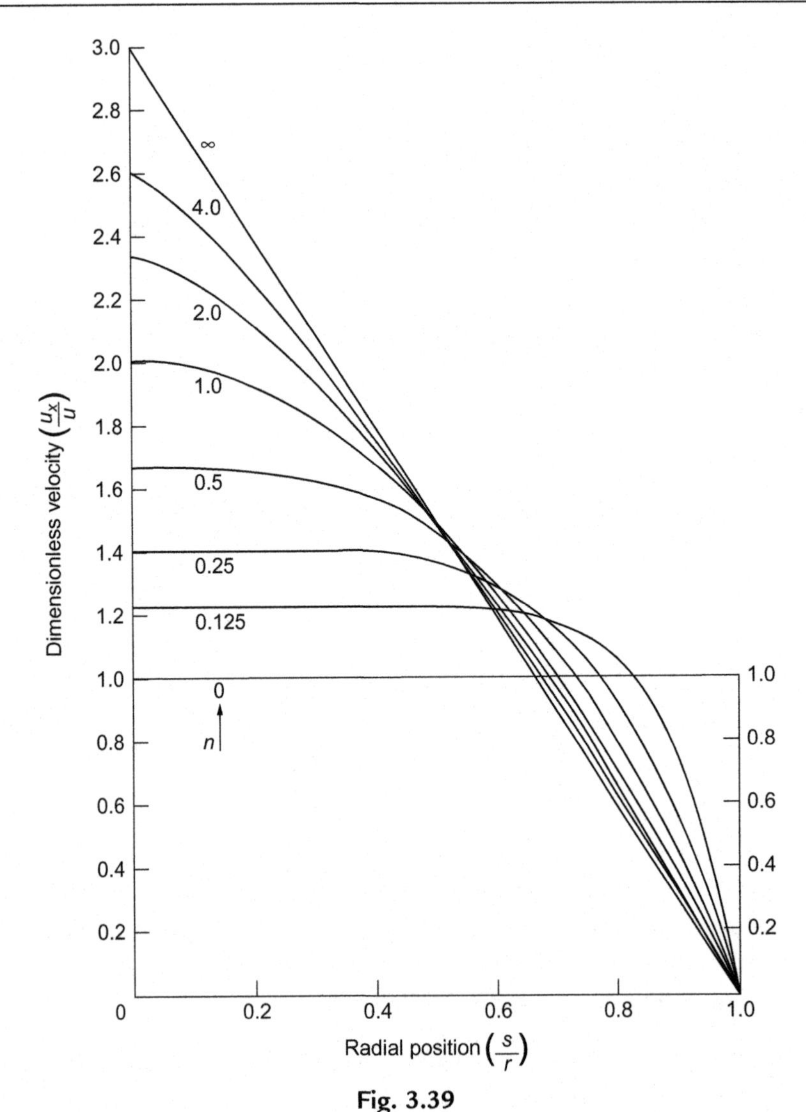

Fig. 3.39
Fully-developed laminar velocity profiles for power-law fluids in a pipe (from Eq. 3.134).

(1) The central unsheared core in which the fluid is all travelling at the centre-line velocity.
(2) The annular region separating the core from the pipe wall, over which the whole of the velocity change occurs.

The boundary between the two regions and the radius r_c of the core is determined by the position in the cross-section at which the shear stress is exactly equal to the yield stress R_Y of the fluid. Since the shear stress is linearly related to the radial position:

$$\frac{R_Y}{R} = \frac{r_c}{r} \quad \text{(Eq. 3.8)}$$

where R is the shear stress at the wall (radius r).

In a given pipe, R is determined solely by the pressure drop $-\Delta P$ and is completely independent of the rheology of the fluid and, from Eq. (3.6):

$$R = -\Delta P \frac{r}{2l} \tag{3.141}$$

and:

$$\frac{r_c}{2l} = \frac{R_Y}{-\Delta P} \tag{3.142}$$

Thus, for a given fluid (R_Y constant), the critical radius r_c is determined entirely by the pressure drop and becomes progressively larger as the pressure drop is reduced. Flow ceases when the shear stress at the wall R falls to a value equal to the yield stress R_Y.

Flow in the annular region ($s > r_c$)

In this region, the relation between the shear stress R_s and the velocity gradient $\mathrm{d}u_x/\mathrm{d}s$ is given by:

$$|R_s| - R_Y = \mu_p \left(-\frac{\mathrm{d}u_x}{\mathrm{d}s}\right)$$

Writing: $R_s = R(s/r)$ and rearranging:

$$\mathrm{d}u_x = \frac{1}{\mu_p}\left(R_Y - \frac{s}{r}R\right)\mathrm{d}s$$

Integrating:

$$u_x = \frac{1}{\mu_p}\left\{R_Y s - R\frac{s^2}{2r}\right\} + \text{constant.}$$

For the no-slip condition at the wall, $u_x = 0$ at $s = r$.

$$\therefore \ \text{constant} = -\frac{1}{\mu_p}\left\{R_Y r - R\frac{r^2}{2r}\right\}$$

Thus:

$$u_x = \frac{1}{\mu_p}\left\{\frac{R}{2r}(r^2 - s^2) - R_Y(r - s)\right\} \tag{3.143}$$

This equation for the velocity profile, reduces to the parabolic form for a Newtonian fluid, when $R_Y = 0$, and applies in the region $r > s > r_c$.

The volumetric flowrate Q_A in the annular region is given by:

$$Q_A = \int_{s=2l(R_Y/-\Delta P)}^{s=r} \left\{\frac{1}{\mu_p}\left[\frac{R}{2r}(r^2 - s^2) - R_Y(r - s)\right]\right\} 2\pi s \, \mathrm{d}s$$

On integration:

$$Q_A = \frac{-\Delta P \pi r^4}{8\mu_p l}\left\{1 - \frac{4}{3}X - 2X^2 + 4X^3 - \frac{5}{3}X^4\right\} \quad (3.144)$$

where $X = \frac{2R_Y l}{-\Delta P r} = \frac{R_Y}{R} = \frac{\text{Yield stress}}{\text{Wall shear stress}}$.

Flow in the centre plug

The velocity u_p of the centre plug is obtained by putting:

$$s = r_c = 2l\frac{R_Y}{-\Delta P}$$

in Eq. (3.143).

Thus:

$$u_p = \frac{1}{\mu_p}\left\{\frac{-\Delta P r}{2l}\frac{1}{2r}\left[r^2 - 4l^2\left(\frac{R_Y}{-\Delta P}\right)^2\right] - R_Y\left(r - 2l\frac{R_Y}{-\Delta P}\right)\right\}$$

This simplifies to:

$$u_p = \frac{-\Delta P r^2}{8\mu_p l}\{2 - 4X + 2X^2\} \quad (3.145)$$

The volumetric flowrate in the plug Q_p is then given by:

$$\begin{aligned} Q_p &= u_p \pi r_c^2 \\ &= \frac{-\Delta P r^2}{8\mu_p l}(2 - 4X + 2X^2)\pi\frac{4l^2 R_Y^2}{(-\Delta P)^2} \\ &= \frac{-\Delta P \pi r^2}{8\mu_p l}\{2X^2 - 4X^3 + 2X^4\} \end{aligned} \quad (3.146)$$

Total flow through the pipe

The total flowrate Q through the pipe is given by:

$$\begin{aligned} Q &= Q_A + Q_p \\ &= \frac{-\Delta P \pi r^4}{8\mu_p l}\left\{1 - \frac{4}{3}X + \frac{1}{3}X^4\right\} \end{aligned} \quad (3.147)$$

Thus, the mean velocity of flow u is given by:

$$\begin{aligned} u &= \frac{Q}{\pi r^2} = \frac{-\Delta P r^2}{8\mu_p l}\left\{1 - \frac{4}{3}X + \frac{1}{3}X^4\right\} \\ &= \frac{-\Delta P d^2}{32\mu_p l}\left\{1 - \frac{4}{3}X + \frac{1}{3}X^4\right\} \end{aligned} \tag{3.148}$$

For a Newtonian fluid both R_Y and X are zero and the mean velocity is:

$$u = \frac{-\Delta P d^2}{32\mu_p l} \quad \text{(cf. Eq. 3.35)}$$

Eq. (3.148) is sometimes known as *Buckingham's equation.*

Example 3.11

The rheological properties of a particular suspension may be approximated reasonably well by either a *power-law* or a *Bingham-plastic* model over the shear rate range of 10–50 s^{-1}. If the consistency coefficient k is 10 N s^n/m^{-2} and the flow behaviour index n is 0.2 in the power law model, what will be the approximate values of the yield stress and of the plastic viscosity in the Bingham-plastic model?

What will be the pressure drop, when the suspension is flowing under laminar conditions in a pipe 200 m long and 40 mm diameter, when the centre line velocity is 1 m/s, according to the power-law model? Calculate the centre-line velocity for this pressure drop for the Bingham-plastic model.

Solution

Using the *power-law model* (Eq. 3.121):

$$|R_y| = k\left(\left|\frac{du_x}{dy}\right|\right)^n = 10\left(\left|\frac{du_x}{dy}\right|\right)^{0.2}$$

When:

$$\left|\frac{du_x}{dy}\right| = 10\ s^{-1}: \quad |R_y| = 10 \times 10^{0.2} = 15.85\ N/m^2$$

$$\left|\frac{du_x}{dy}\right| = 50\ s^{-1}: \quad |R_y| = 10 \times 50^{0.2} = 21.87\ N/m^2$$

Using the *Bingham-plastic model* (Eq. 3.125):

$$|R_y| = R_Y + \mu_p\left|\frac{du_x}{dy}\right|$$

When:

$$\left|\frac{du_x}{dy}\right| = 10\ s^{-1}: \quad 15.85 = R_Y + 10\mu_p$$

$$\left|\frac{du_x}{dy}\right| = 50\ \text{s}^{-1}: \quad \underline{21.87 = R_Y + 50\mu_p}.$$

Subtracting:

$$6.02 = 40\mu_p.$$

Thus:

$$\mu_p = 0.150\ \text{N s/m}^2$$

and:

$$R_Y = 14.35\ \text{N/m}^2$$

Thus, the Bingham-plastic equation is:

$$|R_y| = 14.35 + 0.150\left|\frac{du_x}{dy}\right|$$

For the *power-law fluid*:

Eq. (3.131) gives:

$$u_{CL} = \left(\frac{-\Delta P}{2kl}\right)^{1/n} \frac{n}{n+1} r^{(n+1)/n}$$

Rearranging:

$$-\Delta P = 2klu_{CL}^n \left(\frac{n+1}{n}\right)^n r^{-(n+1)}.$$

The numerical values in SI units are:

$$u_{CL} = 1\ \text{m/s}, \quad l = 200\ \text{m}, \quad r = 0.02\ \text{m}, \quad k = 10\ \text{N s}^n\ \text{m}^{-2}, \quad n = 0.2$$

and:

$$-\Delta P = \underline{\underline{626{,}000\ \text{N/m}^2}}$$

For a Bingham-plastic fluid:

The centre line velocity is given by Eq. (3.145):

$$u_p = \frac{-\Delta Pr^2}{8\mu_p l}(2 - 4X + 2X^2)$$

where:

$$X = \frac{l}{r}\frac{2R_Y}{(-\Delta P)}$$

$$= \frac{200}{0.02} \times \frac{2 \times 14.35}{626{,}000} = 0.458$$

$$\therefore\ 2 - 4X + 2X^2 = 0.589$$

$$u_p = \frac{626{,}000 \times (0.02)^2}{8 \times 0.150 \times 200} \times 0.589$$

$$= \underline{\underline{0.61\ \text{m/s}}}$$

Example 3.12

A Newtonian liquid of viscosity 0.1 N s/m^2 is flowing through a pipe of 25 mm diameter and 20 m in length, and the pressure drop is 10^5 N/m^2. As a result of a process change, a small quantity of polymer is added to the liquid and this causes the liquid to exhibit non-Newtonian characteristics; its rheology is described adequately by the *power-law* model and the *flow index* is 0.33. The apparent viscosity of the modified fluid is equal to the viscosity of the original Newtonian liquid at a shear rate of 1000 s^{-1}.

If the pressure difference over the pipe is unaltered, what will be the ratio of the volumetric flowrates of the two liquids?

Solution

For a power-law fluid:

$$|R_Y| = k\left(\left|\frac{du_x}{dy}\right|\right)^n = \left[k\left(\left|\frac{du_x}{dy}\right|\right)^{n-1}\right]\frac{du_x}{dy}$$

Apparent viscosity

$$\mu_a = k\left(\left|\frac{du_x}{dy}\right|\right)^{-0.67}$$

When:

$$\left|\frac{du_x}{dy}\right| = 1000\ \text{s}^{-1} \quad \mu_a = k(1000)^{-0.67} = \frac{k}{100} = 0.1\ \text{kg/m s}$$

$$\therefore\ k = 10\ \text{N s}^{0.33}/\text{m}^2$$

The rheological equation is

$$|R_y| = 10\left(\left|\frac{du_x}{dy}\right|\right)^{0.33}$$

From Eq. (3.136):

For a power-law fluid:

$$u = \left(\frac{-\Delta P}{4kl}\right)^{1/n}\frac{n}{2(3n+1)}d^{(n+1)/n}.$$

For the polymer solution ($n = 0.33$):

$$u_2 = \left(\frac{10^5}{4\times 10\times 20}\right)^3\frac{1}{12}\cdot 0.025^4$$
$$= 0.0636\ \text{m/s}$$

For the original Newtonian fluid ($n = 1$):

$$u_1 = \left(\frac{10^5}{4\times 0.1\times 20}\right)\cdot\left(\frac{1}{8}\right)\cdot(0.025)^2$$
$$= 0.977\ \text{m/s}$$

and:

$$\frac{u_2}{u_1} = \left(\frac{0.0636}{0.977}\right) = \underline{\underline{0.065}}$$

Example 3.13

Two liquids of equal densities, one Newtonian and the other a non-Newtonian '*power-law*' fluid, flow at equal volumetric rates per unit width down two wide vertical surfaces. The non-Newtonian fluid has a *power-law* index of 0.5 and it has the same apparent viscosity as the Newtonian fluid when its shear rate is 0.01 s^{-1}. Show that, for equal surface velocities of the two fluids, the film thickness of the non-Newtonian fluid is 1.132 times that of the Newtonian fluid.

Solution

For a *power-law* fluid:

$$|R_y| = k\left|\frac{du_x}{dy}\right|^n \quad \text{(Eq. 3.121)}$$

$$= k\left|\frac{du_x}{dy}\right|^{n-1}\left|\frac{du_x}{dy}\right| \quad \text{(Eq. 3.122)}$$

and from Eq. (3.123), the apparent viscosity, $\mu_a = k|du_x/dy|^{n-1}$

For a Newtonian fluid:

$$R_y = \mu|du_x/dy| \quad \text{(Eq. 3.3)}$$

For $n = 0.5$ and $|du_x/dy| = 0.01$ s^{-1}, then, using SI units:

$$\mu_a = \mu = k|du_x/dy|^{n-1}$$

$$\mu = k(0.01)^{-0.5} = 10k \text{ and } k = 0.1\mu.$$

The equation of state of the power-law fluid is therefore in SI units:

$$R_y = 0.1\mu|du_x/dy|^{0.5}$$

For a fluid film of thickness s flowing down a vertical surface of length l and width w, a force balance on the fluid at a distance greater than y from the surface (fluid depth $s - y$) gives:

$$(s-y)wl\rho g = R_y wl = k(du_x/dy)^n wl$$

or:

$$du_x/dy = (\rho g/k)^{1/n}(s-y)^{1/n}$$

and:

$$u_x = (\rho g/k)^{1/n}(s-y)^{(n+1)/n}[-n/(n+1)] + \text{constant}$$

At the surface, $y = 0$, $u_x = 0$ and the constant $= (\rho g/k)^{1/n} s^{(n+1)/n}[n/(n+1]$ and:

$$u_x = \left(\frac{\rho g}{k}\right)^{1/n} \frac{n}{n+1}\left[s^{\frac{n+1}{n}} - (s-y)^{\frac{n+1}{n}}\right]$$

At the free surface where $y = s$:

$$u_s = \left(\frac{\rho g}{k}\right)^{1/n} \frac{n}{n+1} s^{\frac{n+1}{n}} \tag{i}$$

The volumetric flowrate, Q, is given by:

$$\begin{aligned} Q &= \int_0^s w\,dy \left(\frac{\rho g}{k}\right)^{1/n} \frac{n}{n+1}\left[s^{\frac{n+1}{n}} - [s-y]^{\frac{n+1}{n}}\right] \\ &= w\left(\frac{\rho g}{k}\right)^{1/n}\left(\frac{n}{2n+1}\right) s^{\frac{2n+1}{n}} \end{aligned} \tag{ii}$$

For the non-Newtonian fluid, $k = 0.1\mu$, $n = 0.5$ and Eq. (ii) becomes:

$$\begin{aligned} Q &= w(\rho g/(0.1\mu))^2 \times 0.25 s^4 \\ &= 25w\left(\frac{\rho g}{\mu}\right)^2 s^4 \end{aligned} \tag{iii}$$

For the Newtonian fluid, $n = 1$ and $k = \mu$ and substituting in Eq. (ii):

$$Q = w\left(\frac{\rho g}{\mu}\right)\left(\frac{s_N^3}{3}\right) \tag{iv}$$

where s_N is the thickness of the Newtonian fluid.

For equal flowrates, then from Eqs. (iii), (iv):

$$25w(\rho g/\mu)^2 s^4 = 0.33w(\rho g/\mu)s_N^3$$

or:

$$s_N^3 = 75(\rho g/\mu)s^4$$

For equal surface velocities, the term $(\rho g/k)$ may be substituted from Eq. (iv) into Eq. (i) to give, for the non-Newtonian fluid:

$$\begin{aligned} u_s &= (\rho g/0.1\mu)^2 0.33 s^3 \\ &= 100\left(s_N^3/75 s^4\right)^2 0.33 s^3 \\ &= 0.00592 s_N^6/s^5 \end{aligned}$$

for the Newtonian fluid:

$$\begin{aligned} u_s &= (\rho g/\mu)0.5 s_N^2 \\ &= \left(s_N^3/75 s^4\right)0.5 s_N^2 \\ &= 0.0067 s_N^5/s^4 \end{aligned}$$

and hence:

$$s_N/s = (0.0067/0.00592) = \underline{\underline{1.132}}$$

General equations for pipeline flow

Fluids whose behaviour can be approximated by the power-law or Bingham-plastic equation are essentially special cases, and frequently the rheology may be very much more complex so that it may not be possible to fit simple algebraic equations to the flow curves. It is therefore desirable to adopt a more general approach for time-independent fluids in fully-developed flow which is now introduced. For a more detailed treatment and for examples of its application, reference should be made to more specialist sources.[2,25]

If the shear stress is a function of the shear rate, it is possible to invert the relation to give the shear rate, $\dot{\gamma} = -\mathrm{d}u_x/\mathrm{d}s$, as a function of the shear stress, where the negative sign is included here because velocity decreases from the pipe centre outwards.

Thus:

$$\dot{\gamma} = -\frac{\mathrm{d}u_x}{\mathrm{d}s} = f(R_s) = f\left(R\frac{s}{r}\right) \tag{3.149}$$

where R_s is the positive value of the shear stress at radius s.

The advantage of using the relation between shear rate and shear stress in this form is that R_s, unlike $\dot{\gamma}$, is known at all values of s.

On integration, and noting that $u_x = 0$ at $s = r$:

$$u_x = \int_s^r f\left(R\frac{s}{r}\right)\mathrm{d}s \tag{3.150}$$

The total volumetric flowrate Q through the pipe is found by integrating over the whole cross-section to give:

$$\begin{aligned} Q &= \int_0^r 2\pi s u_x \,\mathrm{d}s = \pi \int_0^{r^2} u_x \,\mathrm{d}(s^2) \\ &= \pi \left[s^2 u_x - \int s^2 \,\mathrm{d}u_x\right]_{s=0}^{s=r} \\ &= \pi \int_0^r s^2 f\left(R\frac{s}{r}\right)\mathrm{d}s \quad (\text{since } u_x = 0 \text{ at } s = r) \end{aligned} \tag{3.151}$$

Substituting $s = r(R_s/R)$:

$$Q = \pi \int_0^R \left(r\frac{R_s}{R}\right)^2 f(R_s)\frac{r}{R}\,\mathrm{d}R_s$$

or:

$$\frac{Q}{\pi r^3} = \frac{1}{R^3}\int_0^R R_s^2 f(R_s)\,\mathrm{d}R_s. \tag{3.152}$$

Multiplying Eq. (3.152) by R^3 and differentiating with respect to R gives:

$$\frac{\mathrm{d}}{\mathrm{d}R}\left(R^3\frac{Q}{\pi r^2}\right)=\frac{\mathrm{d}}{\mathrm{d}R}\int_0^R R_s^2 f(R_s)\mathrm{d}R_s.$$

Use of parameters n' and k'

Noting that R is the upper limit of the integral, and using the Leibnitz rule:

$$R^3\frac{\mathrm{d}(Q/\pi r^3)}{\mathrm{d}R}+3R^2\frac{Q}{\pi r^3}=R^2 f(R)$$

Dividing by R^2 and putting $R=-\Delta P\frac{r}{2l}$ gives:

$$-\Delta P\frac{r}{2l}\frac{\mathrm{d}(Q/\pi r^3)}{\mathrm{d}\left(-\Delta P\frac{r}{2l}\right)}+\frac{3Q}{\pi r^3}=f(R)=\left(-\frac{\mathrm{d}u_x}{\mathrm{d}s}\right)_{s=r}$$

$$\therefore\quad \frac{4Q}{\pi r^3}\frac{\mathrm{d}[\ln(Q/\pi r^3)]}{4\mathrm{d}\ \ln\left[-\Delta P\frac{r}{2l}\right]}+\frac{3Q}{\pi r^3}=\left(-\frac{\mathrm{d}u_x}{\mathrm{d}s}\right)_{s=r} \tag{3.153}$$

Writing:

$$n'=\frac{\mathrm{d}\left[\ln\left(-\Delta P\frac{r}{2l}\right)\right]}{\mathrm{d}[\ln(Q/\pi r^3)]} \tag{3.154}$$

where n' is the slope of the log/log plot of $-\Delta P(r/2l)$ against $Q/\pi r^3$ gives:

$$\frac{4Q}{\pi r^3}\left(\frac{1}{4n'}+\frac{3}{4}\right)=\frac{4Q}{\pi r^3}\left(\frac{3n'+1}{4n'}\right)=\left(-\frac{\mathrm{d}u_x}{\mathrm{d}s}\right)_{s=r} \tag{3.155}$$

Eq. (3.155) is frequently referred to as the Rabinowitsch[36]–Mooney[37] relation. Now the rheological data for a fluid may be represented as a curve of $R=-\Delta P(r/2l)$ plotted against $4Q/\pi r^3$ $(=\frac{8u}{d})$ using logarithmic coordinates, and the slope n' may be measured over any small range or even at a point on the curve, irrespective of whether or not an equation may be fitted to the curve. This provides a basis for utilising practical rheological data for *any shear-dependent but time-independent fluid.*

From the definition of n':

$$R=-\Delta P\frac{r}{2l}=\text{constant}\left(\frac{Q}{\pi r^3}\right)^{n'}=k'\left(\frac{4Q}{\pi r^3}\right)^{n'} \tag{3.156}$$

Thus n' and k' are parameters which can be measured for any fluid, and the method may be applied to a wide range of rheological properties.

For a *power-law fluid*, the slope n' (Eq. 3.154) has a constant value n and hence:

$$n' = n \tag{3.157}$$

Thus, using Eq. (3.135) to substitute for $-\Delta P$ in Eq. (3.156):

$$k' = k\left(\frac{3n+1}{4n}\right)^n \tag{3.158}$$

Generalised Reynolds number

The Metzner and Reed Reynolds number Re_{MR} may be expressed in terms of n' and k'. From Eq. (3.140), derived for a power-law fluid:

$$Re_{MR} = 8\left(\frac{n}{6n+2}\right)^n \frac{\rho u^{2-n} d^n}{k}$$

Putting $n = n'$ and $k = k'\left(\frac{4n'}{3n'+1}\right)^{n'}$, then:

$$\begin{aligned} Re_{MR} &= 8\left(\frac{n'}{6n'+2}\right)^{n'}\left(\frac{3n'+1}{4n'}\right)^{n'}\frac{\rho u^{2-n'} d^{n'}}{k'} \\ &= 8^{1-n'}\frac{\rho u^{2-n'} d^{n'}}{k'} \end{aligned} \tag{3.159}$$

Working in terms of the apparent viscosity μ_w at the wall shear rate, by definition:

$$\mu_w = \frac{R}{\left(-\dfrac{du_x}{ds}\right)_{s=r}} \tag{3.160}$$

From Eq. (3.156):

$$R = k'\left(\frac{4Q}{\pi r^3}\right)^{n'} = k'\left(\frac{8u}{d}\right)^{n'} \tag{3.161}$$

and from Eq. (3.155):

$$\left(-\frac{du_x}{ds}\right)_{s=r} = \frac{4Q}{\pi r^3}\frac{3n'+1}{4n'} = \frac{8u}{d}\frac{3n'+1}{4n'} \tag{3.162}$$

Substituting from Eqs. (3.161), (3.162) into Eq. (3.160):

$$\mu_w = \frac{k'\left(\dfrac{8u}{d}\right)^{n'}}{\dfrac{8u}{d}\cdot\dfrac{3n'+1}{4n'}} = k'\left(\frac{8u}{d}\right)^{n'-1}\frac{4n'}{3n'+1} \tag{3.163}$$

Then substituting for k' in Eq. (3.159):

$$Re_{MR} = \frac{8^{1-n'} \rho u^{2-n'} d^{n'}}{\mu_w \left(\frac{8u}{d}\right)^{1-n'} \cdot \frac{3n'+1}{4n'}}$$

or:

$$Re_{MR} = \frac{4n'}{3n'+1} \frac{\rho u d}{\mu_w} \tag{3.164}$$

Velocity-pressure gradient relationships for fluids of specified rheology

Eq. (3.152) provides a method of determining the relationship between pressure gradient and mean velocity of flow in a pipe for fluids whose rheological properties may be expressed in the form of an explicit relation for shear rate as a function of shear stress.

$$\frac{Q}{\pi r^3} = \frac{1}{R^3}\int_0^R R_s^2 f(R_s) \mathrm{d}R_s \quad \text{(Eq. 3.152)}$$

For a *power-law fluid*, from Eq. (3.121):

$$f(R_s) = \dot{\gamma} = \left(\frac{R_s}{k}\right)^{\frac{1}{n}} \tag{3.165}$$

$$\begin{aligned} \frac{Q}{\pi r^3} &= \frac{1}{R^3}\int_0^R R_s^2 \left(\frac{R_s}{k}\right)^{\frac{1}{n}} \mathrm{d}R_s \\ &= \frac{1}{R^3 k^{\frac{1}{n}}}\int_0^R R_s^{\frac{2n+1}{n}} \mathrm{d}R_s \\ &= \frac{1}{R^3 k^{\frac{1}{n}}} \cdot \frac{n}{3n+1} R^{\frac{3n+1}{n}} \\ &= \frac{1}{k^{\frac{1}{n}}} \frac{n}{3n+1} R^{\frac{1}{n}} \end{aligned} \tag{3.166}$$

Writing the mean velocity as $u = \frac{Q}{\pi r^2}$ and $R = -\Delta P \cdot \frac{r}{2l}$, then

$$\begin{aligned} u &= \frac{r}{k^{\frac{1}{n}}} \frac{n}{3n+1} \left(\frac{-\Delta P r}{2l}\right)^{\frac{1}{n}} \\ &= \left(\frac{-\Delta P d}{4kl}\right)^{\frac{1}{n}} \frac{n}{6n+2} d \end{aligned} \tag{3.167}$$

Which is identical to Eq. (3.136).

For a *Bingham-plastic fluid*, from Eq. (3.125):

$$f(R_s)=\dot{\gamma}=\frac{R_s-R_Y}{\mu_p}\quad (R_s>R_Y) \tag{3.168a}$$

$$f(R_s)=\dot{\gamma}=0\quad (R_s\leq R_Y) \tag{3.168b}$$

$$\begin{aligned}
\therefore\quad \frac{Q}{\pi r^3}&=\frac{1}{R^3}\int_{R_Y}^{R} R_s^2\frac{1}{\mu_p}(R_s-R_Y)\mathrm{d}R_s\quad \left(\text{noting that} \int_0^{R_s}=0\right)\\
&=\frac{1}{\mu_p R^3}\int_{R_Y}^{R}(R_s^3-R_s^2R_Y)\mathrm{d}R_s\\
&=\frac{1}{\mu_p R^3}\left[\frac{R_s^4}{4}-\frac{R_s^3R_Y}{3}\right]_{R_Y}^{R}\\
&=\frac{1}{\mu_p R^3}\left(\frac{R^4}{4}-\frac{R^3R_Y}{3}-\frac{R_Y^4}{4}+\frac{R_Y^4}{3}\right)\\
&=\frac{1}{4\mu_p}R\left(1-\frac{4}{3}\frac{R_Y}{R}+\frac{1}{3}\left(\frac{R_Y}{R}\right)^4\right)
\end{aligned} \tag{3.169}$$

Again, noting that $u=\frac{Q}{\pi r^2}$ and $R=-\Delta P\cdot\frac{r}{2l}$ and putting $\frac{R_Y}{R}=X$

$$u=\frac{-\Delta P d^2}{32\mu_p l}\left(1-\frac{4}{3}X+\frac{1}{3}X^4\right) \tag{3.170}$$

As expected, Eq. (3.170) is identical to Eq. (3.148).

The above procedure may be followed using any other equation of state, provided that $\dot{\gamma}$· can be expressed as an explicit function of shear stress R_s.

It may be noted that Eqs. (3.167), (3.170) are identical to Eqs. (3.136), (3.148) derived earlier. Although these derivations are simpler to carry out, the method does not allow the velocity profile in the pipe to be obtained.

For a *Herschel–Bulkley fluid*, from Eq. (3.127):

$$f(R_s)=\dot{\gamma}=\left(\frac{R_s-R_Y}{\mu_p}\right)^{\frac{1}{m}}\quad (R_s>R_Y) \tag{3.171a}$$

$$f(R_s)=\dot{\gamma}=0\quad (R_s\leq R_Y) \tag{3.171b}$$

$$\therefore\quad \frac{Q}{\pi r^3}=\frac{1}{R^3}\int_{R_Y}^{R} R_s^2\left(\frac{R_s-R_Y}{\mu_p}\right)^{\frac{1}{m}}\mathrm{d}R_s$$

Putting $R_s - R_Y = R'$:

$$\frac{Q}{\pi r^3} = \frac{1}{R^3 \mu_p^{\frac{1}{m}}} \int_0^{R-R_Y} (R' + R_Y)^2 R'^{\frac{1}{m}} dR'$$

$$= \frac{1}{R^3 \mu_p^{\frac{1}{m}}} \int_0^{R-R_Y} \left(R'^{\frac{2m+1}{m}} + 2R_Y R'^{\frac{m+1}{m}} + R_Y^2 R'^{\frac{1}{m}} \right) dR'$$

$$= \frac{1}{R^3 \mu_p^{\frac{1}{m}}} \left[\frac{m}{3m+1} R'^{\frac{3m+1}{m}} + R_Y \frac{2m}{2m+1} R'^{\frac{2m+1}{m}} + R_Y^2 \frac{m}{m+1} R'^{\frac{m+1}{m}} \right]_0^{R-R_Y}$$

giving:

$$\frac{Q}{\pi r^3} = \frac{1}{R^3 \mu_p^{\frac{1}{m}}} \left\{ \frac{m}{3m+1} (R-R_Y)^{\frac{3m+1}{m}} + \frac{2m}{2m+1} R_Y (R-R_Y)^{\frac{2m+1}{m}} + \frac{m}{m+1} R_Y^2 (R-R_Y)^{\frac{m+1}{m}} \right\} \tag{3.172}$$

$$\therefore \quad \frac{Q}{\pi r^2} = u = \frac{r}{R^3 \mu_p^{\frac{1}{m}}} (R-R_Y)^{\frac{m+1}{m}} \left\{ \frac{m}{3m+1} (R-R_Y)^2 + \frac{2m}{2m+1} R_Y (R-R_Y) + \frac{m}{m+1} R_Y^2 \right\} \tag{3.173}$$

Eq. (3.172) reduces to Eq. (3.166) for a *power-law fluid* ($R_Y = 0$) and to Eq. (3.169) for a *Bingham-plastic fluid* ($m = 1$)

Expressing Eq. (3.173) in terms of pressure gradient $\left(R = -\frac{\Delta P d}{4l} \right)$ and $X = \frac{R_Y}{R}$:

$$u = \frac{1}{2} \left(-\frac{\Delta P}{4\mu_p l} \right)^{\frac{1}{m}} d^{\frac{m+1}{m}} (1-X)^{\frac{m+1}{m}} \left\{ \frac{m}{3m+1} (1-X)^2 + \frac{2m}{2m+1} X(1-X) + \frac{m}{m+1} X^2 \right\} \tag{3.174}$$

3.4.8 Turbulent flow

Surprising though it may be, there is no completely reliable method of predicting pressure drop for turbulent flow of non-Newtonian fluids in pipes. There is strong evidence that fluids showing similar flow characteristics in laminar flow do not necessarily behave in the same way in turbulent flow. Thus, for instance, a flocculated suspension and a polymer solution following the power-law model and having similar values of n and k may give different results in turbulent flow. Heywood and Cheng[38] and van den Heever et al.[39] have illustrated the difficulties in calculating pressure drops for the turbulent flow of a range of systems including polymer solutions, kaolin slurries, sewage sludge, using the various equations given in the literature.

As indicated earlier, non-Newtonian characteristics have a much stronger influence on flow in the streamline flow region where viscous effects dominate than in turbulent flow where inertial forces are of prime importance. Furthermore, there is a substantial evidence to the effect that, for shear-thinning fluids, the standard friction chart tends to overpredict pressure drop if the Metzner and Reed Reynolds number Re_{MR} is used. Furthermore, the laminar flow can persist for slightly higher Reynolds numbers than that for Newtonian fluids. Overall, therefore, there is a factor of safety involved in treating the fluid as Newtonian when flow is expected to be turbulent, with a notable exception of viscoelastic fluids.

Hartnett and Kostic[40] have recently examined the published correlations for turbulent flow of shear-thinning 'power-law' fluids in pipes and in noncircular ducts, and have concluded that, for smooth pipes, Dodge and Metzner's[42] modification of Eq. (3.11) (to which it reduces for Newtonian fluids) is the most satisfactory.

Dodge and Metzner[41] carried out experimental work using pipes of nominal diameters 1/2 in. (12.7 mm), 1 in. (25.4 mm), and 2 in. (50.8 mm), using polymer gels and solid-liquid suspensions, at Metzner and Reed Reynolds numbers (Re_{MR}) up to 36,000. The flow characteristics of the fluids corresponded approximately to the *power-law* relation, with n values ranging from 0.3 to 1, though only two of the fluids conformed to the power-law relation over the whole range of Reynolds numbers. Fig. 3.40 is a reproduction of Dodge and Metzner's graph of friction factor against Reynolds number (Re_{MR}) with friction factor expressed as ϕ ($=R/\rho u^2$), in order to conform with the standard used elsewhere in this chapter, instead of the Fanning friction factor $f(=2\phi)$ used by the authors. For turbulent flow, the friction factor at a given value of Re_{MR} becomes progressively less as the degree of shear-thinning increases (n decreasing). It will be noted that the experimental results and extrapolated values are separately designated. Extrapolated values should never be used for non-Newtonian fluids; certainly not the extrapolated values for shear-thickening fluids which should be ignored, as virtually no experimental data are available in the literature for n greater than 1. However, the experimental data and the results shown in Fig. 3.40 are in line with some of the recent numerical predictions available in the literature.[42–44] These simulations also suggest that the velocity fluctuations in the flow direction are more intense close to the wall for a shear-thinning fluid than that away from the wall.

Yoo[45] has proposed a simple modification to the Blasius equation for turbulent flow in a pipe, which gives values of the friction factor accurate to within about ±10%. The friction factor is expressed in terms of the Metzner and Reed[35] generalised Reynolds number Re_{MR} and the power-law index n.

$$\frac{f}{2}=\phi=\frac{R}{\rho u^2}=0.0396n^{0.675}\,Re_{MR}^{-0.25} \tag{3.175}$$

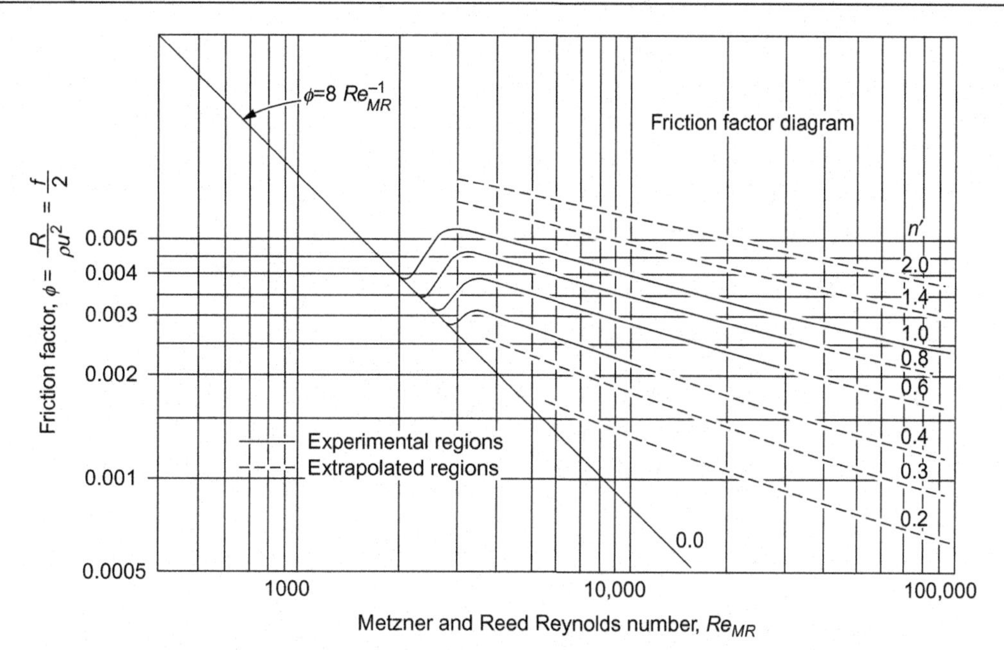

Fig. 3.40
Metzner and Reed correlation of friction factor and Reynolds number.

While Eq. (3.175) reduces to the simple Blasius relation (Eq. 3.10) for a Newtonian fluid ($n=1$), some authors[43] have argued that the two numerical constants (0.0396 and −0.25) are also functions of the power-law index.

Rearranging Eq. (3.175) gives:

$$\frac{R}{\rho u^2} n^{-0.675} = 0.0396 Re_{MR}^{-0.25} \tag{3.176}$$

Thus, the pipe friction chart for a Newtonian fluid (Fig. 3.3) may be used for shear-thinning power-law fluids if Re_{MR} is used in place of Re. In the turbulent region, the ordinate is equal to $(R/\rho u^2)\, n^{-0.615}$. For the streamline region, the ordinate remains simply $R/\rho u^2$, because Re_{MR} has been defined so that it shall be so (see Eq. 3.140). Similarly, Irvine[46] has proposed an improved form of the modified Blasius equation which predicts the friction factor for inelastic shear-thinning polymer-solutions to within ±7%.

$$\phi = \frac{R}{\rho u^2} = \left[\frac{2^{3-2n}}{7^{7n}} \left(\frac{4}{3n+1} \right)^{n(3n+2)} Re_{MR}^{-1} \right]^{1/(3n+1)} \tag{3.177}$$

Both Eqs. (3.176), (3.177) are easy to use as they give $R/\rho u^2$ as explicit functions of Re_{MR} and n. Eq. (3.175) should be used with caution, particularly if the fluid exhibits any elastic properties, bearing in mind the conclusions of Heywood and Cheng[38] and van den Heever et al.[39]

Bowen[47] gives a generalised scale-up method by which the pressure drop in an industrial pipeline may be estimated from the results of laboratory experiments with the same fluid. It involves making measurements on the small-scale using a sample of the fluid. The method obviates the difficulty that erroneous results may be obtained by predicting the pressure drop for flow in the turbulent regime using parameters measured under laminar flow conditions. Reference may also be made to Chhabra and Richardson[27] for details of the procedure.

3.4.9 The Transition From Laminar to Turbulent Flow

The critical value of the Reynolds number (Re_{MR}) for the transition from laminar to turbulent flow may be calculated from the Ryan and Johnson[14] stability parameter, defined earlier by Eq. (3.56). For a *power-law* fluid, this becomes:

$$(Re_{MR})_c = \frac{6464}{(3n+1)^2}(2+n)^{\frac{2+n}{1+n}} \tag{3.178}$$

From Eq. (3.178), the critical Reynolds number for a Newtonian fluid is 2100. As n decreases, $(Re_{MR})_c$ rises to a maximum of 2400, and then falls to 1600 for $n = 0.1$. Most of the experimental evidence suggests, however, that the transition occurs at a value close to 2000.

Example 3.14

A kaolin suspension ($n' = 0.2$, $k' = 1.15$ Pa s$^{n'}$) of density 1210 kg/m^3 is to be pumped through a 50 mm diameter pipe. Calculate the value of the critical velocity for this suspension corresponding to the end of the laminar flow. Due to the varied process requirements, the actual velocity in the pipe can vary on either side of this value by up to 50%, what will be the corresponding range of the pressure gradient in the pipe?

Solution

For a true power-law fluid, one can use Eqs. (3.157), (3.158):

$$n' = n = 0.2;\ \ k = k'\left(\frac{3n+1}{4n}\right)^{-n}$$

$$\therefore\ \ k = 1.15\left(\frac{3\times 0.2+1}{4\times 0.2}\right)^{-0.2} = 1\ \text{Pa s}$$

Now using Eqs. (3.159), (3.178):

$$(Re_{MR})_c = 8^{1-n'}\frac{\rho u^{2-n'} d^{n'}}{k'} = \frac{6464}{(3\times 0.2+1)^2}(2+0.2)^{\frac{2.2}{1.2}}$$

$$\therefore\ \ u_c = \left[\frac{1.15\times 6464\times 2.2^{2.2/1.2}}{1210\times 8^{1-0.2}\times (0.05)^{0.2}\times (1.6)^2}\right]^{\frac{1}{2-0.2}}$$

Or $u_c = 2$ m/s.

Thus, the laminar flow conditions will exist for velocities lower than 2 m/s.

But due to the varied process demand, the velocity in the pipe can range for 1–3 m/s. For $u = 1$ m/s, the flow will be laminar and therefore one can use Eq. (3.136):

$$u = \left(\frac{-\Delta P}{l}\right)^{1/n} \left(\frac{1}{4k}\right)^{1/n} \frac{n}{6n+2} d^{\frac{n+1}{n}}$$

Substituting values and solving for pressure gradient:

$$-\frac{\Delta P}{l} = 4ku^n \left(\frac{6n+2}{n}\right)^n d^{-(n+1)}$$

$$= 4 \times 1 \times 1^{0.2} \left(\frac{6 \times 0.2 + 2}{0.2}\right)^{0.2} (0.05)^{-1.2}$$

i.e., $-\frac{\Delta P}{l} = \underline{\underline{254 \text{ Pa/m}}}$

For the higher value of $u = 3$ m/s, the flow is expected to be turbulent. Evaluating the Reynolds number:

$$Re_{MR} = \frac{8^{1-0.2} \times 1210 \times 3^{2-0.2} \times 0.05^{0.2}}{1.15}$$

$$Re_{MR} = 22040$$

Using Eq. (3.176):

$$R = \left(\frac{-\Delta P}{l}\right)\frac{d}{4} = (\rho u^2)\left(n^{0.675}\right)\left\{0.0396 Re_{MR}^{-0.25}\right\}$$

Substituting values and solving for $-\frac{\Delta P}{l}$:

$$-\frac{\Delta P}{l} = \left(\frac{4}{0.05}\right)(1210 \times 3^2)(0.2)^{0.675} \times 0.0396(22040)^{-0.25}$$

$$-\frac{\Delta P}{l} = \underline{\underline{955 \text{ Pa/m}}}$$

Alternatively, from Fig. 3.40:

$$\frac{R}{\rho u^2} \asymp 0.0011$$

$$\therefore \quad -\frac{\Delta P}{l} = 0.0011 \times 1210 \times 3^2 \times \frac{4}{0.05} = \underline{\underline{960 \text{ Pa/m}}}$$

3.5 Nomenclature

		Units in SI System	Dimensions in M, L, T, θ
A	Area perpendicular to direction of flow	m^2	L^2
B	Width of rectangular channel or notch	m	L
C_c	Coefficient of contraction	–	–
C_f	Coefficient for flow over a bank of tubes	–	–
D	Depth of liquid in channel	m	L
Db	Deborah number	–	–
d	Diameter of pipe	m	L
d_a	Dimension of rectangular duct, or distance separating parallel plates	m	L
d_b	Dimension of rectangular duct	m	L
d_m	Hydraulic mean diameter ($=4\ A/p$)	m	L
E	Eddy kinematic viscosity	m^2/s	$L^2\ T^{-1}$
e	Surface roughness	m	L
F	Energy per unit mass degraded because of irreversibility of process	J/kg	$L^2\ T^{-2}$
f	Fanning friction factor ($=2\ R/\rho u^2$)	–	–
f'	Moody friction factor ($8R/\rho u^2$)	–	–
G	Mass rate of flow	kg/s	$M\ T^{-1}$
$\mathbf{G}$	Young's modulus	N/m^2	$M\ L^{-1}\ T^{-2}$
g	Acceleration due to gravity	m/s^2	$L\ T^{-2}$
h	Depth below surface measured perpendicular to bottom of channel or notch	m	L
h_f	Head lost due to friction	m	L
I_x, I_y, I_z	Normal stresses in x, y, z directions (surfaces A, C, B in Fig. 3.36)	N/m^2	$M\ L^{-1}\ T^{-2}$
i	Hydraulic gradient (h_f/l)	–	–
J	Specific energy of fluid in open channel	J/kg	$L^2\ T^{-2}$
j	Number of banks of pipes in direction of flow	–	–
k	Consistency coefficient in power-law equation	$N\ s^n/m^2$	$M\ L^{-1}\ T^{-2}$
k'	Coefficient defined by Eq. (3.93)	$N\ s^{n'}/m^2$	$M\ L^{-1}\ T^{n'-2}$
L	Characteristic linear dimension	m	L
l	Length of pipe or channel	m	L
m	Index in Eq. (3.127)	–	–
N_1	First normal normal stress difference $I_x - I_y$	N/m^2	$M\ L^{-1}\ T^{-2}$
N_2	Second normal normal stress difference $I_y - I_z$	N/m^2	$M\ L^{-1}\ T^{-2}$
n	Index in power-law equation	–	–

n'	Slope defined by Eq. (3.154)	–	–
P	Pressure	N/m^2	$\mathbf{M\,L^{-1}\,T^{-2}}$
P_f	Pressure due to friction	N/m^2	$\mathbf{M\,L^{-1}\,T^{-2}}$
ΔP	Pressure difference or change	N/m^2	$\mathbf{M\,L^{-1}\,T^{-2}}$
p	Wetted perimeter	m	$\mathbf{L}$
Q	Volumetric rate of flow	m^3/s	$\mathbf{L^3\,T^{-1}}$
Q_A	Volumetric flowrate in sheared annulus	m^3/s	$\mathbf{L^3\,T^{-1}}$
Q_p	Volumetric flowrate in unsheared plug	m^3/s	$\mathbf{L^3\,T^{-1}}$
q	Net heat flow into system	J/kg	$\mathbf{L^2\,T^{-2}}$
R	Shear stress on surface	N/m^2	$\mathbf{M\,L^{-1}\,T^{-2}}$
R_c	Shear stress on surface C of element in Fig. 3.36	N/m^2	$\mathbf{M\,L^{-1}\,T^{-2}}$
Re	Reynolds number with respect to pipe diameter	–	–
Re_{MR}	Generalised (Metzner and Reed) Reynolds number	–	–
$Re_{(MR)c}$	Critical value of Re_{MR} at laminar-turbulent transition	–	–
Re_x	Reynolds number with respect to x	–	–
R_m	Mean value of shear stress at surface	N/m^2	$\mathbf{M\,L^{-1}\,T^{-2}}$
R_S	Shear stress at radius s	N/m^2	$\mathbf{M\,L^{-1}\,T^{-2}}$
R_Y	Yield stress	N/m^2	$\mathbf{M\,L^{-1}\,T^{-2}}$
R_y	Shear stress at some point in fluid distance y from surface	N/m^2	$\mathbf{M\,L^{-1}\,T^{-2}}$
R_0	Shear stress ($-R$) in fluid at boundary surface ($y=0$)	N/m^2	$\mathbf{M\,L^{-1}\,T^{-2}}$
r	Radius of pipe, or outer pipe in case of annulus, or radius of cone	m	$\mathbf{L}$
r_c	Radius of unsheared plug	m	$\mathbf{L}$
r_i	Radius of inner pipe of annulus	m	$\mathbf{L}$
S	Entropy per unit mass	J/kg K	$\mathbf{L^2\,T^{-2}\,\theta^{-1}}$
s	Distance from axis of pipe or from centre-plane or of rotation or thickness of liquid film	m	$\mathbf{L}$
t	Time	s	$\mathbf{T}$
T	Temperature	K	$\boldsymbol{\theta}$
U	Internal energy per unit mass	J/kg	$\mathbf{L^2\,T^{-2}}$
u	Mean velocity	m/s	$\mathbf{L\,T^{-1}}$
u_{CL}	Velocity in pipe at centre line	m/s	$\mathbf{L\,T^{-1}}$
u_{Ex}	Fluctuating velocity component	m/s	$\mathbf{L\,T^{-1}}$
u_i	Instantaneous value of velocity	m/s	$\mathbf{L\,T^{-1}}$
u_p	Velocity of unsheared plug	m/s	$\mathbf{L\,T^{-1}}$
u_s	Velocity at free surface	m/s	$\mathbf{L\,T^{-1}}$
$u_{\max}$	Maximum velocity	m/s	$\mathbf{L\,T^{-1}}$

u_t	Velocity at narrowest cross-section of bank of tubes	m/s	$\mathbf{L\,T^{-1}}$
u_x	Velocity in X-direction at distance y from surface	m/s	$\mathbf{L\,T^{-1}}$
u_w	Velocity of propagation wave	m/s	$\mathbf{L\,T^{-1}}$
v	Volume per unit mass of fluid	m^3/kg	$\mathbf{M^{-1}\,L^3}$
W_s	Shaft work per unit mass	J/kg	$\mathbf{L^2\,T^{-2}}$
w	Width of surface	m	**L**
X	$(2\,l/r)(R_Y/-\Delta P)$	–	–
x	Distance in X-direction or in direction of motion or parallel to surface	m	**L**
y	Distance in Y-direction or perpendicular distance from surface	m	**L**
Z	Stability criterion defined in Eq. (3.56)	–	–
z	Distance in vertical direction	m	**L**
α	Constant in expression for kinetic energy of fluid	–	–
β	Constant in Eq. (3.124)	$s^{2/3}$	$\mathbf{T^{2/3}}$
γ	Strain	–	–
$\dot{\gamma}$	Rate of shear or of strain	s^{-1}	$\mathbf{T^{-1}}$
δ	Boundary layer thickness	m	**L**
λ	Shear force acting on unit length of inner surface of annulus	N/m	$\mathbf{M\,T^{-2}}$
μ	Viscosity of fluid	$N\,s/m^2$	$\mathbf{M\,L^{-1}\,T^{-1}}$
μ_a	Apparent viscosity defined by Eq. (3.117)	$N\,s/m^2$	$\mathbf{M\,L^{-1}\,T^{-1}}$
μ_p	Plastic viscosity	$N\,s/m^2$	$\mathbf{M\,L^{-1}\,T^{-1}}$
μ_0	Apparent viscosity ($\dot{\gamma} \to 0$)	$N\,s/m^2$	$\mathbf{M\,L^{-1}\,T^{-1}}$
μ_∞	Apparent viscosity ($\dot{\gamma} \to \infty$)	$N\,s/m^2$	$\mathbf{M\,L^{-1}\,T^{-1}}$
ϕ	Friction factor ($=R/\rho u^2$)	–	–
ρ	Density of fluid	kg/m^3	$\mathbf{M\,L^{-3}}$
θ	Angle between cone and plate in viscometers or angle	–	–
τ	Shear stress ($=R$)	N/m^2	$\mathbf{M\,L^{-1}\,T^{-2}}$
ω	Angular speed of rotation	s^{-1}	$\mathbf{T^{-1}}$

References

1. Bird RB, Armstrong RC, Hassager O. *Dynamics of polymeric liquids, vol. 1: fluid mechanics*. 2nd ed. New York: Wiley; 1987.
2. Chhabra RP, Richardson JF. *Non-Newtonian flow and applied rheology: engineering applications*. 2nd ed. Oxford: Butterworth-Heinemann; 2008.
3. Barnes HA, Hutton JF, Walters K. *An introduction to rheology*. Amsterdam: Elsevier; 1989.

4. Reynolds O. Papers on Mechanical and Physical Subjects 2 (1881–1901) 51. An experimental investigation of the circumstances which determine whether the motion of water shall be direct or sinuous and the law of resistance in parallel channels. 535. On the dynamical theory of incompressible viscous fluids and the determination of the criterion.
5. Prandtl L. Neure Ergebnisse der Turbulenzforschung. *Z Ver Deut Ing* 1933;**77**:105.
6. Nikuradse J. Strömungsgesetze in rauhen Röhren. *Forsch Ver deut Ing* 1933;**361**.
7. Stanton T, Pannell J. Similarity of motion in relation to the surface friction of fluids. *Phil Trans R Soc* 1914;**214**:199.
8. Moody LF. Friction factors for pipe flow. *Trans Am Soc Mech Eng* 1944;**66**:671.
9. La Violette M. On the history, science and technology included in the Moody diagram. *ASME J Fluids Eng* 2017;https://doi.org/10.1115/1.4035116.
10. Blasius H. Das Ähnlichkeitsgesetz bei Reibungsvorgängen in Flüssigkeiten. *Forsh Ver deut Ing* 1913;**131**.
11. Laufer J. The structure of turbulence in fully developed pipe flow. Report no. 1174. United States National Advisory Committee for Aeronautics; 1955.
12. Hagen G. Ueber die Bewegung des Wassers in engen zylindrischen Röhren. *Ann Phys (Pogg Ann)* 1839;**46**:423.
13. Poiseuille J. Recherches experimentales sur le movement des liquids dans les tubes de très petit diamètre. *Inst de France Acad des Sci Mémoires présentés par divers savants* 1846;**9**:433.
14. Ryan NW, Johnson MA. Transition from laminar to turbulent flow in pipes. *AIChE J* 1959;**5**:433.
15. Tennekes H, Lumley JL. *A first course in turbulence.* Cambridge: MIT Press; 1972.
16. Davidson PA. *Turbulence: an introduction for scientists and engineers.* 2nd ed. Oxford: Oxford University Press; 2015.
17. Shah RK, London AL. *Laminar forced convection in ducts; advances in heat transfer supplement 1.* New York: Academic Press; 1978.
18. White CM. Streamline flow through curved pipes. *Proc Roy Soc A* 1929;**123**:645.
19. Darby R, Chhabra RP. *Chemical engineering fluid mechanics.* 3rd ed. Boca Raton, FL: CRC Press; 2017.
20. Chanson H. *The hydraulics of open channel flow: an introduction.* 2nd ed. London: Elsevier; 2012.
21. British Standard 5168 (BS 5168). *Glossary of rheological terms.* London: British Standards Institution; 1975.
22. Cross MM. Rheology of non-Newtonian fluids: a new flow equation for pseudoplastic systems. *J Colloid Sci* 1965;**20**:417.
23. Mewis J, Wagner NJ. *Colloidal suspension rheology.* New York: Cambridge University Press; 2012.
24. Astarita G, Marrucci G. *Principles of non-Newtonian fluid mechanics.* New York: McGraw-Hill; 1974.
25. Skelland AHP. *Non-Newtonian flow and heat transfer.* New York: Wiley; 1967.
26. Hutton JF, Pearson JRA, Walters K, editors. *Theoretical rheology.* London: Applied Science Publishers; 1975.
27. Nguyen QD, Boger DV. Measuring the flow properties of yield stress fluids. *Annu Rev Fluid Mech* 1992;**24**:47.
28. Paul EL, Atiemo-Obeng VA, Kresta SM. *Handbook of industrial mixing: science and practice.* Hoboken, NJ: Wiley; 2004.
29. Boger DV, Walters K. *Rheological phenomena in focus.* Amsterdam: Elsevier; 1993.
30. Larson RG. *The structure and rheology of complex fluids.* New York: Oxford University Press; 1998.
31. Whorlow RH. *Rheological techniques.* 2nd ed. Chickchester: Wiley; 1992.
32. Macosko CW. *Rheology: principles, measurements and applications.* New York: Wiley; 1994.
33. Farooqi SI, Richardson JF. Rheological behaviour of kaolin suspensions in water and water-glycerol mixtures. *Trans I Chem E* 1980;**58**:116.
34. Metzner AB, White JL, Denn MM. Constitutive equations for viscoelastic fluids for short deformation periods and for rapidly changing flows. *AIChE J* 1966;**12**:836.
35. Metzner AB, Reed JC. Flow of non-Newtonian fluids–correlation of the laminar, transition and turbulent flow regions. *AIChE J* 1955;**1**:434.
36. Rabinowitsch B. Über die Viskosität and Elastizität van Solen. *Z Phys Chem* 1929;**A145**.
37. Mooney M. Explicit formulas for slip and fluidity. *J Rheol* 1931;**2**:210.
38. Heywood NI, Cheng DC-H. Comparison of methods for predicting head loss in turbulent pipe flow of non-Newtonian fluids. *Trans Inst Meas Control* 1984;**6**:33.

39. van den Heever EM, Sutherland APN, Haldenwang R. Influence of the rheological model used in pipe-flow prediction techniques for homogeneous non-Newtonian fluids. *J Hydraul Eng* 2015;**141**:04014084.
40. Hartnett JP, Kostic M. Turbulent friction factor correlations for power-law fluids in circular and non-circular channels. *Int Commun Heat Mass Transfer* 1990;**17**:59.
41. Dodge DW, Metzner AB. Turbulent flow of non-Newtonian systems. *AIChE J* 1959;**5**:189 [see also correction: *AIChE J* 1962;**l.8**:143].
42. Rudman M, Blackburn HM, Graham LJW, Pullum L. Turbulent pipe flow of shear-thinning fluids. *J Non-Newtonian Fluid Mech* 2004;**118**:33.
43. Gnambode PS, Orlandi P, Ould-Rouiss M, Nicolas X. Large-eddy simulation of turbulent pipe flow of power-law fluids. *Int J Heat Fluid Flow* 2015;**54**:196.
44. Gavrilov AA, Rudyak V. Reynolds-averaged modelling of turbulent flows of power-law fluids. *J Non-Newtonian Fluid Mech* 2016;**227**:45.
45. Yoo SS. *Heat transfer and friction factors for non-Newtonian fluids in circular tubes* [PhD thesis]. Chicago: University of Illinois; 1974.
46. Irvine TF. A generalized Blasius equation for power law fluids. *Chem Eng Commun* 1988;**65**:39.
47. Bowen RL. Designing turbulent flow systems. *Chem Eng (Albany)* 1961;**68**:143.

Further Reading

1. Brasch DJ, Whyman D. *Problems in fluid flow*. London: Edward Arnold; 1986.
2. *Flow through valves, fittings and pipes*. Crane: Technical Paper No. 41014th printing New York: Crane; 1974.
3. Kays JM, Nedderman RM. *An introduction to fluid mechanics and heat transfer*. 3rd ed. Cambridge: Cambridge University Press; 1974.
4. Massey BS. *Mechanics of solids*. 5th ed. London: Van Nostrand/Reinhold; 1987.
5. Paterson AR. *A first course in fluid mechanics*. Cambridge: Cambridge University Press; 1985.
6. Pritchard PJ, Mitchell JW. *Fox and McDonald's introduction to fluid mechanics*. 9th ed. New York: Wiley; 2015.
7. Streeter VL, Wylie EB. *Fluid mechanics*. 8th ed. New York: McGraw-Hill; 1985.
8. White FM. *Fluid mechanics*. 8th ed. New York: McGraw Hill; 2015.

CHAPTER 4

Flow of Compressible Fluids

4.1 Introduction

Although all fluids are to some degree compressible, compressibility is sufficiently enough to affect flow under normal conditions only for a gas. Furthermore, if the pressure of the gas does not change by more than about 15%–20%, it is usually satisfactory to treat the gas as an incompressible fluid with a density equal to that at the mean pressure.

When compressibility is taken into account, the equations of flow become much more complex than they are for an incompressible fluid, even if the simplest possible equation of state (the *ideal gas law*) is used to describe their behaviour. In this chapter, attention is confined to consideration of the flow of ideal gases. The physical property of a gas which varies, but which is constant for an incompressible fluid, is the density ρ or specific volume v $(=1/\rho)$. Density is a function of both temperature and pressure and it is necessary therefore to take account of the effects of both of these variables. The relation between the pressure and the density will be affected by the heat transfer to the gas as it passes through the system. Isothermal conditions can be maintained only if there is very good heat transfer to the surroundings and normally exist only at low flowrates in small equipment. At the opposite extreme, in large installations with high flowrates, conditions are much more nearly adiabatic. It should be noted that, except for isothermal flow, the relation between pressure and density is influenced by the way in which the change is caused (for example, the degree of reversibility).

In this chapter consideration is given to the flow of gases through orifices and nozzles, and to flow in pipelines. It is found that, in all these cases, the flow may reach a limiting maximum value which is independent of the downstream pressure; this is a phenomenon which does not arise with incompressible fluids. The discussion presented in this chapter also serves as the starting point for the selection and design of safety release valves.

4.2 Flow of Gas Through a Nozzle or Orifice

This is one of the simplest applications of the flow of a compressible fluid and it can be used to illustrate many of the features of the process. In practical terms, it is highly relevant to the design of relief valves or bursting discs, which are often incorporated into pressurised systems in order to protect the equipment and personnel from dangers which may arise if the equipment

Coulson and Richardson's Chemical Engineering. https://doi.org/10.1016/B978-0-08-101099-0.00004-5

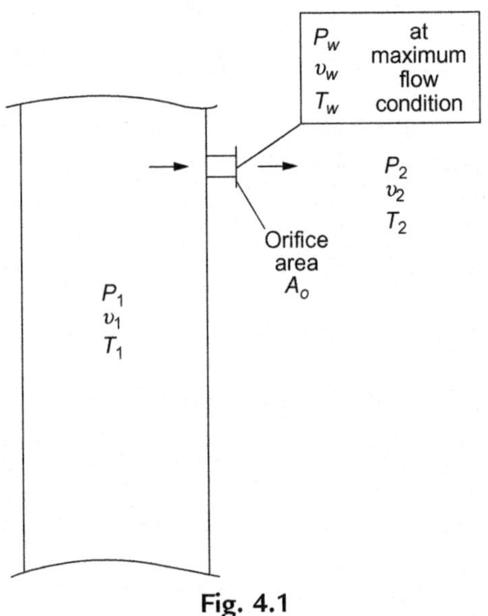

Fig. 4.1
Discharge through an orifice.

is subjected to pressures in excess of design values. In many cases it is necessary to vent gases evolved during a chemical reaction.

For this purpose, the gas flowrate at an aperture through which it is discharged from a vessel maintained at a constant pressure P_1 to surroundings at a pressure P_2 (Fig. 4.1) is considered.

The energy balance, Eq. (2.54), is used here as follows:

$$\frac{1}{\alpha}u\mathrm{d}u + g\mathrm{d}z + v\mathrm{d}P + \delta W_s + \delta F = 0 \tag{4.1}$$

If the flow is horizontal ($dz=0$), the velocity distribution is assumed to be flat ($\alpha=1$) and friction neglected ($\delta F=0$) then, as no external work is performed, Eq. (4.1) reduces to:

$$u\mathrm{d}u + v\mathrm{d}P = 0 \tag{4.2}$$

If the velocity in the vessel at which the gas approaches the outlet is negligible ($u_1=0$), integration of Eq. (4.2) gives:

$$\frac{u_2^2}{2} = -\int_{P_1}^{P_2} v\mathrm{d}P \tag{4.3}$$

The value of the integral depends on the pressure–volume relation. In some cases, this relationship can be approximated by the ideal gas law while in other situations, this integral can be evaluated numerically provided sufficient data on pressure–density (or specific volume) is available for the system of interest.

4.2.1 Isothermal Flow

For the isothermal flow of an ideal gas, then from Eq. (2.69):

$$-\int_{P_1}^{P_2} v\mathrm{d}P = -P_1 v_1 \ln\frac{P_2}{P_1} = P_1 v_1 \ln\frac{P_1}{P_2} \tag{4.4}$$

Substituting into Eq. (4.3), gives the discharge velocity u_2:

$$\frac{u_2^2}{2} = P_1 v_1 \ln\frac{P_1}{P_2}$$

or:

$$u_2 = \sqrt{2P_1 v_1 \ln\frac{P_1}{P_2}} \tag{4.5}$$

The mass rate of flow is given by:

$$\begin{aligned} G &= u_2 A_2 \rho_2 = \frac{u_2 A_2}{v_2} \\ &= \frac{A_2}{v_2}\sqrt{2P_1 v_1 \ln\frac{P_1}{P_2}} \end{aligned} \tag{4.6}$$

It is shown in Chapter 6 that the minimum flow area A_2 tends to be somewhat smaller than the area A_0 of the aperture because the gas leaves with a small radial inwards velocity component. Furthermore, there will be some reduction of discharge rate because of the frictional effects, which have been neglected in Eq. (4.2). Grouping these factors together by means of a coefficient of discharge C_D, where $C_D < 1$, gives:

$$G = \frac{C_D A_0}{v_2}\sqrt{2P_1 v_1 \ln\frac{P_1}{P_2}} \tag{4.7}$$

Substituting for isothermal flow:

$$v_2 = v_1 \frac{P_1}{P_2}$$

$$\begin{aligned} G &= C_D A_0 \frac{P_2}{P_1 v_1}\sqrt{2P_1 v_1 \ln\frac{P_1}{P_2}} \\ &= C_D A_0 P_2 \sqrt{\frac{2}{P_1 v_1} \ln\frac{P_1}{P_2}} = C_D A_0 P_2 \sqrt{\frac{2M}{\mathbf{R}T} \ln\frac{P_1}{P_2}} \end{aligned} \tag{4.8}$$

where $\mathbf{R}$ is the universal gas constant, and M is the molecular weight of the gas.

Maximum flow conditions

Eq. (4.8) gives $G=0$ for $P_2=P_1$, and also for $P_2=0$. As it is a continuous function, it must yield a maximum value of G at some intermediate pressure P_w, where $0<P_w<P_1$. Differentiating both sides of Eq. (4.8) with respect to P_2:

$$\begin{aligned}\frac{dG}{dP_2} &= C_D A_0 \sqrt{\frac{2}{P_1 v_1}} \frac{d}{dP_2}\left\{P_2\left[\ln\left(\frac{P_1}{P_2}\right)\right]^{1/2}\right\} \\ &= C_D A_0 \sqrt{\frac{2}{P_1 v_1}}\left\{\left[\ln\left(\frac{P_1}{P_2}\right)\right]^{1/2} + P_2 \frac{1}{2}\left[\ln\left(\frac{P_1}{P_2}\right)\right]^{-1/2}\left(\frac{P_2}{P_1}\right)\left(\frac{P_1}{-P_2^2}\right)\right\}\end{aligned}$$

For a maximum value of G, $dG/dP_2=0$, and:

Then:

$$\ln\frac{P_1}{P_2} = \frac{1}{2} \tag{4.9}$$

or:

$$\frac{P_1}{P_2} = 1.65 \tag{4.10}$$

and the critical pressure ratio,

$$\frac{P_2}{P_1} = w_c = 0.607 \tag{4.11}$$

Substituting into Eq. (4.8) to give the maximum value of G (G_{max}):

$$\begin{aligned}G_{max} &= C_D A_0 P_2 \sqrt{\frac{2}{P_1 v_1} \cdot \frac{1}{2}} \\ &= C_D A_0 \sqrt{\frac{P_2}{v_2}} \\ &= C_D A_0 \rho_2 \sqrt{P_2 v_2}\left(\text{where } \rho_2 = \frac{1}{v_2}\right)\end{aligned} \tag{4.12}$$

Thus the velocity,

$$u_w = \sqrt{P_2 v_2} = \sqrt{P_0 v_0} = \sqrt{\frac{\mathbf{R}T}{M}} \tag{4.13}$$

It is shown later in Section **4.3** that this corresponds to the velocity at which a small pressure wave will be propagated under isothermal conditions, sometimes referred to as the 'isothermal sonic velocity', though the heat transfer rate will not generally be sufficient to maintain truly isothermal conditions.

Writing Eq. (4.12) in terms of the upstream conditions (P_1, v_1):

$$G_{max} = C_D A_0 \frac{1}{v_1}\frac{P_2}{P_1}\sqrt{P_1 v_1}$$

or, on substitution from Eq. (4.11):

$$G_{max} = 0.607 C_D A_0 \sqrt{\frac{P_1}{v_1}} \tag{4.14}$$

At any given temperature, $P_1 v_1 = P_0 v_0 = \text{constant}$, where v_0 is the value of v at some reference pressure P_0.

Then:

$$G_{max} = 0.607 C_D A_0 P_1 \sqrt{\frac{1}{P_0 v_0}} = 0.607 C_D A_0 P_1 \sqrt{\frac{M}{\mathbf{R}T}} \tag{4.15}$$

Thus, G_{max} is linearly related to P_1.

It will be seen that when the pressure ratio P_2/P_1 is less than the critical value ($w_c = 0.607$), the flow rate becomes independent of the downstream pressure P_2. The fluid at the orifice is then flowing at the velocity of a small pressure wave and the velocity of the pressure wave *relative to the orifice* is zero. That is the upstream fluid cannot be influenced by the pressure in the downstream reservoir. Thus, the pressure falls to the critical value at the orifice, and further expansion to the downstream pressure takes place in the reservoir with the generation of a *shock wave*, as discussed in Section **4.6**.

4.2.2 Nonisothermal Flow

If the flow is nonisothermal, it may be possible to represent the relationship between pressure and volume by an equation of the form:

$$Pv^k = \text{constant} \tag{4.16}$$

Then from Eq. (2.73):

$$-\int_{P_1}^{P_2} v\,dP = \frac{k}{k-1} P_1 v_1 \left[1 - \left(\frac{P_2}{P_1}\right)^{(k-1)/k}\right] \tag{4.17}$$

Substituting from Eq. (4.17) in Eq. (4.2), the discharge velocity u_2 is then given by:

$$\frac{u_2^2}{2}=\frac{k}{k-1}P_1v_1\left[1-\left(\frac{P_2}{P_1}\right)^{(k-1)/k}\right]$$

$$\text{i.e.}\quad u_2=\sqrt{\frac{2k}{k-1}P_1v_1\left[1-\left(\frac{P_2}{P_1}\right)^{(k-1)/k}\right]} \tag{4.18}$$

Allowing for a discharge coefficient, the mass rate of flow is given by:

$$G=\frac{C_DA_0}{v_2}\sqrt{\frac{2k}{k-1}P_1v_1\left[1-\left(\frac{P_2}{P_1}\right)\right]^{(k-1)/k}} \tag{4.19}$$

$$=\frac{C_DA_0}{v_1}\left(\frac{P_2}{P_1}\right)^{1/k}\sqrt{\frac{2k}{k-1}P_1v_1\left[1-\left(\frac{P_2}{P_1}\right)^{(k-1)/k}\right]} \tag{4.20}$$

Maximum flow conditions

This equation also gives $G=0$ for $P_2=0$ and for $P_2=P_1$:

Differentiating both sides of Eq. (4.20) with respect to P_2/P_1:

$$\frac{dG}{d(P_2/P_1)}=\frac{C_DA_0}{v_1}\sqrt{\frac{2k}{k-1}P_1v_1}\left\{\left(\frac{P_2}{P_1}\right)^{1/k}\frac{1}{2}\left[1-\left(\frac{P_2}{P_1}\right)^{(k-1)/k}\right]^{-1/2}\right.$$

$$\left.\times\left(-\frac{k-1}{k}\right)\left(\frac{P_2}{P_1}\right)^{-1/k}+\left[1-\left(\frac{P_2}{P_1}\right)^{(k-1)/k}\right]^{1/2}\frac{1}{k}\left(\frac{P_2}{P_1}\right)^{(1/k)-1}\right\}$$

Putting $dG/d(P_2/P_1)=0$ for the maximum value of G (G_{max}):

$$1-\left(\frac{P_2}{P_1}\right)^{(k-1)/k}=\frac{1}{2}(k-1)\left(\frac{P_2}{P_1}\right)\left(\frac{P_2}{P_1}\right)^{-1/k}$$

Which simplifies to give:

$$\frac{P_2}{P_1}=\left(\frac{2}{k+1}\right)^{k/_{k}-1} \tag{4.21}$$

Substituting Eq. (4.21) into Eq. (4.19):

$$\begin{aligned} G_{\max} &= \frac{C_D A_0}{v_2}\sqrt{\frac{2k}{k-1}P_1 v_1 \frac{1}{2}(k-1)\frac{P_2}{P_1}\left(\frac{P_2}{P_1}\right)^{-1/k}} \\ &= C_D A_0 \rho_2 \sqrt{k P_2 v_2} \end{aligned} \tag{4.22}$$

Hence:

$$u_w = \sqrt{k P_2 v_2} \tag{4.23}$$

The velocity $u_w = \sqrt{kP_2v_2}$ is shown to be the velocity of a small pressure wave if the pressure–volume relation is given by Pv^k=constant. If the expansion approximates to a reversible adiabatic (isentropic) process, $k \approx \gamma$, the ratio of the specific heats of the gases, as indicated in Eq. (2.30).

Eq. (4.22) then becomes:

$$G_{\max} = C_D A_0 \rho_2 \sqrt{\gamma P_2 v_2} \tag{4.24}$$

and:

$$u_w = \sqrt{\gamma P_2 v_2} = \sqrt{\frac{\gamma \mathbf{R} T_2}{M}} \tag{4.25}$$

where $u_w = \sqrt{\gamma P_2 v_2}$ is the velocity of propagation of a small pressure wave under isentropic conditions.

As for isothermal conditions, when maximum flow is occurring, the velocity of a small pressure wave, *relative to the orifice* is zero, and the fluid at the orifice is not influenced by the pressure further downstream.

Substituting the critical pressure ratio $w_c = P_2/P_1$ in Eq. (4.21):

$$1 + w_c^{(k-1)/k} = \frac{k-1}{2} w_c^{(k-1)/k}$$

giving:

$$w_c = \left[\frac{2}{k+1}\right]^{k/(k-1)} \tag{4.26}$$

For isentropic conditions:

$$w_c = \left[\frac{2}{\gamma+1}\right]^{\gamma/(\gamma-1)} \tag{4.26a}$$

For a diatomic gas at approximately atmospheric pressure, $\gamma = 1.4$, and $w_c = 0.53$. From Eq. (4.22):

$$\begin{aligned} G_{max} &= C_D A_0 \sqrt{\frac{kP_2}{v_2}} \\ &= C_D A_0 \sqrt{\frac{k}{v_1}\left(\frac{P_2}{P_1}\right)^{(k+1)/k} P_2} \\ &= C_D A_0 \sqrt{\frac{kP_1}{v_1}\left(\frac{P_2}{P_1}\right)^{(k+1)/k}} \end{aligned} \tag{4.27}$$

Substituting from Eq. (4.26):

$$G_{max} = C_D A_0 \sqrt{\frac{kP_1}{v_1}\left[\frac{2}{k+1}\right]^{(k+1)/(k-1)}} \tag{4.28}$$

For isentropic conditions, $k = \gamma$, and:

$$G_{max} = C_D A_0 \sqrt{\frac{\gamma P_1}{v_1}\left[\frac{2}{\gamma+1}\right]^{(\gamma+1)/(\gamma-1)}} \tag{4.28a}$$

For a given upstream temperature T_1, $P_1 v_1 = P_0 v_0 =$ constant where v_0 is the value of v at some reference pressure P_0 and temperature T_1.

Thus:

$$G_{max} = C_D A_0 P_1 \sqrt{\frac{k}{P_0 v_0}\left[\frac{2}{k+1}\right]^{(k+1)/(k-1)}} = C_D A_0 P_1 \sqrt{\frac{M}{\mathbf{R}T_1} k \left[\frac{2}{k+1}\right]^{(k+1)/(k-1)}} \tag{4.29}$$

For isentropic conditions $k = \gamma$, and:

$$G_{max} = C_D A_0 P_1 \sqrt{\frac{\gamma}{P_0 v_0}\left[\frac{2}{\gamma+1}\right]^{(\gamma+1)/(\gamma-1)}} = C_D A_0 P_1 \sqrt{\frac{M}{\mathbf{R}T_1} \gamma \left[\frac{2}{\gamma+1}\right]^{(\gamma+1)/(\gamma-1)}} \tag{4.30}$$

For a diatomic gas, $\gamma \asymp 1.4$ for pressures near atmospheric and:

$$G_{max} = 0.685 C_D A_0 P_1 \sqrt{\frac{1}{P_0 v_0}} = 0.685 C_D A_0 P_1 \sqrt{\frac{M}{\mathbf{R}T_1}} \tag{4.31}$$

It may be noted that Eqs (4.29)–(4.31) give a linear relation between G_{max} and P_1. Comparison with Eq. (4.15) (for isothermal conditions) shows that the maximum flowrate G_{max} is $(0.685/0.607) = 1.13$ times greater than that for isothermal flow of a diatomic gas.

In Fig. 4.2, the mass flowrate is plotted as a function of cylinder pressure for discharge through an orifice to an atmosphere at a constant downstream pressure P_2 – for the conditions given in Example 4.1. In Fig. 4.3, the cylinder pressure P_1 is maintained constant and the effect of the downstream pressure P_2 on the flowrate is shown.

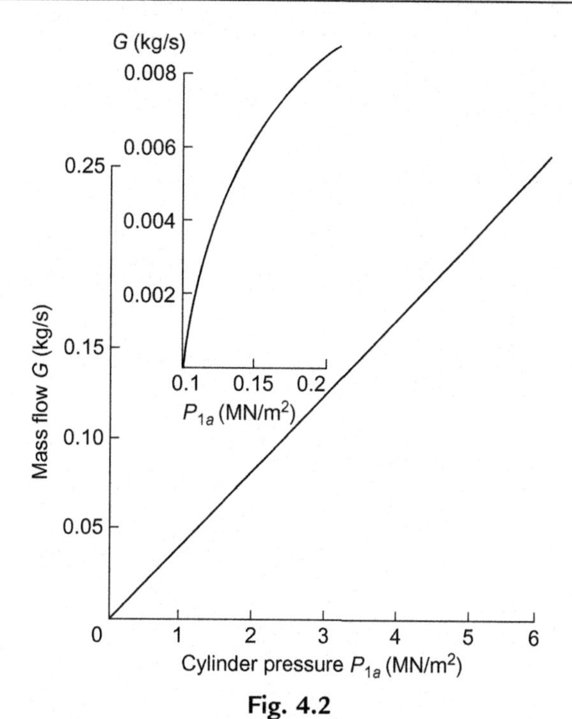

Fig. 4.2
Mass rate of discharge of gas as function of cylinder pressure for constant downstream pressure.

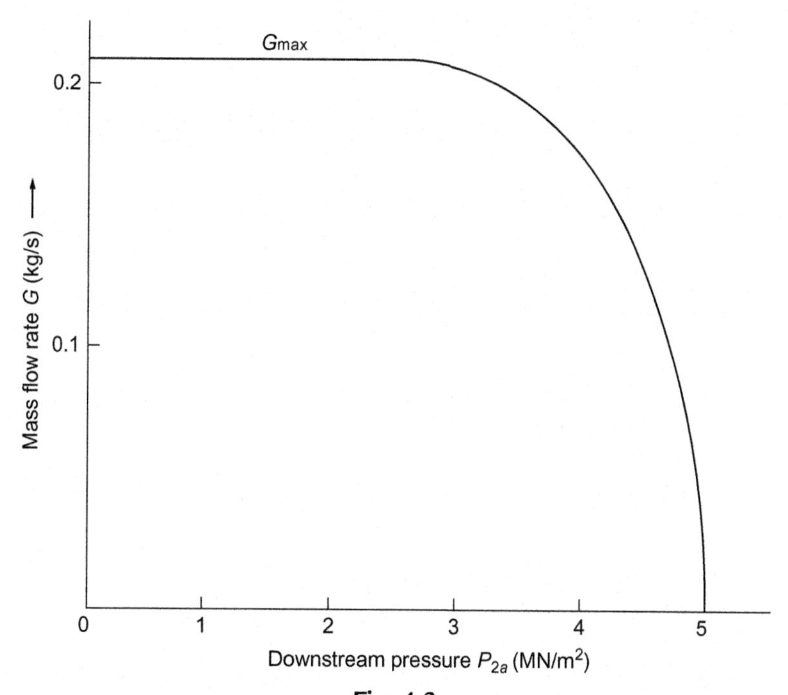

Fig. 4.3
Mass rate of discharge of gas as function of downstream pressure for constant cylinder pressure.

Example 4.1

A cylinder contains air at a pressure of 6.0 MN/m^2 and discharges to atmosphere through a valve which may be taken as equivalent to a sharp-edged orifice of 6 mm diameter (coefficient of discharge $=0.6$). (i) Plot the rate of discharge of air from the cylinder against cylinder pressure. Assume that the expansion at the valve is approximately isentropic and that the temperature of the air in the cylinder remains unchanged at 273 K. (ii) For a constant pressure of 5 MN/m^2 in the cylinder, plot the discharge rate as a function of the pressure P_2 of the surroundings.

Solution

The critical pressure ratio for discharge through the valve is given by Eq. (4.26a):

$$=\left[\frac{2}{\gamma+1}\right]^{[\gamma/(\gamma-1)]}$$

where γ varies from 1.40 to 1.54 over the pressure range 0.1–6.0 MN/m^2. Critical conditions occur at a relatively low cylinder pressure. Taking $\gamma=1.40$, then:

$$\text{Critical ratio}=W_c=\left(\frac{2}{2.40}\right)^{1.4/1.4-1}=0.53$$

(i) Sonic velocity will occur until the pressure in the cylinder falls to $\left(\frac{P_2}{W_c}\right)$, i.e. $(101.3/0.53)=$ 191.1 kN/m^2.

The rate of discharge for cylinder pressures greater than 191.1 kN/m^2 is given by Eq. (4.30):

$$G_{max}=C_D A_0 P_1\sqrt{\frac{\gamma M}{\mathbf{R}T_1}\left(\frac{2}{\gamma+1}\right)^{(\gamma+1)/(\gamma-1)}}$$

Taking a mean value for γ of 1.47, then:

$$\gamma\left(\frac{2}{\gamma+1}\right)^{(\gamma+1)/(\gamma-1)}=0.485$$

$$A_0=\frac{\pi}{4}\times(0.006)^2=2.83\times10^{-5}\,m^2$$

$$\frac{1}{P_1 v_1}=\frac{M}{\mathbf{R}T_1}=\frac{29}{8314\times273}=1.28\times10^{-5}\,s^2/m^2$$

$$P_1 v_1=\frac{\mathbf{R}T_1}{M}=78{,}100\,s^2/m^2$$

$$\begin{aligned}G_{max}&=0.6\times\left(2.83\times10^{-5}\right)\times P_1\sqrt{\left(1.28\times10^{-5}\right)\times0.485}\\&=4.23\times10^{-8}P_1\ \text{kg/s}\quad\left(P_1\ \text{in}\ N/m^2\right)\\&=4.23\times10^{-2}P_{1a}\ \text{kg/s}\quad\left(P_{1a}\ \text{in}\ MN/m^2\right)\end{aligned}$$

For cylinder pressures below 191.1 kN/m^2, the mass flowrate is given by Eq. (4.20). Putting $k=\gamma$ for isentropic conditions:

$$G = \frac{C_D A_0}{v_1}\left(\frac{P_2}{P_1}\right)^{1/\gamma}\sqrt{\frac{2\gamma}{\gamma-1}P_1 v_1\left[1-\left(\frac{P_2}{P_1}\right)^{(\gamma-1)/\gamma}\right]}$$

Using a value of 1.40 for γ in this low pressure region, and writing:

$$\frac{1}{v_1} = \frac{1}{P_1 v_1}\cdot P_1$$

$$G = \frac{0.6\times(2.83\times10^{-5})}{78{,}100}P_1\left(\frac{P_2}{P_1}\right)^{0.714}\sqrt{\frac{2\times1.4}{1.4-1}\times78{,}100\left[1-\left(\frac{P_2}{P_1}\right)^{0.286}\right]}$$

$$= 1.608\times10^{-7}P_1\left(\frac{P_2}{P_1}\right)^{0.714}\sqrt{\left[1-\left(\frac{P_2}{P_1}\right)^{0.286}\right]}$$

For discharge to atmospheric pressure, $P_2 = 101{,}300\ \text{N/m}^2$, and:

$$G = 6.030\times10^{-4}P_1^{0.286}\sqrt{1-27P_1^{-0.286}}\ \text{kg/s}\quad (P_1\text{ in N/m}^2)$$

Putting pressure P_{1a} in MN/m^2, then:

$$G = 0.0314P_{1a}^{0.286}\sqrt{1-0.519P_{1a}^{-0.286}}\ \text{kg/s}\quad (P_{1a}\text{ in MN/m}^2)$$

The discharge rate is plotted in Fig. 4.2.

(ii)

G vs P_{1a} data

(i) Above $P_{1a} = 0.19$ MN/m^2		**(ii) Below $P_{1a} = 0.19$ MN/m^2**	
P_{1a} (MN/m^2)	**G (kg/s)**	**P_{1a} (MN/m^2)**	**G (kg/s)**
0.19	0.0080		
0.2	0.0084	0.10	0
0.5	0.021	0.125	0.0042
1.0	0.042	0.15	0.0060
2.0	0.084	0.17	0.0070
3.0	0.126	0.19	0.0079
4.0	0.168		
5.0	0.210		
6.0	0.253		

Note: The slight mismatch in the two columns for $P_{1a} = 0.19$ MN/m^2 is attributable to the use of an average value of 1.47 for γ in column 2 and 1.40 for γ in column 4.

From this table, for a constant upstream pressure of 5 MN/m^2, the mass flowrate G remains at 0.210 kg/s for all pressures P_2 below $5\times0.53 = 2.65$ MN/m^2.

For higher values of P_2, the flowrate is obtained by substitution of the constant value of P_1 (5 MN/m^2) in Eq. (4.20).

$$G = \left(1.608 \times 10^{-7}\right) \times (82.40) \times P_2^{0.714} \sqrt{1 - 0.01214 P_2^{0.286}} \quad \left(P_2 \text{ in N/m}^2\right)$$

$$= 0.2548 P_{2a}^{0.714} \sqrt{1 - 0.631 P_{2a}^{0.286}} \text{ kg/s} \quad \left(P_2 \text{ in N/m}^2\right)$$

This relationship is plotted in Fig. 4.3 and values of G as a function of P_{2a} are:

P_{2a} MN/m^2	G kg/s
<2.65	0.210
3.0	0.206
3.5	0.194
4.0	0.171
4.5	0.123
4.9	0.061
4.95	0.044
5	0

4.3 Velocity of Propagation of a Pressure Wave

When the pressure at some point in a fluid is changed, the new condition takes a finite time to be transmitted to some other point in the fluid because the state of each intervening element of fluid has to be changed. The velocity of propagation is a function of the bulk modulus of elasticity ε, where ε is defined by the relation:

$$\varepsilon = \frac{\text{increase of stress within the fluid}}{\text{resulting volumetric strain}} = \frac{\mathrm{d}P}{-(\mathrm{d}v/v)} = -c\frac{\mathrm{d}P}{\mathrm{d}v} \tag{4.32}$$

If a pressure wave is transmitted at a velocity u_w over a distance $\mathrm{d}l$ in a fluid of cross-sectional area A, from section A to section B, as shown in Fig. 4.4, it may be brought to rest by causing the fluid to flow at a velocity u_w in the opposite direction. The pressure and specific volume are at plane B P and v, and at plane A $(P+\mathrm{d}P)$ and $(v+\mathrm{d}v)$, respectively. As a result of the change in pressure, the velocity of the fluid changes from u_w at B to $(u_w+\mathrm{d}u_w)$ at A and its mass rate of flow is:

$$G = \frac{u_w A}{v} = \frac{(u_w + \mathrm{d}u_w)A}{v + \mathrm{d}v}$$

The net force acting on the fluid between sections A and B is equal to the rate of change of momentum of the fluid, or:

$$PA - (P + \mathrm{d}P)A = G\,\mathrm{d}u_w$$

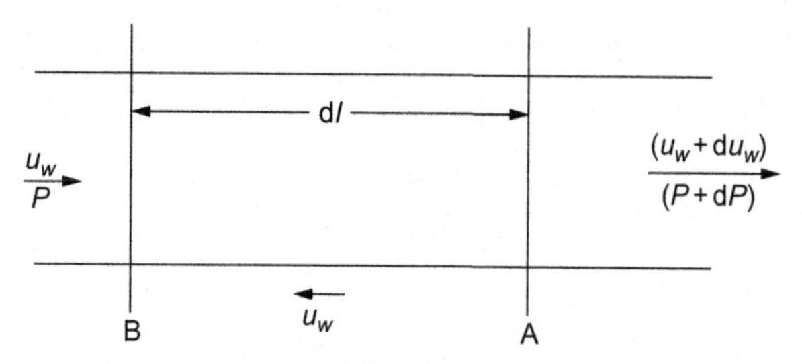

Fig. 4.4
Propagation of a pressure wave.

Substituting for

$$du_w = \left(\frac{G}{A}\right) dv$$

and:

$$-A\,\mathrm{d}P = G\frac{G}{A}dv$$

$$-\frac{\mathrm{d}P}{\mathrm{d}v} = \frac{G^2}{A^2}$$

$$= \frac{\varepsilon}{v} \quad \text{(from Eq. 4.32)}$$

$$\therefore \quad u_w = \sqrt{\varepsilon v} \tag{4.33}$$

For an ideal gas, ε may be calculated from the equation of state. Under isothermal conditions:

$$Pv = \text{constant}$$

$$\therefore \quad -\frac{\mathrm{d}P}{\mathrm{d}v} = \frac{P}{v}$$

This together with Eq. (4.33) leads to:

$$\therefore \quad \varepsilon = P \tag{4.34}$$

and:

$$u_w = \sqrt{Pv} = \sqrt{\frac{\mathbf{R}T}{M}} \tag{4.35}$$

Under isentropic conditions:

$$Pv^{\gamma} = \text{constant}$$

$$-\frac{\mathrm{d}P}{\mathrm{d}v} = \frac{\gamma P}{v}$$

$$\therefore \quad \varepsilon = \gamma P \tag{4.36}$$

and:

$$u_w = \sqrt{\gamma P v} = \sqrt{\gamma \frac{\mathbf{R}T}{M}} \tag{4.37}$$

This value of u_w corresponds closely to the velocity of sound in the fluid. That is, for normal conditions of transmission of a small pressure wave, the process is almost isentropic. When the relation between pressure and volume is $Pv^k = \text{constant}$, then:

$$u_w = \sqrt{kPv} = \sqrt{k \frac{\mathbf{R}T}{M}} \tag{4.38}$$

4.4 Converging-Diverging Nozzles for Gas Flow

Converging-diverging nozzles, as shown in Fig. 4.5A, sometimes known as Laval nozzles, are used for the expansion of gases where the pressure drop is large. If the nozzle is carefully designed so that the contours closely follow the lines of flow, the resulting expansion of the gas is almost reversible. Because the flow rate is large for high-pressure differentials, there is little time for heat transfer to take place between the gas and surroundings and the expansion is effectively isentropic. In the analysis of the nozzle, the change in flow is examined for various pressure differentials across the nozzle.

The specific volume v_2 at a downstream pressure P_2, is given by:

$$v_2 = v_1 \left(\frac{P_1}{P_2}\right)^{1/\gamma} = v_1 \left(\frac{P_2}{P_1}\right)^{-1/\gamma} \tag{4.39}$$

If gas flows under turbulent conditions from a reservoir at a pressure P_1, through a horizontal nozzle, the velocity of flow u_2, at the pressure P_2 is given by:

$$\frac{u_2^2}{2} + \int_1^2 v\,dP = 0 \quad \text{(from Eq. 2.42)}$$

Thus:

$$u_2^2 = \frac{2\gamma}{\gamma - 1} P_1 v_1 \left[1 - \left(\frac{P_2}{P_1}\right)^{(\gamma-1)/\gamma}\right] \tag{4.40}$$

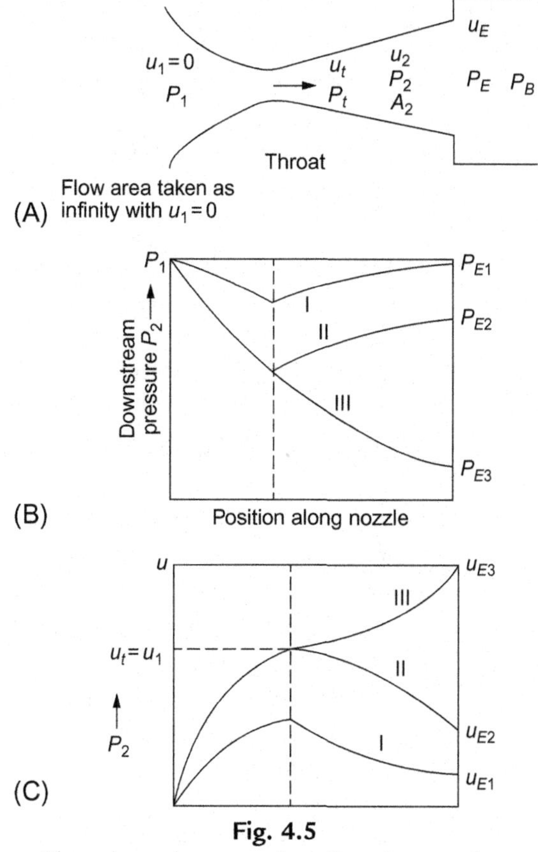

Fig. 4.5
Flow through converging-diverging nozzles.

Since:

$$A_2 = \frac{Gv_2}{u_2} \quad \text{(from Eq. 2.36)} \tag{4.41}$$

the required cross-sectional area for flow when the pressure has fallen to P_2 may be found.

4.4.1 Maximum Flow and Critical Pressure Ratio

In the flow of a gas through a nozzle, the pressure falls from its initial value P_1 to a value P_2 at some point along the nozzle; at first the velocity rises more rapidly than the specific volume and therefore the area required for flow decreases. For low values of the pressure ratio P_2/P_1, however, the velocity changes much less rapidly than the specific volume so that the area for flow must increase again. The effective area for flow presented by the nozzle must therefore pass through a minimum. It is shown that this occurs if the pressure ratio P_2/P_1 is less than the critical pressure ratio (usually approximately 0.5) and that the velocity at the throat is then equal to the velocity of sound. For expansion to pressures below the critical value the flowing stream must be diverging. Thus in a converging nozzle the velocity of the gas stream will never exceed the sonic velocity though in a converging-diverging nozzle supersonic velocities may be obtained in the diverging section. The flow changes which occur as the backpressure P_B is steadily reduced are shown in Fig. 4.5B and 4.5C. The pressure at the exit of the nozzle is denoted by P_E, which is not necessarily the same as the backpressure P_B. In the following three cases, the exit and backpressures correspond and the flow is approximately isentropic. The effect of a mismatch in the two pressures is considered later.

Case I Backpressure P_B is quite high. Curves I show how pressure and velocity change along the nozzle. The pressure falls to a minimum at the throat and then rises to a value $P_{E1}=P_B$ The velocity increases to maximum at the throat (less than sonic velocity) and then decreases to a value of u_{E1} at the exit of the nozzle. This situation corresponds to conditions in a venturi operating entirely at subsonic velocities.

Case II Backpressure is reduced (curves II). The pressure falls to the critical value at the throat where the velocity is sonic. The pressure then rises to $P_{E2}=P_B$ at the exit. The velocity rises to the sonic value at the throat and then falls to u_{E2} at the outlet.

Case III Backpressure is low, with pressure less than critical value at the exit. The pressure falls to the critical value at the throat and continues to fall to give an exit pressure $P_{E3}=P_B$. The velocity increases to sonic at the throat and continues to increase to supersonic in the diverging cone to a value u_{E3}.

With a converging-diverging nozzle, the velocity increases beyond the sonic velocity only if the velocity at the throat is sonic and the pressure at the outlet is lower than the throat pressure. For a converging nozzle the rate of flow is independent of the downstream pressure, provided the critical pressure ratio is reached and the throat velocity is sonic.

4.4.2 The Pressure and Area for Flow

As indicated in Section **4.4.1**, the area required at any point depends upon the ratio of the downstream to the upstream pressure P_2/P_1 and it is helpful to establish the minimum value of $A_2 \cdot A_2$ may be expressed in terms of P_2 and w [$=(P_2/P_1)$] using Eqs (4.39)–(4.41).

Thus:

$$\begin{aligned} A_2^2 &= G^2\frac{\gamma-1}{2\gamma}\frac{v_1^2(P_2/P_1)^{-2/\gamma}}{P_1 v_1\left[1-(P_2/P_1)^{(\gamma-1)/\gamma}\right]} \\ &= \frac{G^2 v_1(\gamma-1)}{2P_1\gamma}\frac{w^{-2/\gamma}}{1-w^{(\gamma-1)/\gamma}} \end{aligned} \tag{4.42}$$

For a given rate of flow G, A_2 decreases from an effectively infinite value at pressure P_1 at the inlet to a minimum value given by:

$$\frac{dA_2^2}{dw} = 0$$

or, when:

$$1 - w^{(\gamma-1)/\gamma}\frac{-2}{\gamma}w^{-1-2/\gamma} - w^{-2/\gamma}\frac{1-\gamma}{\gamma}w^{-1/\gamma} = 0$$

or:

$$w = \left(\frac{2}{\gamma+1}\right)^{\gamma/(\gamma-1)} \tag{4.43}$$

The value of w given by Eq. (4.43) is the critical pressure ratio w_c given by Eq. (4.26a). Thus the velocity at the throat is equal to the sonic velocity. Alternatively, Eq. (4.42) may be put in terms of the flowrate (G/A_2) as:

$$\left(\frac{G}{A_2}\right)^2 = \frac{2\gamma}{\gamma-1}\left(\frac{P_2}{P_1}\right)^{2/\gamma}\frac{P_1}{v_1}\left[1 - \left(\frac{P_2}{P_1}\right)^{(\gamma-1)/\gamma}\right] \tag{4.44}$$

and the flowrate G/A_2 may then be shown to have a maximum value of $G_{max}/A_2 = \sqrt{\gamma P_2/v_2}$.

A nozzle is correctly designed for any outlet pressure between P_1 and $P_{E1,}$ in Fig. 4.5. Under these conditions the velocity will not exceed the sonic velocity at any point, and the flowrate will be independent of the exit pressure $P_E = P_B$. It is also correctly designed for supersonic flow in the diverging cone for an exit pressure of P_{E3}.

It has been shown above that when the pressure in the diverging section is greater than the throat pressure, subsonic flow occurs. Conversely, if the pressure in the diverging section is less than the throat pressure, the flow will be supersonic beyond the throat. Thus at a given point in the diverging cone where the area is equal to A_2 the pressure may have one of two values for isentropic flow.

As an example, when γ is 1.4, using Eq. (4.39):

$$v_2 = v_1 w^{-0.71} \tag{4.45}$$

and:

$$u_2^2 = 7P_1 v_1 \left(1 - w^{0.29}\right) \tag{4.46}$$

Thus:

$$A_2^2 = \frac{v_1 w^{-1.42}}{7P_1(1 - w^{0.29})}G^2 \tag{4.47}$$

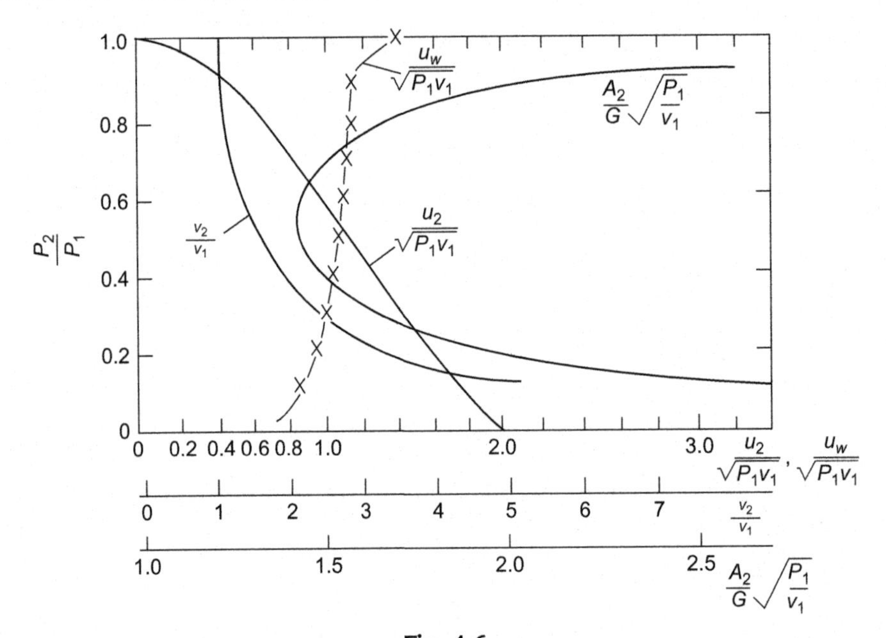

Fig. 4.6
Specific volume, velocity, and nozzle area as a function of pressure.

In Fig. 4.6 values of $v_2/v_1, u_2/\sqrt{P_1 v_1}$, and $(A_2/G)\ \sqrt{P_1/v_1}$ which are proportional to v_2, u_2, and A_2 respectively are plotted as abscissae against P_2/P_1. It is seen that the area A_2 decreases to a minimum and then increases again. At the minimum cross-section, the velocity is equal to the sonic velocity and P_2/P_1 is the critical ratio. $u_w/\sqrt{P_1 v_1}$ is also plotted and by comparing with the curve for $u_2/\sqrt{P_1 v_1}$, it is easy to compare the corresponding values of u_2 and u_w. The flow is seen to be subsonic for P_2/P_1 <0.53 and supersonic for P_2/P_1 <0.53.

4.4.3 Effect of Backpressure on Flow in Nozzle

It is of interest to study how the flow in the nozzle varies with the backpressure P_B.

(1) $P_B > P_{E2}$. The flow is subsonic throughout the nozzle and the rate of flow G is determined by the value of the backpressure P_B. Under these conditions $P_E = P_B$. It is shown as P_{E1} in Fig. 4.5.
(2) $P_B = P_{E2}$. The flow is subsonic throughout the nozzle, except at the throat where the velocity is sonic. Again $P_E = P_B$.
(3) $P_{E3} < P_B < P_{E2}$. The flow in the converging section and at the throat is exactly as for Case 2. For this range of exit pressures, the flow is not isentropic through the whole nozzle and the area at the exit section is greater than the design value for a backpressure P_B. One of two things may occur. Either the flow will be supersonic but not occupy the

whole area of the nozzle, or a shock wave will occur at some intermediate point in the diverging cone, giving rise to an increase in pressure and a change from supersonic to subsonic flow.

(4) $P_B = P_{E3}$. The flow in the converging section and at the throat is again the same as for Cases 2, 3, and 4. The flow in the diverging section is supersonic throughout and the pressure of P_B is reached smoothly. P_{E3} is therefore the design pressure for supersonic flow at the exit.

(5) $P_B < P_{E3}$. In this case, the flow throughout the nozzle is exactly as for Case 4. The pressure at the exit will again be P_{E3} and the pressure will fall beyond the end of the nozzle from P_{E3} to P_B by continued expansion of the gas.

It may be noted that the flowrate through the nozzle is a function of backpressure only for Case 1.

A practical example of flow through a converging-diverging nozzle is given in Example 4.4 after a discussion of the generation of shock waves.

4.5 Flow in a Pipe

Compressibility of a gas flowing in a pipe can have significant effect on the relation between flowrate and the pressures at the two ends of the pipe. Changes in fluid density can arise as a result of changes in either temperature or pressure, or in both, and the flow will be affected by the rate of heat transfer between the pipe and the surroundings. Two limiting cases of particular interest are for isothermal and adiabatic conditions.

Unlike in the case of an orifice or nozzle, the pipeline maintains the area of flow constant and equal to its cross-sectional area. There is no possibility therefore of the gas expanding laterally. Supersonic flow conditions can be reached in pipeline installations in a manner similar to that encountered in flow through a nozzle, but *not within the pipe itself* unless the gas enters the pipe at a supersonic velocity. If a pipe connects two reservoirs and the upstream reservoir is maintained at constant pressure P_1, the following pattern will occur as the pressure P_2 in the downstream reservoir is reduced.

(1) Starting with $P_2 = P_1$ there is, of course, no flow and $G = 0$.

(2) Reduction of P_2 initially results in an increase in G. G increases until the gas velocity at the outlet of the pipe reaches the velocity of propagation of a pressure wave ('sonic velocity'). This value of P_2 will be denoted as P_w.

(3) Further reduction of P_2 has no effect on the flow in the pipeline. The pressure distribution along the length of the pipe remains unaltered and the pressure at its outlet remains at P_w. The gas, on leaving the pipe, expands laterally and its pressure falls to the reservoir pressure P_2.

In considering the flow in a pipe, the differential form of the general energy balance equation (2.54) is used, and the friction term δF will be written in terms of the energy dissipated per unit mass of fluid for flow through a length dl of pipe. In the first instance, isothermal flow of an ideal gas is considered and the flowrate is expressed as a function of the upstream and downstream pressures. Nonisothermal and adiabatic flow are discussed later.

4.5.1 Energy Balance for Flow of Ideal Gas

In Chapter 2, the general energy equation for the flow of a fluid through a pipe has been expressed in the form:

$$\frac{u\mathrm{d}u}{\alpha}+g\mathrm{d}z+v\mathrm{d}P+\delta W_s+\delta F=0 \quad \text{(Eq. 2.54)}$$

For a fluid flowing through a length dl of pipe of constant cross-sectional area A:

$$\delta W_s=0$$

$$\delta F=4\left(\frac{R}{\rho u^2}\right)u^2\frac{\mathrm{d}l}{d} \quad \text{(from Eq. 3.19)}$$

$$\therefore \quad \frac{u\mathrm{d}u}{\alpha}+g\mathrm{d}z+v\mathrm{d}P+4\frac{R}{\rho u^2}u^2\frac{\mathrm{d}l}{d}=0 \tag{4.48}$$

This equation cannot be integrated directly because the velocity u increases as the pressure falls and is, therefore, a function of l (Fig. 4.7). It is therefore, convenient to work in terms of the mass flow G which remains constant throughout the length of pipe.

The velocity,

$$u=\frac{Gv}{A} \quad \text{(from Eq. 2.36)}$$

and hence:

$$\frac{1}{\alpha}\left(\frac{G}{A}\right)^2 v\mathrm{d}v+g\mathrm{d}z+v\mathrm{d}P+4\left(\frac{R}{\rho u^2}\right)\left(\frac{G}{A}\right)^2 v^2\frac{\mathrm{d}l}{d}=0 \tag{4.49}$$

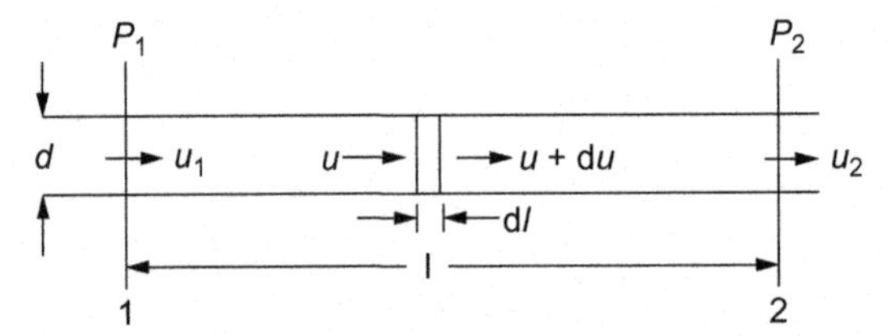

Fig. 4.7
Flow of compressible fluid through a pipe.

For turbulent flow, which is usual for a gas, $\alpha=1$.

Then, for flow in a horizontal pipe ($dz=0$):

$$\left(\frac{G}{A}\right)^2 v\,dv + v\,dP + 4\left(\frac{R}{\rho u^2}\right)\left(\frac{G}{A}\right)^2 v^2\frac{dl}{d}=0 \tag{4.50}$$

Dividing by v^2:

$$\left(\frac{G}{A}\right)^2 \frac{dv}{v} + \frac{dP}{v} + 4\left(\frac{R}{\rho u^2}\right)\left(\frac{G}{A}\right)^2 \frac{dl}{d}=0 \tag{4.51}$$

Now the friction factor $R/\rho u^2$ is a function of the Reynolds number Re and the relative roughness e/d of the pipe surface which will normally be constant along a given pipe.

The Reynolds number is given by:

$$Re=\frac{ud\rho}{\mu}=\frac{Gd}{A\mu}=\frac{4G}{\pi d\mu} \tag{4.52}$$

Since G is constant over the length of the pipe, Re varies only as a result of changes in the viscosity μ. Although μ is a function of temperature, and to some extent of pressure, it is not likely to vary widely over the length of the pipe. Furthermore, the friction factor $R/\rho u^2$ is only a weak function of the Reynolds number when Re is high, and little error will therefore arise by regarding it as constant for a fixed value of (e/d).

Thus, integrating Eq. (4.51) over a length l of pipe:

$$\left(\frac{G}{A}\right)^2 \ln\frac{v_2}{v_1} + \int_{P_1}^{P_2}\frac{dP}{v} + 4\left(\frac{R}{\rho u^2}\right)\frac{l}{d}\left(\frac{G}{A}\right)^2=0 \tag{4.53}$$

The integral term in Eq. (4.53) will depend on the P–v relationship during the expansion of the gas in the pipe, and several cases are now considered.

4.5.2 Isothermal Flow of an Ideal Gas in a Horizontal Pipe

For isothermal changes in an ideal gas:

$$\int_{P_1}^{P_2}\frac{dP}{v}=\frac{1}{P_1v_1}\int_{P_1}^{P_2}P\,dP=\frac{P_2^2-P_1^2}{2P_1v_1} \tag{4.54}$$

and, therefore, substituting in Eq. (4.53):

$$\left(\frac{G}{A}\right)^2 \ln\frac{P_1}{P_2} + \frac{P_2^2-P_1^2}{2P_1v_1} + 4\left(\frac{R}{\rho u^2}\right)\frac{l}{d}\left(\frac{G}{A}\right)^2=0 \tag{4.55}$$

Since v_m, the specific volume at the mean pressure in the pipe, is given by:

$$\frac{P_1+P_2}{2}v_m=P_1v_1$$

and:

$$\left(\frac{G}{A}\right)^2\ln\frac{P_1}{P_2}+(P_2-P_1)\frac{1}{v_m}+4\left(\frac{R}{\rho u^2}\right)\frac{l}{d}\left(\frac{G}{A}\right)^2=0 \tag{4.56}$$

If the pressure drop in the pipe is a small proportion of the inlet pressure, the first term is negligible and the fluid may be treated as an incompressible fluid at the mean pressure in the pipe.

It is sometimes convenient to substitute $\mathbf{R}T/M$ for P_1v_1 in Eq. (4.55) to give:

$$\left(\frac{G}{A}\right)^2\ln\frac{P_1}{P_2}+\frac{P_2^2-P_1^2}{2\mathbf{R}T/M}+4\left(\frac{R}{\rho u^2}\right)\frac{l}{d}\left(\frac{G}{A}\right)^2=0 \tag{4.57}$$

Eqs (4.55), (4.57) are the most convenient forms for the calculation of gas flowrate as a function of P_1 and P_2 under isothermal conditions. Some additional refinement can be added if a compressibility factor is introduced as defined by the relation $P_v=Z\mathbf{R}T/M$, for conditions where there are significant deviations from the *ideal gas law* (Eq. 2.15).

Maximum flow conditions

Eq. (4.55) expresses G as a continuous function of P_2, the pressure at the downstream end of the pipe for a given upstream pressure P_1. If there is no pressure change over the pipe then, $P_2=P_1$, and substitution of P_1 for P_2 in Eq. (4.55) gives $G=0$, as would be expected. For some intermediate value of P_2 ($=P_w$, say), where $0<P_w<P_1$, the flowrate G is a maximum.

Multiplying Eq. (4.55) by $(A/G)^2$:

$$-\ln\left(\frac{P_2}{P_1}\right)+\left(\frac{A}{G}\right)^2\frac{(P_2^2-P_1^2)}{2P_1v_1}+4\left(\frac{R}{\rho u^2}\right)\frac{l}{d}=0 \tag{4.58}$$

Differentiating with respect to P_2, for a constant value of P_1:

$$-\frac{P_1}{P_2}\frac{1}{P_1}+\left(\frac{A}{G}\right)^2\frac{2P_2}{2P_1v_1}+\frac{A^2}{2P_1v_1}(P_2^2-P_1^2)\frac{-2\,\mathrm{d}G}{G^3\,\mathrm{d}P_2}=0$$

The rate of flow is maximum when $\mathrm{d}G/\mathrm{d}P_2=0$. Denoting conditions at the downstream end of the pipe by suffix w, when the flow is a maximum:

$$\frac{1}{P_w}=\left(\frac{A}{G}\right)^2\frac{P_w}{P_1v_1}$$

or:

$$\left(\frac{G}{A}\right)^2 = \frac{P_w^2}{P_1 v_1} = \frac{P_w^2}{P_w v_w} = \frac{P_w}{v_w} \tag{4.59}$$

or:

$$u_w = \sqrt{P_w v_w} \tag{4.60}$$

It has been shown in Eq. (4.35) that this velocity v_w is equal to the velocity of transmission of a small pressure wave in the fluid at the pressure P_w if heat could be transferred rapidly enough to maintain isothermal conditions. If the pressure at the downstream end of the pipe were P_w, the fluid there would then be moving with the velocity of a pressure wave, and therefore a wave would not be transmitted through the fluid in the opposite direction because its velocity relative to the pipe would be zero. If, at the downstream end, the pipe were to be connected to a reservoir in which the pressure was reduced below P_w, the flow conditions within the pipe would be unaffected and the pressure at the exit of the pipe would remain at the value P_w as shown in Fig. 4.8. The drop in pressure from P_w to P_2 would then take place by virtue of lateral expansion of the gas beyond the end of the pipe. If the pressure P_2 in the reservoir at the downstream end were gradually reduced from P_1, the rate of flow would increase until the pressure reached P_w and then remain constant at this maximum value as the pressure was further reduced.

Thus, with compressible flow there is a maximum mass flowrate G_w which can be attained by the gas for a given upstream pressure P_1, and further reduction in pressure in the downstream reservoir below P_w will not give any further increase in G.

The maximum flowrate G_w is given by Eq. (4.59):

$$G_w = A\sqrt{\frac{P_w}{v_w}}$$

Substituting $v_w = v_1(P_1/P_w)$:

$$G_w = AP_w\sqrt{\frac{1}{P_1 v_1}} \tag{4.61}$$

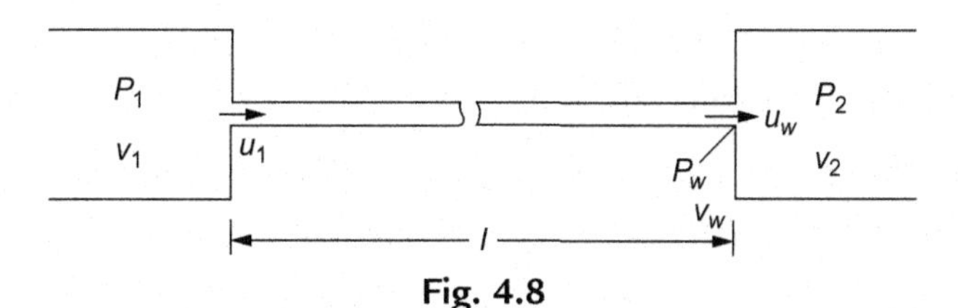

Fig. 4.8
Maximum flow conditions.

P_w is given by substituting in Eq. (4.58), that is:

$$\ln\left(\frac{P_1}{P_w}\right)+\frac{v_w}{P_w}\frac{1}{P_w v_w}\frac{P_w^2-P_1^2}{2}+4\left(\frac{R}{\rho u^2}\right)\frac{l}{d}=0$$

or:

$$\ln\left(\frac{P_1}{P_w}\right)^2+1-\left(\frac{P_1}{P_w}\right)^2+8\left(\frac{R}{\rho u^2}\right)\frac{l}{d}=0 \quad (4.62)$$

or:

$$8\left(\frac{R}{\rho u^2}\right)\frac{l}{d}=\left(\frac{1}{w_c}\right)^2-\ln\left(\frac{1}{w_c}\right)^2-1 \quad (4.63)$$

where $w_c=P_w/P_1$ (the critical value of the pressure ratio $w=(P_2/P_1)$).

Eqs (4.62), (4.63) are very important in that they give:

(a) The maximum value of the pressure ratio P_1/P_2 ($=P_1/P_w$) for which the whole of the expansion of the gas can take place in the pipe.
If $P_1/P_2>P_1/P_w$, the gas expands from P_1 to P_w in the pipe, and from P_w to P_2 in the downstream reservoir. Eqs (4.55)–(4.57) may be used to calculate what is happening in the pipe only provided $P_1/P_2 \ngtr P_1/P_w$ as given by Eq. (4.62).
(b) The minimum value of $8(R/\rho u^2)(l/d)$ for which, for any pressure ratio P_1/P_2, the fall in gas pressure will take place entirely within the pipe.

In Table 4.1, the relation between P_1/P_w and $8(R/\rho u^2)(l/d)$ is given for a series of numerical values of w_c.

It is seen that P_1/P_w increases with $8(R/\rho u^2)(l/d)$; in other words, in general, the longer the pipe (assuming $R/\rho u^2 \asymp$ constant) the greater is the ratio of the upstream to downstream pressure P_1/P_2 which can be accommodated in the pipe itself.

For an assumed average pipe friction factor $(R/\rho u^2)$ of 0.0015, this has been expressed as an l/d ratio. For a 25-mm diameter pipeline, the corresponding pipelengths are given in metres. In addition, for an upstream pressure P_1 of 100 bar, the *average* pressure gradient in the 25-mm pipe is given in the last column. At first sight it might seem strange that limiting conditions should be reached in such very short pipes for values of $P_1/P_w \asymp 1$, but it will be seen that the *average pressure gradients* then become very high.

In Fig. 4.9, values of P_1/P_w and w_c are plotted against $8(R/\rho u^2)(l/d)$. This curve gives the limiting value of P_1/P_2 for which the whole expansion of the gas can take place within the pipe.

Table 4.1 Limiting pressure ratios for pipe flow

					For $\frac{R}{\rho u}=0.0015$	For $d=25$ mm	For $P_1=100$ bar	
$\frac{P_1}{P_w}$	$\frac{P_w}{P_1}$	$\left(\frac{1}{w_c}\right)^2$	$\ln\left(\frac{1}{w_c}\right)^2$	$8\frac{R}{\rho u^2}(l/d)$	$\frac{l}{d}$	l(m)	P_1-P_w (bar)	Average $\frac{P_1-P_w}{l}$ (bar/m)
1	1	1	0	0	0	0	0	–
1.1	0.9091	1.21	0.1906	0.0194	1.62	0.0404	9.1	225
1.78	0.562	3.16	1.151	1.01	84.2	2.11	43.8	20.8
3.16	0.316	10	2.302	6.70	558	14.0	68.4	4.89
5.62	0.178	31.6	3.453	27.1	2258	56.5	82.2	1.45
10	0.1	100	4.605	94.4	7867	197	90	0.457
17.8	0.0562	316	5.756	309.2	25,770	644	94.4	0.146
31.6	0.0316	1000	6.901	992	82,700	2066	96.8	0.046
56.2	0.0178	3160	8.058	3151	263,000	6575	98.2	0.015
100	0.01	10,000	9.210	9990	832,500	20,800	99	0.005

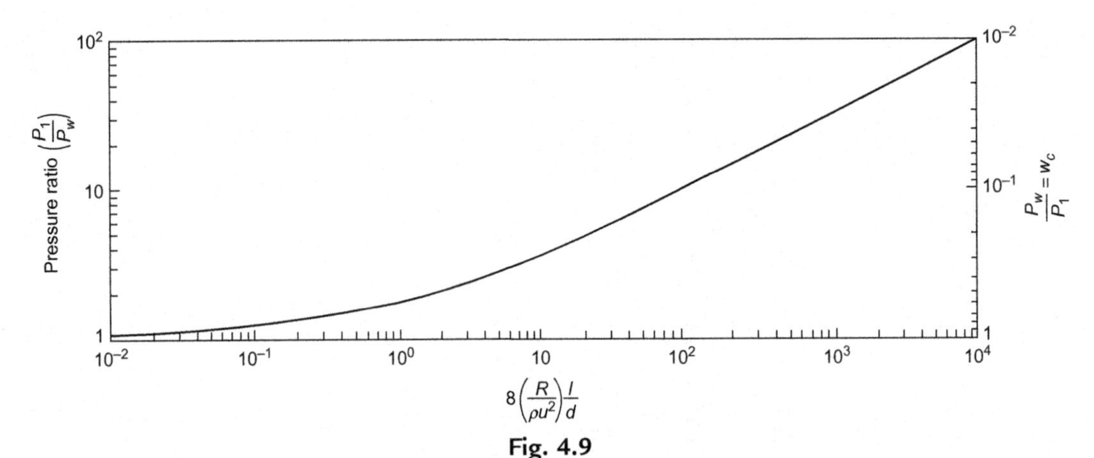

Fig. 4.9
Critical pressure ratio P_1/P_w (maximum value of P_1/P_2 which can occur in a pipe) as function of $8(R/\rho u^2)(l/d)$.

It is seen in Table 4.1, for instance, that for a 25-mm pipeline and an assumed value of 0.0015 for $R/\rho u^2$ that for a pipe of length 14 m, the ratio of the pressure at the pipe inlet to that at the outlet cannot exceed 3.16 ($w_c = 0.316$). If $P_1/P_2 > 3.16$, the gas expands to the pressure P_w ($= 0.316P_1$) inside the pipe and then it expands down to the pressure P_2 *within the downstream reservoir*.

Flow with fixed upstream pressure and variable downstream pressure

As an example, the flow of air at 293 K in a pipe of 25 mm diameter and length 14 m is considered, using the value of 0.0015 for $R/\rho u^2$ employed in the calculation of the figures in Table 4.1; $R/\rho u^2$ will, of course, show some variation with Reynolds number, but this effect will be neglected in the following calculation. The variation in flowrate G is examined, for a given upstream pressure of 10 MN/m^2, as a function of downstream pressure P_2. As the critical value of P_1/P_2 for this case is 3.16 (see Table 4.1), the maximum flowrate will occur at all values of P_2 less than $10/3.16 = 3.16$ MN/m^2. For values of P_2 greater than 3.16 MN/m^2, Eq. (4.57) applies:

$$\left(\frac{G}{A}\right)^2 \ln\frac{P_1}{P_2} + \frac{P_2^2 - P_1^2}{2\mathbf{R}T/M} + 4\left(\frac{R}{\rho u^2}\right)\frac{l}{d}\left(\frac{G}{A}\right)^2 = 0$$

Multiplying by $(A/G)^2$ and inserting the numerical values with pressures P_1 and P_2 expressed in MN/m^2 (P_{1a} and P_{2a}):

$$\ln\frac{P_{1a}}{P_{2a}} - 5.95 \times 10^6 \left(P_{1a}^2 - P_{2a}^2\right)\left(\frac{A}{G}\right)^2 + 3.35 = 0$$

For $P_{1a} = 10$ MN/m^2:

$$2.306 - \ln P_{2a} - 5.95 \times 10^6 \left(100 - P_{2a}^2\right) \left(\frac{A}{G}\right)^2 + 3.35 = 0$$

giving:

$$\left(\frac{G}{A}\right)^2 = \frac{5.95 \times 10^6 \left(100 - P_{2a}^2\right)}{5.656 - \ln P_{2a}} \quad (\text{kg}^2/\text{m}^4\text{s}^2)$$

Values of G calculated from this relation are given in Table 4.2 and plotted in Fig. 4.10.

In the last column of Table 4.2, values are given for G^*/A calculated by ignoring the effect of changes in the kinetic energy of the gas, that is by neglecting the log term in Eq. (4.57). It will be noted that the effects are small ranging from about 2% for $P_{2a} = 9$ MN/m^2 to about 13% for $P_{2a} = 4$ MN/m^2.

Values of G_{max} for the maximum flow conditions can also be calculated from Eq. (4.61):

$$G_{max} = AP_w \sqrt{\frac{1}{P_1 v_1}}$$

Substitution of the numerical values gives:

$G_{max} = 5.4$ kg/s, which is in agreement with the value in Table 4.2.

For $G/A = 10^4$ kg/m^2 s, the Reynolds number $Re \approx \frac{\left(10^4 \times 25 \times 10^{-3}\right)}{10^{-5}} \approx 2.5 \times 10^7$

The chosen value of 0.0015 for $R/\rho u^2$ is reasonable for a smooth pipe over most of the range of Reynolds numbers encountered in this exercise.

Table 4.2 Mass flowrate as function of downstream pressure P_{2a} for constant upstream pressure of 10 MN/m^2

Pressure P_{2a} (MN/m^2)	$\frac{G}{A} \times 10^{-4}$ (kg/m^2 s)	G (kg/s)	$\frac{G^*}{A} \times 10^{-4}$ (kg/m^2 s)
10	0	0	0
9.5	0.41	2.0	0.42
9	0.57	2.8	0.58
8	0.78	3.8	0.80
7	0.90	4.4	0.95
6	0.99	4.9	1.07
5	1.05	5.2	1.15
4	1.08	5.3	1.22
<3.16	1.09	5.4	1.26

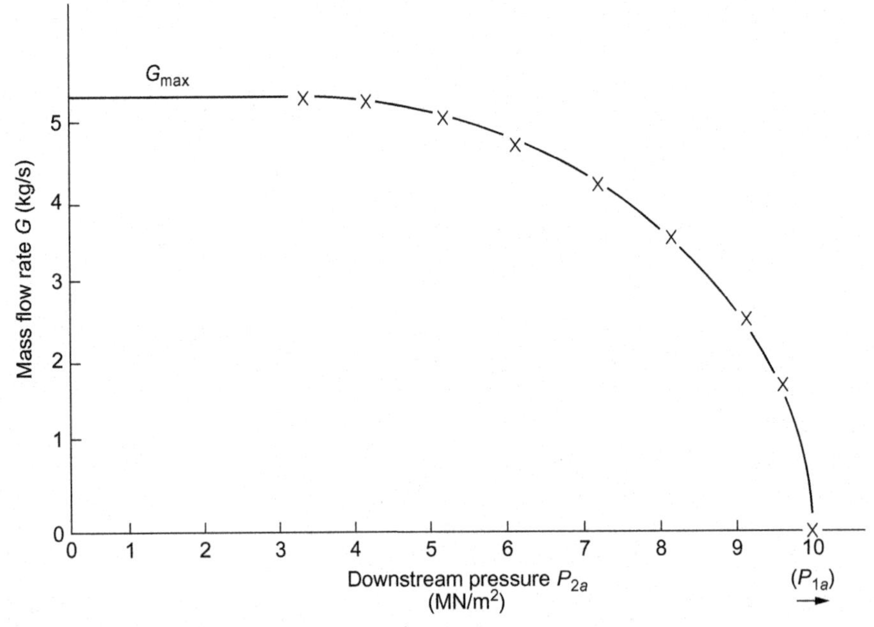

Fig. 4.10
Mass flowrate as function of downstream pressure for isothermal flow of air in pipe (l=14 m, d=25 mm).

Example 4.2

Over a 30 m length of a 150 mm vacuum line carrying air at 295 K, the pressure falls from 0.4 to 0.13 kN/m^2. If the relative roughness of the pipe e/d is 0.003 what is the approximate flowrate? It may be noted that the flow of gases in vacuum systems is discussed fully by Griffiths.[1]

Solution

Since the friction factor is not known, an iterative solution is required. Using $(R/gu^2)=0.004$, we can now use Eq. (4.55):

$$\left(\frac{G}{A}\right)^2 \ln\frac{P_1}{P_2} + \frac{P_2^2 - P_1^2}{2P_1v_1} + 4\left(\frac{R}{\rho u^2}\right)\frac{l}{d}\left(\frac{G}{A}\right)^2 = 0 \quad \text{(Eq. 4.55)}$$

$$\left(\frac{G}{A}\right)^2 ln\frac{0.4}{0.13} + \frac{(0.13^2 - 0.40^2)\times 10^6}{2\times(8314\times 295/29)} + 4\times 0.004\times\frac{30}{0.150}\times\left(\frac{G}{A}\right)^2 = 0$$

$$1.124\left(\frac{G}{A}\right)^2 - 0.846 + 3.2\left(\frac{G}{A}\right)^2 = 0$$

$$\frac{G}{A} = 0.44\,\text{kg/m}^2\,\text{s}$$

The viscosity of air at 295 K = 1.8×10^{-5} N s/m^2

Reynolds number, $Re = \dfrac{0.44\times 0.15}{1.8\times 10^{-5}} = 3670$ (from Eq. 4.52)

From Fig. 3.7, $\dfrac{R}{\rho u^2} = 0.005$

Inserting new values:

$$1.124\left(\frac{G}{A}\right)^2 - 0.846 + 4.0\left(\frac{G}{A}\right)^2 = 0$$

and:

$$\frac{G}{A} = \underline{\underline{0.41\,\text{kg/m}^2\,\text{s}}}$$

A check is made that this flowrate is possible.

$$8\frac{R}{\rho u^2}\frac{l}{d} = \frac{8 \times 0.005 \times 30}{0.15} = 8.0$$

From Fig. 4.9:

$$\frac{P_1}{P_w} = 3.5$$

Actual value:

$$\frac{P_1}{P_2} = \left(\frac{0.4}{0.13}\right) = 3.1$$

Thus, the flow condition just falls within the possible range.

Example 4.3

A flow of 50 m^3/s methane, measured at 288 K and 101.3 kN/m^2, has to be delivered along a 0.6 m diameter line, 3.0 km long with a relative roughness of 0.0001, linking a compressor and a processing unit. The methane is to be discharged at the plant at 288 K and 170 kN/m^2 and it leaves the compressor at 297 K. What pressure must be developed at the compressor in order to achieve this flowrate?

Solution

Taking the mean temperature of $(288 + 297)/2 = 293$ K and using the ideal gas law,

$$P_1 v_1 = \frac{\mathbf{R}T}{M} = \frac{8314 \times 293}{16} = 1.5225 \times 10^5\,\text{m}^2/\text{s}^2$$

At 288 K and 101.3 kN/m^2,

$$v = \frac{1.5225 \times 10^5}{1.013 \times 10^5} \times \frac{288}{293} = 1.477\,\text{m}^3/\text{kg}$$

Mass flowrate of methane,

$$G = \frac{50}{1.497} = 33.85\,\text{kg/s}$$

Cross-sectional area of pipe

$$A=\frac{\pi}{4}\times(0.6)^2=0.283\,\text{m}^2$$

$$\frac{G}{A}=\frac{33.85}{0.283}=119.6\,\text{kg/m}^2\,\text{s}$$

$$\left(\frac{G}{A}\right)^2=1.431\times10^4\,(\text{kg/m}^2\,\text{s})^2$$

Viscosity of methane at 293 K$\asymp 0.01\times10^{-3}$ N s/m^2

$$\text{Reynolds number } Re=\frac{119.6\times0.6}{10^{-5}}=7.18\times10^6$$

$$\text{For } \frac{e}{d}=0.0001,\ \frac{R}{\rho u^2}=0.0015\ \text{(from Fig. 3.7)}$$

The upstream pressure is calculated using Eq. (4.55):

$$\left(\frac{G}{A}\right)^2\ln\frac{P_1}{P_2}+\frac{P_2^2-P_1^2}{2P_1v_1}+4\left(\frac{R}{\rho u^2}\right)\frac{l}{d}\left(\frac{G}{A}\right)^2=0$$

Substituting values:

$$1.431\times10^4\ \ln\frac{P_1}{1.7\times10^5}-\frac{P_1^2-(1.7\times10^5)^2}{2\times1.5225\times10^5}+4\times0.0015\times\frac{3000}{0.6}\times1.431\times10^4=0$$

Dividing by 1.431×10^4 gives:

$$\ln P_1-12.04-2.29\times10^{-1}P_1^2+6.63+30.00=0$$

$$2.29\times10^{-10}P_1^2-\ \ln P_1=24.59$$

A trial and error solution gives: $P_1=4.05\times10^5\,\text{N/m}^2$.

It is necessary to check that this degree of expansion is possible *within* pipe

$$\frac{P_1}{P_2}=\frac{4.05\times10^5}{1.7\times10^5}=2.38$$

By reference to Fig. 4.9:

$$8\frac{R}{\rho u^2}\frac{l}{d}=60\ \text{ and }\ \frac{P_1}{P_w}=8.1$$

Thus, the pressure ratio is within the possible range.

Heat flow required to maintain isothermal conditions

As the pressure in a pipe falls, the kinetic energy of the fluid increases at the expense of the internal energy and the temperature tends to fall. The maintenance of isothermal conditions therefore depends on the transfer of an adequate amount of heat from the surroundings. For a small change in the system, the energy balance is given in Chapter 2 as:

$$\delta q - \delta W_S = dH + g dz + u du \quad \text{(from Eq. 2.51)}$$

For a horizontal pipe, $dz = 0$, and for isothermal expansion of an ideal gas $dH = 0$. Thus if the system does no work on the surroundings:

$$\delta q = u du \tag{4.64}$$

and the required transfer of heat (in mechanical energy units) per unit mass is $\Delta u^2/2$. Thus the amount of heat required is equivalent to the increase in the kinetic energy of the fluid. If the mass rate of flow is G, the total heat to be transferred per unit time is $G\Delta u^2/2$.

In cases where the change in the kinetic energy is small, the required flow of heat is correspondingly small, and conditions are almost adiabatic.

4.5.3 Nonisothermal Flow of an Ideal Gas in a Horizontal Pipe

In general, when an ideal gas expands or is compressed, the relation between the pressure P and the specific volume v can be represented approximately as:

$$Pv^k = \text{a constant} = P_1 v_1^k$$

where k will depend on the heat transfer to the surroundings.

Evaluation of the integral gives:

$$\int_{P_1}^{P_2} \frac{dP}{v} = \frac{k}{k+1} \frac{P_1}{v_1} \left[\left(\frac{P_2}{P_1} \right)^{(k+1)/k} - 1 \right] \tag{4.65}$$

Inserting this value in Eq. (4.53):

$$\left(\frac{G}{A}\right)^2 \frac{1}{k} \ln\left(\frac{P_1}{P_2}\right) + \frac{k}{k+1} \frac{P_1}{v_1} \left[\left(\frac{P_2}{P_1} \right)^{(k+1)/k} - 1 \right] + 4\left(\frac{R}{\rho u^2}\right) \frac{l}{d} \left(\frac{G}{A}\right)^2 = 0 \tag{4.66}$$

For a given upstream pressure P_1, the maximum flow rate occurs when $u_2 = \sqrt{kP_2 v_2}$, the velocity of transmission of a pressure wave under these conditions (Eq. 4.38). Flow under adiabatic conditions is considered in detail in the next section, although an approximate result may be obtained by putting k equal to γ in Eq. (4.66); this is only approximate because equating k to γ implies reversibility.

4.5.4 Adiabatic Flow of an Ideal Gas in a Horizontal Pipe

The conditions existing during the adiabatic flow in a pipe may be calculated using the approximate expression $Pv^k = $ a constant to give the relation between the pressure and the specific volume of the fluid. In general, however, the value of the index k may not be known for an irreversible adiabatic process. An alternative approach to the problem is therefore desirable.[2,3]

For a fluid flowing under turbulent conditions in a pipe, $\delta W_s = 0$ and:

$$\delta q = \mathrm{d}H + g\,\mathrm{d}z + u\,\mathrm{d}u \quad \text{(from Eq. 2.51)}$$

In an adiabatic process, $\delta q = 0$, and the equation may then be written for the flow in a pipe of constant cross-sectional area A to give:

$$\left(\frac{G}{A}\right)^2 v\,\mathrm{d}v + g\,\mathrm{d}z + \mathrm{d}H = 0 \tag{4.67}$$

Now:

$$\begin{aligned} \mathrm{d}H &= \mathrm{d}U + \mathrm{d}(Pv) \\ &= C_v\,\mathrm{d}T + \mathrm{d}(Pv) \end{aligned}$$

for an ideal gas (from Eq. 2.24)

Further:

$$C_p\,\mathrm{d}T = C_v\,\mathrm{d}T + \mathrm{d}(Pv)$$

for an ideal gas (from Eq. 2.26)

$$\therefore\ \mathrm{d}T = \frac{\mathrm{d}(Pv)}{C_p - C_v} \tag{4.68}$$

so that:

$$\mathrm{d}H = \mathrm{d}(Pv)\left(\frac{C_v}{C_p - C_v} + 1\right) \tag{4.69}$$

$$= \frac{\gamma}{\gamma - 1}\mathrm{d}(Pv) \tag{4.70}$$

Substituting this value of $\mathrm{d}H$ in Eq. (4.67) and writing $g\ \mathrm{d}z = 0$ for a horizontal pipe:

$$\left(\frac{G}{A}\right)^2 v\,\mathrm{d}v + \frac{\gamma}{\gamma - 1}\mathrm{d}(Pv) = 0 \tag{4.71}$$

Integrating, a relation between P and v for adiabatic flow in a horizontal pipe is obtained:

$$\frac{1}{2}\left(\frac{G}{A}\right)^2 v^2 + \frac{\gamma}{\gamma - 1}Pv = \frac{1}{2}\left(\frac{G}{A}\right)^2 v_1^2 + \frac{\gamma}{\gamma - 1}P_1 v_1 = \text{constant, } K \text{ (say)} \tag{4.72}$$

From Eq. (4.72):

$$P = \frac{\gamma - 1}{\gamma}\left[\frac{K}{v} - \frac{1}{2}\left(\frac{G}{A}\right)^2 v\right] \tag{4.73}$$

$$\mathrm{d}P = \frac{\gamma - 1}{\gamma}\left[-\frac{K}{v^2} - \frac{1}{2}\left(\frac{G}{A}\right)^2\right]\mathrm{d}v$$

$$\frac{\mathrm{d}P}{v} = \frac{\gamma - 1}{\gamma}\left[-\frac{K}{v^3} - \frac{1}{2}\left(\frac{G}{A}\right)^2\frac{1}{v}\right]\mathrm{d}v \tag{4.74}$$

$$\int_{P_1}^{P_2}\frac{\mathrm{d}P}{v} = \frac{\gamma - 1}{\gamma}\left[\frac{K}{2}\left(\frac{1}{v_2^2} - \frac{1}{v_1^2}\right) - \frac{1}{2}\left(\frac{G}{A}\right)^2 \ln\frac{v_2}{v_1}\right] \tag{4.75}$$

Substituting for K from Eq. (4.72):

$$\begin{aligned}\int_{P_1}^{P_2}\frac{\mathrm{d}P}{v} &= \frac{\gamma - 1}{\gamma}\left[\left(\frac{G}{A}\right)^2\frac{v_1^2}{4}\left(\frac{1}{v_2^2} - \frac{1}{v_1^2}\right) + \frac{\gamma P_1 v_1}{2(\gamma - 1)}\left(\frac{1}{v_2^2} - \frac{1}{v_1^2}\right) - \frac{1}{2}\left(\frac{G}{A}\right)^2 \ln\frac{v_2}{v_1}\right]\\ &= \frac{\gamma - 1}{4\gamma}\left(\frac{G}{A}\right)^2\left(\frac{v_1^2}{v_2^2} - 1 - 2\ln\frac{v_2}{v_1}\right) + \frac{P_1 v_1}{2}\left(\frac{1}{v_2^2} - \frac{1}{v_1^2}\right)\end{aligned} \tag{4.76}$$

Inserting the value of $\int_{P_1}^{P_2}\mathrm{d}P/v$ from Eq. (4.76) into Eq. (4.53):

$$\left(\frac{G}{A}\right)^2 \ln\frac{v_2}{v_1} + \frac{\gamma - 1}{4\gamma}\left(\frac{G}{A}\right)^2\left(\frac{v_1^2}{v_2^2} - 1 - 2\ln\frac{v_2}{v_1}\right) + \frac{P_1 v_1}{2}\left(\frac{1}{v_2^2} - \frac{1}{v_1^2}\right) + 4\left(\frac{R}{\rho u^2}\right)\frac{l}{d}\left(\frac{G}{A}\right)^2 = 0$$

Simplifying:

$$8\left(\frac{R}{\rho u^2}\right)\frac{l}{d} = \left[\frac{\gamma - 1}{2\gamma} + \frac{P_1}{v_1}\left(\frac{A}{G}\right)^2\right]\left[1 - \left(\frac{v_1}{v_2}\right)^2\right] - \frac{\gamma + 1}{\gamma}\ln\frac{v_2}{v_1} \tag{4.77}$$

This expression enables v_2, the specific volume at the downstream end of the pipe, to be calculated for the fluid flowing at a mass rate G from an upstream pressure P_1.

Alternatively, the mass rate of flow G may be calculated in terms of the specific volume of the fluid at the two pressures P_1 and P_2.

The pressure P_2 at the downstream end of the pipe is obtained by substituting the value of v_2 in Eq. (4.72).

For constant upstream conditions, the maximum flow through the pipe is found by differentiating with respect to v_2 and putting (dG/dv_2) as equal to zero. The maximum flow is thus shown to occur when the velocity at the downstream end of the pipe is the sonic velocity $\sqrt{\gamma P_2 v_2}$ (Eq. 4.37).

The rate of flow of gas under adiabatic conditions is never more than 20% greater than that obtained for the same pressure difference with isothermal conditions. For pipes of length at least 1000 diameters, the difference does not exceed about 5%. In practice the rate of flow may be limited, not by the conditions in the pipe itself, but by the development of sonic velocity at some valve or other constriction in the pipe. Care should, therefore be taken in the selection of fittings for pipes conveying gases at high velocities.

Analysis of conditions for maximum flow

It will now be shown from purely thermodynamic considerations that for adiabatic conditions, supersonic flow cannot develop in a pipe of constant cross-sectional area because the fluid is in a condition of maximum entropy when flowing at the sonic velocity. The condition of the gas at any point in the pipe where the pressure is P is given by the equations:

$$Pv = \frac{1}{M}\mathbf{R}T \quad \text{(Eq. 2.16)}$$

and:

$$\frac{\gamma}{\gamma-1}Pv + \frac{1}{2}\left(\frac{G}{A}\right)^2 v^2 = K \quad \text{(Eq. 4.72)}$$

It may be noted that if the changes in the kinetic energy of the fluid are small, the process is almost isothermal.

Eliminating v, an expression for T is obtained as:

$$\frac{\gamma}{\gamma-1}\frac{\mathbf{R}T}{M} + \frac{1}{2}\left(\frac{G}{A}\right)^2 \frac{\mathbf{R}^2T^2}{P^2M^2} = K \tag{4.78}$$

The corresponding value of the entropy is now obtained:

$$dH = TdS + vdP = C_p dT \text{ for an ideal gas (from Eqs 2.28 and 2.26)}$$

$$\therefore \quad dS = C_p\frac{dT}{T} - \frac{\mathbf{R}}{MP}dP$$

$$\therefore \quad S = C_p \ln\frac{T}{T_0} - \frac{\mathbf{R}}{M}\ln\frac{P}{P_0} \quad \text{(if } C_p \text{ is constant)} \tag{4.79}$$

where T_0, P_0 represents the datum condition of the gas at which the entropy is arbitrarily taken as zero.

The temperature or enthalpy of the gas may then be plotted to a base of entropy to give a *Fanno line*.[4] This line shows the condition of the fluid as it flows along the pipe. If the velocity at entrance is subsonic (the normal condition), then the enthalpy will decrease along the pipe and the velocity will increase until sonic velocity is reached. If the flow is supersonic at the entrance, the velocity will decrease along the duct until it becomes sonic. The entropy has a maximum value corresponding to sonic velocity as shown in Fig. 4.11. (Mach number $Ma < 1$ represents subsonic conditions; $Ma > 1$ supersonic.)

Fanno lines are also useful in presenting conditions in nozzles, turbines, and other units where supersonic flow arises.[5]

For small changes in pressure and entropy, the kinetic energy of the gas increases only very slowly, and therefore the temperature remains almost constant. As the pressure is further reduced, the kinetic energy changes become important and the rate of fall of temperature increases and eventually $\mathrm{d}T/\mathrm{d}S$ becomes infinite. Any further reduction of the pressure would cause a decrease in the entropy of the fluid and is, therefore, impossible.

The condition of maximum entropy occurs when $\mathrm{d}S/\mathrm{d}T = 0$, where:

$$\frac{\mathrm{d}S}{\mathrm{d}T} = \frac{C_p}{T} - \frac{\mathbf{R}}{MP}\frac{\mathrm{d}P}{\mathrm{d}T} \quad \text{(from Eq. 4.79)}$$

The entropy is, therefore, a maximum when:

$$\frac{\mathrm{d}P}{\mathrm{d}T} = \frac{MPC_p}{\mathbf{R}T} \tag{4.80}$$

For an ideal gas:

$$C_p - C_v = \frac{\mathbf{R}}{M} \quad \text{(Eq. 2.27)}$$

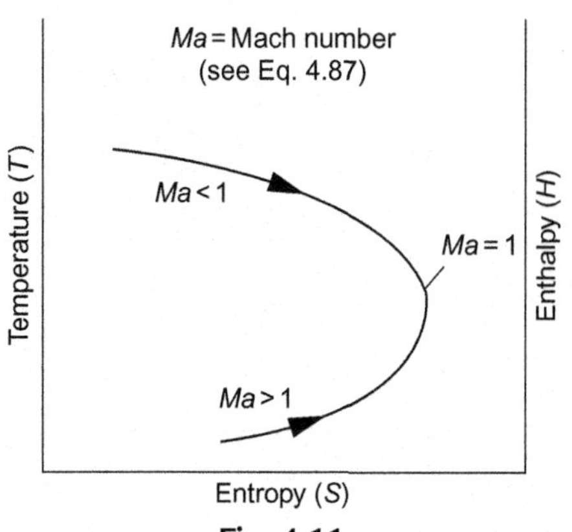

Fig. 4.11
The Fanno line.

Substituting in Eq. (4.80):

$$\frac{dP}{dT}=\frac{PC_v}{(C_p-C_v)T}=\frac{P}{T}\frac{\gamma}{\gamma-1} \tag{4.81}$$

The general value of dP/dT may be obtained by differentiating Eq. (4.78) with respect to T, or:

$$\frac{\mathbf{R}}{M}+\frac{\gamma-1}{2\gamma}\left(\frac{G}{A}\right)^2\frac{\mathbf{R}^2}{M^2}\left(\frac{P^2 2T-T^2 2P(dP/dT)}{P^4}\right)=0$$

$$\therefore\quad 1+\frac{\gamma-1}{\gamma}\left(\frac{G}{A}\right)^2\frac{\mathbf{R}}{M}\left(\frac{T}{P^2}-\frac{T^2}{P^3}\frac{dP}{dT}\right)=0$$

$$\therefore\quad \frac{dP}{dT}=\frac{P}{T}+\frac{\gamma}{\gamma-1}\left(\frac{A}{G}\right)^2\frac{MP^3}{\mathbf{R}T^2} \tag{4.82}$$

The maximum value of the entropy occurs when the values of dP/dT given by Eqs (4.81), (4.82) are the same, that is when:

$$\frac{\gamma}{\gamma-1}\frac{P}{T}=\frac{P}{T}+\frac{\gamma}{\gamma-1}\left(\frac{A}{G}\right)^2\frac{MP^3}{\mathbf{R}T^2}$$

and

$$\left(\frac{G}{A}\right)^2=\gamma\left(\frac{M}{\mathbf{R}}\right)\frac{P^2}{T}=\frac{\gamma MP^2}{\mathbf{R}\,T}$$

or

$$\left(\frac{u}{v}\right)^2=\frac{\gamma P}{v}$$

i.e. when:

$$u=\sqrt{\gamma Pv}=u_w \quad \text{(from Eq. 4.37)}$$

which has already been shown to be the velocity of propagation of a pressure wave. This represents the limiting velocity which can be reached in a pipe of constant cross-sectional area.

4.5.5 Flow of Nonideal Gases

In the preceding sections, methods have been given for the calculation of the pressure drop for the flow of an incompressible fluid and for a compressible fluid which behaves as an ideal gas. If the fluid is compressible and deviations from the ideal gas law are appreciable, one of the approximate equations of state, such as van der Waals' equation, may be used in place of the law $PV = n\mathbf{R}T$ to give the relation between temperature, pressure, and volume. Alternatively, if the enthalpy of the gas is known over a range of temperature and pressure, the energy balance, Eq. (2.56), which involves a term representing the change in the enthalpy, may be employed:

$$\Delta\frac{u^2}{2\alpha} + g\Delta z + \Delta H = q - W_s \quad \text{(Eq. 2.56)}$$

This method of approach is useful in considering the flow of steam at high pressures.

4.6 Shock Waves

It has been seen while deriving Eqs (4.33)–(4.38) that for a small disturbance, the velocity of propagation of the pressure wave is equal to the velocity of sound. If the changes are much larger and the process is not isentropic, the wave developed is known as a shock wave, and the velocity may be much greater than the velocity of sound. Material and momentum balances must be maintained and the appropriate equation of state for the fluid must be followed. Furthermore, any change which takes place must be associated with an increase, never a decrease, in entropy. For an ideal gas in a uniform pipe under adiabatic conditions a material balance gives:

$$\frac{u_1}{v_1} = \frac{u_2}{v_2} \quad \text{(Eq. 2.62)}$$

Momentum balance:

$$\frac{u_1^2}{v_1} + P_1 = \frac{u_2^2}{v_2} + P_2 \quad \text{(Eq. 2.64)}$$

Equation of state:

$$\frac{u_1^2}{2} + \frac{\gamma}{\gamma - 1}P_1 v_1 = \frac{u_2^2}{2} + \frac{\gamma}{\gamma - 1}P_2 v_2 \quad \text{(from Eq. 4.72)}$$

Substituting from Eq. (2.62) into Eq. (2.64):

$$v_1 = \frac{u_1^2 - u_1 u_2}{P_2 - P_1} \tag{4.83}$$

$$v_2 = \frac{u_2^2 - u_1 u_2}{P_1 - P_2} \tag{4.84}$$

Then, from Eq. (4.72):

$$\frac{u_1^2 - u_2^2}{2} + \frac{\gamma}{\gamma - 1}\frac{P_1}{P_2 - P_1}u_1(u_1 - u_2) - \frac{\gamma}{\gamma - 1}\frac{P_2}{P_1 - P_2}u_2(u_2 - u_1) = 0$$

that is: $(u_1 - u_2) = 0$, representing no change in conditions, or:

$$\frac{u_1 + u_2}{2} + \frac{\gamma}{\gamma - 1}\frac{1}{P_2 - P_1}(u_1 p_1 - u_2 P_2) = 0$$

Hence:

$$\frac{u_2}{u_1} = \frac{(\gamma - 1)(P_2/P_1) + (\gamma + 1)}{(\gamma + 1)(P_2/P_1) + (\gamma - 1)} \tag{4.85}$$

and:

$$\frac{P_2}{P_1} = \frac{(\gamma + 1) - (\gamma - 1)(u_2/u_1)}{(\gamma + 1)(u_2/u_1) - (\gamma - 1)} \tag{4.86}$$

Eq. (4.86) gives the pressure changes associated with a sudden change of velocity. In order to understand the nature of a possible velocity change, it is convenient to work in terms of Mach numbers. The Mach number (Ma) is defined as the ratio of the velocity at a point to the corresponding velocity of sound where:

$$Ma_1 = \frac{u_1}{\sqrt{\gamma P_1 v_1}} \tag{4.87}$$

and:

$$Ma_2 = \frac{u_2}{\sqrt{\gamma P_2 v_2}} \tag{4.88}$$

From Eqs (4.83), (4.84), (4.87), (4.88):

$$v_1 = \frac{u_1^2}{Ma_1^2}\frac{1}{\gamma P_1} = \frac{u_1(u_1 - u_2)}{P_2 - P_1} \tag{4.89}$$

$$v_2 = \frac{u_2^2}{Ma_2^2}\frac{1}{\gamma P_2} = \frac{u_2(u_2 - u_1)}{P_1 - P_2} \tag{4.90}$$

Giving:

$$\frac{Ma_2^2}{Ma_1^2} = \frac{u_2 P_1}{u_1 P_2} \tag{4.91}$$

From Eq. (4.89):

$$\frac{1}{\gamma Ma_1^2} = \frac{1 - (u_2/u_1)}{(P_2/P_1) - 1} \tag{4.92}$$

Also,

$$\frac{1}{\gamma Ma_1^2} = \frac{2}{(\gamma + 1)(P_2/P_1) + (\gamma - 1)} \quad \text{(from Eq. 4.85)} \tag{4.93}$$

Thus:

$$\frac{P_2}{P_1} = \frac{2\gamma Ma_1^2 - (\gamma - 1)}{\gamma + 1} \tag{4.94}$$

and:

$$\frac{u_2}{u_1} = \frac{(\gamma - 1)Ma_1^2 + 2}{Ma_1^2(\gamma + 1)} \tag{4.95}$$

From Eq. (4.91):

$$\frac{Ma_2^2}{Ma_1^2} = \frac{(\gamma - 1)Ma_1^2 + 2}{Ma_1^2(\gamma + 1)} \frac{(\gamma + 1)}{2\gamma Ma_1^2 - (\gamma - 1)}$$

or:

$$Ma_2^2 = \frac{(\gamma - 1)Ma_1^2 + 2}{2\gamma Ma_1^2 - (\gamma - 1)} \tag{4.96}$$

For a sudden change or normal shock wave to occur, the entropy change per unit mass of fluid must be positive.

From Eq. (4.79), the change in entropy is given by:

$$\begin{aligned} S_2 - S_1 &= C_p \ln \frac{T_2}{T_1} - \frac{\mathbf{R}}{M} \ln \frac{P_2}{P_1} \\ &= C_p \ln \frac{P_2}{P_1} + C_p \ln \frac{v_2}{v_1} - \frac{R}{M} \ln \frac{P_2}{P_1} \\ &= C_v \ln \frac{P_2}{P_1} + C_p \ln \frac{u_2}{u_1} \quad \text{(from Eqs 2.62 and 2.27)} \\ &= C_v \ln \frac{2\gamma Ma_1^2 - (\gamma - 1)}{\gamma + 1} - C_p \ln \frac{Ma_1^2(\gamma + 1)}{(\gamma - 1)Ma_1^2 + 2} \end{aligned} \tag{4.97}$$

$S_2 - S_1$ is positive when $Ma_1 > 1$. Thus a normal shock wave can occur only when the flow is supersonic. From Eq. (4.96), if $Ma_1 > 1$, then $Ma_2 < 1$, and therefore the flow necessarily changes from supersonic to subsonic. If $Ma_1 = 1$, $Ma_2 = 1$ also, from Eq. (4.96), and no change therefore takes place. It should be noted that there is no change in the energy of the fluid as it passes through a shock wave, though the entropy increases and therefore the change is irreversible.

For flow in a pipe of constant cross-sectional area, therefore, a shock wave can develop only if the gas enters at supersonic velocity. It cannot occur spontaneously.

Example 4.4

A reaction vessel in a building is protected by means of a bursting disc and the gases are vented to the atmosphere through a stack pipe having a cross-sectional area of 0.07 m^2. The ruptured disc has a flow area of 4000 mm^2 and the gases expand to the full area of the stack pipe in a divergent section. If the gas in the vessel is at a pressure of 10 MN/m^2 and a temperature of 500 K, calculate: (a) the initial rate of discharge of gas, (b) the pressure and Mach number immediately upstream of the shock wave, and (c) the pressure of the gas immediately downstream of the shock wave.

Assume that isentropic conditions exist on either side of the shock wave and that the gas has a mean molecular weight of 40 kg/kmol, a ratio of specific heats of 1.4, and obeys the ideal gas law.

Solution

The pressure ratio w_c at the throat is given by Eq. (4.43):

$$w_c = \frac{P_c}{P_1} = \left[\frac{2}{\gamma+1}\right]^{\gamma/(\gamma-1)} = \left(\frac{2}{2.4}\right)^{1.4/0.4} = 0.53$$

Thus, the throat pressure $= (10 \times 0.53) = 5.3\ \text{MN/m}^2$.

Specific volume of gas in reactor:

$$v_1 = \left(\frac{22.4}{40}\right)\left(\frac{500}{273}\right)\left(\frac{101.3}{10{,}000}\right) = 0.0103\,\text{m}^3/\text{kg}$$

Specific volume of gas at the throat $= (P_1/P_2)^{\frac{1}{\gamma}} v_1 = (0.0103)(1/0.53)^{1/1.4} = (0.0103 \times 1.575) = 0.0162\,\text{m}^3/\text{kg}$.

$$\begin{aligned} &\therefore\ \text{velocity at the throat} \\ &= \sqrt{\gamma P v}\ \ (\text{from Eq. 4.37}) \\ &= \sqrt{1.4 \times 5.3 \times 10^6 \times 0.0162} = 347\,\text{m/s} \end{aligned}$$

Initial rate of discharge of gas;

$$\begin{aligned} G &= uA/v\ \ (\text{at throat}) \\ &= \frac{347 \times 4000 \times 10^{-6}}{0.0162} \\ &= \underline{\underline{85.7\,\text{kg/s}}} \end{aligned}$$

The gas continues to expand isentropically and the pressure ratio w is related to the flow area by Eq. (4.47). If the cross-sectional area of the exit to the divergent section is such that $w^{-1} = (10{,}000/101.3) = 98.7$, the pressure here will be atmospheric and the expansion will be entirely isentropic. The duct area, however, has nearly twice this value, and the flow is *over-expanded*, atmospheric pressure being reached within the divergent section. In order to satisfy the boundary conditions, a shock wave occurs further along the divergent section across which the pressure increases. The gas then expands isentropically to atmospheric pressure.

If the shock wave occurs when the flow area is A, then the flow conditions at this point can be calculated by solution of the equations for:

(1) *The isentropic expansion from conditions at the vent.* The pressure ratio w (pressure/pressure in the reactor) is given by Eq. (4.47) as:

$$\begin{aligned} w^{-1.42}/(1 - w^{0.29}) &= \left(\frac{A}{G}\right)^2 \left(\frac{7P_1}{v_1}\right) \\ &= \left(\frac{A}{85.7}\right)^2 \frac{(7 \times 10{,}000 \times 10^3)}{0.0103} = 9.25 \times 10^5 A^2 \end{aligned} \quad (1)$$

The pressure at this point is $10 \times 10^6 w$ N/m^2.

Specific volume of gas at this point is given by Eq. (4.45) as:

$$v = v_1 w^{-0.71} = 0.0103 w^{-0.71}$$

The velocity is given by Eq. (4.46) as:

$$\begin{aligned} u^2 &= 7P_1 v_1 (1 - w^{0.29}) \\ &= (7 \times 10 \times 10^6 \times 0.0103)(1 - w^{0.29}) \end{aligned}$$

$$\therefore \quad u = 0.849 \times 10^3 (1 - w^{0.29})^{0.5} \text{ m/s}$$

Velocity of sound at a pressure of $10 \times 10^6 w$ N/m^2

$$= \sqrt{1.4 \times 10 \times 10^6 w \times 0.0103 w^{-0.71}}$$

$$= 380 w^{0.145} \text{ m/s}$$

$$\text{Mach number} = \frac{0.849 \times 10^3 (1 - w^{0.29})^{0.5}}{380 w^{0.145}} = 2.23 \left(w^{-0.29} - 1\right)^{0.5} \quad (2)$$

(2) *The nonisentropic compression across the shock wave.* The velocity downstream from the shock wave (suffix s) is given by Eq. (4.95) as:

$$\begin{aligned} u_s &= u_1 \frac{(\gamma - 1) Ma_1^2 + 2}{Ma_1^2 (\gamma + 1)} \\ &= \frac{0.849 \times 10^3 (1 - w^{0.29})^{0.5} [1.4 - 1 \times 4.97 (w^{-0.29} - 1) + 2]}{4.97 (w^{-0.29} - 1) \times (1.4 + 1)} \quad (3) \\ &= 141 (1 - w^{0.29})^{-0.5} \text{ m/s} \end{aligned}$$

The pressure downstream from the shock wave P_s is given by Eq. (4.94):

$$\frac{P_s}{10^7 w} = \frac{2\gamma Ma_1{}^2 - (\gamma - 1)}{\gamma + 1} \quad (4)$$

Substituting for Ma_1 from Eq. (2):

$$P_s = \left\{56.3 w \left(w^{-0.29} - 1\right) - 1.66\right\} \times 10^6 \text{ N/m}^2$$

(3) *The isentropic expansion of the gas to atmospheric pressure.* The gas now expands isentropically from P_s to P_a (=101.3 kN/m^2) and the flow area increases from A to the full bore of 0.07 m^2. Denoting conditions at the outlet by suffix a, then from Eq. (4.46):

$$u_a^2 - u_s^2 = 7 P_s v_s \left[1 - \left(\frac{P_a}{P_s}\right)^{0.25}\right] \quad (5)$$

$$\frac{u_a}{v_a} = \frac{85.7}{0.07} = 1224 \text{ kg/m}^2\text{ s} \quad (6)$$

$$\frac{u_s}{v_s} = \frac{85.7}{A} \text{ kg/m}^2\text{ s} \quad (7)$$

$$\frac{v_a}{v_s} = \left(\frac{P_a}{P_s}\right)^{-0.71} \tag{8}$$

Eqs (1), (3)–(8), involving seven unknowns, may be solved by trial and error to give $w = 0.0057$.

Thus the pressure upstream from the shock wave is:

$$(10 \times 10^6 \times 0.0057) = 0.057 \times 10^6 \,\text{N/m}^2$$

or:

$$57 \,\text{kN/m}^2$$

The Mach number, from Eq. (2) = $\underline{\underline{4.15}}$.

The pressure downstream from shock wave P_s, from Eq. (4),

$$= \underline{\underline{1165 \,\text{kN/m}^2}}$$

In closing this chapter, two comments are in order. Many authors prefer to express the various forms of the energy equation obtained in this chapter in dimensionless forms involving Mach numbers, specific heat ratio (k), isentropic constant (γ), etc. These are available in several text books, e.g. see Refs. 6–8. Secondly, many authors have presented graphs which facilitate the calculations of compressible pipe flow applications, but all such charts are really based (together with some simplifications) on the basic equations presented in this chapter, e.g. see Refs. 9–11.

4.7 Nomenclature

		Units in SI System	Dimensions in M, L, T, θ
A	Cross-sectional area of flow	m^2	$\mathbf{L}^2$
A_0	Area of orifice	m^2	$\mathbf{L}^2$
C_D	Coefficient of discharge	–	–
C_p	Specific heat capacity at constant pressure	J/kg K	$\mathbf{L^2T^{-2}}\boldsymbol{\theta}^{-1}$
c_v	Specific heat capacity at constant volume	J/kg K	$\mathbf{L^2T^{-2}}\boldsymbol{\theta}^{-1}$
d	Pipe diameter	m	$\mathbf{L}$
e	Pipe roughness	m	$\mathbf{L}$
F	Energy dissipated per unit mass	J/kg	$\mathbf{L^2T^{-2}}$
G	Mass flowrate	kg/s	$\mathbf{MT^{-1}}$
G_{max}	Mass flowrate under conditions of maximum flow	kg/s	$\mathbf{MT^{-1}}$
G^*	Value of G calculated ignoring changes in kinetic energy	kg/s	$\mathbf{MT^{-1}}$
g	Acceleration due to gravity	m/s^2	$\mathbf{LT^{-2}}$
H	Enthalpy per unit mass	J/kg	$\mathbf{L^2T^{-2}}$
K	Energy per unit mass	J/kg	$\mathbf{L^2T^{-2}}$

k	Gas expansion index	–	–
l	Pipe length	m	$\mathbf{L}$
M	Molecular weight	kg/kmol	$\mathbf{MN^{-1}}$
n	Number of moles	k mol	$\mathbf{N}$
P	Pressure	N/m^2	$\mathbf{ML^{-1}T^{-2}}$
P_B	Backpressure at nozzle	N/m^2	$\mathbf{ML^{-1}T^{-2}}$
P_E	Exit pressure of gas	N/m^2	$\mathbf{ML^{-1}T^{-2}}$
P_w	Downstream pressure P_2 at maximum flow condition	N/m^2	$\mathbf{ML^{-1}T^{-2}}$
q	Heat added per unit mass	J/kg	$\mathbf{L^2T^{-2}}$
R	Shear stress at pipe wall	N/m^2	$\mathbf{ML^{-1}T^{-2}}$
R	Universal gas constant	8314 J/ kmol K	$\mathbf{MN^{-1}L^2T^{-2}}\boldsymbol{\theta}^{-1}$
S	Entropy per unit mass	J/kg K	$\mathbf{L^2T^{-2}}\boldsymbol{\theta}^{-1}$
T	Temperature (absolute)	K	$\boldsymbol{\theta}$
U	Internal energy per unit mass	J/kg	$\mathbf{L^2T^{-2}}$
u	Velocity	m/s	$\mathbf{LT^{-1}}$
u_E	Velocity at exit of nozzle	m/s	$\mathbf{LT^{-1}}$
u_w	Velocity of pressure wave	m/s	$\mathbf{LT^{-1}}$
V	Specific volume ($=\rho^{-1}$)	m^3/kg	$\mathbf{M^{-1}L^3}$
W_s	Shaft work per unit mass	J/kg	$\mathbf{L^2T^{-2}}$
w	Pressure ratio P_2/P_1	–	–
W_c	Pressure ratio P_w/P_1	–	–
Z	Vertical height	m	$\mathbf{L}$
α	Kinetic energy correction factor	–	–
y	Specific heat ratio (C_p/C_v)	–	–
ε	Bulk modulus of elasticity	N/m^2	$\mathbf{ML^{-1}T^{-2}}$
μ	Viscosity	Ns/m^2	$\mathbf{ML^{-1}T^{-1}}$
ρ	Density	kg/m^3	$\mathbf{ML^{-3}}$
Ma	Mach number	–	–
Re	Reynolds number	–	–

Suffixes

0	Reference condition
1	Upstream condition
2	Downstream condition
W	Maximum flow condition

*Value obtained neglecting kinetic energy changes.

References

1. Griffiths H. Some problems of vacuum technique from a chemical engineering standpoint. *Trans Inst Chem Eng* 1945;**23**:113.
2. Saad MA. *Compressible fluid flow.* 2nd ed. Englewood Cliffs, NJ: Prentice Hall; 1998.
3. Oosthuizen PH, Carscallen WE. *Introduction to compressible fluid flow.* 2nd ed. Boca Raton, FL: CRC Press; 2013.
4. Stodola A, Lowenstein LC. *Steam and gas turbines.* New York: McGraw-Hill; 1945.
5. Sears FW, Salinger GL. *Thermodynamics, kinetic theory and statistical thermodynamics.* 3rd ed. Reading, MA: Addison-Wesley; 1975.
6. White FM. *Fluid mechanics.* 8th ed. New York: McGraw-Hill; 2015.
7. Shames IH. *Mechanics of fluids.* 3rd ed. New York: McGraw-Hill; 1992.
8. Darby R, Chhabra RP. *Chemical engineering fluid mechanics.* 3rd ed. Boca Raton, FL: CRC Press; 2017.
9. Yu FC. Compressible fluid pressure drop calculation—isothermal versus adiabatic. *Hydrocarbon Processing* 1999;**78**(5):89.
10. Walters T. Gas flow calculations: don't choke. *Chem Eng* 2000;**107**(1):70.
11. Teng F, Medina P, Heigold M. Compressible fluid flow calculation methods. *Chem Eng* 2014;**121**(2):32.

Further Reading

1. Anderson JD. *Modern compressible flow: with historical perspective.* 3rd ed. New York: McGraw-Hill; 2012.
2. Becker E. *Gas dynamics.* New York: Academic Press; 1968.
3. Liepmann HW, Roshko A. *Elements of gas dynamics.* New York: Dover; 2002.
4. Mayhew YR, Rogers GFC. *Thermodynamics and transport properties of fluids.* 5th ed. Oxford: Wiley-Blackwell; 1995.
5. Shames IH. *Mechanics of fluids.* 3rd ed. New York: McGraw-Hill; 1992.
6. Shapiro AH. The dynamics and thermodynamics of compressible fluid flow, vols. I and II. New York: Ronald; 1953 and 1954.

CHAPTER 5

Flow of Multiphase Mixtures

5.1 Introduction

The flow problems considered in previous chapters are concerned with homogeneous fluids, either single phase or suspensions of fine particles whose settling velocities are sufficiently low for the solids to be completely suspended in the fluid. Consideration is now given to the far more complex problem of the flow of multiphase systems in which the composition of the mixture may vary over the cross-section of the pipe or channel; furthermore, the components may be moving at different velocities to give rise to the phenomenon of "slip" between the phases.

Multiphase flow is important in many areas of chemical and process engineering and the behaviour of the material will depend on the properties of the components, the flowrates, and the geometry of the system. In general, the complexity of the flow is so high that design methods depend very much on an analysis of the behaviour of such systems in practice and, only to a limited extent, on theoretical predictions. Some of the more important systems to be considered are:

Mixtures of liquids with gas or vapour.
Liquids mixed with solid particles (hydraulic transport).
Gases carrying solid particles wholly or partly in suspension (pneumatic transport).
Mixtures of immiscible liquids like oil and water.
Multiphase systems containing solids, liquids, and gases.

Mixed materials may be transported horizontally, vertically, or at an inclination to the horizontal in pipes and, in the case of liquid–solid mixtures, in open channels (flumes, launders for instance). Although there is some degree of common behaviour between the various systems, the range of physical properties is so high that each different type of system must be considered separately. Liquids may have densities up to three orders of magnitude greater than gases but they do not exhibit any significant compressibility. Liquids themselves can range from simple Newtonian liquids such as water, to non-Newtonian fluids with very high apparent viscosities. These very large variations in density and viscosity are responsible for the large differences in behaviour of solid–gas and solid–liquid mixtures that must, in practice, be considered separately. For all multiphase flow systems, however, it is important to understand the nature of the interactions between the phases and how these influence the

Coulson and Richardson's Chemical Engineering. https://doi.org/10.1016/B978-0-08-101099-0.00005-7

flow patterns—the ways in which the two or three phases are distributed over the cross-section of the pipe or duct. In design it is necessary to be able to predict *pressure drop* which, usually, depends not only on the flow pattern, but also on the relative velocity of the phases; this *slip velocity* will influence the *hold-up*, the fraction of the pipe volume which is occupied by a particular phase. It is important to note that, in the flow of a two-component mixture, the hold-up (or in situ concentration) of a component will differ from that in the mixture discharged at the end of the pipe because, as a result of *slip* of the phases relative to one another, their residence times in the pipeline will not be the same. Special attention is therefore focused on three aspects of the flow of these complex mixtures.

(1) The flow patterns.
(2) The hold-up of the individual phases and their relative velocities.
(3) The relationship between pressure gradient in a pipe and the flowrates and physical properties of the phases.

The difference in density between the phases is important in determining flow pattern. In gas–solid and gas–liquid mixtures, the gas will always be the lighter phase, and in liquid–solid systems it will be usual for the liquid to be less dense than the solid. In vertical upward flow, therefore, there will be a tendency for the lighter phase to rise more quickly than the denser phase giving rise to a slip velocity. For a liquid–solid or gas–solid system, this *slip* velocity will be close to the terminal falling velocity of the particles. In a liquid–gas system, the slip velocity will depend on the flow pattern in a complex way. In all cases, there will be a net upward force resulting in a transference of energy from the faster to the slower moving phase, and a vertically downwards gravitational force will be balanced by a vertically upwards drag force. There will be axial symmetry of flow in vertical ducts.

In horizontal flow, the flow pattern will inevitably be more complex because the gravitational force will act perpendicular to the pipe axis, the direction of flow, and will cause the denser component to flow preferentially near the bottom of the pipe. Energy transfer between the phases will again occur as a result of the difference in velocity, but the net force will be horizontal and the suspension mechanism of the particles, or the dispersion of the fluid will be a more complex process. In this case, the flow will not be symmetrical about the pipe axis, except at very high velocities and/or when the difference in the two densities is very small.

In practice, many other considerations will affect the design of an installation. For example, wherever solid particles are present, there is the possibility of *blockage* of the pipe and it is therefore important to operate under conditions where the probability of this occurring is minimised if not eliminated altogether. Solids may be *abrasive* and cause undue wear if the velocities are too high or changes in direction of flow are too sudden. Choice of suitable materials of construction, associated equipment like pumps and valves, flow meters, for instance and operating conditions is therefore important. In pneumatic transport, *electrostatic charging* may take place, which can cause considerable increase in pressure gradient.

5.2 Two-Phase Gas (Vapour)-Liquid Flow

5.2.1 Introduction

Some of the important features of the flow of two-phase mixtures composed of a liquid together with a gas or vapour are discussed in this section. There are many applications in the chemical and process industries, ranging from the flow of mixtures of oil and gas from well heads to the flow of vapour–liquid mixtures in boilers, condensers, and evaporators.

Because of the presence of the two phases, there are considerable complications in describing and quantifying the nature of the flow compared with conditions with a single phase. The lack of knowledge of the velocities at a point in the individual phases makes it impossible to give any real picture of the velocity distribution. In most cases, the gas phase, which may be flowing with a much greater velocity than the liquid, continuously accelerates the liquid thus involving a transfer of energy or momentum. Either phase may be in *streamline* or in *turbulent* flow, though the most important case is that in which both phases are turbulent. The criterion for streamline or turbulent flow of a phase is whether the Reynolds number for its flow at the same rate on its own is less or $>$1000–2000. This distinction is to some extent arbitrary in that injection of a gas into a liquid initially in streamline flow may result in turbulence developing.

If there is no heat transfer to the flowing mixture, the mass rate of flow of each phase will remain substantially constant, though the volumetric flowrates (and velocities) will increase progressively as the gas expands with the falling pressure in the direction of flow along the length of pipe. In a boiler or evaporator, there will be a progressive vapourisation of the liquid leading to a decreased mass flowrate of liquid and corresponding increase for the vapour, with the total mass rate of flow remaining constant. The volumetric flowrate will increase very rapidly as a result of the combined effects of falling pressure and increasing vapour/liquid ratio.

A gas–liquid mixture will have a lower density than the liquid alone. Therefore, if in a U-tube one limb contains liquid and the other a liquid–gas mixture, the equilibrium height in the second limb will be higher than in the first. If two-phase mixture is discharged at a height less than the equilibrium height, a continuous flow of liquid will take place from the first to the second limb, provided that a continuous feed of liquid and gas is maintained. This principle is used in the design of the airlift pump described in Chapter 8.

Consideration will now be given to the various flow regimes that may exist and how they may be represented on a 'Flow Pattern Map'; to the calculation and prediction of hold-up of the two phases during flow; and to the calculation of pressure gradients for gas–liquid flow in pipes. In addition, when gas–liquid mixtures flow at high velocities, serious erosion problems can arise and it is necessary for the designer to restrict flow velocities to avoid serious damage to equipment.

A more detailed treatment of the subject is given by Govier and Aziz,[1] Chisholm,[2] Hewitt[3], and more recently by Michaelides et al.[4]

5.2.2 Flow Regimes and Flow Patterns

Horizontal flow

The flow pattern is complex and is influenced by the diameter of the pipe, the physical properties of the fluids, and their flowrates. In general, as the velocities are increased and as the gas–liquid ratio increases, changes will take place from 'bubble flow' through to 'mist flow' as shown in Fig. 5.1[1–8]; the principal characteristics of the flow patterns are described in Table 5.1. At high liquid–gas ratios, the liquid forms the continuous phase and at low values it forms the disperse phase. In the intervening region, there is generally some instability; and sometimes several flow regimes are lumped together. In plug flow and slug flow, the gas is flowing faster than the liquid and liquid from a slug tends to become detached, to move as a relatively slow moving film along the surface of the pipe and then to be reaccelerated when the next liquid slug catches it up. This process can account for a significant proportion of the total energy losses. Particularly in short pipelines, the flow develops an oscillating pattern arising largely from discontinuities associated with the expulsion of successive liquid slugs in an intermittent manner.

The regions over which the different types of flow can occur are conveniently shown on a 'Flow Pattern Map' in which a function of the gas flowrate is plotted against a function of the

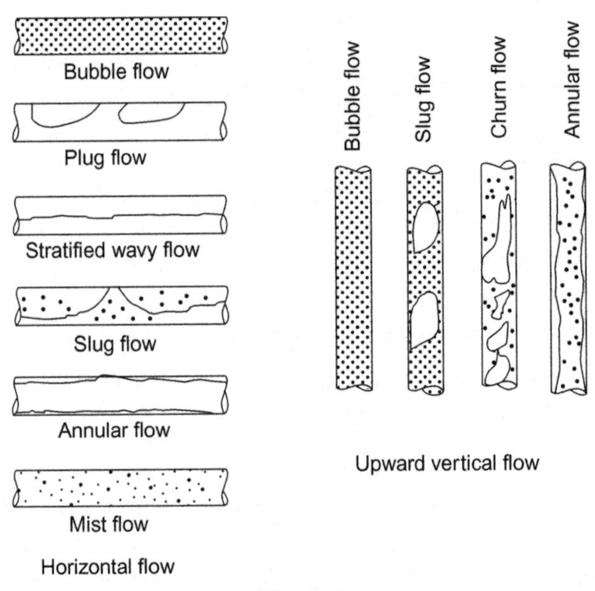

Fig. 5.1
Flow patterns in two-phase flow.

Table 5.1 Flow regimes in horizontal two-phase flow

Regime	Description	Typical Velocities (m/s)	
		Liquid	Vapour
1. Bubble flow[a]	Bubbles of gas dispersed throughout the liquid	1.5–5	0.3–3
2. Plug flow[a]	Plugs of gas in liquid phase	0.6	<1.0
3. Stratified flow	Layer of liquid with a layer of gas above	<0.15	0.6–3
4. Wavy flow	As stratified but with a wavy interface due to higher velocities	<0.3	>5
5. Slug flow[a]	Slug of gas in liquid phase	Occurs over a wide range of velocities	
6. Annular flow[b]	Liquid film on inside walls with gas in centre		>6
7. Mist flow[b]	Liquid droplets dispersed in gas		>60

[a]Frequently grouped together as *intermittent* flow.
[b]Sometimes grouped as *annular/mist* flow.

liquid flowrate and boundary lines are drawn to delineate the various regions. It should be borne in mind that the distinction between any two flow patterns is not clear-cut and that these divisions are only approximate as each flow regime tends to merge in with its neighbours; in any case, the whole classification is based on highly subjective observations. Several researchers have produced their own flow pattern maps.[5–9]

Most of the data used for compiling such maps have been obtained for the flow of water and air at near atmospheric temperature and pressure, and scaling factors have been introduced to extend their applicability to other systems. However, bearing in mind the diffuse nature of the boundaries between the regimes and the relatively minor effect of changes in physical properties, such a refinement does not appear to be justified. The flow pattern map for horizontal flow illustrated in Fig. 5.2 which has been prepared by Chhabra and Richardson[10] is based on those previously presented by Mandhane et al.[9] and Weisman et al.[8] The axes of this diagram are superficial liquid velocity u_L and superficial gas velocity u_G (in each case the volumetric flowrate of the phase divided by the total cross-sectional area of the pipe). While some authors have introduced the multiplication factors involving the ratios of density, viscosity, and surface tension using air and water as the reference fluids for the gas and liquid phase respectively, these do not seem to be particularly justified in view of their minor influence. Also, each physical property does not necessarily influence each transition or in the same way.[11] It must however, be borne in mind that the classification of flow regimes based on visual observations, or even on the signal processing methods intrinsically tends to be

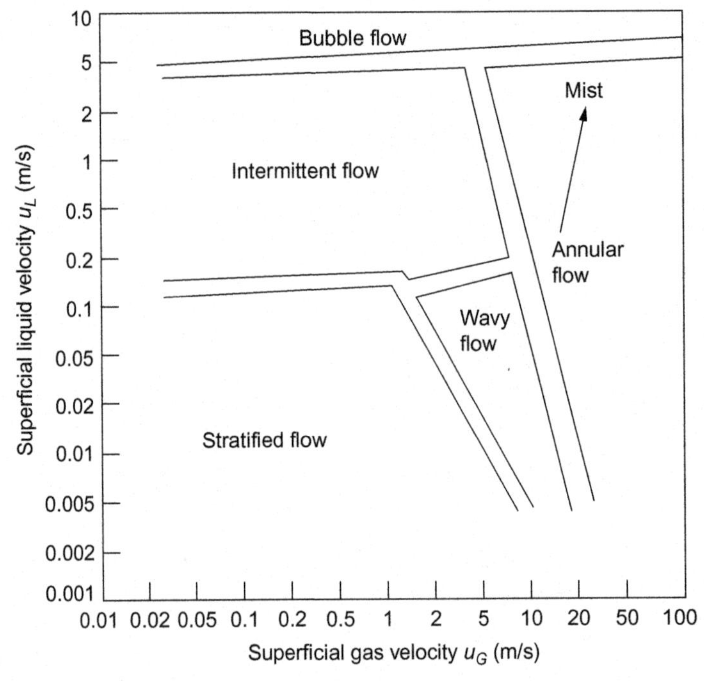

Fig. 5.2
Flow pattern map.

somewhat subjective and arbitrary.[12] However, some of the recent experimental data[13,14] is also consistent with the flow pattern map shown in Fig. 5.2.

Slug flow should be avoided when it is necessary to obviate unsteady conditions, and it is desirable to design so that annular flow still persists at loadings down to 50% of the normal flow rates. Even though in many applications both phases should be turbulent, excessive gas velocity will lead to a high pressure drop, particularly in small pipes.

Although most of the data relate to flow in pipes of small diameters (<42 mm), results of experiments carried out in a 205 mm pipe fit well on the diagram. The flow pattern map, shown in Fig. 5.2, also gives a good representation of results obtained for the flow of mixtures of gas and shear-thinning non-Newtonian liquids, including very highly shear-thinning suspensions (power law index $n \asymp 0.1$) and viscoelastic polymer solutions.

Vertical flow

In vertical flow, axial symmetry exists and flow patterns tend to be somewhat more stable. However, with slug flow in particular, oscillations in the flow can occur as a result of sudden changes in pressure as liquid slugs are discharged from the end of the pipe.

The principal flow patterns are shown in Fig. 5.1. In general, the flow pattern map (Fig. 5.2) is also applicable to vertical flow. Further reference to flow of gas–liquid mixtures in vertical pipes is made in Section 8.4.1 with reference to the operation of the air-lift pump.

5.2.3 Hold-Up

Because the gas always flows at a velocity greater than that of the liquid, the *in situ* volumetric fraction of liquid at any point in a pipeline will be greater than the input volume fraction of liquid; furthermore it will progressively change along the length of the pipe as a result of expansion of the gas with the falling pressure.

There have been several experimental studies of two-phase flow in which the hold-up has been measured, either directly or indirectly. The direct method of measurement involves suddenly isolating a section of the pipe by means of quick-acting valves and then determining the quantity of liquid trapped.[15,16] Such methods are cumbersome and are subject to errors arising from the fact that the valves cannot operate instantaneously. Typical of the indirect methods is that in which the pipe cross-section is scanned by γ-rays and the hold-up is determined from the extent of their attenuation.[17–19] Other techniques based on impedance measurements, tomographic methods, etc. are well described by Chaouki et al.[20]

Lockhart and Martinelli[21] expressed hold-up in terms of a parameter X, characteristic of the relative flowrates of liquid and gas on their own at the same flowrate, defined as:

$$X = \sqrt{\frac{-\Delta P_L}{-\Delta P_G}} \tag{5.1}$$

where $-\Delta P_L$ and $-\Delta P_G$ are the frictional pressure drops which would arise from the flow of the respective phases on their own at the same flow rates. Their correlation is reproduced in Fig. 5.3. As a result of more recent work it is now generally accepted that the correlation overpredicts values of liquid hold-up. Thus Farooqi and Richardson,[22] the results of whose work are also shown in Fig. 5.3, have given the following expression for liquid hold-up for cocurrent flow of air and Newtonian liquids in horizontal pipes:

$$\left.\begin{array}{ll} \in_L = 0.186 + 0.0191X & 1 < X < 5 \\ \in_L = 0.143X^{0.42} & 5 < X < 50 \\ \in_L = \dfrac{1}{0.97 + 19/X} & 50 < X < 500 \end{array}\right\} \tag{5.2}$$

It should be noted that, for the turbulent flow of each phase when the friction factor is nearly constant, pressure drop is approximately proportional to the square of velocity and X is then equal to the ratio of the superficial velocities of the liquid and gas.

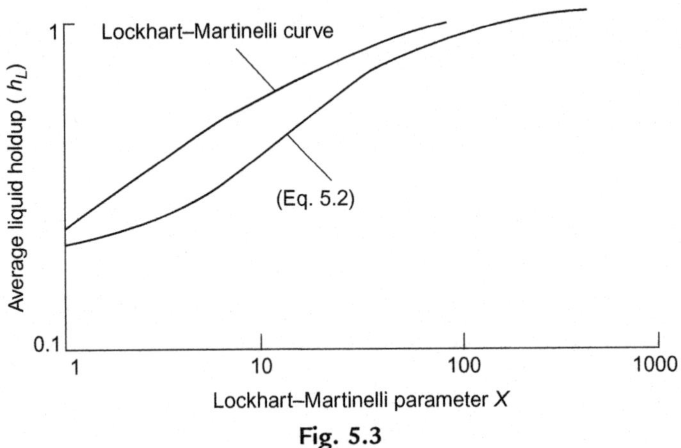

Fig. 5.3
Correlation for average liquid hold-up,

Eq. (5.2) is found to hold well for non-Newtonian shear-thinning suspensions as well, provided that the liquid flow is turbulent. However, for the laminar flow of the liquid, Eq. (5.2) considerably over predicts the liquid hold-up $\in_L$. The extent of over prediction increases as the degree of shear-thinning increases and as the liquid Reynolds number becomes progressively less. A modified parameter X' has therefore been defined[22,23] for a power-law fluid (Chapter 3) in such a way that it reduces to X both at the superficial velocity u_L equal to the transitional velocity $(u_L)_c$ from streamline to turbulent flow and when the liquid exhibits Newtonian properties. The parameter X' is defined by the relation

$$X' = X\left(\frac{u_L}{(u_L)_c}\right)^{1-n} \tag{5.3}$$

where n is the *power-law* index. It will be seen that the correction factor becomes progressively more important as n deviates increasingly from unity and as the velocity deviates from the critical velocity. Eq. (5.2) may then be applied provided X' is used in place of X in the equation.

Thus, in summary, liquid hold-up can be calculated using Eq. (5.2) for:

Newtonian fluids in laminar or turbulent flow.
Non-Newtonian fluids in *turbulent flow only.*

Eq. (5.2), with the modified parameter X' used in place of X, may be used for the laminar flow of shear-thinning fluids whose behaviour can be described by the *power-law* model.

A knowledge of hold-up is particularly important for vertical flow since the hydrostatic pressure gradient, which is frequently the major component of the total pressure gradient, is directly proportional to liquid hold-up. However, in slug flow, the situation is complicated by the fact that any liquid, which is in the form of an annular film surrounding the gas slug does not contribute to the hydrostatic pressure.[16] A good overview of the experimental and theoretical/numerical developments is available in the literature.[24]

5.2.4 Pressure, Momentum, and Energy Relations

Methods for determining the drop in pressure start with a physical model of the two-phase system, and the analysis is developed as an extension of that used for single-phase flow. In the *separated flow* model, the phases are first considered to flow separately; and their combined effect is then examined.

The total pressure gradient in a horizontal pipe, $(-dP_{TPF}/dl)$, consists of two components that represent the frictional and the acceleration contributions respectively, or:

$$\frac{-dP_{TPF}}{dl} = \frac{-dP_f}{dl} + \frac{-dP_a}{dl} \tag{5.4}$$

A momentum balance for the flow of a two-phase fluid through a horizontal pipe and an energy balance may be written in an expanded form of that applicable to single-phase fluid flow. These equations for two-phase flow cannot be used in practice since the individual phase velocities and local densities are not known. Some simplification is possible if it is assumed that the two phases flow separately in the channel occupying fixed fractions of the total area, but even with this assumption of separated flow regimes, progress is difficult due to the uncertainty in the modelling of the interfacial stress. It is important to note that, as in the case of single-phase flow of a compressible fluid, it is no longer possible to relate the shear stress to the pressure drop in a simple form since the pressure drop now covers both frictional and acceleration losses. The shear stress at the wall is proportional to the total rate of momentum transfer, arising from friction and acceleration, so that the total drop in pressure $-\Delta P_{TPF}$ is given by:

$$-\Delta P_{TPF} = (-\Delta P_f) + (-\Delta P_a) \tag{5.5}$$

The pressure drop due to acceleration is important in two-phase flow because the gas is normally flowing much faster than the liquid, and therefore as it expands the liquid phase will accelerate with consequent transfer of energy. For flow in a vertical direction, an additional term $-\Delta P_{gravity}$ must be added to the right hand side of Eq. (5.5) to account for the hydrostatic pressure attributable to the liquid in the pipe, and this may be calculated approximately provided that the in situ liquid hold-up is known.

Analytical solutions for the equations of motion are not possible because of the difficulty of specifying the flow pattern and of defining the precise nature of the interaction between the phases. Rapid fluctuations in flow frequently occur and these cannot readily be taken into account. For these reasons, it is necessary for design purposes to use correlations which have been obtained using experimental data combined with dimensional considerations.

Great care should be taken however, if these are used outside the limits used in the experimental work.

Practical methods for evaluating pressure drop

Probably the most widely used method for estimating the drop in pressure due to friction is that proposed by Lockhart and Martinelli[21] and later modified by Chisholm.[25] This is based on the physical model of separated flow in which each phase is considered separately and then a combined effect formulated. The two-phase pressure drop due to friction $-\Delta P_{TPF}$ is taken as the pressure drop $-\Delta P_L$ or $-\Delta P_G$ that would arise for either phase flowing alone in the pipe at the stated rate, multiplied by some factor $\Phi_L{}^2$ or $\Phi_G{}^2$. This factor is presented as a function of the ratio of the individual single-phase pressure drops and:

$$\frac{-\Delta P_{TPF}}{-\Delta P_G} = \Phi_G^2 \tag{5.6}$$

$$\frac{-\Delta P_{TPF}}{-\Delta P_L} = \Phi_L^2 \tag{5.7}$$

The relation between Φ_G and Φ_L and X (defined by Eq. 5.1) is shown in Fig. 5.4, where it is seen that separate curves are given according to the nature of the flow of the two phases. This relation was developed from studies on the flow in small tubes of up to 25 mm diameter with water, oils, and hydrocarbons using air at a pressure of up to 400 kN/m^2. For mass flowrates per unit area of L' and G' for the liquid and gas, respectively, Reynolds numbers $Re_F(L'd/\mu_L)$ and $Re_G(G'd/\mu_G)$ may be used as criteria for defining the flow regime; values

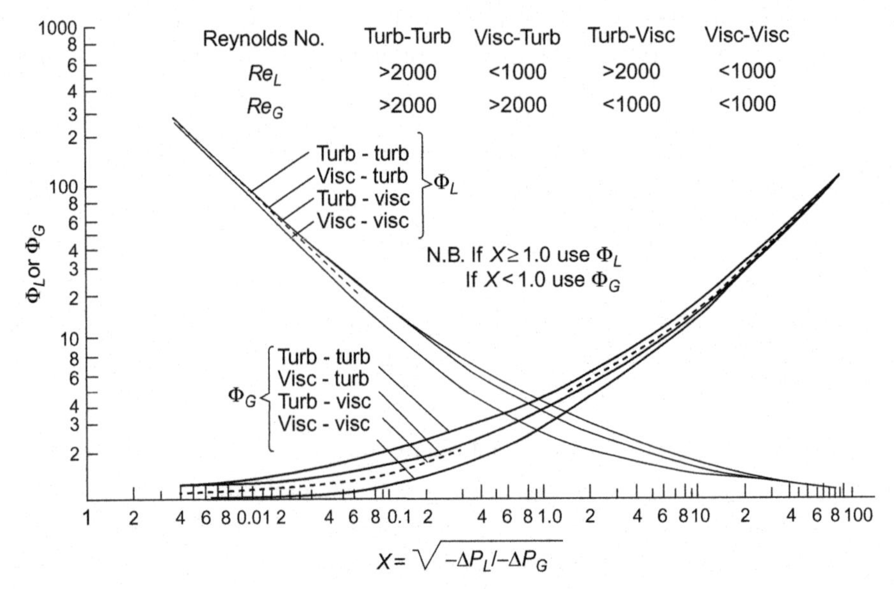

Fig. 5.4
Relation between Φ and X for two-phase flow.

<1000 to 2000, however, do not necessarily imply that the fluid is in truly laminar flow. Later experimental work showed that the total pressure has an influence and data presented by Griffith[26] may be consulted where pressures are in excess of 3 MN/m^2. Chisholm[25] has developed a relation between Φ_L and X which he puts in the form:

$$\Phi_L^2 = 1 + \frac{c}{X} + \frac{1}{X^2} \tag{5.8}$$

where c has a value of 20 for turbulent/turbulent flow, 10 for turbulent liquid/streamline gas, 12 for streamline liquid/turbulent gas, and 5 for streamline/streamline flow. If the densities of the fluids are significantly different from those of water and air at atmospheric temperature and pressure, the values of c are somewhat modified.

Chenoweth and Martin[27,28] have presented an alternative method for calculating the drop in pressure, which is empirical and based on experiments with pipes of 75 mm and pressures up to 0.7 MN/m^2. They have plotted the volume fraction of the inlet stream that is liquid as abscissa against the ratio of the two-phase pressure drop to that for liquid flowing at the same volumetric rate as the mixture. An alternative technique has been described by Baroczy.[29] If heat transfer gives rise to evaporation then reference should be made to work by Dukler et al.[30] A good discussion concerning the relative merits and demerits of various methods is available in the literature.[24]

An illustration of the method of calculation of two-phase pressure drop is included here as Example 5.1.

Critical flow

For the flow of a compressible fluid, conditions of sonic velocity may be reached, thus limiting the maximum flowrate for a given upstream pressure. This situation can also occur with two-phase flow, and such critical velocities may sometimes be reached with a drop in pressure of only 30% of the inlet pressure.

Example 5.1

Steam and water flow through a 75 mm i.d. pipe at flowrates of 0.05 and 1.5 kg/s, respectively. If the mean temperature and pressure are 330 K and 120 kN/m^2, what is the pressure drop per unit length of pipe assuming adiabatic conditions?

Solution

$$\text{Cross-sectional area for flow} = \frac{\pi(0.075)^2}{4} = 0.00442\ \text{m}^2$$

$$\text{Flow rate of water} = \frac{1.5}{1000} = 0.0015\ \text{m}^3/\text{s}$$

$$\text{Water velocity} = \frac{0.0015}{0.00442} = 0.339\ \text{m/s}$$

Density of steam at 330 K and 120 kN/m^2 (assuming ideal gas behaviour):

$$=\left(\frac{18}{22.4}\right)\left(\frac{273}{330}\right)\left(\frac{120}{101.3}\right)=0.788\ \text{kg/m}^3$$

$$\text{Flow rate of steam}=\frac{0.05}{0.788}=0.0635\ \text{m}^2/\text{s}$$

$$\text{Steam velocity}=\frac{0.0635}{0.00442}=14.37\ \text{m/s}$$

Viscosities at 330 K and 120 kN/m^2:

$$\text{steam}=0.0113\times10^{-3}\,\text{Ns/m}^2;\ \ \text{water}=0.52\times10^{-3}\,\text{Ns/m}^2$$

Therefore:

$$Re_L=\frac{0.075\times0.339\times1000}{0.52\times10^{-3}}=4.89\times10^4$$

$$Re_G=\frac{0.075\times14.37\times0.788}{0.0113\times10^{-3}}=7.52\times10^4$$

That is, both the gas and liquid are in turbulent flow.

From the friction chart (Fig. 3.7), assuming $e/d=0.00015$:

$$\left(\frac{R}{\rho u^2}\right)_L=0.0025\ \ \text{and}\ \ \left(\frac{R}{\rho u^2}\right)_G=0.0022$$

∴ From Eq. (3.18):

$$-\Delta P_L=4\left(\frac{R}{\rho u^2}\right)_L\frac{l}{d}\rho u^2=4\times0.0025\left(\frac{1}{0.075}\right)(1000\times0.339^2)=15.32\,(\text{N/m}^2)/\text{m}$$

$$-\Delta P_G=4\times0.0022\left(\frac{1}{0.075}\right)(0.778\times14.37^2)=18.35(\text{N/m}^2)/\text{m}$$

$$\therefore\ \frac{-\Delta P_L}{-\Delta P_G}=\frac{15.32}{18.85}=0.812$$

and:

$$X^2=0.812\ \ \text{and}\ \ X=0.901$$

From Fig. 5.4, for turbulent-turbulent flow,

$$\Phi_L=4.35\ \ \text{and}\ \ \Phi_G=3.95$$

Therefore:

$$\frac{-\Delta P_{TPF}}{-\Delta P_G} = 3.95^2 = 15.60$$

and:

$$-\Delta P_{TPF} = 15.60 \times 18.85 = 294\,(\mathrm{N/m^2})/\mathrm{m}$$

$$\underline{\underline{-\Delta P_{TPF} = 0.29\,(\mathrm{kN/m^2})/\mathrm{m}}}$$

One can also use Eq. (5.8) in this instance. Since both phases are in turbulent flow, $c = 20$. Substituting values:

$$\phi_L^2 = 1 + \frac{C}{X} + \frac{1}{X^2} \tag{5.8}$$

$$= 1 + \frac{20}{0.901} + \frac{1}{(0.901)^2}$$

i.e.

$$\phi_L^2 = 24.43$$

$$\therefore -\Delta P_{TPF} = \phi_L^2 \times (-\Delta P_L)$$

$$= 24.43 \times 15.32$$

or

$$-\Delta P_{TPF} = 374\,\mathrm{Pa/m}$$

This value is about 30% larger than the one calculated above. This is an indication of the typical accuracy of such predictions.

Non-Newtonian flow

When a liquid exhibits non-Newtonian characteristics, the above procedures for Newtonian fluids are valid provided that the liquid flow is turbulent.

For streamline flow of non-Newtonian liquids, the situation is completely different and the behaviour of two-phase mixtures in which the liquid is a shear-thinning fluid is now examined.

The injection of air into a shear-thinning liquid in laminar flow may result in a substantial reduction in the pressure drop[31] and values of the *drag ratio* ($\Phi_L^2 = -\Delta P_{TPF}/-P_L$) may be substantially below unity. For a constant flowrate of liquid, the drag ratio gradually falls from unity as the gas rate is increased. At a critical air flowrate, the drag ratio passes through a minimum $(\Phi_L^2)_{min}$ and then increases, reaching values in excess of unity at high gas flowrates. This effect has been observed with shear-thinning solutions of polymers and with flocculated suspensions of fine kaolin and anthracite coal, and is confined to conditions where the liquid flow would be laminar in the absence of air.

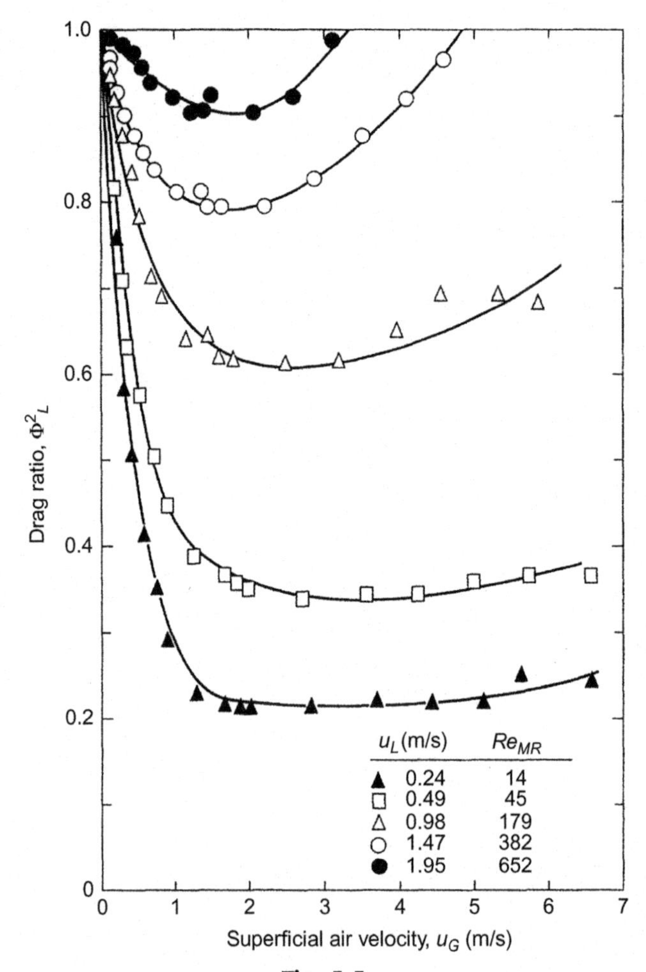

Fig. 5.5
Drag ratio as function of superficial gas velocity (liquid velocity as parameter).

A typical graph of drag ratio as a function of superficial air velocity is shown in Fig. 5.5 in which each curve refers to a constant superficial liquid velocity. The liquids in question exhibited power law rheology and the corresponding values of the Metzner and Reed Reynolds numbers Re_{MR} based on the superficial liquid velocity u_L (see Chapter 3) are given. The following characteristics of the curves may be noted:

(1) For a given liquid, the value of the minimum drag ratio $(\Phi_L^2)_{min}$ decreases as the superficial liquid velocity is decreased.
(2) The superficial air velocity required to give the minimum drag ratio increases as the liquid velocity decreases.

For a more shear-thinning liquid, the minimum drag ratio becomes even smaller although more air must be added to achieve this condition.

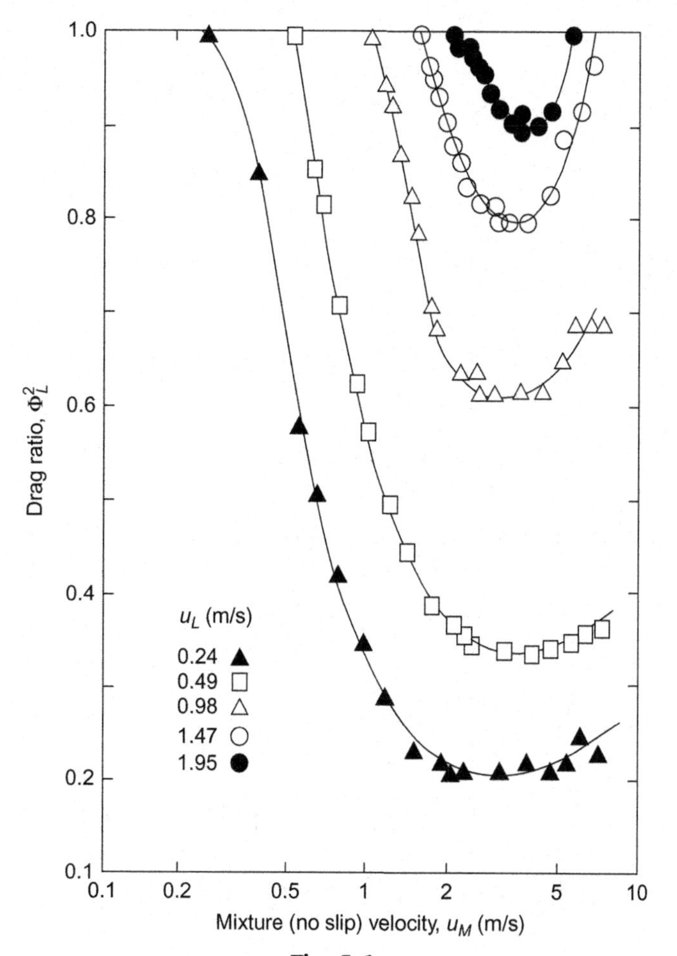

Fig. 5.6
Drag ratio as function of mixture velocity.

If, for a given liquid, drag ratio is plotted (Fig. 5.6) against the mixture velocity (superficial air velocity + superficial liquid velocity) as opposed to the superficial gas velocity, it is found that the minima all occur at about the same mixture velocity, irrespective of the liquid flowrate. For liquids of different rheological properties, the minimum occurs at the same Reynolds number Re_{MR} (based on mixture as opposed to superficial liquid velocity)—about 2000 which corresponds to the limit of laminar flow. This suggests that the extent of drag reduction increases progressively until the liquid flow ceases to be laminar. Thus, at low flowrates more air can be injected before the liquid flow becomes turbulent, as indicated previously.

At first sight, it seems anomalous that, on increasing the total volumetric flowrate by injection of air, the pressure drop can actually be reduced. Furthermore, the magnitude of the effect can be very large with values of drag ratio being as low as 0.2, i.e. the pressure drop

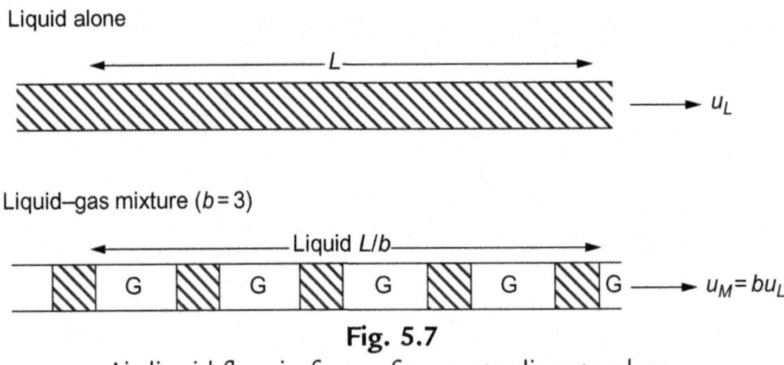

Fig. 5.7
Air-liquid flow in form of separate discrete plugs.

can be reduced by a *factor of 5* by air injection. How this can happen can be illustrated by means of a highly simplified model. If the air and liquid flow as a series of separate plugs, as shown in Fig. 5.7, the total pressure drop can then be taken as the sum of the pressure drops across the liquid and gas slugs; the pressure drop across the gas slugs however will be negligible compared with that contributed by the liquid.

For a 'power-law' fluid in laminar flow at a velocity u_L in a pipe of length l, the pressure drop $-\Delta P_L$ will be given by:

$$-\Delta P_L = K u_L^n l \tag{5.9}$$

If air is injected so that the mixture velocity is increased to bu_L, then the total length of liquid slugs in the pipe will be $(1/b)l$. Then, neglecting the pressure drop across the air slugs, the two-phase pressure drop $-\Delta P_{TP}$ will be given by:

$$\begin{aligned} -\Delta P_{TPF} &= K(bu_L)^n \left(\frac{1}{b} l\right) \\ &= K b^{n-1} {u_L}^n l \end{aligned} \tag{5.10}$$

Thus, dividing Eq. (5.10) by Eq. (5.9):

$$\Phi_L^2 = \frac{-\Delta P_{TPF}}{-\Delta P_L} = b^{n-1} \tag{5.11}$$

Because $n < 1$ for a shear-thinning fluid, Φ_L^2 will be less than unity and a reduction in pressure drop occurs. The lower the value of n and the larger the value of b, the greater the effect will be. It will be noted that the effects of expansion of the air as the pressure falls have not been taken into account.

It has been found experimentally that Eq. (5.11) does apply for small rates of injection of air. At higher injection rates, the model is not realistic and the pressure drop is reduced by a smaller amount than predicted by the equation.

It is seen that for a fluid with a power-law index of 0.2 (a typical value for a flocculated kaolin suspension) and a flowrate of air equal to that of liquid ($b = 2$), $\Phi_L^2 = 0.57$, that is, there is a pressure drop reduction of 43%. For $n = 0.1$ and $b = 5$, $\Phi_L^2 = 0.23$, corresponding to a 77% reduction.

It will be noted that for a Newtonian fluid ($n = 1$) Eq. (5.11) gives $\Phi_L^2 = 1$ for all values of b. In other words, the pressure drop will be unaffected by air injection provided that the liquid flow remains laminar. In practice, because of losses not taken into account in the simplified model, the pressure drop for a Newtonian fluid always increases with air injection.

Furthermore, it can be seen that for turbulent flow when $n \rightarrow 2$, air injection would result in substantial increases in pressure drop.

Air injection can be used, in practice, in two ways:

(1) To reduce the pressure drop, and hence the upstream pressure in a pipeline, for a given flowrate of shear-thinning liquid.
(2) To increase the flowrate of liquid for any given pipeline pressure drop.

It may be noted that energy will be required for compressing the air to the injection pressure, which must exceed the upstream pressure in the pipeline. The conditions under which power-saving is achieved have been examined by Dziubinski and Richardson,[32] who has shown that the relative efficiency of the liquid pump and the air compressor are critically important factors.

The pressure drop for a fluid exhibiting a yield stress, such as a *Bingham plastic* material, can be similarly reduced by air injection.

In a practical situation, air injection can be beneficial in that, when a pipeline is shut down, it may be easier to start up again if the line is not completely full of a slurry. On the other hand, if the pipeline follows an undulating topography, difficulties can arise if air collects at the high spots.

Air injection may sometimes be an alternative to deflocculation. In general, deflocculated suspensions flow more readily but they tend to give much more highly consolidated sediments, which can be difficult to re-suspend on starting up following a shutdown. Furthermore, deflocculants are expensive and may adversely affect the suitability of the solids for subsequent use.

5.2.5 Erosion

The flow of two-phase systems often causes erosion, and many empirical relationships have been suggested to define exactly when the effects are likely to be serious. Since high velocities may be desirable to avoid the instability associated with slug flow, there is a danger that any increase in throughput above the normal operating condition will lead to a situation where erosion may become a serious possibility.

An indication of the velocity at which erosion becomes significant may be obtained from:

$$\rho_M u_M^2 = 15{,}000$$

where ρ_M is the mean density of the two-phase mixture (kg/m^3) and u_M is the mean velocity of the two-phase mixture (m/s). Here:

$$\rho_M = [L' + G'] \bigg/ \left[\frac{L'}{\rho_L} + \frac{G'}{\rho_G}\right]$$

and:

$$u_M = u_L + u_G$$

where u_L and u_G are the superficial velocities of the liquid and gas respectively.

It is apparent that some compromise may be essential between avoiding a slug-flow condition and velocities, which are likely to cause erosion.

5.3 Flow of Liquid–Liquid Mixtures

5.3.1 Introduction

There are numerous situations where two immiscible liquids flow together in pipes, the commonest example being that of water-crude oil. Other examples are found in liquid–liquid extraction, transportation of viscous media like grease, bitumen, heavy oils, in pipes with a thin layer of water close to the wall to reduce friction.[33–35] However, such drag reduction is observed only in a specific flow regime.

The flow of liquid–liquid mixtures in pipes is somewhat similar to that of gas–liquid mixtures discussed in Section 5.2, especially when the surface tension effects are minor. However, the two also differ in significant ways as well. The density differential in liquid–liquid systems is not as large as in the case of gas–liquid systems ($\rho_G/\rho_L \sim 10^{-3}$) and this reduces the influence of gravity in determining the prevailing flow pattern in a pipe for given conditions. Thus, the interface between the two phases tends to be relatively smooth in liquid–liquid systems. Similarly, the difference in the viscosity of the two phases is not always as large here as that in the case of gas–liquid systems. On the other hand, the material of construction is an influencing factor on the flow behaviour via the wetting/nonwetting wall characteristics of the liquid, and thus surface tension plays a role here, as demonstrated by the experimental results of Angelli and Hewitt.[36]

From a practical standpoint, liquid–liquid systems can broadly be classified into two categories: emulsions in which one phase is nearly uniformly dispersed in the form of droplets in the second continuous phase. Depending upon the physical properties of the two phases, their relative properties and flow rates, one can encounter oil-in-water or water-in-oil emulsions as well as the same system can undergo a phase inversion, i.e. the dispersed and continuous phases can switch their roles. Depending upon the size of the droplets, their physical properties especially interfacial tension and relative flow rates, such emulsions are generally stable and can be treated as pseudo-homogeneous single phase fluids which may exhibit non-Newtonian (shear-thinning, yield stress, etc.) characteristics.[37–39] Emulsion rheology is strongly dependent on the volume fraction and viscosity of the dispersed phase, size distribution of the dispersed phase, shear rate, temperature, type of emulsifying agent, etc. In the second category are the systems, which are characterized by the two distinct phases present in the pipe. While one can employ the methods presented in Chapter 3 for estimating the pipe frictional pressure drop for the flow of emulsions, consideration is given here to the flow of liquid–liquid mixtures in pipes as two distinct phases. The treatment here follows the structure similar that used in Section 5.2 for the gas–liquid systems, though the currently available information on liquid–liquid systems is neither as extensive nor cohesive as that for gas–liquid systems. An excellent overview of the developments in this field is available in the literature.[40]

5.3.2 Flow Patterns

Just as in the case of gas–liquid systems, a wide variety of flow patterns has also been reported in liquid–liquid systems. While the early studies[1,41,42] on the delineation of flow regimes are based on visual observations by naked eye and/or on photographic methods, recent works are based on the measurements of local holdup, pressure fluctuations, etc. via the use of conductivity and impedance probes, gamma-ray, laser-induced fluorescence, based methods, etc. Irrespective of the method used, the classification of the flow patterns entails a degree of subjectivity and arbitrariness. Notwithstanding the larger number of possible minor variations, one can discern four main flow patterns: stratified with smooth or wavy interface, spherical or elongated slugs of one liquid in the other (similar to intermittent flow in Section 5.2), dispersed flow (fine drops of one liquid in the other), and the annular film flow wherein one of the liquids flows in the form of a thin film and the other liquid constitutes the central core. Of course, the boundaries between the two adjacent flow regimes are often blurred and therefore, mixed flow patterns are observed in most practical situations. Fig. 5.8 shows schematically the aforementioned flow patterns in horizontal flow.[43]

At relatively low flow rates, the two phases are completely stratified with a smooth interface. At such flow rates, the resulting shear forces are relatively small and the gravity dominates. With the gradual increase of the flow rates, the interfacial stress increases and the interface becomes wavy with a mushy zone involving drops on both sides of the interface. In the case of oil–water systems, oil-in-water dispersion is obtained at very high flow rates of water. Likewise, at very large oil flowrates, the water phase becomes dispersed leading to

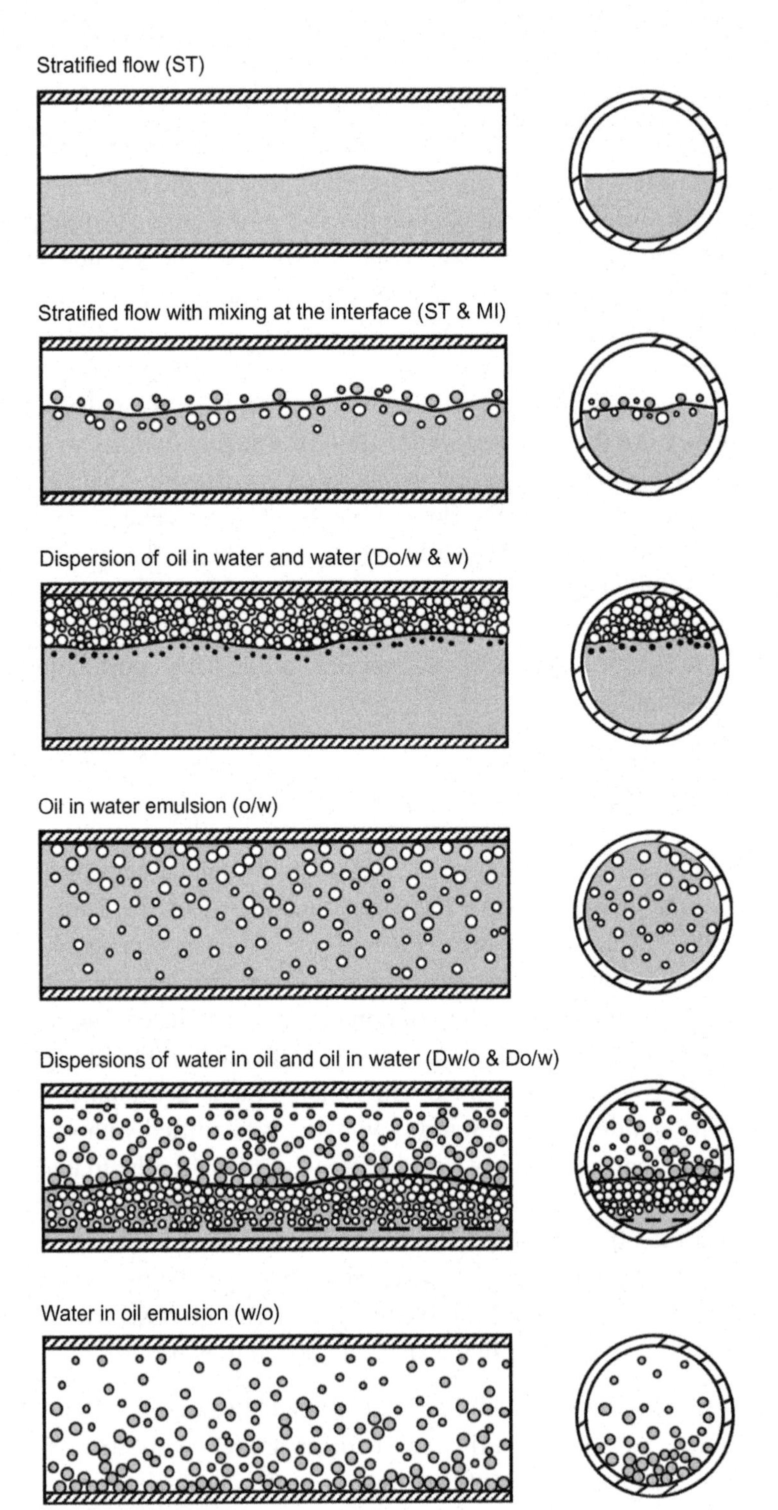

Fig. 5.8

Horizontal oil/water flow pattern sketches. *Based on Trallero JL, Sarica C, Brill JP. A study of oil/water flow patterns in horizontal pipes. SPE Prod Facil 1997;12:165–72.*

water-in-oil dispersion. As in the case of gas–liquid systems, for a given liquid–liquid combination (their densities, viscosities, interfacial tension), the flow patterns are influenced by their flow rates, orientation of pipe (horizontal, inclined, vertical), type of flow (cocurrent, countercurrent, upward, downward), and pipe diameter. Thus, for instance, in vertical upward low viscosity oil–water flow,[44] one can encounter oil bubbles or slugs in water, churn transitional, water drops in oil, and eventually at very high flow rates oil/water or water/oil emulsions, as shown schematically in Fig. 5.9. Of course, stratified flow pattern is not possible in vertical and inclined (>30° from horizontal) pipes. More detailed description of the various flow patterns and of conditions for transition from one to another is available in the literature.[40]

Owing to a degree of similarity between the flow pattern behaviour of the gas–liquid and liquid–liquid systems in horizontal flow, it is natural to wonder if the flow pattern maps developed for gas–liquid systems (e.g., Fig. 5.2) can be adapted for the liquid–liquid systems. In this context, Brauner[40] has proposed that the gas–liquid flow pattern maps also work for liquid–liquid systems characterized by large values of the familiar Eotvös number, $Eo = (\Delta \rho g D^2/8\sigma) \gg 1$ (gravity dominated flow). In this case, the density difference ($\Delta\rho$) and inclination (component of gravity) determine the transition from one flow pattern to another. On the other hand, the flow pattern transitions in liquid–liquid systems characterized by small values of the Eotvös number are influenced by surface tension, wettability characteristics, the manner in which the second phase is introduced in the pipe, etc. In this case, if the pipe diameter is lower than a critical value ($\sim(\sigma/\Delta\rho g)^{1/2}$), one may not observe stratified flow even at low flow rates.[45] The dramatic effect of Eotvös number on flow patterns in vertical upward flow is shown in Fig. 5.10. Thus, as of now, no completely satisfactory flow pattern map is

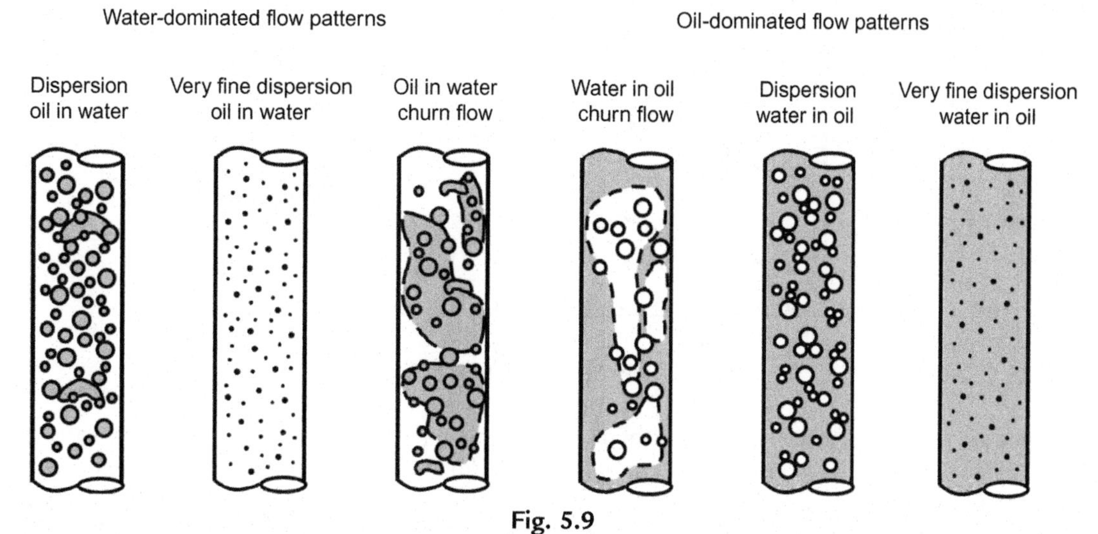

Fig. 5.9
Schematic representation of the vertical oil-water flow patterns. *Based on Flores JG, Chen XT, Sarica C, Brill JP. Characterization of oil-water flow patterns in vertical and deviated wells. SPE Prod Facil 1999;14:102–9.*

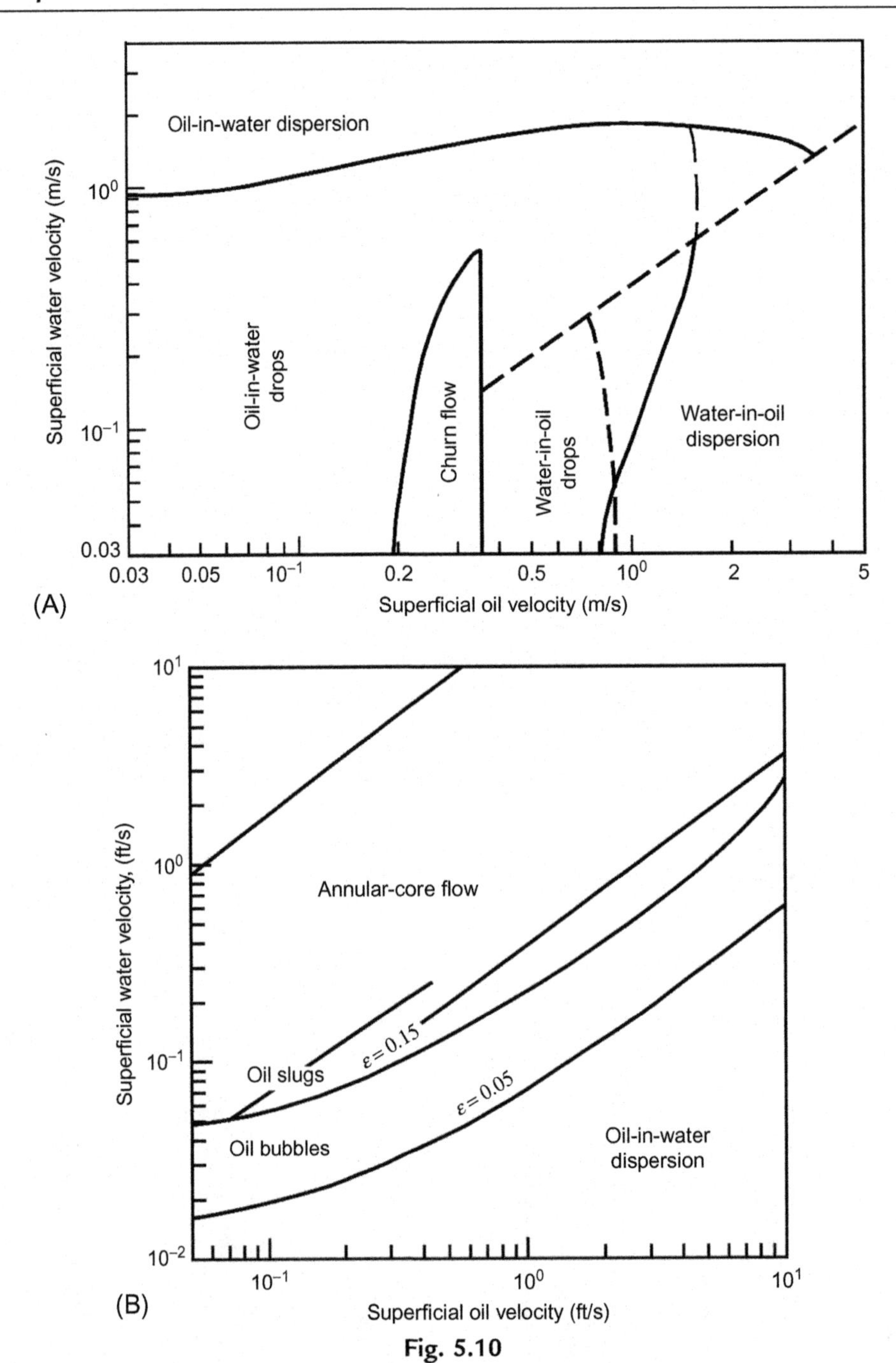

Fig. 5.10

Flow pattern map for vertical upward flow of oil/water system. (A) $\mu_{oil} = 20$ mPa.s; $\rho_{oil} = 850$ kg/m^3; $E_0 = 14$; $D = 50$ mm. (B) $\mu_{oil} = 600$ mPa.s; $\rho_{oil} = 910$ kg/m^3; $D = 9.5$ mm; $E_0 = 1.17$. Here ε is the hold up of less viscous phase. *Based on Brauner N. In: Bertola V, editor. Modelling and control of two-phase flow phenomena: liquid-liquid two-phaseflow systems. Udine, Italy: International Center for Mechanical Sciences; 2004.*

available for the flow of liquid–liquid mixtures in pipes. Many investigators[40] have developed flow pattern maps using the superficial velocities of the two phases, while others[46] have used the input volume fraction of one phase and the mixture velocity as the coordinates. Undoubtedly, the wetting characteristics and physical properties of the two phases do seem to influence the transitions from one flow pattern to another, it is not yet clear as to how to incorporate these effects into the prediction of flow patterns.

5.3.3 Average Holdup

Only limited data is available on the average holdup of the two phases in horizontal[43,47,48] and vertical[44,49] pipes. Figs 5.11 and 5.12 show representative results on holdup as functions of the superficial velocities of the two phases. In vertical flows, the total pressure gradient includes a significant contribution from the static head and therefore, it is important to estimate or measure the holdup of each phase in this case. Early attempts on the prediction of the average holdup are based on the assumption of "no-slip" between the two-phase (homogeneous models) and in this case the *in situ* holdup is equal to the volume fraction at the inlet to pipe. However, in practice, due to the density difference, there will be a slip (particularly in the stratified flow regime) between the two phases and the so-called homogeneous model is not expected to work well. In vertical flow, it is possible to estimate in situ holdup of a phase by using the idea of Zuber and Findlay.[50] This approach necessitates the estimation of the terminal falling velocity of the droplets of the dispersed phase,

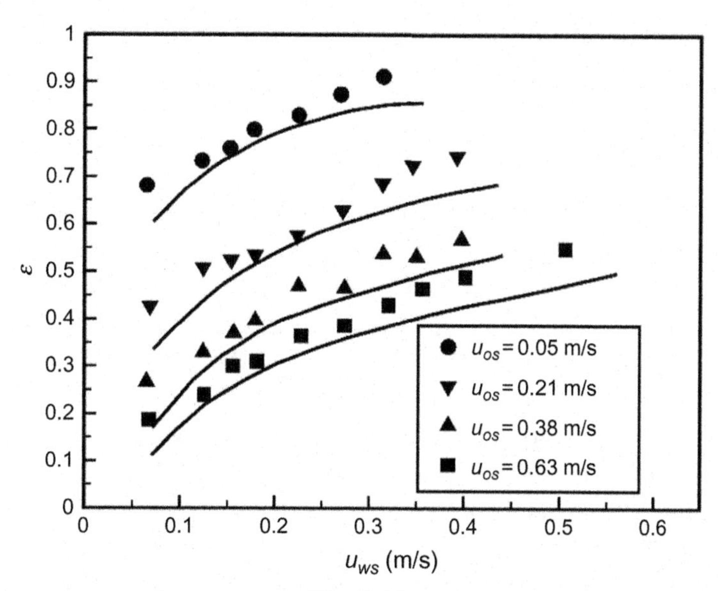

Fig. 5.11

Representative water hold up data. *From Liu Y, Zhang H, Wang S, Wang J. Prediction of pressure gradient and holdup in horizontal liquid-liquid segregated flow with small Eotvos number. Chem Eng Commun 2009;196:697–714.*

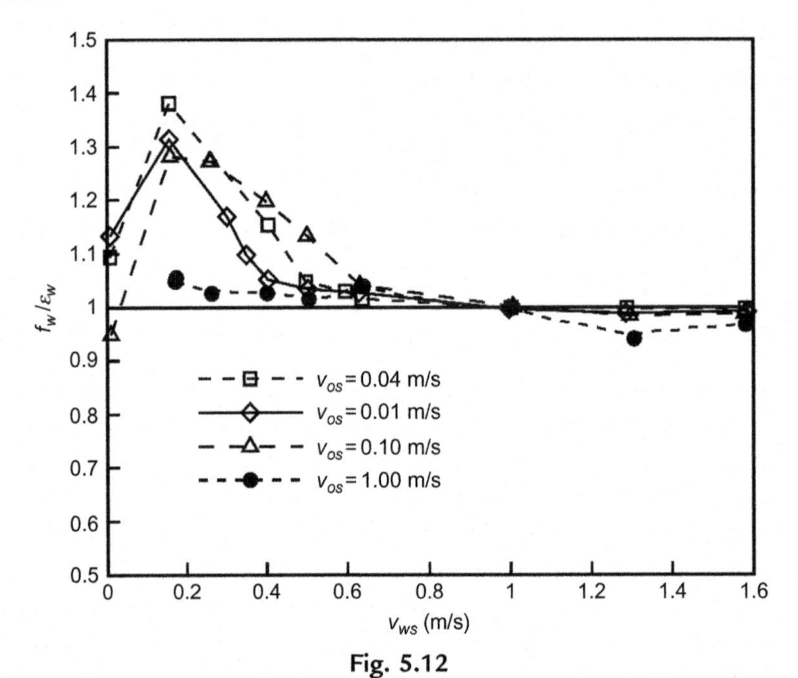

Fig. 5.12

Experimental hold up ratio. f_w and ε_w are the inlet fraction and hold up pf water. *From Trallero JL, Sarica C, Brill JP. A study of oil/water flow patterns in horizontal pipes. SPE Prod Facil 1997;12:165–72.*

as has been demonstrated in reference 40. In horizontal flow, energetic considerations (minimizing the surface and potential energy of the interface), Brauner et al.[51] and Liu et al.[47] have obtained implicit expressions for holdup and these predictions are seen to be in line with the scant experimental data.

5.3.4 Pressure Gradient

Just as in the case of gas–liquid flows, the pressure drop for the flow of liquid–liquid mixtures is determined by a large number of kinematic variables and physical properties of the two phases, flow pattern, and flow configuration (concurrent, countercurrent, pipe inclination, etc.). Many authors have attempted to develop semitheoretical expressions for the prediction of pressure drop for a specific flow regime, e.g. Brauner,[40] Al-Wahibi,[52] Liu et al.[47] for stratified flows with smooth or wavy interface, core annular flows[40,49] etc. On the other hand, some attempts have also been made to develop a general correlation along the lines of Lockhart–Martinelli approach presented in the preceding section on gas–liquid flows. Fig. 5.13 shows representative experimental data for an oil–water system in a horizontal pipe.[47] Dasari et al.[53] have attempted to adapt the formulation of Lockhart–Martinelli for liquid–liquid systems in horizontal pipes. Treating the less viscous phase (water in most cases) as the 'gas' and the other more viscous liquid (oils of high viscosity) as the 'liquid' phase, one can use the same definitions as that given by Eqs (5.1), (5.6), and (5.7). Based on the available experimental data, Dasari et al.[53] put forward the following correlation for ϕ_G^2:

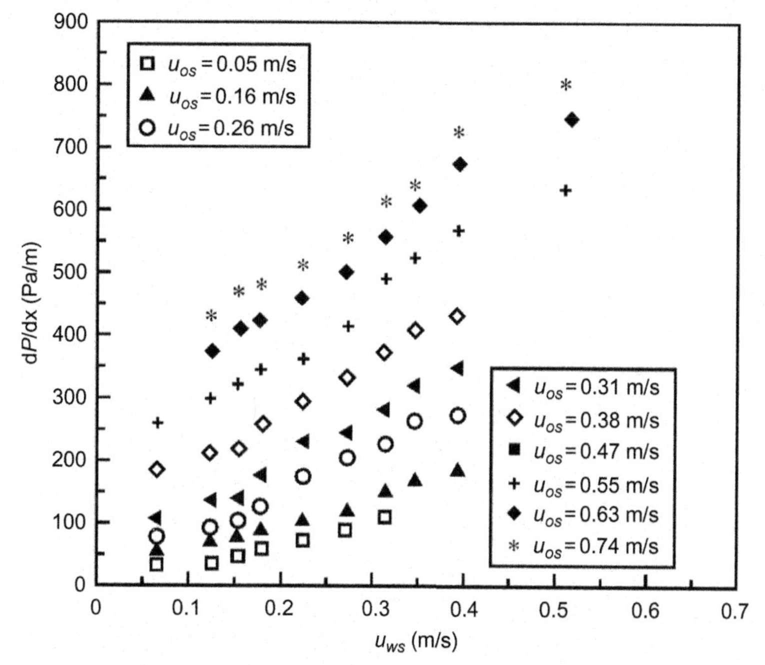

Fig. 5.13

Experimental mean pressure gradient data. *From Liu Y, Zhang H, Wang S, Wang J. Prediction of pressure gradient and holdup in horizontal liquid-liquid segregated flow with small Eotvos number. Chem Eng Commun 2009;196:697–714.*

$$\frac{-\Delta P_{TP}}{-\Delta P_W} = \phi_G^2 = 1.4 - \frac{0.006}{X} + \frac{0.06}{X^2} + \frac{0.019}{X^3} \tag{5.12}$$

Eq. (5.12) was stated to predict the two-phase pressure drop with an uncertainty of $\pm 25\%$ over the range $\sim 0.1 \leq X \leq 20$.

5.4 Flow of Solids–Liquid Mixtures

5.4.1 Introduction

Hydraulic transport is the general name frequently given to the transportation of solid particles in liquids—the term *hydraulic* relates to the fact that most of the earlier applications of the technique involved water as the carrier fluid. Today there are many industrial plants, particularly in the mining industries, where particles are transported in a variety of liquids. Transport may be in vertical or horizontal pipes and, as is usually the case in long pipelines, it may follow the undulations of the land over which the pipeline is constructed. The diameter and length of the pipeline and its inclination, the properties of the solids and of the liquid, and the flowrates all influence the nature of the flow and the pressure gradient. Design methods are, in general, not very reliable, particularly for the transportation of coarse particles in horizontal pipelines and calculated values of pressure drop and energy requirements should be treated with caution. In practice, it may be more important to ensure that the system operates reliably, and without the risk of blockage and without excessive erosion, than to achieve optimum conditions in relation to power requirements.

The most important variables, which must be considered in estimating power consumption and pressure drop are:

(a) The pipeline length, diameter, inclination to the horizontal, necessity for bends, valves, etc.
(b) The liquid its physical properties, including density, viscosity, and rheology, and its corrosive nature, if any.
(c) The solids their particle size and size distribution, shape, and density—all factors which affect their behaviour in a fluid.
(d) The concentration of particles and the flowrates of both solids and liquid.

Suspensions are conveniently divided into two broad classes—*fine suspensions* in which the particles are reasonably uniformly distributed in the liquid; and *coarse suspensions* in which particles tend to travel predominantly in the bottom part of a horizontal pipe at a lower velocity than the liquid, and to have a significantly lower velocity than the liquid in a vertical pipe. This is obviously not a sharp classification and furthermore, is influenced by the flowrate and concentration of solids. However, it provides a useful initial basis on which to consider the behaviour of solid–liquid mixtures.

5.4.2 Homogeneous Nonsettling Suspensions

Fine suspensions are reasonably homogeneous and segregation of solid and liquid phases does not occur to any significant extent during flow. The settling velocities of the particles are low in comparison with the liquid velocity and the turbulent eddies within the fluid are responsible for the suspension of the particles. In practice, turbulent flow will always be used, except when the liquid has a very high viscosity or exhibits non-Newtonian characteristics. The particles may be individually dispersed in the liquid or they may be present as flocs.

In disperse, or deflocculated, suspensions, the particles generally all have like charges and therefore repel each other. Such conditions exist when the pH range and the concentration of ions in the liquid is appropriate; this subject is discussed in Volume 2. Disperse suspensions tend to exhibit Newtonian behaviour. High fractional volumetric concentrations (0.4–0.5) are achievable and pressure drops for pipeline flow are comparatively low and correspond closely with those calculated for homogeneous fluids of the same density as the suspension. The particles do not have a very large effect on the viscosity of the liquid except at very high concentrations, when non-Newtonian shear-thickening characteristics may be encountered as a result of the build up of a 'structure'. There are two possible reasons for not transporting particles in the deflocculated state. First, they tend to form dense coherent sediments and, if the flow in the pipe is interrupted, there may be difficulties in restarting operation. Secondly, the cost of the chemicals, which need to be added to maintain the particles in the dispersed state may be considerable.

In most large-scale pipelines, the suspensions of fine particles are usually transported in the flocculated state. Flocs consist of large numbers of particles bound loosely together with liquid occluded in the free space between them. They therefore tend to behave as relatively large particles of density intermediate between that of the liquid and the solid. Because some of the liquid is immobilised within the flocs, the maximum solids concentration obtainable is less than that for deflocculated suspensions. The flocs are fragile and can break down and deform in the shear fields, which exist within a flowing fluid. They therefore tend to exhibit non-Newtonian behaviour. At all but the highest concentrations they are *shear-thinning* and frequently exhibit a *yield stress* (Chapter 3). Their behaviour, over limited ranges of shear rates, can usually be reasonably well-described by the *power law* or *Bingham plastic* models (see Eqs 3.120 and 3.122). Thixotropic and viscoelastic effects are usually negligible under the conditions existing in hydraulic transport lines.

Because concentrated flocculated suspensions generally have high apparent viscosities at the shear rates existing in pipelines, they are frequently transported under laminar flow conditions. Pressure drops are then readily calculated from their rheology, as described in Chapter 3. When the flow is turbulent, the pressure drop is difficult to predict accurately and will generally be somewhat *less* than that calculated assuming Newtonian behaviour. As the Reynolds number becomes greater, the effects of non-Newtonian behaviour become progressively less. There is thus a safety margin for design purposes if the suspensions are treated as Newtonian fluids when in turbulent flow.

Since, by definition, the settling velocity of the particles is low in a fine suspension, its behaviour is not dependent on its direction of flow and, if allowance is made for the hydrostatic pressure, pressure gradients are similar in horizontal and vertical pipelines.

In a series of experiments on the flow of flocculated kaolin suspensions in laboratory and industrial scale pipelines,[54–56] measurements of pressure drop were made as a function of flowrate. Results were obtained using a laboratory capillary-tube viscometer, and pipelines of 42 and 205 mm diameter arranged in a recirculating loop. The rheology of all of the suspensions was described by the *power-law* model with a power law index less than unity, that is they were all shear-thinning. The behaviour in the laminar region can be described by the equation:

$$|R_y| = k\left|\frac{du_x}{dy}\right|^n \tag{5.13}$$

(see Eq. 3.121) where,

Table 5.2 Power-law parameters for flocculated kaolin suspensions

Solids Volume Fraction, *C*	Flow Index, *n*	Consistency, *k* (Ns^n/m^2)
0.086	0.23	0.89
0.122	0.18	2.83
0.142	0.16	4.83
0.183	0.15	15.3
0.220	0.14	32.4
0.234	0.13	45.3

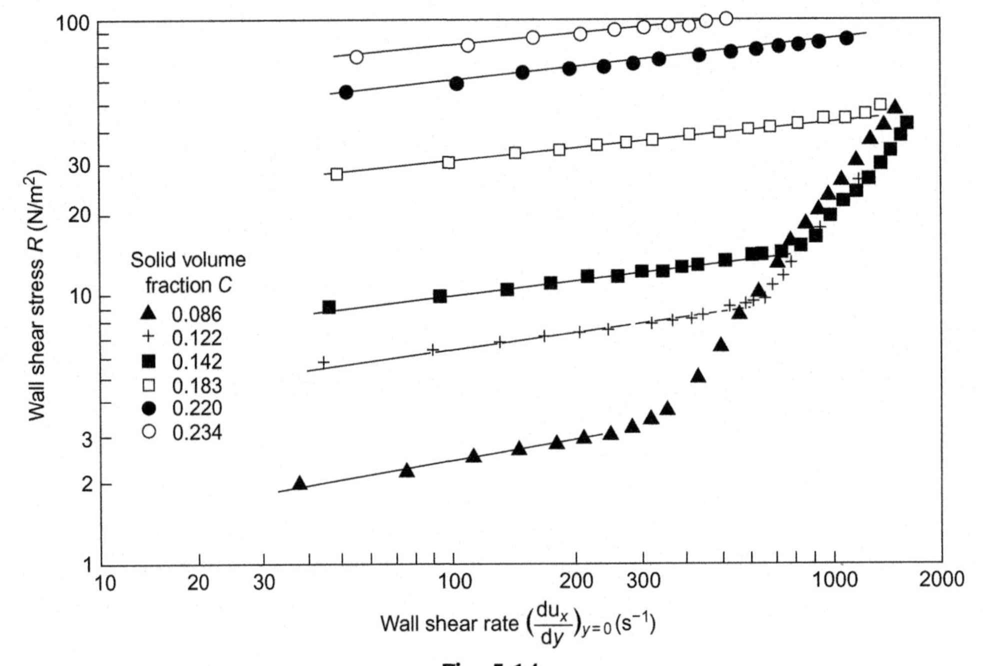

Fig. 5.14
Rheograms for flocculated kaolin suspensions.

R_y is the shear stress at a distance y from the wall,
u_x is the velocity at that position,
n is the *flow index*, and
k is the *consistency*.

Values of n and k for the suspensions used are given in Table 5.2. Experimental results are shown in Fig. 5.14 as wall shear stress R as a function of wall shear rate $(du_x/dy)_{y=0}$ using logarithmic coordinates.

It is shown in Chapter 3 (Eq. 3.137) that:

$$\left[\frac{du_x}{dy}\right]_{y=0} = \frac{6n+2}{n}\left(\frac{u}{d}\right) \tag{5.14}$$

Fig. 5.14 shows clearly the transition point from laminar to turbulent flow for each of the suspensions when flowing in the 42 mm diameter horizontal pipe.

5.4.3 Coarse Solids

The flow behaviour of suspensions of coarse particles is completely different in horizontal and vertical pipes. In horizontal flow, the concentration of particles increases towards the bottom of the pipe, the degree of nonuniformity increasing as the velocity of flow is decreased. In vertical transport however, axial symmetry is maintained with the solids evenly distributed over the cross-section. The two cases are therefore considered separately.

5.4.4 Coarse Solids in Horizontal Flow

Only with fine solids and/or at uneconomically high velocities are the particles uniformly distributed over the cross-section of a horizontal pipe. For coarse particles, the following principal types of flow are observed as the velocity is decreased:

(a) Heterogeneous suspension with all the particles suspended but with a significant concentration gradient vertically.
(b) Heterogeneous suspension in the upper part of the bed but a sliding bed moving along the bottom of the pipe.
(c) A similar pattern to (b), but with the bed composed of moving layers at the top and a stationary deposit at the bottom.
(d) Transport as a bed with the lower layers stationary and a few particles moving over the surface of the bed in intermittent suspension.

In addition, it is possible to obtain what is known as *dense phase flow*[57] with the particles filling the whole bore of the pipe and sliding with little relative movement between the particles.

In all cases where the two phases are moving with different velocities, it is important to differentiate between the concentration of particles in the pipe (their holdup ϵ_S) and the volume fraction of particles (C) in the discharge. The implications of this will now be considered, together with possible means of experimentally determining the holdup.

Hold-up and slip velocity

In any two-phase flow system in which the two phases are flowing at different velocities, the in-line concentration of a component will differ from that in the stream, which leaves the end of the pipe. The in-line concentration of the component with the *lower* velocity will be greater than its concentration in the exit stream because it will have a longer residence time. It is important to understand what is happening within the pipe because the relative velocity between the phases results in energy transfer from the faster to the slower moving component. Thus, in hydraulic transport the liquid will transfer energy to the solid particles at a rate, which is a function of the *slip velocity*. The solid particles will, in turn, be losing energy as a result of impact with the walls and frictional effects. In the steady state, the rate of gain and of loss of energy by the particles will be equal.

It is not possible to calculate the in-line concentrations and slip velocity from purely *external* measurements on the pipe, i.e. knowledge of the rates at which the two components are delivered from the end of the pipe provides no evidence for what is happening within the pipe. It is thus necessary to measure *one* or more of the following variables:

The absolute linear velocity of the particles u'_S.
The absolute linear velocity of the liquid u'_L.
The slip velocity $U_R = u'_L - u'_S$.
The hold-up of the solids ϵ_S.
The hold-up of the liquid $\epsilon_L = 1 - \epsilon_S$.

In an extensive experimental study of the transport of solids by liquid in a 38 mm diameter pipe,[58] the following variables were measured:

u is the mixture velocity (by electromagnetic flowmeter)
u'_L is the linear velocity of liquid (by salt injection method)
ϵ_S is the hold-up of solid particles in the pipe (by γ-ray absorption method).

In the salt injection method,[59] a pulse of salt solution is injected into the line and the time is measured for it to travel between two electrode pairs positioned a known distance apart, downstream from the injection point.

The γ-ray absorption method of determining in-line concentration (hold-up) of particles depends on the different degree to which the solid and the liquid attenuate γ-rays; details of the method are given in the literature.[18,19]

All the other important parameters of the systems can be determined from a series of material balances as follows:

The superficial velocity of the liquid:

$$u_L = u'_L(1 - \epsilon_S) \tag{5.15}$$

The superficial velocity of the solids: $u_s = u - u_L$

$$= u - u'_L(1 - \epsilon_S) \tag{5.16}$$

where the superficial velocity of a component is defined as the velocity it would have at the same volumetric flowrate if it occupied the total cross-section of the pipe.

The absolute velocity of the solids:

$$u_S' = \frac{u_S}{\epsilon_s} = u_L' - \frac{1}{\epsilon_S}(u_L' - u) \tag{5.17}$$

The slip, or relative, velocity:

$$u_R = u_L' - u_S' = \frac{1}{\epsilon_S}(u_L' - u) \tag{5.18}$$

The fractional volumetric concentration of solids C in the mixture emerging from the end of the pipe can then be obtained simply as the ratio of the superficial velocity of the solids (u_S) to the mixture flowrate.

$$u = (u_s + u_L)$$

or:

$$C = \frac{u_S}{u} = \frac{u_s' \epsilon_S}{u} = 1 - \frac{u_L'}{u}(1 - \epsilon_S)$$

or:

$$\frac{u_R}{u} = \frac{\epsilon_S - C}{(1 - \epsilon_S)\epsilon_S} \tag{5.19}$$

The consistency of the data can be checked by comparing values calculated using Eq. (5.19) with the measured values for samples collected at the outlet of the pipe.

When an industrial pipeline is to be designed, there will be no *a priori* way of knowing what the in-line concentration of solids or the slip velocity will be. In general, the rate at which solids are to be transported will be specified and it will be necessary to predict the pressure gradient as a function of the properties of the solid particles, the pipe dimensions, and the flow velocity. The main considerations will be to select a pipeline diameter, such that the liquid velocity and concentrations of solids in the discharged mixture will give acceptable pressure drops and power requirements and will not lead to conditions where the pipeline is likely to block.

It is found that the major factor, which determines the behaviour of the solid particles is their terminal falling velocity in the liquid. This property gives a convenient way of taking account of particle size, shape, and density.

In the experimental study, which has just been referred to, it was found that the slip velocity was of the same order as the terminal falling velocity of the particles. Although there is no theoretical basis, this assumption does provide a useful working guide and enables all the internal parameters, including holdup, to be calculated for a given mixture velocity u and delivered concentration C using Eqs (5.15)–(5.19). It is of interest to note that, in pneumatic conveying (discussed in Section 5.5), slip velocity is again found to approximate terminal falling velocity.

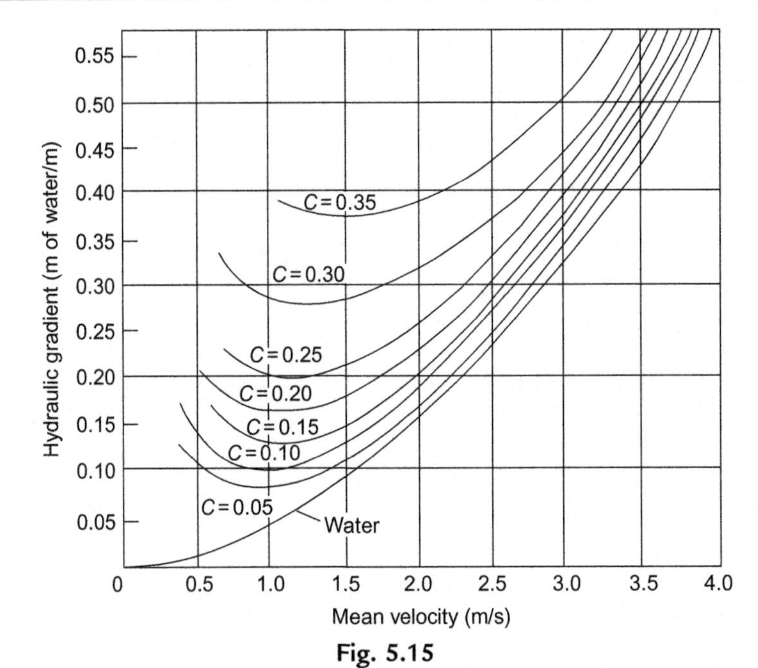

Fig. 5.15
Hydraulic gradient-velocity curves for 200 μm sand in 25 mm diameter hydraulic conveying line.

Some of the experimental results of different researchers and methods of correlating results are now described.

Predictive methods for pressure drop

A typical curve for the conveying of solids in water is shown in Fig. 5.15 which refers to the transport of *ca.* 200 μm sand in a small pipeline of 25 mm diameter.[60] It shows hydraulic gradient i (head lost per unit length of pipe) as a function of the mean velocity, with delivered concentration (C) as parameter. Each curve shows a minimum, which corresponds approximately to the transition between flow with a bed and suspended flow. The economic operating condition is frequently close to the critical velocity. It will be noted that as concentration is increased, the minimum occurs at a progressively high velocity. The contribution of the solids to hydraulic gradient is seen to be greatest at low velocities. Great care should be exercised in operating at velocities below the critical values as the system is then unstable and blockage can easily occur; it is a region in which the pressure drop increases as the velocity is reduced. The prediction of conditions under which blockage is likely to occur is a complex area and the subject has been discussed in depth by Hisamitsu et al.[61] and Wilson et al.[62]

There were several studies of hydraulic transport in the 1950s, sparked off particularly by an interest in the economic possibilities of transportation of coal and other minerals over long distances. Newitt et al.[60] working with solids of a range of particle sizes (up to 5 μm) and densities (1180–4600 kg/m^3) in a 25 mm diameter pipe, suggested separate correlations for flow with a bed deposit and for conditions where the particles were predominantly in heterogeneous suspension.

For flow where a bed deposit tends to form:

$$\frac{i-i_w}{Ci_w}=66\frac{gd}{u^2}(s-1)=66\left[\frac{u^2}{gd(s-1)}\right]^{-1} \tag{5.20}$$

and for heterogeneous suspensions of particles of terminal falling velocity u_0:

$$\frac{i-i_w}{Ci_w}=1100\frac{gd}{u^2}(s-1)\frac{u_0}{u}=1100\frac{u_0}{u}\left[\frac{u^2}{gd(s-1)}\right]^{-1} \tag{5.21}$$

The following features of Eqs (5.20) and (5.21) should be noted.

a. They both represent pressure gradient in the form of the quotient of a dimensionless excess hydraulic gradient $(i - i_w)/i_w$ and the delivered concentration (C). This implies that the excess pressure gradient is linearly related to concentration. Subsequent work casts doubt on the validity of this assumption, particularly for flow in suspension.[63]
b. The excess pressure gradient is seen to be inversely proportional to a modified Froude number $u^2/[gd(s-1)]$ in which s is the ratio of the densities of the solids and the liquid. The pipe diameter d has been included to make the right hand side dimensionless but its effect was not studied.
c. Particle characteristics do not feature in the correlation for flow with a bed. This is because the additional hydraulic gradient is calculated using a force balance in which the contribution of the particles to pressure drop is attributed to solid–solid friction at the walls of the pipe. This equation does not include the coefficient of friction, the importance of which will be referred to in a later section.

In a comprehensive study carried out at roughly the same time by Durand[64–66] the effect of pipe diameter was examined using pipes of large diameter (40–560 mm) and a range of particle sizes d_p. The experimental data were correlated by:

$$\frac{i-i_w}{Ci_w}=121\left\{\frac{gd}{u^2}(s-1)\frac{u_0}{\left[gd_p(s-1)\right]^{1/2}}\right\}^{1.5} \tag{5.22}$$

Reference to Volume 2 (Chapter 3) shows that at low particle Reynolds numbers (Stokes' Law region), $u_0 \propto d_p^2$ and that at high Reynolds number $u_0 \propto d_p^{1/2}$—at intermediate Reynolds numbers the relation between u_0 and d_p is complex, but over a limited range $u_0 \propto d_p^m$ where $\frac{1}{2} < m < 2$. It will be seen, therefore, that the influence of particle diameter is greatest for small particles (low Reynolds numbers) and becomes progressively less as particle size increases, becoming independent of size at high Reynolds numbers—as in Eq. (5.20), which refers to flow with a moving bed. Durand and Condolios[67] found in their experiments using large diameter pipes that particle size when in excess of 20 mm did not affect the pressure difference. James and Broad[68] also used pipelines ranging from 102 to 207 mm diameter for the transportation of coarse particles under conditions, which tended to give rise to the formation of a bed deposit. Their results suggest that the coefficient in Eq. (5.20) is dependent on pipe diameter and that the constant value of 66 should be replaced by $(60+0.24d)$ where d is in mm. Most of their results related to conditions of heterogeneous suspension.

Eqs (5.21) and (5.22) give results, which are reasonably consistent, and they both give $(i - i_w)/Ci_w$ proportional to u^{-3}.

The term $u_0/[gd_p(s - 1)]^{1/2}$ is shown in Volume 2 (Chapter 3) to be proportional to the reciprocal square root of the *drag coefficient* (C_D) for a particle settling at its terminal falling velocity.

Substituting $\frac{4}{3}(s-1)\left(gd_p/u_0^2\right)=C_D$ into Eq. (5.22) gives:

$$\frac{i-i_w}{Ci_w}=150\left\{\frac{gd}{u^2}(s-1)\frac{1}{\sqrt{C_D}}\right\}^{1.5} \tag{5.23}$$

If the concentration term C is transferred to the right-hand side of Eq. (5.23) (taking account of the fact that $(i - i_w)$ is not necessarily linearly related to C), it may be written as:

$$\frac{i-i_w}{i_w}=f\left[\frac{u^2\sqrt{C_D}}{gd(s-1)C}\right] \tag{5.24}$$

In Volume 2A, the drag coefficient $C'_D(=C_D/2)$ is used in the calculation of the behaviour of single particles. However, C_D is used in this chapter to facilitate comparison with the results of other researchers in the field of Hydraulic Transport.

In Fig. 5.16,[69] results of a number of researchers[61,63,68–73] covering a wide range of experimental conditions, are plotted as:

$$\frac{i-i_w}{i_w}\text{versus}\frac{u^2\sqrt{C_D}}{gd(s-1)C}$$

The scatter of the results is considerable but this arrangement of the groups seems to be the most satisfactory. The best line through all the points is given by:

$$\frac{i-i_w}{i_w}=30\left[\frac{u^2\sqrt{C_D}}{gd(s-1)C}\right]^{-1} \tag{5.25}$$

Zandi and Govatos[74] suggest that the transition from flow with bed formation to flow as a heterogeneous suspension occurs at the condition where:

$$x=\frac{u^2\sqrt{C_D}}{Cgd(s-1)}=40 \tag{5.26}$$

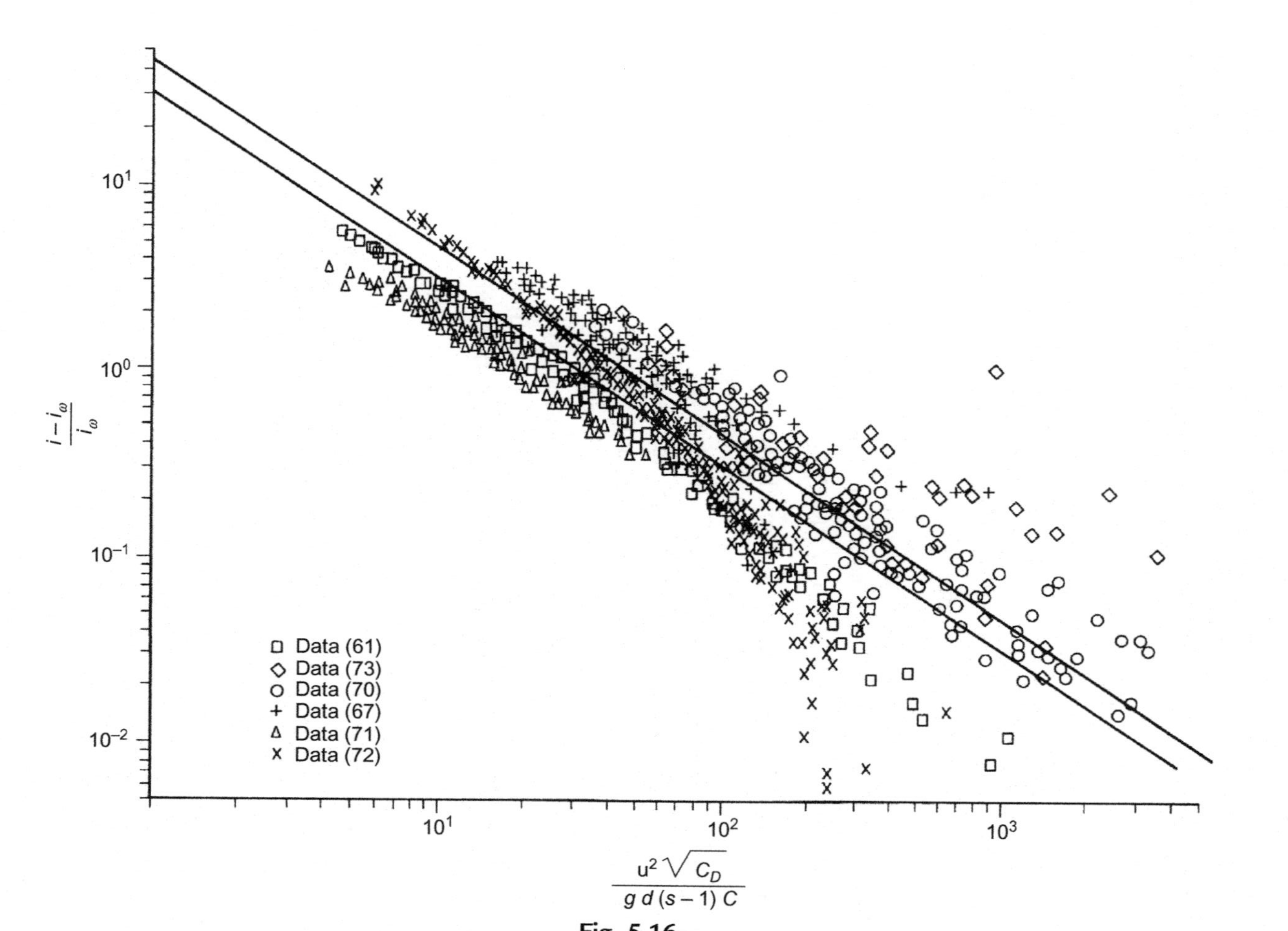

Fig. 5.16
Excess hydraulic gradient due to solids as function of modified Froude number. Comparison of results of different researchers.

For $x < 40$, corresponding to flow with a bed, the best value for the slope of the line in Fig. 5.16 is about -1. For higher values of x, corresponding to heterogeneous suspension, there is some evidence for a slope nearer -1.5, in line with Eqs (5.22) and (5.23) of Durand. However, in this region, values of $(i - i_w)/i_w$ are so low that the experimental errors will be very high, for example at $x = 300$, $(i - i_w)/i_w \asymp 0.1$ and friction losses differ from those for water alone by only about 10%. What the figure does show is that the excess pressure gradient due to the solids becomes very large only under conditions where bed formation tends to occur ($x < 40$). As this is also the region of greatest practical interest, further consideration will be confined to this region.

It will be noted that in this region ($x < 40$) the experimental data show an approximately fourfold spread of ordinate at any given value of x.

Even taking into account the different experimental conditions of the various researchers and errors in their measurements, and the generally unstable nature of solid–liquid flow, Eq. (5.25) is completely inadequate as a design equation.

An alternative approach to the representation of results for solid–liquid flow is to use the *two-layer model* that will be described in the following section. It will be seen that the coefficient of friction between the particles and the wall of the pipe is an important parameter in the model. It is suggested that its complete absence in Eq. (5.25) may be an important reason for the extent of the scatter. Unfortunately, it is a quantity which has been measured in only a very few investigations. It is interesting to note that the form of Eq. (5.20) was obtained by Newitt et al.[60] using a force balance similar to that employed in the two-layer model and it is implicit in their analysis that the coefficient (66 in Eq. (5.20)) will be a function of the coefficient of friction between the particles at the wall of the pipe.

Roco and Shook[75,76] do not accept that the suspended solids and the bed constitute two identifiable regions in the pipe and have developed a model for the prediction of concentrations and velocities as continuous functions of position in the cross-section. The prediction of pressure drop from their theory involves complex numerical calculations and there is some difficulty in assessing the range of conditions over which it can be used because of the paucity of experimental data available. Some of the earlier experiments were carried out by Shook and Daniel[77] who used a γ-ray absorption system for measuring in line concentration profiles. An extensive programme of experimental work was carried out at the Saskatchewan Research Council with a view to providing much needed data on large diameter pipelines (250 mm).

The two-layer model

It is seen that the models put forward do not adequately explain the behaviour of hydraulic conveying systems and they poorly correlate the results of different researchers.

A two-layer model was first proposed by Wilson[78] in 1976 to describe the flow of solid–liquid mixtures in pipes under conditions where part of the solids are present in a moving bed and part are in suspended flow in the upper part of the pipe. He carried out a steady-state force balance on the two layers in order to calculate the pressure drop in the pipeline. The original model has been modified considerably by Wilson himself and by others.[79,80] However, it has not been used widely for design purposes, partly because of its complexity and the difficulty of obtaining convergent solutions to the iterative calculations, which are necessary. Furthermore, it is necessary to be able to predict both the proportion of the solids, which are present in each layer and the interfacial shear at the upper surface of the bed. The former is dependent on the experimental measurement of solids distribution and the generation of a reliable method of predicting it in terms of system properties and flowrates. The latter, the interfacial shear stress, cannot readily be measured and is calculated on the basis that the upper liquid layer is moving over the surface of a bed whose effective roughness is a function of the size of the particles.

The force balance is as follows (Fig. 5.17):

$$\text{For the bed: } F_B = -\Delta P A_B = (\mu_F \Sigma F_N + R_B S_B - R_i S_i) l \tag{5.27}$$

$$\text{For the upper layer: } F_L = -\Delta P A_L = (R_i S_i + R_L S_L) l \tag{5.28}$$

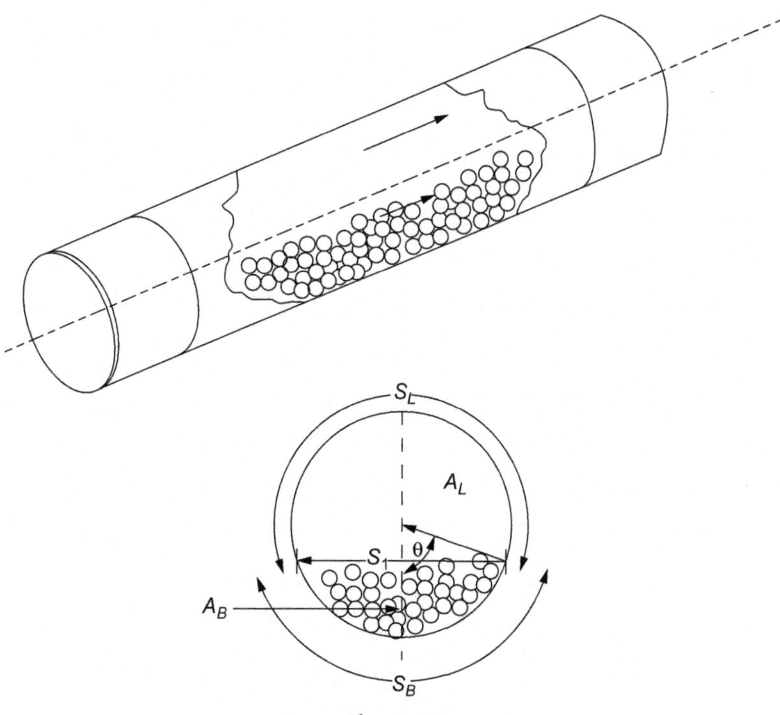

Fig. 5.17
Solids distribution over pipe section in the two-layer model.

where

$-\Delta P$ is the pressure drop over the pipe of length l,
A_B is the cross-section occupied by bed,
A_L is the cross-section occupied by the upper layer,
R_B is the shear stress at the wall in the liquid in the bed,
R_i is the interfacial shear stress,
R_L is the shear stress at the wall in the liquid above the bed,
S_B is the perimeter of the cross-section in contact with the bed,
S_i is the length of the chord between the two layers, and.
S_L is the perimeter of the cross-section in contact with the upper layer.

The term $\mu_F \Sigma F_N$ is the product of the coefficient of friction μ_F and the sum ΣF_N of the normal forces per unit length between the particles and the wall.

Adding Eqs (5.27) and (5.28) and dividing by A, the total cross-section area of the pipe, then:

$$\frac{-\Delta P}{l} = \frac{R_L S_L + \mu_F \Sigma F_N + R_B S_B}{A} \tag{5.29}$$

It may be noted that the term $R_i S_i$ cancels out, the interfacial drag forces on the two layers being equal and opposite.

The two perimeters S_L and S_B are functions of the depth of the bed and the angle subtended by the bed surface to the centre of the pipe. In order to calculate their values, it is necessary to know the in-line concentration of solids in the pipe and the voidage of the bed.

The shear stress R_L at the pipe wall in the upper portion of the pipe may be calculated on the assumption that the liquid above the bed is flowing through a noncircular duct, bounded at the top by the wall of the pipe and at the bottom by the upper surface of the bed. The hydraulic mean diameter may then be used in the calculation of wall shear stress. However, this does not take account of the fact that the bottom boundary, the top surface of the bed, is not stationary and not impenetrable, and will have a greater effective roughness than the pipe wall. The effects of these two assumptions do, however, operate in opposite directions and therefore partially cancel each other out. The shear stress R_B due to the flow of liquid in the bed, relative to the pipe wall, is calculated on the assumption that it is unaltered by the presence of the particles, since their size is likely to be considerably greater than the thickness of the laminar sublayer for turbulent flow in a pipe.

Before the pressure drop can be calculated from Eq. (5.29), it is still necessary to know the in-line concentration of solids in the pipe for a given flowrate and discharge concentration. Furthermore, it is necessary to know how the liquid and solid are distributed between the two layers.

The *two-layer model* is being progressively updated as fresh experimental results and correlations become available. The most satisfactory starting-point for anyone wishing to use the model to calculate pressure gradients for flow of solids–liquid mixtures in a pipeline is the text of Shook and Roco[81] and that of Wilson[62] which have worked examples. However, there are many pitfalls to be avoided in this area, and there is no substitute for practical experience gained by working in the field.

In an extensive study of the transport of coarse solids in a horizontal pipeline of 38 mm diameter,[58] measurements were made of pressure drop, as a function not only of mixture velocity (determined by an electromagnetic flowmeter) but also of in-line concentration of solids and liquid velocity. The solids concentration was determined using a *γ-ray absorption* technique, which depends on the difference in the attenuation of γ-rays by solid and liquid. The liquid velocity was determined by a *salt injection* method,[59] in which a pulse of salt solution was injected into the flowing mixture, and the time taken for the pulse to travel between two electrode pairs a fixed distance apart was measured. It was then possible, using Eq. (5.18), to calculate the relative velocity of the liquid to the solids. This relative velocity was found to increase with particle size and to be of the same order as the terminal falling velocity of the particles in the liquid.

The values of R_L, S_L, R_B, and S_B were then calculated based on the following assumptions:

1. That the relative velocity u_R was equal to the terminal falling velocity of the particles. This assumption does not necessarily apply to large pipes.
2. That the particles all travelled in a bed of porosity equal to 0.5. It had previously been shown that, as soon as bed formation occurred, the major part of the contribution of the solids was attributable to those present within the bed.
3. That the particles and liquid in *the bed* were travelling at the same velocity, i.e. there was no slip within the bed.

In addition, off-line experiments were carried out to measure the coefficient of solid friction μ_F between the particles and the surface of the pipe. The term $\mu_F \Sigma F_N$ was then calculated on the assumption that ΣF_N was equal to the net weight (actual minus buoyancy) of the particles in the pipeline. This assumption was reasonable provided that the bed did not occupy more than about 30% of the pipe cross-section.

A sensitivity analysis showed that the calculated value of pressure gradient was very sensitive to the value of the coefficient of friction μ_F, but was relatively insensitive to the slip velocity u_R and to the bed voidage.

Some examples of calculated pressure gradients are shown in Fig. 5.18 in which solids concentration is a parameter. Experimental points are given, with separate designations for each concentration band. The difference between the experimental and predicted value does not generally exceed about 15%.

The sensitivity of the pressure drop to the coefficient of solids-surface friction μ_F may well account for the wide scatter in the results shown earlier in Fig. 5.16. Unfortunately this quantity has been measured by only very few investigators. It must be emphasised that in the design

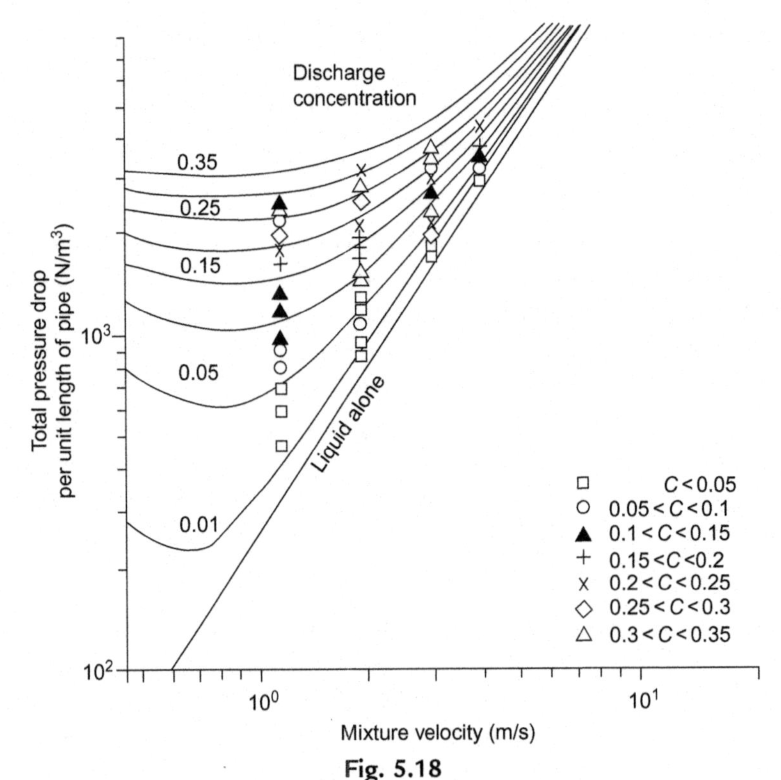

Fig. 5.18

Comparison of predicted and experimental results for simplified "two-layer" model (3.5 mm gravel in 38 mm pipe[58]).

of any hydraulic transport system it is extremely important to have a knowledge of the coefficient of friction.

Transport of coarse particles in heavy and shear-thinning media

There have been several studies involving the use of media consisting of fine dense particles suspended in water for transporting coarse particles. The fine suspension behaves as a homogeneous fluid of increased density, but its viscosity is not sufficiently altered to have a significant effect on the pressure drop during turbulent flow, the normal condition for hydraulic transport. The cost of the dense particles may however, be appreciable and their complete separation from the coarse particles may be difficult.

More recently, attention has been focused on the use of suspensions of fines at very high concentrations; the desired increase in density can then be achieved using lighter particles. Furthermore, if the particles are flocculated, the suspension will generally exhibit shear-thinning rheology conforming approximately to the power-law or the Bingham-plastic model (see Chapter 3). Contamination of the product can frequently be eliminated by using fines of the same material as the coarse particles. For example, coarse particles of coal can be transported in suspensions of fine coal with beneficial results.

Concentrated (20%–30% by volume) flocculated suspensions of coal and of china clay are highly shear-thinning (power-law index $0.1 < n < 0.2$) and, at the shear rates encountered in pipeline flow, have high apparent viscosities, as a result of which the flow is frequently laminar. In these circumstances, the suspension has a very high apparent viscosity in the low-shear region in the core of the flow, and therefore suspension of coarse particles is facilitated. At the same time, the apparent viscosity in the high-shear region near the pipe wall is much lower and pressure drops are not excessive. Thus, in effect, a shear-thinning suspension is a tailor-made fluid for hydraulic transport giving a highly desirable variation of apparent viscosity over the cross-section of the pipe. Similar benefits also accrue if the conveying medium exhibits viscoplastic behaviour.

Solutions of some polymeric materials also have similar rheological properties and behave in the same way as suspensions in pipe flow. However, they do in time break down and they do not give any appreciable increase in buoyancy as their densities differ little from that of water. Then, of course, there is the issue of contamination.

One of the earlier studies was carried out in 1971 by Charles and Charles[82] who investigated the feasibility of transporting coarse materials in heavy media (sand in flocculated suspensions of clay in water). They concluded that the power requirement for the transport of one million tonnes/year of solids in a 200 mm diameter pipe could be reduced by a factor of about 6 by using heavy media in place of water. Kenchington,[83] who also studied the transport of coarse particles in china clay suspensions, concluded that a significant proportion of the coarse particles may be suspended in the dense medium without a noticeable increase in pressure drop. Duckworth et al.[84–86] showed that it is possible to transport coal with a maximum particle size of 19 mm in a 250 mm diameter pipeline under laminar conditions using a fine coal slurry as the carrier fluid. Work has also been carried out on characterising suspensions of coarse particles in fines by means of a form of rotary viscometer, a 'shearometer', and designing pipelines on the basis of the resulting measurements.[87]

Chhabra and Richardson[88] transported coarse particles in a 42 mm pipe in a wide range of fluids, including Newtonian liquids of high viscosity, shear-thinning polymer solutions, and shear-thinning suspensions of fine flocculated particles. They correlated their own and Kenchington's experimental results by means of linear regression to give:

$$\frac{i - i_w}{i_w}\frac{\phi}{C} = 0.20\left[\frac{u^2}{gd(s-1)}\right]^{-1} \tag{5.30}$$

A somewhat improved correlation is obtained by nonlinear regression as:

$$\frac{i - i_w}{i_w}\frac{\phi}{C} = 0.28\left[\frac{u^2}{gd(s-1)}\right]^{-1.25} \tag{5.31}$$

ϕ is the friction factor, $R/\rho u^2$, for the flow of the liquid alone at the same superficial velocity u. Because most of the work was carried out in the laminar-flow region (high apparent viscosity of the medium), ϕ is very sensitive to the flowrate. This contrasts with the work described in the previous section using water. The flow was then turbulent and the friction factor did not vary significantly over the range of velocities studied; in effect, the value of ϕ was implicitly incorporated in the coefficient in the equations. It is of interest to note that Eq. (5.30) is consistent with Eq. (5.20) for heterogeneous suspension in water for a value of ϕ equal to 0.003, corresponding to a Reynolds number of about 22,000 for flow in a smooth pipe.

Transport of particles of low density

Hydraulic transport has been extensively used in the minerals industry over a long period. A characteristic of mineral particles is that they have higher densities than the transporting fluid and always travel at a lower velocity than the liquid. Recently, developments in the continuous processing of foodstuffs[89–91] have led to the use of shear-thinning non-Newtonian fluids (usually starch-based) to transport food particles through heat exchangers for the purpose of sterilisation. Laminar flow conditions usually prevail, but the particles are readily maintained in suspension because they are almost neutrally-buoyant in the carrier fluid which usually has a very high apparent viscosity, and in many cases a yield stress. As the particles tend to migrate away from the walls of the heat exchanger to the core region where the velocity of the fluid is greatest, their velocities are frequently greater than the mean velocity of the fluid, and the slip velocity is then negative. The residence time of the particles will then be lower than that calculated from the mean velocity of flow, and it is important that this difference be taken into account in order to avoid under-sterilisation of the foodstuff.

5.4.5 Coarse Solids in Vertical Flow

Durand[92] has also studied vertical transport of sand and gravel of particle size ranging between 0.18 and 4.57 mm in a 150 mm diameter pipe, and Worster and Denny[93] conveyed coal and gravel in vertical pipes of diameters 75, 100, and 150 mm. They concluded that the pressure drop for the slurry was the same as for the water alone, if due allowance was made for the static head attributable to the solids in the pipe.

Newitt et al.[94] conveyed particles of densities ranging from 1190 to 4560 kg/m^3, of sizes 0.10 to 3.8 mm, in a 25 mm diameter pipe 12.8 m tall and in a 50 mm diameter pipe 6.7 m tall. The particles used had a 30-fold variation in terminal falling velocity. It was found that the larger particles had little effect on the frictional losses, provided the static head due to the solids was calculated on the assumption that the particles had a velocity relative to the liquid equal to their terminal falling velocities. Furthermore, it was shown photographically that at high velocities these particles travel in a central core, and thus the frictional forces at the wall will be unaffected by their presence. Very fine particles of sand give suspensions which behave as homogeneous fluids, and the hydraulic gradient due to friction is the same as for horizontal flow. When the settling velocity cannot be neglected in comparison with the liquid velocity, the hydraulic gradient was found to be given by:

$$\frac{i-i_w}{Ci_w}=0.0037\left(\frac{gd}{u^2}\right)^{1/2}\frac{d}{d_p}u_0^2 \tag{5.32}$$

For the transport of coarse particles, the relative velocity between the liquid and solids is an important factor determining the hold-up, and hence the in-line concentration of solids. Cloete et al.[95] who conveyed sand and glass ballotini particles through vertical pipes of 12.5 and 19 mm diameter at velocities up to 3 m/s, obtained values of the in-line concentration by means of a γ-ray technique similar to that discussed earlier. The vertical pressure gradient in the pipe was measured and the component due to the hydrostatic pressure of the mixture was subtracted in order to obtain the frictional component. It was found that the frictional component was similar to that for water alone for velocities up to 0.7 m/s, but tended to increase at higher velocities.

In the absence of a direct measurement of the in-line concentration of solids ϵ_S, it is necessary to make an estimate of its value in order that the hydrostatic pressure gradient in the pipe may be calculated. This can be done for a given mixture velocity u and delivered concentration C, provided that the velocity of the particles relative to the liquid u_R is known.

Following the same argument as used for horizontal flow (Eqs 5.15–5.19):

For unit area of pipe cross-section:

Mixture flowrate = Solids flowrate + Liquid flowrate

$$u=u'_S\epsilon_S+u'_L(1-\epsilon_S) \tag{5.33}$$

where u'_S and u'_L are the linear velocities of solid and liquid, respectively.

and

$$u_R=u'_L-u'_S. \tag{5.34}$$

Considering the flow of solids:

$$uC=u'_S\epsilon_S \tag{5.35}$$

From Eqs (5.33)–(5.35), the relation between ϵ_S and C is given by:

$$C=\epsilon_S\left[1-\frac{u_R}{u}(1-\epsilon_S)\right] \tag{5.36}$$

For the transport of a dilute suspension of solids, u_R will approximate to the free-falling velocity u_0 of the particles in the liquid. For concentrated suspensions, a correction must be applied to take into account the effect of neighbouring particles. This subject is considered in detail in Volume 2A (Chapter 5) from which it will be seen that the simplest form of correction takes the form:

$$\frac{u_R}{u_0} = (1 - \epsilon_S)^{m-1} \tag{5.37}$$

where m ranges from about 4.8 for fine particles to 2.4 for coarse particles. Eqs (5.36) and (5.37) are then solved simultaneously to give the value of ϵ_S, from which the contribution of the solids to the hydrostatic pressure may be calculated.

Kopko et al.[96] made measurements of the pressure drop for the flow of suspensions of iron shot (0.0734–0.131 mm diameter) in a vertical pipe 61 mm diameter and 4.3 m tall. The frictional pressure drop was obtained by subtracting the hydrostatic component from the total measured pressure drop and it was found that it constituted only a very small proportion of the total measured. Subsequently, Al-Salihi[97] has confirmed that, if allowance is made for the hydrostatic component using the procedure outlined above, the frictional pressure drop is largely unaltered by the presence of the solids.

In summary, therefore, it is recommended that the frictional pressure drop for vertical flow be calculated as follows:

For *nonsettling suspensions*: The standard equation for a single phase fluid is used with the physical properties of the suspension in place of those of the liquid.

For *suspensions of coarse particles*: The value calculated for the carrier fluid flowing alone at the mixture velocity is used.

Example 5.2

Sand with a mean particle diameter of 0.2 mm is to be conveyed in water flowing at 0.5 kg/s in a 25 mm internal diameter horizontal pipe 100 m long. Assuming fully suspended flow, what is the maximum amount of sand that may be transported in this way if the head developed by the pump is limited to 300 kN/m^2?

The terminal falling velocity of the sand particles in water may be taken as 0.0239 m/s. This value may be confirmed using the method given in Volume 2.

Solution

Assuming the mean velocity of the suspension is equal to the water velocity, that is, neglecting slip, then:

$$u_m = 0.5/(1000 \times \pi \times 0.025^2/4) = 1.02\,\text{m/s}$$

For water alone, flowing at 1.02 m/s:

$$Re = (0.025 \times 1.02 \times 1000)/0.001 = 25,500$$

Assuming $e/d = 0.008$, then, from Fig. 3.7:

$$\phi = 0.0046 \quad \text{or} \quad f = 0.0092$$

From, Eq. (3.20), the head loss is:

$$h_f = (8 \times 0.0046)(100/0.025)\left(1.02^2/(2 \times 9.81)\right) = 7.8\,\text{m water}$$

and the hydraulic gradient is:

$$i_w = (7.8/100) = 0.078\,\text{m water/m}$$

A pressure drop of 300 kN/m^2 is equivalent to:

$(300 \times 1000)/(1000 \times 9.81) = 30.6\,\text{m water}$

and hence $i = (30.6/100) = 0.306\,\text{m water/m}$

Substituting in Eq. (5.21):

$$(0.306 - 0.078)/0.078C = 1100(9.81 \times 0.025/1.02^2)(0.0239/1.02)(2.6 - 1)$$

from which: $C = 0.30$

[Eq. (5.22) may also be used, in which case:

$$(0.306 - 0.078)/0.078C = 121[9.81 \times 0.025(2.6 - 1)0.0239]/[1.02^2(9.81 \times 0.0002(2.6 - 1)^{0.5}])^{1.5}$$

from which:

$$C = 0.36$$

which is a very similar result.]

If G kg/s is the mass flow of sand, then:

$$\text{Volumetric flow of sand} = (G/2600) = 0.000385G\ \text{m}^3/\text{s}$$
$$\text{Volumetric flow of water} = (0.5/1000) = 0.0005\,\text{m}^3/\text{s}$$

and:

$$0.000385G/(0.000385G + 0.0005) = 0.30$$

from which:

$$\underline{\underline{G = 0.56\ \text{kg/s}}} = \underline{\underline{2\,\text{tonne/h}}}$$

5.5 Flow of Gas–Solids Mixtures

5.5.1 General Considerations

Pneumatic conveying involves the transport of particulate materials by air or other gases. It is generally suitable for the transport of particles in the size range 20 μm to 50 mm. Finer particles cause problems arising from their tendency to adhere together and to the walls of the pipe and ancillary equipment. Sticky and moist powders are the worst of all. Large particles may need excessively high velocities in order to maintain them in suspension or to lift them from the bottom of the pipe in horizontal systems. Pneumatic conveying lines may be horizontal, vertical, or inclined and may incorporate bends, valves, and other fittings, all of which may exert a considerable influence on the behaviour of the whole installation. Whereas it is possible to make reasonable predictions of pressure gradients in long straight pipes, the effects of such fittings can be ascertained reliably only from the results of practical studies under conditions close to those that will be experienced in the final installation. The successful operation of a pneumatic conveyor may well depend much more on the need to achieve reliable operation, by removing the risks of blockage and of damage by erosion, than on achieving conditions which optimise the performance of the straight sections of the pipeline. It is important to keep changes in direction of flow as gradual as possible, to use suitable materials of construction (polyurethane lining is frequently employed) and to use velocities of flow sufficiently high to keep the particles moving, but not so high as to cause serious erosion. Whenever possible, it is desirable to carry out pilot scale tests with a sample of the solids.

Two characteristics of the conveying fluid result in considerable differences in the behaviour of pneumatic and hydraulic conveying systems: fluid density and compressibility. In hydraulic conveying, the densities of the solids and fluid are of the same order of magnitude, with the solids usually having a somewhat higher density than the liquid. Practical flow velocities are commonly in the range of 1–5 m/s. In pneumatic transport, the solids may have a density two to three orders of magnitude greater than the gas and velocities will be considerably greater—up to 20–30 m/s. In a mixture with a volume fraction of solids, of say 0.05, the mass ratio of solids to fluid will be about 0.15–0.20 for hydraulic conveying compared with about 50 for pneumatic conveying. Compressibility of the gas is important because it expands as it flows down the pipe and if, for example, the ratio of the pressures at the upstream and downstream ends is 4, the velocity will increase by this factor and it would be necessary to double the diameter of the pipe to maintain the same linear gas velocity as at the beginning of the pipe. In practice, 'stepped' pipelines are commonly used with progressively larger pipes used towards the downstream end. Horizontal pipelines up to 500 m long are in common use and a few 2000 m lines now exist; vertical lifts usually do not exceed about 50 m.

There are three basic modes of transport that are employed. The first and most common, is termed *dilute phase* or *lean phase* transport in which the volume fraction of solids in the suspension does not exceed about 0.05 and a high proportion of the particles spend most of their time in suspension. The second is transport which takes place largely in the form of a *moving bed* in which the solids volume fraction may be as high as 0.6; this is relevant only for horizontal or slightly inclined pipelines. The third form is *dense phase* transport in which fairly close packed 'slugs' of particles, with volume fractions of up to 0.5, alternate with slugs of gas and are propelled along the pipe. Because the velocities (~3 m/s) are very much lower than in dilute phase transport, there is less attrition of particles and less pipe wear. The gas usage is reduced but the risk of blockage is serious in horizontal pipelines.

Considerably more work has been carried out on horizontal as opposed to vertical pneumatic conveying. A useful review of relevant work and of correlations for the calculation of pressure drops has been given by Klinzing et al.[98,99] Some consideration will now be given to horizontal conveying, with particular reference to dilute phase flow, and this is followed by a brief analysis of vertical flow.

5.5.2 Horizontal Transport

Flow patterns

In a horizontal pipeline, the distribution of the solids over the cross-section becomes progressively less uniform as the velocity is reduced. The following flow patterns which are commonly encountered in sequence at decreasing gas velocities have been observed in pipelines of small diameter.

1. *Uniform suspended flow*
   The particles are evenly distributed over the cross-section across the whole length of pipe.
2. *Nonuniform suspended flow*
   The flow is similar to that described above but there is a tendency for particles to flow preferentially in the lower portion of the pipe. If there is an appreciable size distribution, the larger particles are found predominantly at the bottom.
3. *Slug flow*
   As the particles enter the conveying line, they tend to settle out before they are fully accelerated. They form dunes, which are then swept bodily downstream giving an uneven longitudinal distribution of particles along the pipeline.
4. *Dune flow*
   The particles settle out as in slug flow but the dunes remain stationary with particles being conveyed above the dunes and also being swept from one dune to the next.
5. *Moving bed*
   Particles settle out near the feed point and form a continuous bed on the bottom of the pipe. The bed develops gradually throughout the length of the pipe and moves slowly

forward. There is a velocity gradient in the vertical direction in the bed, and conveying continues in suspended form above the bed.

6. *Stationary bed*

   The behaviour is similar to that of a moving bed, except that there is virtually no movement of the bed particles. The bed can build up until it occupies about three-quarters of the cross-section. Further reduction in velocity quickly gives rise to a complete blockage of the pipe.

7. *Plug flow*

   Following slug flow, the particles, instead of forming stationary dunes, gradually build up over the cross-section until they eventually cause a blockage. This type of flow is less common than dune flow.

Suspension mechanisms

The mechanism of suspension is related to the type of flow pattern prevailing in the pipe. Suspended types of flow are usually attributable to dispersion of the particles by the action of the turbulent eddies in the fluid. In turbulent flow, the vertical component of the eddy velocity will lie between one-seventh and one-fifth of the forward velocity of the fluid and, if this is more than the terminal falling velocity of the particles, they will tend to be supported in the fluid. In practice, it is found that this mechanism is not as effective as might be thought because there is a tendency for the particles to damp out the eddy currents.

If the particles tend to form a bed, they will be affected by the lateral dispersive forces described by Bagnold.[100,101] A fluid in passing through a loose bed of particles exerts a dilating action on the system. This gives rise to a dispersion of the particles in a direction at right angles to the flow of fluid.

If a particle presents a face inclined at an angle to the direction of motion of the fluid, it may be subjected to an upward lift due to the *aerofoil* effect. In Fig. 5.19, a flat plate is shown at an angle to a stream of fluid flowing horizontally. The fluid pressure acts normally at the surface and thus produces forces with vertical and horizontal components as shown.

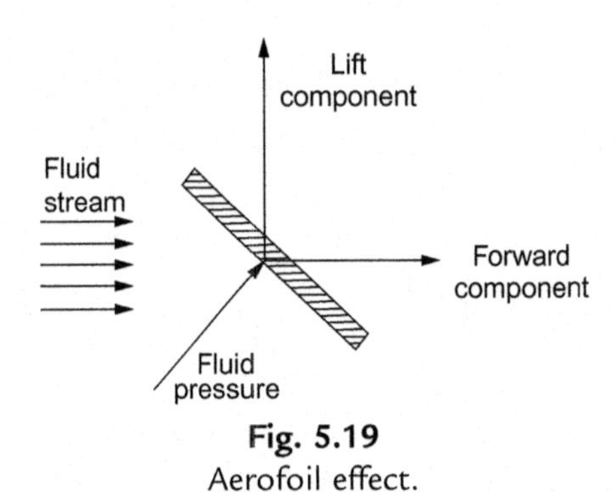

Fig. 5.19
Aerofoil effect.

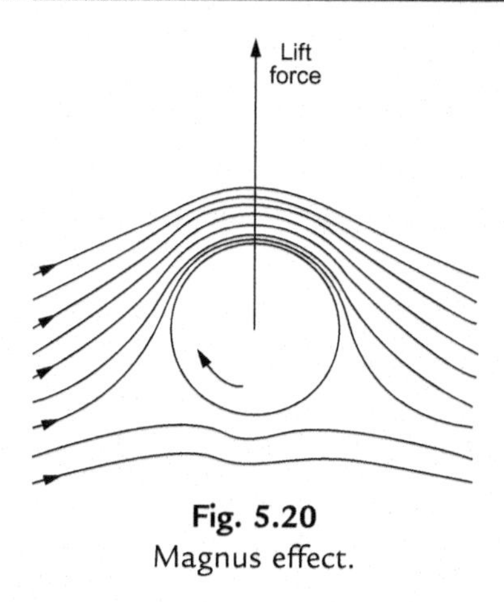

Fig. 5.20
Magnus effect.

If the particle rotates in the fluid, it will be subjected to a lift on the *Magnus Principle*. Fig. 5.20 shows a section through a cylinder rotating in a fluid stream. At its upper edge, the cylinder and the fluid are both moving in the same direction, but at its lower edge they are moving in opposite directions. The fluid above the cylinder is therefore accelerated, and that below the cylinder is retarded. Thus the pressure is greater below the cylinder and an upward force is exerted.

The above processes, other than the action of the dispersive forces, can result in an upward or downward displacement of an individual particle, there being approximately equal chances of the net force acting upwards or downwards. However, because the gravitational force gives rise to a tendency for the concentration of particles to be greater at the bottom of the pipe, the overall effect is to lift the particles.

The principal characteristic of a particle, which determines the dominant suspension mechanism is its terminal falling velocity. Particles with low falling velocities will be readily suspended by the action of the eddies, whereas the dispersive forces will be most important with particles of high falling velocities. In a particular case, of course, the fluid velocity will also be an important factor, full suspension of a given particle occurring more readily at high velocities.

Energy requirements for dilute phase conveying

The energy required for conveying can conveniently be considered in two parts: that required for the flow of the air alone, and the additional energy necessitated by the presence of the particles, similar to the idea used in the context of hydraulic transport in Section 5.4. It should be noted, however, that the fluid friction will itself be somewhat modified for the following reasons: the total cross-sectional area will not be available for the flow of fluid; the pattern of turbulence will be affected by the solids; and the pressure distribution through the pipeline will be different, and hence the gas density at a given point will be affected by the solids.

The presence of the solids is responsible for an increased pressure gradient for a number of reasons. If the particles are introduced from a hopper, they will have a lower forward velocity than the fluid and therefore have to be accelerated. Because the relative velocity is greatest near the feed point and progressively falls as the particles are accelerated, their velocity will initially increase rapidly and, as the particles approach their limiting velocities, the acceleration will become very small. The pressure drop due to acceleration is therefore greatest near the feed point. Similarly, when solids are transported round a bend, they are retarded and the pressure gradient in the line following the bend is increased as a result of the need to accelerate the particles again. In pneumatic conveying, the air is expanding continuously along the line and therefore the solid velocity is also increasing. Secondly, work must be done against the action of the earth's gravitational field because the particles must be lifted from the bottom of the pipe each time they drop. Finally, particles will collide with one another and with the walls of the pipe, and therefore their velocities will fall and they will need to be accelerated again. Collisions between particles will be less frequent and result in less energy loss than impacts with the wall, because the relative velocity is much lower in the former case.

The transference of energy from the gas to the particles arises from the existence of a relative velocity. The particles will always be travelling at a lower velocity than the gas. The loss of energy by a particle will generally occur on collision and thus be a discontinuous process. The acceleration of the particle will be a gradual process occurring after each collision, the rate of transfer of energy falling off as the particle approaches the gas velocity.

The accelerating force exerted by the fluid on the particle will be a function of the properties of the gas, the shape and size of the particle, and the relative velocity. It will also depend on the dispersion of the particles over the cross-section and the shielding of individual particles. The process is complex and therefore it is not possible to develop a precise analytical treatment, but it is obviously important to know the velocity of the particles.

Determination of solid velocities

The determination of the velocity of individual particles can be carried out in a number of ways. First, the particles in a given section of pipe can be isolated using two rapidly acting shutters or valves, and the quantity of particles trapped measured by removing the intervening section of pipe. This method was used by Segler,[102] Clark et al.,[103] and Mitlin.[104] It is, however, cumbersome, very time-consuming, and dependent upon extremely good synchronisation of the shutters. Another method is to take two photographs of the particles in the pipeline at short time-intervals and to measure the distance the particles have travelled in the time.

This method is restricted to very low concentrations of relatively large particles, and does not permit the measurement of more than a few particles.[105]

A third method consists of measuring the time taken for a "tagged" particle (e.g. radioactive or magnetic) to travel between two points.[106] The method gives results applicable only to an isolated particle, which may not be representative of the bulk of the particles. These techniques can readily be used in experimental equipment but are not practicable for industrial plants.

Cross-correlation methods

Because of the general oscillations in the flow condition in pneumatic conveying, it is possible to use sensing probes to measure the high frequency fluctuations at two separate locations at a specified distance apart and then to determine the mean time interval between a given pattern appearing at the two locations. Klinzing and Mathur[107] used a dielectric measuring device for this purpose. Such methods are readily applied to both laboratory and large-scale plants.

Richardson and McLeman[108] used a method which was really a precursor of the cross-correlation technique. They injected a pulse of air into the conveying line as a result of which there was a very short period during which the walls at any particular point were not subject to bombardment by particles, and the noise level was substantially reduced. By placing two transducers in contact with the wall of the pipe at a known distance apart and connecting each to a thyratron, it was possible to arrange for the first to start a frequency counter and for the second to stop the counter. A very accurate method of timing the air pulse was thus provided. It was found that the pulse retained its identity over a long distance, and this suggested that the velocities of all the particles tended to be the same. The method enabled extremely rapid and accurate measurements of the solids velocity to be obtained. Over a 16 m distance the error was $<1\%$.

The importance of obtaining accurate measurements of solid velocity is associated with the fact that the drag exerted by the fluid on the particle is approximately proportional to the square of the relative velocity. As the solid velocity frequently approaches the air velocity, the necessity for very accurate values is apparent.

Pressure drops and solid velocities for dilute phase flow

When solid particles are introduced into an air stream a large amount of energy is required to accelerate the particles, and the acceleration period occupies a considerable length of pipe. In order to obtain the values of pressure gradients and solid velocities under conditions approaching equilibrium, measurements must be made at a considerable distance from the feed point. Much of the earlier experimental work suffered from the facts that conveying lines were much too short and that the pressure gradients were appreciably influenced by the acceleration of the particles. A typical curve obtained by Clark et al.[103] for the pressure

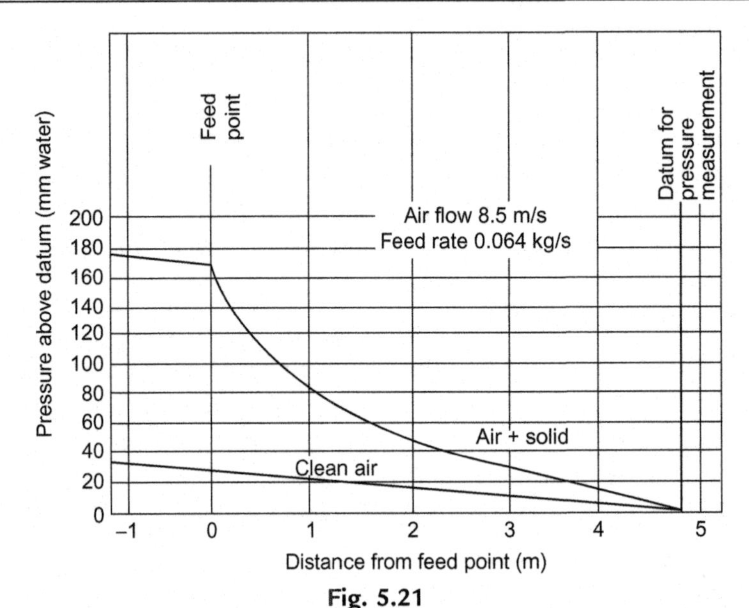

Fig. 5.21
Pressure in horizontal 25 mm conveying line for transport of cress seed.

distribution along a 25 mm diameter conveying line, is shown in Fig. 5.21. It may be noted that the pressure gradient gradually diminishes from a very high value in the neighbourhood of the feed point to an approximately constant value at distances greater than about 2 m. It has been confirmed experimentally that the solid velocity is increasing in the region of decreasing pressure gradient and that the length of the acceleration period increases with the mass of the particles, as might be expected. When the pressure gradient is approximately constant, so is the solid velocity. A further factor, which makes it necessary to obtain measurements over a long section of pipe is that the pressure gradient does in many cases exhibit a wave form.[108] This appears to be associated with the tendency for dune formation to occur within the pipe, and thus the measured value of the pressure gradient may be influenced by the exact location of the pressure tappings. It is therefore suggested that measurements should be made over a length of at least 15 m of pipe, and remote from the solids feed point.

If the pressure drop is plotted against airflow rate as shown in Fig. 5.22, it is seen that at a given feed rate the curve always passes through a minimum. At air velocities above the minimum of the curve, the solids are in suspended flow, but at lower velocities particles are deposited at the bottom of the pipe and there is a serious risk of blockage occurring.

Accurate measurements of solid velocities and pressure gradients have been made by Richardson and McLeman[108] using a continuously operating system in which the solids were separated from the discharged air in a cyclone separator and introduced again to the feed hopper at the high pressure end of the system by means of a specially constructed rotary valve.

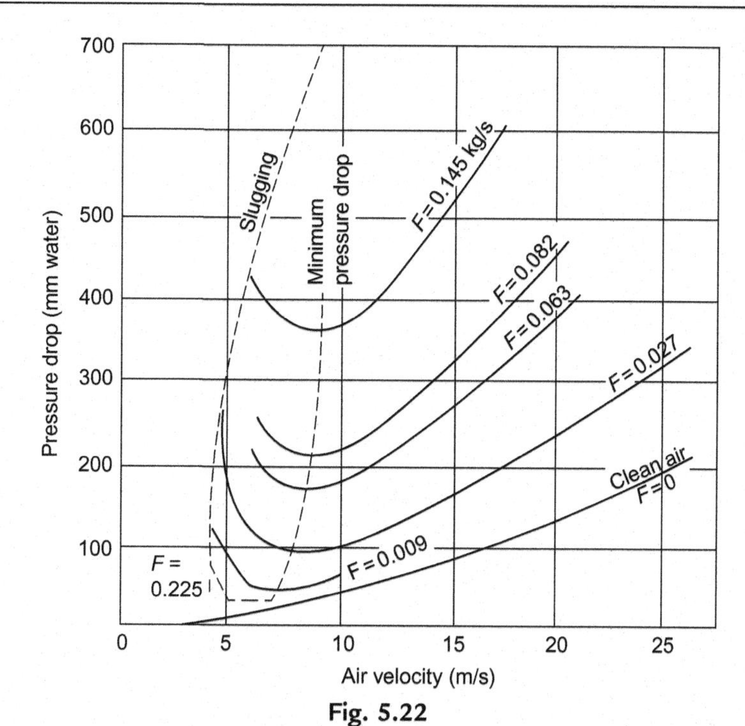

Fig. 5.22
Effect of air rate on pressure drop for transport of cress seed in 4.8 m horizontal length of 25 mm pipe.

A 25 mm diameter pipeline was used with two straight lengths of about 35 m joined by a semicircular bend. Experiments were carried out with air velocities up to 35 m/s. The solids used and their properties are listed in Table 5.3.

The velocities of solid particles u_s (m/s) are represented in terms of the air velocity u_G (m/s), the free-falling velocity of the particles u_0 (m/s), and the density of the solid particles ρ_s (kg/m^3) by the equation:

$$u_G - u_S = \frac{u_0}{0.468 + 7.25\sqrt{u_0/\rho_S}} \tag{5.38}$$

Over the range studied, the slip velocity u_R ($=u_G - u_s$) is close to the terminal falling velocity of the particles in air ($0.67 < u_R/u_0 < 1.03$).

Deviations from Eq. (5.38) are noted only at high loadings with fine solids of wide size distribution. Experimental results are plotted in Fig. 5.23.

The additional pressure drop due to the presence of solids in the pipeline ($-\Delta P_x$) could be expressed in terms of the solid velocity, the terminal falling velocity of the particles, and the feed rate of solids F (kg/s). The experimental results for a 25 mm pipe are correlated to within ±10% by:

Table 5.3 Physical properties of the solids used for pneumatic conveying[108]

Material	Shape	Particle Size (mm)		Density (kg/m³)	Free-Falling Velocity u_0 (m/s)
		Range	Mean		
Coal A	Rounded	1.5–Dust	0.75	1400	2.80
Coal B	Rounded	1.3–Dust	0.63		2.44
Coal C	Rounded	1.0–Dust	0.50		2.13
Coal D	Rounded	2.0–Dust	1.0		3.26
Coal E	Rounded	4.0–Dust	2.0		3.72
Perspex A	Angular	2.0–1.0	1.5	1185	3.73
Perspex B	Angular	5.0–2.5	3.8		5.00
Perspex C	Spherical	1.0–0.5	0.75		2.35
Polystyrene	Spherical	0.4–0.3	0.36	1080	1.62
Lead	Spherical	1.0–0.15	0.30	11,080	8.17
Brass	Porous, feathery filings	0.6–0.2	0.40	8440	4.08
Aluminium	Rounded	0.4–0.1	0.23	2835	3.02
Rape seed	Spherical	2.0–1.8	1.91	1080	5.91
Radish seed	Spherical	2.8–2.3	2.5	1065	6.48
Sand	Nearly spherical	1.5–1.0	1.3	2610	4.66
Manganese dioxide	Rounded	1.0–0.25	0.75	4000	5.27

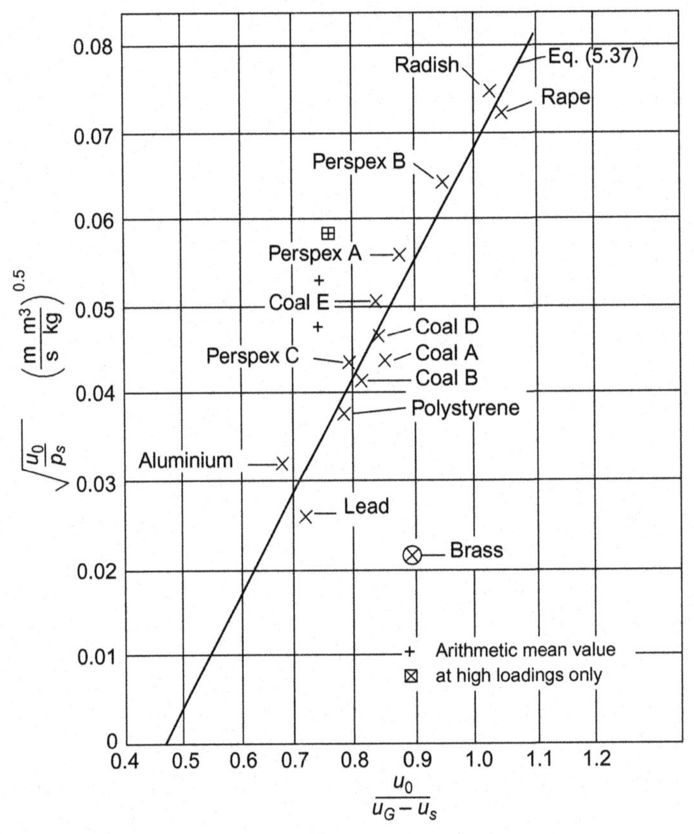

Fig. 5.23
Slip velocities $u_G - u_S$ for various materials.

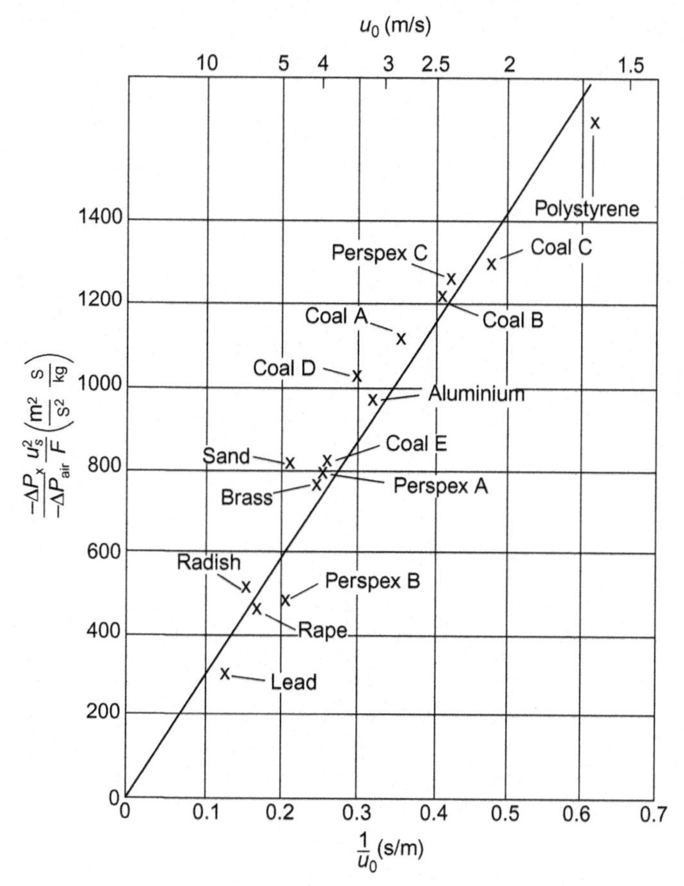

Fig. 5.24
Pressure drop for flow of solids in pneumatic conveying.

$$\frac{-\Delta P_x}{-\Delta P_{air}} \frac{u_S^2}{F} = \frac{2805}{u_0} \tag{5.39}$$

In Fig. 5.24 $(-\Delta P_x/-\Delta P_{air})\, u_S^2/F$ is plotted against $1/u_0$ for a 21 m length of pipe. $-\Delta P_{air}$ is the pressure drop for the flow of air alone at the pressure existing within the pipe.

The effect of pipe diameter on the pressure drop in a conveying system is seen by examining the results of Segler[102] who conveyed wheat grain in pipes up to 400 mm diameter. Taking a value of the solids velocity for wheat from the work of Gasterstädt,[109] $[(-\Delta P_x/-\Delta P_{air})\, u_S^2\, u_0]/F$ is plotted against pipe diameter in Fig. 5.25. The constant in Eq. (5.39) decreases with pipe diameter and is proportional to $d^{-0.7}$. Eq. (5.39) may thus be written (with d in m) as:

$$\frac{-\Delta P_x}{-\Delta P_{air}} \frac{u_S^2}{F} = \frac{210}{u_0 d^{0.7}} \tag{5.40}$$

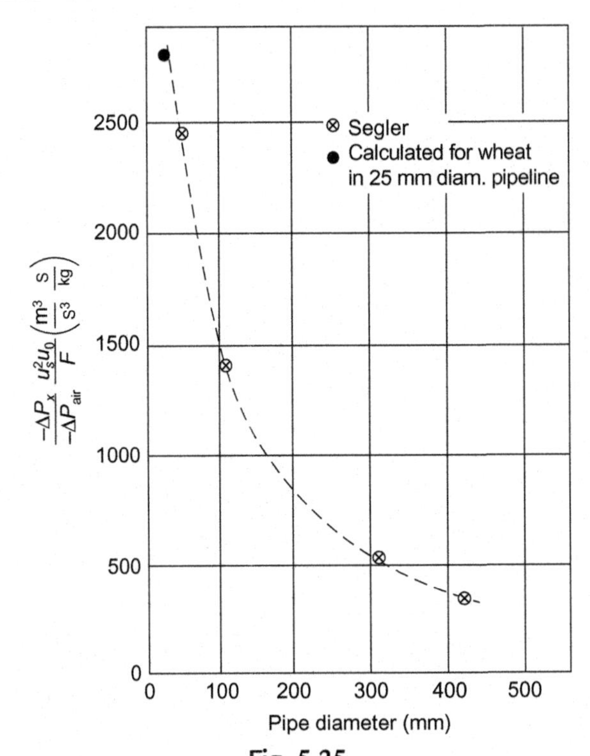

Fig. 5.25
Effect of pipe diameter on pressure drop for transport of wheat.

The above correlations all apply in straight lengths of pipe with the solids suspended in the gas stream. Under conditions of slug or dune flow the pressure gradients will be considerably greater. It is generally found that the economic velocity corresponds approximately with the minimum of the curves in Fig. 5.22. If there are bends in the conveying line, the energy requirement for conveying will be increased because the particles are retarded and must be reaccelerated following the bend. Bends should be eliminated wherever possible because, in addition, the wear tends to be very high.

Electrostatic charging

Eqs (5.38)–(5.40) for solid velocity and pressure drop are applicable only in the absence of electrostatic charging of the particles. Many materials, including sand, become charged during transport and cause the deposition of a charged layer on the surface of the pipe. The charge remains on the earthed pipeline for long periods but can be removed by conveying certain materials, including coal and perspex, through the pipe. The charging is thought to be associated with attrition of the particles and therefore to be a relatively slow process, while the discharging is entirely an electrical phenomenon and therefore is more rapid. The effect of electrostatic charging is to increase the frequency of collisions between the particles and the wall. As a result, the velocity of the solids is substantially reduced and the excess pressure difference ($-\Delta P_x$) may be increased by as much as tenfold. The results included in Fig. 5.24 are those obtained before an appreciable charge has built up. Increasing the humidity of the air will substantially reduce the electrostatic effects.

Dense phase conveying

There are several attractive features of dense phase conveying which warrant its use in some circumstances. In general, pipe wear is less and separation of the product is easier. Konrad[110] has carried out a comprehensive review of the current state of the art and reference should also be made to excellent books available on this subject.[99,111] He points out that the technique is not yet well established and that there are considerable design difficulties, not least a satisfactory procedure for predicting pressure drop. The risk of blocking the pipeline if the operating conditions are not correctly chosen is a deterrent to the widescale application of dense phase conveying.

5.5.3 Vertical Transport

Vertical pneumatic conveying has received considerably less attention than horizontal conveying and interest is usually confined to transport over much smaller distances. It is used for transferring particulate materials between plant units at different levels and for unloading into storage vessels. Another important application is in the recycling of solids in fluidised bed reactor systems (See Volume 2A, Chapter 6). As in hydraulic transport, vertical upward flows are inherently simpler than horizontal flows in that gravitational forces are directly balanced by forces due to the pressure gradient in the pipeline and the drag forces of the fluid on the particles. Furthermore, the flow is axisymmetric and there is no tendency for particles to settle out on to the pipe walls.

Detailed consideration of the interaction between particles and fluids is given in Volume 2A to which reference should be made. Briefly, however, if a particle is introduced into a fluid stream flowing vertically upwards it will be transported by the fluid provided that the fluid velocity exceeds the terminal falling velocity u_0 of the particle; the relative or *slip* velocity will be approximately u_0. As the concentration of particles increases, this slip velocity will become progressively less and, for a *slug* of fairly close packed particles, will approximate to the *minimum fluidising velocity* of the particles. (See Volume 2A, Chapter 6.)

The evidence from practical studies of vertical transport[106] is that the contribution of the solids to the pressure gradient in the pipeline is attributable predominantly to the weight of the particles. Because the density of the particles ρ_s is much greater than that of the gas, this additional pressure gradient can be written as:

$$-\frac{dP}{dl}=(1-\in_G)\rho_S g \tag{5.41}$$

where $\in_G$ is the voidage (gas holdup) of the mixture in the pipeline.

The conveying of fine particles in vertical pipes of diameters 25, 50, and 75 mm has been studied, amongst others, by Boothroyd.[106] He measured the pressure gradient in the pipeline, and found that the frictional pressure drop was less than that for air alone in the 25 mm pipe, but was greater in the larger pipes. This effect was attributed to the fact that the extent to which the fluid turbulence was affected by the presence of the particles was markedly influenced by pipe size.

Klinzing et al.[98] reported the results of experiments on dense-phase plug-flow of a cohesive coal. Pressure drops were measured over plugs up to 0.6 m in length and were found to be linearly related to plug length. There was little wall friction and pressure drops were almost entirely attributable to the gravitational contribution, given by Eq. (5.41).

Li and Kwauk[112] made a theoretical and practical study of dense phase vertical flow in a tapered rectangular transport line designed to maintain an approximately constant gas velocity. The voidage of the bed was measured at three points along its height, employing a capacitance void detector, and the normal wall shear stress arising from interparticle contact pressure was measured with a probe incorporating a diaphragm and strain gauge. Typically, pressure gradient along the line was about 1.5 times the gravitational component calculated using Eq. (5.41).

Sinclair and Jackson[113] have presented a theoretical relation between pressure gradient and the flowrates of gas and solids over the whole range of possible conditions for both cocurrent and countercurrent flow. It predicts marked segregation of gas and particles in the radial direction.

Miller and Gidaspow[114] have studied transport of 75 μm catalyst particles in a 75 mm diameter vertical pipe. The flow was characterized by a dilute rising core and a dense annular region at the walls that tended to move downwards. The fractional volumetric concentration of the solids was from 0.007 to 0.04 in the core and up to 0.25 in the annular region.

More extensive reviews of the pertinent studies are available in the literature.[99,111]

5.5.4 Practical Applications

Although pressure drop and energy requirements are important considerations in the design of pneumatic conveying installations, the over-riding factor must be that the plant will operate safely and without giving rise to serious operating and maintenance problems. There is always a considerable reluctance to operate at velocities below that corresponding to the minimum pressure drop shown in Fig. 5.22 for horizontal transport because this constitutes an unstable operating region, with the pressure drop increasing as the velocity falls, and the risk of blockage of the line is considerable. As a rule of thumb, it is undesirable to use gas velocities less than about twice the terminal falling velocity of the particles—a figure that is equally valid for vertical transport. Excessive velocities should be avoided, however, not only because of the consequent high costs of power, but also because of the increased effects of abrasion from the solids when moving at high velocity. The effects of abrasion are particularly marked at bends and fittings and changes of flow direction should be as gradual as possible. With combustible solids, there are further risks of dust explosions and it may be necessary to convey in an inert gas, particularly where there is likely to be a buildup of electrostatic charges.

The solids to gas ratio should be kept as high as possible to avoid the costs of pumping unnecessarily large amounts of gas and of separating the excess gas from the solids at the pipe outlet. However, because fluctuating flow conditions frequently occur it is desirable to avoid high concentrations because of the risk of blockage.

At the downstream end of the pipeline, it is necessary to disengage the solids from the gas and this is most usually carried out in cyclone separators (See Volume 2A, Chapter 5). However, there may be a carry-over of fine particles which must be eliminated before the gas is vented, and gas filters or electrostatic precipitators may be used for this purpose (See again Volume 2A, Chapter 5). At the upstream end, the particles must be introduced using some form of positive feeder, such as a rotary valve or a blow tank.

The design of pneumatic conveying installations presents a number of problems associated with the behaviour of the particles to be transported—particularly their tendency to agglomerate and to electrostatic charging. These factors are very difficult to define, depending as they do on the exact form of solids and the humidity of the air used for conveying. Reliable design methods have therefore not been developed for any but the simplest of problems. It should be borne in mind that variations in excess pressure drops up to 2–10-fold may arise from some of these ill-definable effects. This has led Wypych and Arnold[115] to suggest that it is desirable wherever possible to use a test rig to determine the conveying characteristics of the particles, using standardised procedures, before finalising the design.

Example 5.3

Sand of particle size 1.25 mm and density 2600 kg/m^3 is to be transported in air at the rate of 1 kg/s through a horizontal pipe 200 m long. Estimate the pipe diameter, the pressure drop in the pipeline and the airflow required.

Solution

Assuming a solids: gas mass ratio of 5, then:

$$\text{mass flow of air} = (1/5) = 0.20\,\text{kg/s}$$

and, assuming an air density of 1.0 kg/m^3:

$$\text{Volumetric flow of air} = (0.20/1.0) = \underline{\underline{0.20\,\text{m}^3/\text{s}}}$$

In order to avoid an excessive pressure drop, an air velocity of 30 m/s is acceptable.

Ignoring the volume occupied by the sand—about 0.2% of that occupied by the air—the cross-sectional area of the pipe required is (0.20/30) = 0.0067 m^2.

and thus:

$$\text{pipe diameter} = (4 \times 0.0067/\pi)^{0.5} = 0.092\,\text{m or } 92\,\text{mm}$$

The nearest standard size is 100 mm.

For sand of particle size 1.25 mm and density 2600 kg/m^3, the free-falling velocity is given in Table 5.3 as:

$$u_0 = 4.7\,\text{m/s}$$

In Eq. (5.38):

$$(u_G - u_s) = 4.7/\left[0.468 + 7.25(4.7/2600)^{0.5}\right] = 6.05\,\text{m/s}$$

The cross-sectional area of a 100 mm ID. pipe = ($\pi \times 0.100^2/4$) = 0.00786 m^2.

and hence:

$$\text{air velocity } u_G = (0.20/0.00786) = 25.5\,\text{m/s}$$

and:

$$\text{solids velocity, } u_s = (25.5 - 6.05) = 19.4\,\text{m/s}$$

Taking the viscosity and the density of air as 1.7×10^{-5} Ns/m^2 and 1.0 kg/m^3 respectively then, for the air alone:

$$Re = \left(0.100 \times 25.5 \times 1.0/\left(1.7 \times 10^{-5}\right)\right) = 150{,}000$$

and from Fig. 3.7:

$$\text{friction factor, } \phi = 0.004$$

Assuming isothermal conditions and incompressible flow, then, in Eq. (3.18):

$$\begin{aligned} -\Delta P_{air} &= [4 \times 0.004(200/0.100] \times 1.0 \times 25.5^2)/2 \\ &= 10404\,\text{N/m}^2 \text{ or } 10.4\,\text{kN/m}^2 \end{aligned}$$

and in Eq. (5.39):

$$(-\Delta P_x/-\Delta P_{air})\left(u_S^2/F\right) = (2805/u_0)$$

or

$$\begin{aligned} -\Delta P_x &= (2805\Delta P_{air}F)/(u_0 u_s)^2 \\ &= (2805 \times 10.4 \times 1.0)/\left(4.7 \times 19.4^2\right) \\ &\underline{\underline{16.5\,\text{kN/m}^2}} \end{aligned}$$

The total pressure drop = $(-\Delta P_a) + (-\Delta P_x)$ = 10.4 + 16.5 = 26.8 kN/m^2.

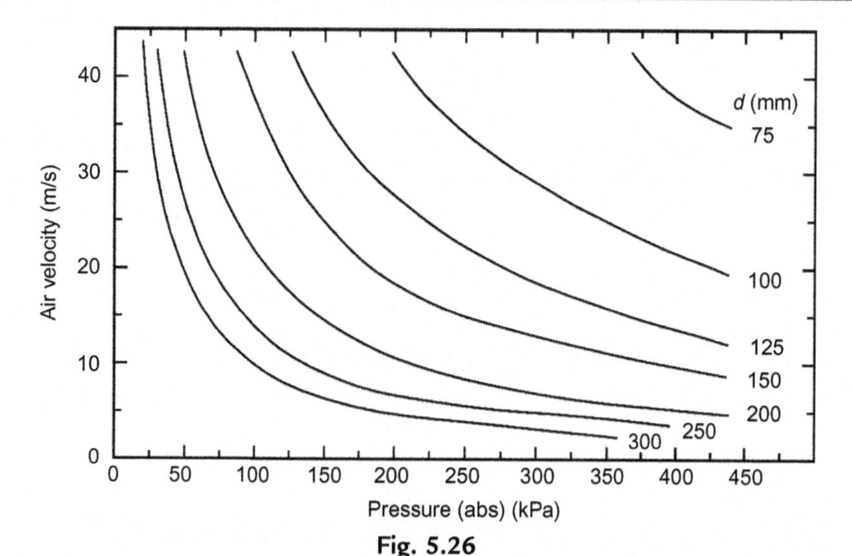

Fig. 5.26
The influence of air pressure and pipeline bore on conveying air velocity for a free airflow rate of 2400 m^3/h.

Mills[111] and others[116] have alluded to some of the other practical difficulties and possible remedies for the design and operation of pneumatic transport pipelines. For instance, with the decreasing pressure in the pipe in the direction of flow, the gas expands and therefore the mean air velocity progressively increases. For a constant diameter pipe, the air velocity at the exit can thus be 3–4 times or even higher than that at the inlet. For instance, a pipeline operating with an inlet pressure of 2 atm and discharging at 1 atm, the gas velocity at the exit will be twice the value at the inlet. For the sake of explanation and disregarding the effect of solids on the air velocity, Fig. 5.26 shows the mean air velocity as a function of pipe diameter and air pressure under isothermal conditions (corresponding to the airflow at standard temperature and pressure conditions of about 2500 m^3/h and assuming the ideal gas law behaviour). As the pressure decreases, not only the air velocity increases but these curves also become steeper. Bear in mind that not only the frictional pressure drop varies as a square of the air velocity, but such high velocities can also cause severe erosion problems. Indeed this problem is quite acute with constant diameter pipes, especially at high pressures (or large negative pressures), as is the case for long distance pipelines and/or those operating in the so-called dense phase regime. A common practice to limit the increase in air velocity on this count is to progressively increase the pipe diameter, as shown in Fig. 5.27, for a pipe transporting 100 tonnes/h of fine grade fly ash from a coal-based thermal power plant over a distance of about 1 mile with the absolute inlet air pressure of 4 atm. However, caution is needed in positioning the step and the next pipe diameter. Naturally, a minimum value of the air velocity is required which itself depends upon the size, density, and loading of solids. Thus, if the stepping up of the pipe diameter is done too early, the velocity may fall below this value

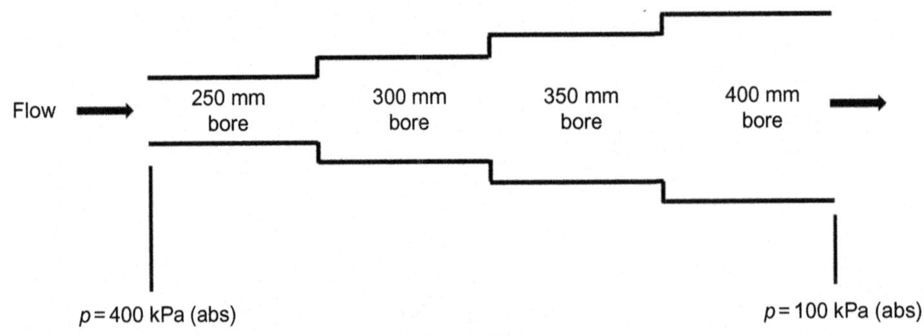

Fig. 5.27
Stepped pipeline parameters for the conveying of a fine grade of fly ash at 100 tonne/h over a distance of 1.5 km.

thereby leading to the blockage of line. As an illustration, taken from Mills,[111] let us say for a specific application, the minimum conveying velocity is 15 m/s and 2400 m^3/h of air (at standard temperature and pressure) is available for this duty. The inlet pressure is about 5 atm (absolute). Preliminary design calculations show that a 125 mm diameter pipe is needed for this duty, which gives the inlet air velocity of 16.5 m/s. As can be seen in Fig. 5.28, if this line were to continue all the way (when the pressure drops to atmospheric, the exit velocity will be of the

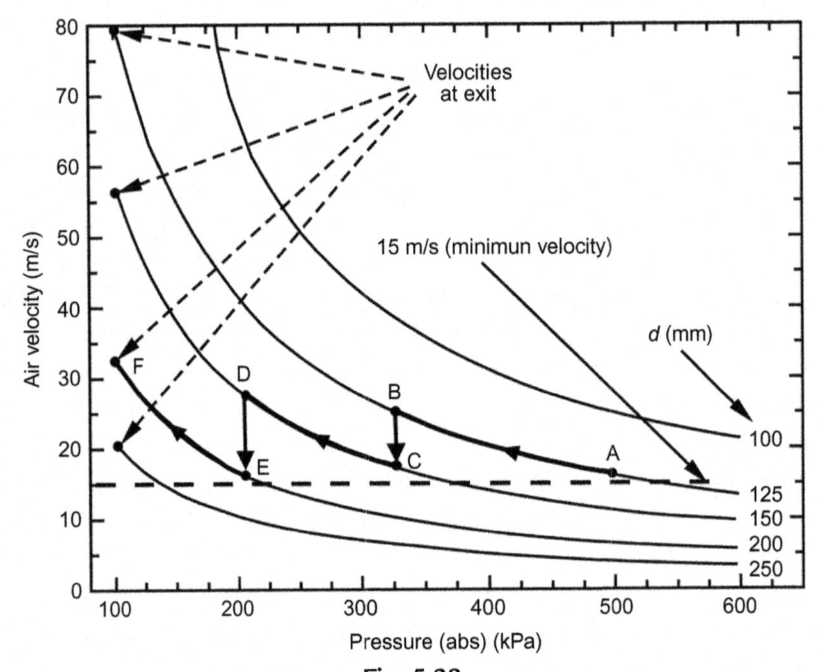

Fig. 5.28
Stepped pipeline velocity profile for high-pressure dilute phase system using 2500 m^3/h of air at free air conditions.

order of 81 m/s. The mean velocity for a possible combination of 125, 150, and 200 mm diameter pipes is shown in Fig. 5.28, but even with this the exit velocity is of the order of 32 m/s (reduced by the area factor, i.e. $(125/200)^2$). Thus, perhaps the next pipe diameter of 250 mm can be used towards the end of the pipe to limit the increase in air velocity. More detailed discussions are given by Mills.[111]

5.6 Nomenclature

		Units in SI System	Dimensions in M, L, T
A	Total cross-sectional area of pipe	m^2	L^2
A_B	Cross-sectional area of bed deposit in pipe	m^2	L^2
A_L	Cross-sectional area of pipe over which suspended flow occurs	m^2	L^2
b	Ratio of volumetric flow rates of mixture and liquid	–	–
C	Fractional volumetric concentration of solids in *flowing* mixture	–	–
C_D	Drag coefficient for particle settling at its terminal falling velocity $\left[=\frac{4}{3}(s'-1)\frac{d_p g}{u_0^2}\right]$	–	–
c	Constant in Eq. (5.8)	–	–
d	Pipe diameter	m	L
d_p	Particle diameter	m	L
e	Roughness of pipe wall	m	L
F_B	Force on bed layer in length l of pipe	N	MLT^{-2}
F_L	Force on material in suspended flow in length l of pipe	N	MLT^{-2}
ΣF_N	Sum of normal forces between particles and wall in length l of pipe	N	MLT^{-2}
G	Mass flowrate	kg/s	MT^{-1}
G'	Mass flowrate per unit area for gas	kg/m^2s	$ML^{-2}T^{-1}$
g	Acceleration due to gravity	m/s^2	LT^{-2}
h_f	Friction head	L	m
i	Hydraulic gradient for two-phase flow	–	–
i_w	Hydraulic gradient for liquid (water) in two-phase flow	–	–
K	Proportionality constant in Eq. (5.9)	–	–

k	Consistency coefficient for power-law fluid	N s^n/m^2	$ML^{-1}T^{n-2}$
L'	Mass flowrate per unit area for liquid	kg/m^2s	$ML^{-2}T^{-1}$
l	Pipe length	m	L
M	Mass rate of feed of solids	kg/s	MT^{-1}
m	Exponent in Eq. (5.37)	–	–
n	Flow index for power-law fluid	–	–
P	Pressure	N/m^2	$ML^{-1}T^{-2}$
$-\Delta P$	Pressure drop	N/m^2	$ML^{-1}T^{-2}$
$-\Delta P_a$	Pressure drop associated with acceleration of fluid	N/m^2	$ML^{-1}T^{-2}$
$-\Delta P_{air}$	Pressure drop for flow of air alone in pipe	N/m^2	$ML^{-1}T^{-2}$
$-\Delta P_f$	Pressure drop associated with friction	N/m^2	$ML^{-1}T^{-2}$
$-\Delta P_G$	Pressure drop for gas flowing *alone* at same superficial velocity as in two-phase flow	N/m^2	$ML^{-1}T^{-2}$
$-\Delta P_{gravity}$	Pressure drop associated with hydrostatic effects	N/m^2	$ML^{-1}T^{-2}$
$-\Delta P_L$	Pressure drop for liquid flowing *alone* at same superficial velocity as in two-phase flow	N/m^2	$ML^{-1}T^{-2}$
$-\Delta P_{TPF}$	Pressure drop for two-phase flow over length l of pipe	N/m^2	$ML^{-1}T^{-2}$
$-\Delta P_x$	Excess pressure drop due to solids in pneumatic conveying	N/m^2	$ML^{-1}T^{-2}$
R	Shear stress at pipe wall	N/m^2	$ML^{-1}T^{-2}$
R_B	Shear stress at pipe wall in fluid within bed	N/m^2	$ML^{-1}T^{-2}$
R_i	Interfacial shear stress at surface of bed	N/m^2	$ML^{-1}T^{-2}$
R_L	Shear stress at pipe wall in fluid above bed	N/m^2	$ML^{-1}T^{-2}$
R_y	Shear stress at distance y from wall	N/m^2	$ML^{-1}T^{-2}$
S_B	Perimeter of pipe in contact with bed	m	L
S_i	Length of chord at top surface of bed	m	L
S_L	Perimeter of pipe in contact with suspended flow	m	L
s	Ratio of density of solid to density of liquid (ρ_s/ρ)	–	–
u	Mean velocity in pipe or mixture velocity	m/s	LT^{-1}
u_G	Superficial velocity of gas	m/s	LT^{-1}
u_L	Superficial velocity of liquid	m/s	LT^{-1}
u_m	Mixture velocity	m/s	LT^{-1}
$(U_L)_C$	Critical value of u_L for laminar-turbulent transition	m/s	LT^{-1}
u_R	Relative velocity of phases; slip velocity	m/s	LT^{-1}
u_S	Superficial velocity of solids	m/s	LT^{-1}

u_X	Velocity in x-direction at distance y from surface	m/s	$\mathbf{LT^{-1}}$
u_0	Terminal falling velocity of particle in fluid	m/s	$\mathbf{LT^{-}1}$
u'_L	Actual linear velocity of liquid	m/s	$\mathbf{LT^{-1}}$
u'_S	Actual linear velocity of solids	m/s	$\mathbf{LT^{-1}}$
X	Lockhart and Martinelli parameter $\sqrt{-\Delta P_L / -\Delta P_G}$	–	–
X'	Modified parameter X, defined by Eq. (5.3)	–	–
y	Distance perpendicular to surface	m	**L**
Z	Vertical height	m	**L**
Re	Reynolds number	–	–
Re_G	Reynolds number for gas flowing alone	–	–
Re_L	Reynolds number for liquid flowing alone	–	–
Re_{MR}	Metzner and Reed Reynolds number for power-law fluid (see Chapter 3)	–	–
Re'_0	Particle Reynolds number for terminal falling conditions	–	–
ϵ_G	Average hold-up for gas in two-phase flow	–	–
ϵ_L	Average hold-up for liquid in two-phase flow	–	–
ϵ_S	Average hold-up for solids in two-phase flow	–	–
μ_F	Coefficient of friction between solids and pipe wall	–	–
μ_G	Viscosity of gas	Ns/m^2	$\mathbf{ML^{-1}T^{-1}}$
μ_L	Viscosity of liquid	Ns/m^2	$\mathbf{ML^{-1}T^{-1}}$
ρ	Density	kg/m^3	$\mathbf{ML^{-3}}$
ρ_G	Density of gas	kg/m^3	$\mathbf{ML^{-3}}$
ρ_L	Density of liquid	kg/m^3	$\mathbf{ML^{-3}}$
ρ_M	Density of mixture	kg/m^3	$\mathbf{ML^{-3}}$
ρ_S	Density of solid	kg/m^3	$\mathbf{ML^{-3}}$
Φ_G^2	Drag ratio for two-phase flow $(-\Delta P_{TPF}/-\Delta P_G)$	–	–
Φ_L^2	Drag ratio for two-phase flow $(-\Delta P_{TPF}/-\Delta P_L)$	–	–

References

1. Govier GW, Aziz K. *The flow of complex mixtures in pipes.* Malabar, FL: Krieger; 1982.
2. Chisholm D. *Two-phase flow in pipelines and heat exchangers.* London: George Goodwin; 1983.
3. Hewitt GF. In: Hetsroni G, editor. *Handbook of multiphase systems.* New York: McGraw-Hill; 1982.
4. Michaelides EE, Crowe CT, Schwarzkopf JD, editors. *Multiphase flow handbook.* 2nd ed. Boca Raton, FL: CRC Press; 2017.
5. Baker O. Simultaneous flow of gas and oil. *Oil Gas J* 26 July 1954;**53**:185. Multiphase flow in pipelines; 56 (10 Nov. 1958) 156.
6. Hoogendoorn CJ. Gas-liquid flow in horizontal pipes. *Chem Eng Sci* 1959;**9**:205.

7. Griffith P, Wallis GB. Two-phase slug flow. *J Heat Transf* 1961;**83**:307.
8. Weisman J, Duncan D, Gibson J, Crawford T. Effects of fluid properties and pipe diameter on two phase flow patterns in horizontal lines. *Int J Multiphase Flow* 1979;**5**:437–62.
9. Mandhane JM, Gregory GA, Aziz K. A flow pattern map for gas-liquid flow in horizontal pipes. *Int J Multiphase Flow* 1974;**1**:537–53.
10. Tzotzi C, Bontozoglou V, Andritsos N, Vlachogiannis M. Effect of fluid properties on flow patterns in two-phase gas-liquid flow in horizontal and downward pipes. *Ind Eng Chem Res* 2011;**50**:645–55.
11. Chhabra RP, Richardson JF. Prediction of flow patterns for the cocurrent flow of gas and non-Newtonian liquid in horizontal pipes. *Can J Chem Eng* 1984;**62**:449.
12. Ferguson MEG, Spedding PL. Measurement and prediction of pressure drop in two-phase flow. *J Chem Technol Biotechnol* 1995;**62**:262–78.
13. Luo D, Ghiaasiaan SM. Interphase mass transfer in concurrent vertical two-phase channel flows with non-Newtonian liquids. *Int Commun Heat Mass Transfer* 1997;**24**:1.
14. Xu JY, Wu YX, Shi ZH, Lao LY, Li DH. Studies on two-phase co-current air/non-Newtonian shear-thinning fluid flows in inclined smooth pipes. *Int J Multiphase Flow* 2007;**33**:948–69.
15. Hewitt GF, King I, Lovegrove PC. Holdup and pressure drop measurements in the two phase annular flow of air-water mixtures. *Brit Chem Eng* 1963;**8**:311–8.
16. Chen JJJ, Spedding PL. An analysis of holdup in horizontal two phase gas-liquid flow. *Intl J Multiphase Flow* 1983;**9**:147–59.
17. Petrick P, Swanson BS. Radiation attenuation method of measuring density of a two phase fluid. *Rev Sci Instrum* 1958;**29**:1079–85.
18. Heywood NI, Richardson JF. Slug flow of air-water mixtures in a horizontal pipe: determination of liquid holdup by γ-ray absorption. *Chem Eng Sci* 1979;**34**:17–30.
19. Khatib Z, Richardson JF. Vertical co-current flow of air and shear thinning suspensions of kaolin. *Chem Eng Res Des* 1984;**62**:139.
20. Chaouki J, Larachi F, Dudukovic MP. *Non-invasive monitoring of multiphase flows.* New York: Elsevier; 1997.
21. Lockhart RW, Martinelli RC. Proposed correlation of data for isothermal two-phase, two-component flow in pipes. *Chem Eng Prog* 1949;**45**:39.
22. Farooqi SI, Richardson JF. Horizontal flow of air and liquid (Newtonian and non-Newtonian) in a smooth pipe: Part I: correlation for average liquid holdup. *Trans Inst Chem Eng* 1982;**60**:292–305.
23. Chhabra RP, Farooqi SI, Richardson JF. Isothermal two phase flow of air and aqueous polymer solutions in a smooth horizontal pipe. *Chem Eng Res Des* 1984;**62**:22–32.
24. Ghajar AJ, Bhagwat SM. Gas-liquid flow in ducts. In: Michaelides EE, Crowe CT, Schwarzkopf JD, editors. *Multiphase flow handbook.* 2nd ed. Boca Raton, FL: CRC Press; 2016 [chapter 3].
25. Chisholm D. A theoretical basis for the Lockhart-Martinelli correlation for two-phase flow. *Int J Heat Mass Transf* 1967;**10**:1767.
26. Griffith P. Section 14. Two-phase flow. In: Rohsenow WM, Hartnett JP, editors. *Handbook of heat transfer.* New York: McGraw-Hill; 1973.
27. Chenoweth JM, Martin MW. Turbulent two-phase flow. *Pet Refin Eng* 1955;**34**(10):151.
28. Chenoweth JM, Martin MW. A pressure drop correlation for turbulent two-phase flow of gas-liquid mixtures in horizontal pipes. *Trans Am Soc Mech Eng* 1955. Paper 55 PET-9.
29. Baroczy CJ. A systematic correlation for two-phase pressure drop. *Chem Eng Prog Symp Ser* 1966;**64**(62):232.
30. Dukler AE, Wicks III M, Cleveland RG. Frictional pressure drop in two-phase flow. A comparison of existing correlations for pressure loss and holdup. *AIChE J* 1964;**10**:38.
31. Farooqi SI, Richardson JF. Horizontal flow of air and liquid (Newtonian and non-Newtonian) in a smooth pipe: Part II: average pressure drop. *Trans Inst Chem Eng* 1982;**60**:323.
32. Dziubinski M, Richardson JF. Two-phase flow of gas and non-Newtonian liquids in horizontal pipes – superficial velocity for maximum power saving. *J Pipeline Syst Eng Pract* 1985;**5**:107.
33. Charles ME. The reduction of pressure gradient in oil pipelines. *Trans Can Inst Min* 1960;**63**:306–10.

34. Joseph DD, Bai R, Mata C, Sury K, Grant C. Self-lubricated transport of bitumen froth. *J Fluid Mech* 1999;**386**:127–48.
35. McKibben MJ, Gillies RG, Shook CA. Predicting pressure gradients in heavy oil-water pipelines. *Can J Chem Eng* 2000;**78**:752–6.
36. Angelli P, Hewitt GF. Pressure gradient in horizontal liquid-liquid flows. *Int J Multiphase Flow* 1998;**24**:1183.
37. Pal R. *Rheology of particulate dispersions and composites.* Boca Raton, FL: CRC Press; 2006.
38. Schramm LL. *Emulsions, foams, suspensions, and aerosols.* 2nd ed. Berlin: VCH-Wiley; 2014.
39. Mewis JM, Wagner NJ. *Colloidal suspension rheology.* Cambridge: Cambridge University Press; 2012.
40. Brauner N. In: Bertola V, editor. *Modelling and control of two-phase flow phenomena: liquid-liquid two-phase flow systems.* Udine, Italy: International Center for Mechanical Sciences; 2004.
41. Charles ME, Govier GW, Hodgson GW. The horizontal flow of equal density oil-water mixtures. *Can J Chem Eng* 1961;**39**:27–36.
42. Russell TWF, Hodgson GW, Govier GW. Horizontal pipeline flow of oil and water. *Can J Chem Eng* 1959;**37**:9–17.
43. Trallero JL, Sarica C, Brill JP. A study of oil/water flow patterns in horizontal pipes. *SPE Prod Facil* 1997;**12**:165–72.
44. Flores JG, Chen XT, Sarica C, Brill JP. Characterization of oil-water flow patterns in vertical and deviated wells. *SPE Prod Facil* 1999;**14**:102–9.
45. Brauner N, Maron DM. Flow pattern transitions in two-phase liquid-liquid horizontal pipes. *Int J Multiphase Flow* 1992;**18**:123–40.
46. Angelli P, Hewitt GF. Flow structure in horizontal oil-water flow. *Int J Multiphase Flow* 2000;**26**:1117–40.
47. Liu Y, Zhang H, Wang S, Wang J. Prediction of pressure gradient and holdup in horizontal liquid-liquid segregated flow with small Eötvös number. *Chem Eng Commun* 2009;**196**:697–714.
48. Oliemans R. Oil-water liquid flow rate determined from measured pressure drop and water hold-up in horizontal pipes. *J Braz Soc Mech Sci Eng* 2011;**33**:259–64.
49. Jana AK, Ghoshal P, Das G, Das PK. An analysis of pressure drop and holdup for liquid-liquid upflow through vertical pipes. *Chem Eng Technol* 2007;**30**:920–5.
50. Zuber N, Findlay I. Average volumetric concentration in two phase flow systems. *J Heat Transf* 1965;**87**:453–68.
51. Brauner N, Rovinsky J, Moalem Maron D. Determination of the Interface curvature in stratified two-phase systems by energy considerations. *Int J Multiphase Flow* 1996;**22**:1167–85.
52. Al-Wahaibi T. Pressure gradient correlation for oil-water separated flow in horizontal pipes. *Exp Thermal Fluid Sci* 2012;**42**:196–203.
53. Dasari A, Desamala AB, Ghosh UK, Dasmahapatra AK, Mandal TK. Correlations for prediction of pressure gradient of liquid-liquid flow through a circular horizontal pipe. *J Fluids Eng* 2014;**136**:1–12 071302.
54. Heywood NI, Richardson JF. Rheological behavior of flocculated and dispersed aqueous kaolin suspensions in pipe flow. *J Rheol* 1978;**22**:599.
55. Farooqi SI, Richardson JF. Rheological behavior of kaolin suspensions in water and water-glycerol mixtures. *Trans Inst Chem Eng* 1980;**58**:116.
56. Chhabra RP, Farooqi SI, Richardson JF, Wardle AP. Cocurrent flow of air and china clay suspensions in large diameter pipes. *Chem Eng Res Des* 1983;**61**:56–61.
57. Streat, M.: Dense phase flow of solid-water mixtures in pipelines: a state of the art review. Hydrotransport 10 (BHRA, Fluid Engineering, Innsbruck, Austria), (October 1986). Paper B1.
58. Pirie RL. *Transport of coarse particles in water and shear-thinning suspensions in horizontal pipes. PhD Thesis.* University of Wales; 1990.
59. Pirie RL, Davies T, Khan AR, Richardson JF. Measurement of liquid velocity in multiphase flow by salt injection method. In: *Proceedings of the 2nd International Conf. on Flow Measurement—BHRA, London*; 1988 Paper F3, 187.
60. Newitt DM, Richardson JF, Abbott M, Turtle RB. Hydraulic conveying of solids in horizontal pipes. *Trans Inst Chem Eng* 1955;**33**:93.

61. Hisamitsu N, Ise T, Takeishi Y. Blockage of slurry pipeline. *Hydrotransport-7 (BHRA Fluid Engineering, Sendai, Japan)* 1980;**B4**:71.
62. Wilson KC, Addie GR, Sellgren A, Clift R. *Slurry transport using centrifugal pumps*. 3rd ed. New York: Springer; 2006.
63. Chhabra RP, Richardson JF. Hydraulic transport of coarse gravel particles in a smooth horizontal pipe. *Chem Eng Res Des* 1983;**61**:313.
64. Durand R. Transport hydraulique des materiaux solides en conduite. *Houille Blanche* 1951;**6**:384.
65. Durand R. Transport hydraulique de graviers et galets en conduite. *Houille Blanche* 1951;**6**:609.
66. Durand R. Basic relationships of transportation of solids in pipes—experimental research. In: *Proceedings of the Minnesota International Hydraulic Convention*; 1953. p. 89.
67. Durand R, Condolios E. The hydraulic transportation of coal and solid materials in pipes. In: *Proceedings of a Colloquium on the Hydraulic Transport of Coal, National Coal Board, London*; 1952. Paper IV.
68. James JG, Broad BA. Conveyance of coarse particle solids by hydraulic pipeline: Trials with limestone aggregates in 102, 156 and 207 mm diameter pipes. *Transport and Road Research Laboratory, TRRL Supplementary Report* 1980;**635**.
69. Khan AR, Pirie RL, Richardson JF. Hydraulic transport of solids in horizontal pipelines—predictive methods for pressure gradient. *Chem Eng Sci* 1987;**42**:767–77.
70. Abbott M. *The hydraulic conveying of solids in pipe lines. PhD Thesis*. University of London; 1955.
71. Turtle RB. *The hydraulic conveying of granular material. PhD Thesis*. University of London; 1952.
72. Haas DB, Gillies R, Small M, Husband WHW. *Study of the hydraulic properties of coarse particles of metallurgical coal when transported in slurry form through pipelines of various diameters*. Saskatchewan Research Council Publication No. E-835-1-C80; March 1980.
73. Gaessler H. *Experimentelle und theoretische Untersuchungen über die Strömungsvorgänge beim Transport von Feststoffen in Flüssigkeiten durch horizontale Rohrleitungen*. Karlsruhe, Germany: Dissertation Technische Hochschule; 1966.
74. Zandi I, Govatos G. Heterogeneous flow of solids in pipelines. *J Hydr Div Amer Soc Civ Engrs* 1967;**93**:145–69.
75. Roco M, Shook CA. Modelling of slurry flow: the effect of particle size. *Can J Chem Eng* 1983;**61**:494–503.
76. Roco M, Shook CA. A model for turbulent slurry flow. *J Pipelines* 1984;**4**:3–13.
77. Shook CA, Daniel SM. Flow of suspensions of solids in pipelines. Part I. Flow with a stable stationary deposit. *Can J Chem Eng* 1965;**43**:56–62.
78. Wilson KC. *A unified physically-based analysis of solid-liquid pipeline flow. Hydrotransport 4*. Banff, Alberta, Canada: BHRA Fluid Engineering; May 1976. Paper A1.1.
79. Wilson KC, Brown NP. Analysis of fluid friction in dense-phase pipeline flow. *Can J Chem Eng* 1982;**60**:83–91.
80. Wilson KC, Pugh FJ. Dispersive-force modelling of turbulent suspensions in heterogeneous slurry flow. *Can J Chem Eng* 1988;**66**:721–30.
81. Shook CA, Roco MC. *Slurry flow. Principles and practice*. Oxford: Butterworth-Heinemann; 1991.
82. Charles ME, Charles RA. The use of heavy media in the pipeline transport of particulate solids. In: Zandi I, editor. *Advances in solid-liquid flow in pipes and its application*. Pergamon Press; 1971, chapter 12.
83. Kenchington JM. *Prediction of pressure gradient in dense phase conveying. Hydrotransport 5*. Hanover, Germany: BHRA Fluid Engineering; May 1978. Paper D7.
84. Duckworth RA, Pullum L, Lockyear CF. The hydraulic transport of coarse coal at high concentration. *J Pipelines* 1983;**3**:251.
85. Duckworth RA, Pullum L, Addie GR, Lockyear CF. *The pipeline transport of coarse materials in a non-Newtonian carrier fluid. Hydrotransport 10*. Innsbruck, Austria: BHRA Fluid Engineering; October 1986. Paper C2.
86. Duckworth RA, Addie GR, Maffett J. Minewaste disposal by pipeline using a fine slurry carrier. In: *Proceedings of the 11th International Conference on Slurry Technology—The Second Decade*, organised by STA Washington, DC, Hilton Head, SC; March 1986. p. 187.

87. Tatsis A, Jacobs BEA, Osborne B, Astle RD. A comparative study of pipe flow prediction for high concentration slurries containing coarse particles. In: *Proceedings of the 13th International Conference on Slurry Technology,* organised by STA Washington, DC, Hilton Head, SC; March 1988.
88. Chhabra RP, Richardson JF. Hydraulic transport of coarse particles in viscous Newtonian and non-Newtonian media in a horizontal pipe. *Chem Eng Res Des* 1985;**63**:390–7.
89. Lareo C, Fryer PJ, Barigou M. The fluid mechanics of solid-liquid food flows. *Trans Inst Chem Eng* 1997;**75**:73. Part C.
90. Tucker G, Heydon C. Food particle residence time measurement for the design of commercial tubular heat exchangers suitable for processing suspensions of solids in liquids. *Trans Inst Chem Eng* 1998;**76**:208. Part C.
91. Berk Z. *Food process engineering and technology.* New York: Academic Press; 2008.
92. Durand R. Ecoulements de mixture en conduites verticales – influence de la densite des materiaux sur les caracteristiques de refoulement en conduite horizontale. *Houille Blanche* 1953;**8**:124.
93. Worster RC, Denny DF. The hydraulic transport of solid material in pipes. *Proc Inst Mech Eng H* 1955;**169**:563.
94. Newitt DM, Richardson JF, Gliddon BJ. Hydraulic conveying of solids in vertical pipes. *Trans Inst Chem Eng* 1961;**39**:93.
95. Cloete FLD, Miller AI, Streat M. Dense phase flow of solid-water mixtures through vertical pipes. *Trans Inst Chem Eng* 1967;**45**:T392.
96. Kopko RJ, Barton P, McCormick RH. Hydrodynamics of vertical liquid-solids transport. *Ind Eng Chem Process Des Dev* 1975;**14**:264.
97. Al-Salihi, L.: Hydraulic transport of coarse particles in vertical pipelines. (Ph.D. Thesis), University of Wales (1989).
98. Klinzing GE, Rohatgi ND, Zaltash A, Myler CA. Pneumatic transport – a review (generalized phase diagram approach to pneumatic transport). *Powder Technol* 1987;**51**:135.
99. Klinzing GE, Rizk F, Marcus R, Leung LS. *Pneumatic conveying of solids.* 3rd ed. New York: Springer; 2010.
100. Bagnold RA. Some flume experiments on large grains but little denser than the transporting fluid, and their implications. *Proc Inst Civ Eng* 1955;**4**(iii):174.
101. Bagnold RA. The flow of cohesionless grains in fluids. *Philos Trans* 1957;**249**:235.
102. Segler G. Untersuchungen an Kornergeblasen und Grundlagen fur ihre Berechnung. [Pneumatic Grain Conveying (1951), National Institute of Agricultural Engineering] *Z Ver Dtsch Ing* 1935;**79**:558.
103. Clark RH, Charles DE, Richardson JF, Newitt DM. Pneumatic conveying. Part I. The pressure drop during horizontal conveyance. *Trans Inst Chem Eng* 1952;**30**:209.
104. Mitlin, L.: A study of pneumatic conveying with special reference to solid velocity and pressure drop during transport. (Ph.D. Thesis), University of London, (1954).
105. Jones C, Hermges G. The measurement of velocities for solid-fluid flow in a pipe. *Br J Appl Phys* 1952;**3**:283.
106. Boothroyd RG. Pressure drop in duct flow of gaseous suspensions of fine particles. *Trans Inst Chem Eng* 1966;**44**:T306.
107. Klinzing GE, Mathur MP. The dense and extrusion flow regime in gas-solid transport. *Can J Chem Eng* 1981;**59**:590.
108. Richardson JF, McLeman M. Pneumatic conveying. Part II. Solids velocities and pressure gradients in a one-inch horizontal pipe. *Trans Inst Chem Eng* 1960;**38**:257.
109. Gasterstadt J. Die experimentelle Untersuchung des pneumatischen Fordervorganges. *Forsch Arb Geb Ing Wes* 1924;**265**:1–76.
110. Konrad K. Dense-phase pneumatic conveying: a review. *Powder Technol* 1986;**49**:1.
111. Mills D. *Pneumatic conveying design guide.* 3rd ed. Oxford, UK: Butterworth-Heinemann; 2015.
112. Li H, Kwauk M. Vertical pneumatic moving bed transport. I. Analysis of flow dynamics. II Experimental findings. *Chem Eng Sci* 1989;**44**:249, 261.
113. Sinclair JL, Jackson R. Gas-particle flow in a vertical pipe with particle – particle interactions. *AIChE J* 1989;**35**:1473.
114. Miller A, Gidaspow D. Dense, vertical gas-solid flow in a pipe. *AIChE J* 1992;**38**:1801.

115. Wypych PW, Arnold PC. On improving scale-up procedures for pneumatic conveying design. *Powder Technol* 1987;**50**:281.
116. Wilms H, Dhodapkar S. Pneumatic conveying: optimal system design, operation and control. *Chem Eng* October 2014;**121**:59.

Further Reading

1. Bergles AE, Collier JG, Delhaye JM, Hewitt GF, Mayinger F. *Two-phase flow and heat transfer in the power and process industries.* New York: Hemisphere Publishing Corporation and McGraw-Hill; 1981.
2. Brown NP, Heywood NI, editors. *Slurry handling. Design of solid-liquid systems.* Amsterdam: Elsevier; 1991.
3. Chhabra RP. Hydraulic transport of solids in horizontal pipes. In: Cheremisinoff PN, Cheremisinoff NP, Cheng SL, editors. *Civil engineering practice.* vol. 2. PA: Technomic Pub. Press; 1988.
4. Chhabra RP, Richardson JF. Co-current horizontal and vertical upwards flow of gas and non-Newtonian liquid. In: Cheremisinoff NP, editor. *Encyclopedia of fluid mechanics.* vol. 3. Houston, TX: Gulf Publishing Co; 1986. Gas-liquid flow.
5. Hewitt GF. Two-phase flow studies in the United Kingdom. *Int J Multiphase Flow* 1983;**9**:715–49.
6. Shook, C. A.: In Handbook of fluids in motion by Cheremisinoff, N. P. and Gupta, R. (editors) (Ann Arbor, New York, 1983), pp. 929–943: Pipeline flow of coarse particle slurries.
7. Williams OA. *Pneumatic and hydraulic conveying of solids.* New York: Dekker; 1983.
8. Zenz FA, Othmer DF. *Fluidization and fluid-particle systems.* New York: Reinhold; 1960.
9. Chhabra RP, Richardson JF. Hydraulic transport of coarse particles in viscous Newtonian and non-Newtonianmedia in a horizontal pipe. *Chem Eng Res Des* 1985;**63**:390.

CHAPTER 6

Flow and Pressure Measurement

6.1 Introduction

The most important parameters measured to provide information on the operating conditions of a plant are flowrates, pressures, and temperatures for single phase systems, while additional measurements are required for the flow of multiphase systems in terms of in-situ velocity and distribution of each phase, as discussed in Chapter 5. The instruments used may give either an instantaneous reading or, in the case of flow, may be arranged to give a cumulative flow over any given period. In either case, the instrument may be required to give a signal to some control unit which will then govern one or more parameters on the plant. It should be noted that in industrial plants it is usually more important to have information on the change in the value of a given parameter than to use meters that give particular absolute accuracy. To maintain the value of a parameter at a desired value a control loop is used.

A simple control system, or *loop*, is illustrated in Fig. 6.1. The temperature T_0 of the water at Y is measured by means of a sensor, the output of which is fed to a controller mechanism. The latter can be divided into two sections (normally housed in the same unit). In the first (the *comparator*), the *measured value* (T_0) is compared with the desired value (T_d) to produce an *error* (e), where:

$$e = T_d - T_0 \tag{6.1}$$

The second section of the mechanism (the *controller*) produces an output which is a function of the magnitude of e. This is fed to a control valve in the steam line, so that the valve closes when T_0 increases and vice versa. The system as shown may be used to counteract fluctuations in temperature due to extraneous causes such as variations in water flowrate or upstream temperature—termed *load* changes. It may also be employed to change the water temperature at Y to a new value by adjustment of the desired value.

It is very important to note that in this loop system the parameter T_0, which must be kept constant, is measured, though all subsequent action is concerned with the magnitude of the error and not with the actual value of T_0. This simple loop will frequently be complicated by there being several parameters to control, which may necessitate considerable instrumental analysis and the control action will involve operation of several control valves.

Coulson and Richardson's Chemical Engineering. https://doi.org/10.1016/B978-0-08-101099-0.00006-9

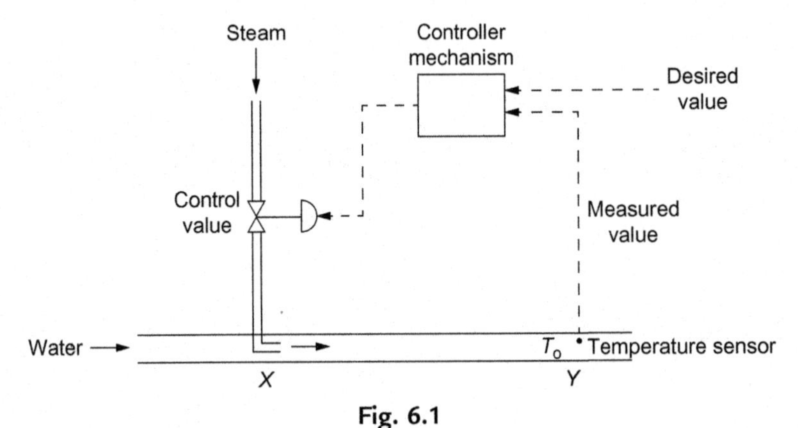

Fig. 6.1
Simple feedback control system.

This represents a simple form of control for a single variable, though in a modern plant many parameters are controlled simultaneously from various measuring instruments, and the variables in a plant such as a distillation unit are frequently linked together, thus increasing the complexity of control that is required.

In industrial plants, the instruments are therefore required not only to act as indicators but also to provide some link, which can be used to help in the control of the plant. In this chapter, pressure measurement is briefly described and methods of measurement of flowrate are largely confined to those that depend on the application of the energy balance equation (2.57). For further detail, reference should be made to Wightman[1] and to other sources.[2–4]

6.2 Fluid Pressure

In a stationary fluid, the pressure is exerted equally in all directions and is referred to as the *static pressure*. In a moving fluid, the static pressure is exerted on any plane parallel to the direction of motion. The pressure exerted on a plane at right angles to the direction of flow is greater than the static pressure because the surface has, in addition, to exert sufficient force to bring the fluid to rest. This additional pressure is proportional to the kinetic energy of the fluid; it cannot be measured independent of the static pressure.

6.2.1 Static Pressure

The energy balance equation can be applied between any two sections in a continuous fluid. If the fluid is not moving, the kinetic energy and the frictional loss are both zero, and therefore:

$$v\,dP + g\,dz = 0 \quad \text{(from Eq. 2.57)}$$

For an incompressible fluid:

$$\begin{aligned} v(P_2 - P_1) + g(z_2 - z_1) &= 0 \\ \text{or: } (P_2 - P_1) &= -\rho g(z_2 - z_1) \end{aligned} \tag{6.2}$$

Thus the pressure difference can be expressed in terms of the height of a vertical column of fluid.

If the fluid is compressible and behaves as an ideal gas, for isothermal conditions:

$$\begin{aligned} P_1 v_1 \ln\frac{P_2}{P_1} + g(z_2 - z_1) = 0 \quad &(\text{from Eq. 2.69}) \\ \therefore \frac{P_2}{P_1} = \exp\left[\frac{-gM}{RT}(z_2 - z_1)\right] \quad &(\text{from Eq. 2.16}) \end{aligned} \tag{6.3}$$

This expression enables the pressure distribution within an ideal gas to be calculated for isothermal conditions.

When the static pressure in a moving fluid is to be determined, the measuring surface must be parallel to the direction of flow so that no kinetic energy is converted into pressure energy at the surface. If the fluid is flowing in a circular pipe, the measuring surface must be perpendicular to the radial direction at any point. The pressure connection, which is known as a *piezometer tube*, should terminate flush with the wall of the pipe so that the flow is not disturbed: the pressure is then measured near the walls where the velocity is a minimum and the reading would be subject only to a small error if the surface were not quite parallel to the direction of flow. A piezometer tube of narrow diameter is used for accurate measurements.

The static pressure should always be measured at a distance of not less than 50 diameters from bends or other obstructions, so that the flow lines are almost parallel to the walls of the tube. If there are likely to be large cross-currents or eddies, a piezometer ring should be used. This consists of four pressure tappings equally spaced at 90 degrees intervals round the circumference of the tube; they are joined by a circular tube which is connected to the pressure measuring device. By this means, false readings due to irregular flow are avoided. If the pressure on one side of the tube is relatively high, the pressure on the opposite side is generally correspondingly low; with the piezometer ring providing a mean value. The cross-section of the piezometer tubes and ring should be small to prevent any appreciable circulation of the fluid.

6.2.2 Pressure Measuring Devices

(a) *The simple manometer*, shown in Fig. 6.2A, consists of a transparent U-tube containing the fluid **A** of density p whose pressure is to be measured and an immiscible fluid **B** of higher density ρ_m. The limbs are connected to the two points between which the

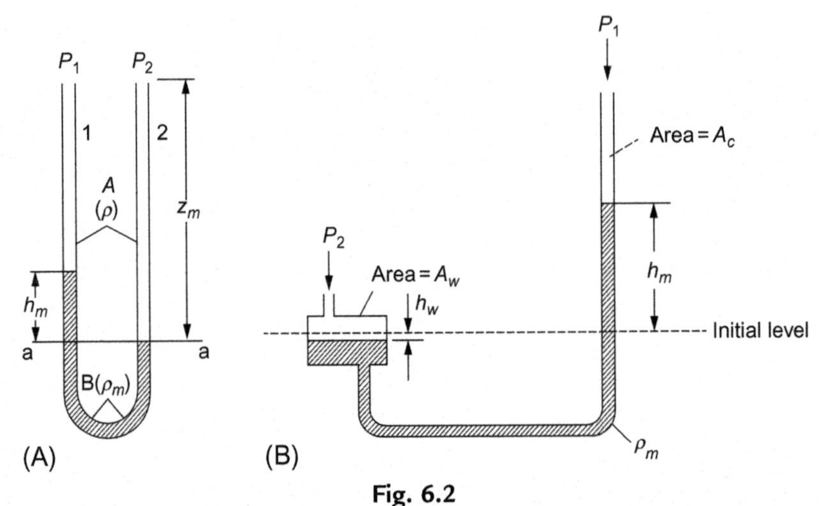

Fig. 6.2
(A) The simple manometer. (B) The well-type manometer.

pressure difference $(P_2 - P_1)$ is required; the connecting leads should be completely full of fluid **A**. If P_2 is greater than P_1, the interface between the two liquids in limb 2 will be decreased by a distance h_m (say) below that in limb 1. The pressure at the level $a-a$ (joined by the same fluid and at the same horizontal level) must be the same in each of the limbs and, therefore (assuming constant density):

$$P_2 + z_m \rho g = P_1 + (z_m - h_m)\rho g + h_m \rho_m g$$
$$\text{and}: \Delta P = P_2 - P_1 = h_m(\rho_m - \rho)g \quad (6.4)$$

If fluid **A** is a gas, its density ρ will normally be small compared with the density of the manometer fluid ρ_m so that:

$$\Delta P = h_m \rho_m g \quad (6.5)$$

(b) In order to avoid the inconvenience of having to read two limbs, the *well-type manometer* shown in Fig. 6.2B can be used. If A_w and A_c are the cross-sectional areas of the well and the column and h_m is the increase in the level of the column and h_w the decrease in the level of the well, then:

$$P_2 = P_1 + \rho_m g(h_m + h_w)$$
$$\text{or}: P_2 - P_1 = \rho_m g(h_m + h_w)$$

The quantity of liquid expelled from the well is equal to the quantity pushed into the column so that:

$$A_w h_w = A_c h_w$$
$$\text{and}: h_w = \frac{A_c}{A_w} h_m$$

Substituting:

$$P_2 - P_1 = \rho_m g h_m \left(1 + \frac{A_c}{A_w}\right) \tag{6.6}$$

If the well is large in comparison to the column then:

$$P_2 - P_1 = \rho_m g h_m \tag{6.7}$$

(c) The *inclined manometer* shown in Fig. 6.3 enables the sensitivity of the manometers described previously to be increased by measuring the length of the column of liquid. If θ is the angle of inclination of the manometer (typically about 10–20 degrees) and L is the movement of the column of liquid along the limb, then:

$$L = \frac{h_m}{\sin\theta} \tag{6.8}$$

and if $\theta = 10$ degrees, the manometer reading L is increased by about 5.7 times compared to the reading h_m which would have been obtained from a simple manometer.

(d) The *inverted manometer* (Fig. 6.4) is used for measuring pressure differences in liquids. The space above the liquid in the manometer is filled with air, which can be admitted or expelled through the tap A in order to adjust the level of the liquid in the manometer.

(e) *The two-liquid manometer.* Small differences in pressure in gases are often measured with a manometer of the form shown in Fig. 6.5. The reservoir at the top of each limb is of a sufficiently large cross-section for the liquid level to remain approximately the same on each side of the manometer. The difference in pressure is then given by:

$$\Delta P = (P_2 - P_1) = h_m(\rho_{m1} - \rho_{m2})g \tag{6.9}$$

where ρ_{m1} and ρ_{m2} are the densities of the two manometer liquids. The sensitivity of the instrument is very high if the densities of the two liquids are nearly the same. To obtain accurate readings it is necessary to choose liquids which give sharp interfaces: paraffin oil and industrial alcohol are commonly used. According to Ower and Pankhurst,[5] benzyl alcohol (specific gravity 1.048) and calcium chloride solutions give the most satisfactory results. The difference in density can be varied by altering the concentration of the calcium chloride solution.

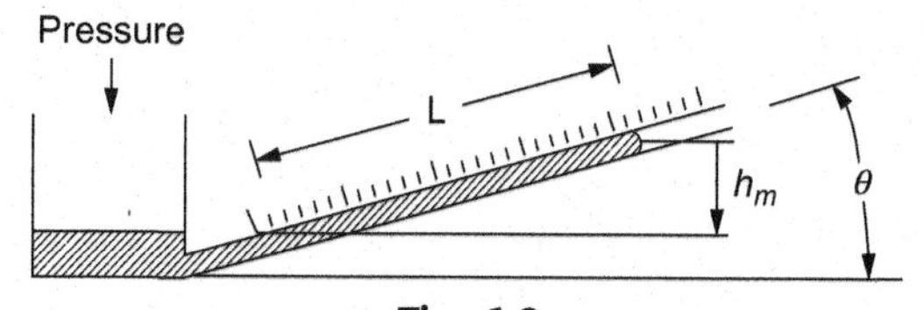

Fig. 6.3
An inclined manometer.

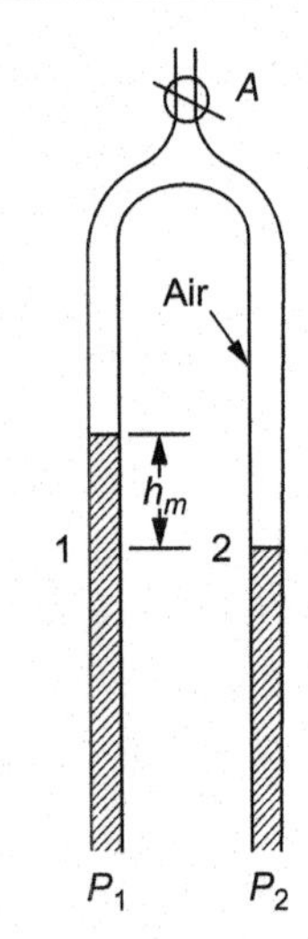

Fig. 6.4
Inverted manometer.

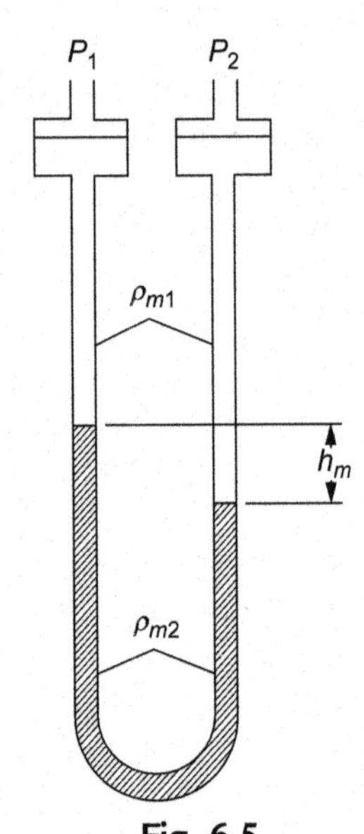

Fig. 6.5
Two-liquid manometer.

(f) *The Bourdon gauge* (Fig. 6.6). The pressure to be measured is applied to a curved tube, oval in cross-section, and the deflection of the end of the tube is communicated through a system of levers to a recording needle. This gauge is widely used for steam and compressed gases, and frequently forms the indicating element of flow controllers. The simple form of the gauge is illustrated in Fig. 6.6A and B. Fig. 6.6C shows a Bourdon sensing element in the form of a helix; this element has a very greater sensitivity and is suitable for very high pressures.

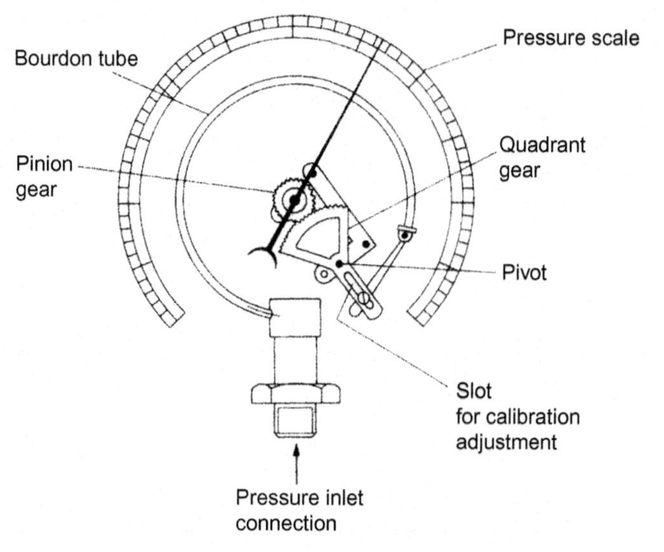

(A) standard mechanism

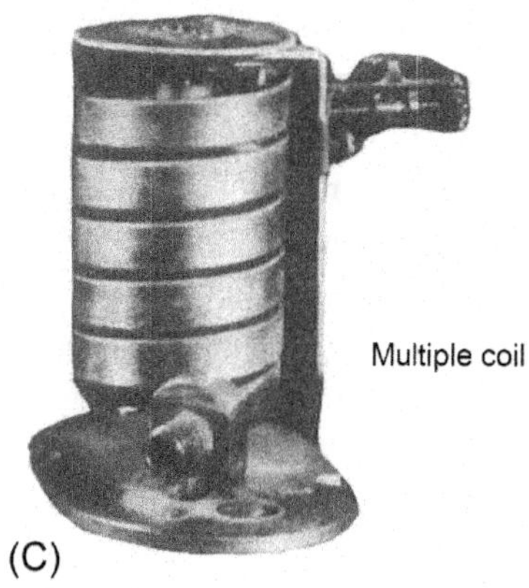

(C)

(B) Single coil

Fig. 6.6
Bourdon gauge.

It may be noted that the pressure measuring devices (a) to (e) all measure a pressure *difference* ΔP ($=P_2 - P_1$). In the case of the Bourdon gauge (f), the pressure indicated is the difference between that communicated by the system to the tube and the external (ambient) pressure, and this is usually referred to as the *gauge* pressure. It is then necessary to add on the ambient pressure in order to obtain the (absolute) pressure. Even the mercury barometer measures, not atmospheric pressure, but the difference between atmospheric pressure and the vapour pressure of mercury which, of course, is negligible. Gauge pressures are however not used in the SI System of units.

6.2.3 Pressure Signal Transmission—The Differential Pressure Cell

The meters described so far provide a measurement usually in the form of a pressure differential, though in most modern plants these readings must be transmitted to a central control facility where they form the basis for either recording or automatic control. The quantity being measured is converted into a signal using a device consisting of a sensing element and a conversion or control element. The sensor may respond to movement, heat, light, magnetic, or chemical effects and these physical quantities may be converted to change in an electrical parameter such as voltage, resistance, capacitance, or inductance. The transmission is most conveniently effected by pneumatic or electrical methods, and one typical arrangement known as a differential pressure (d.p.) cell system, shown in Fig. 6.7, operates on the *force balance* system illustrated in Fig. 6.8.

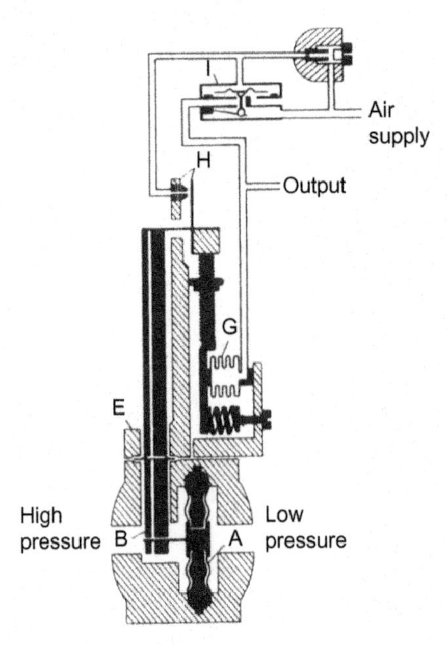

Fig. 6.7
Pneumatic differential pressure cell.

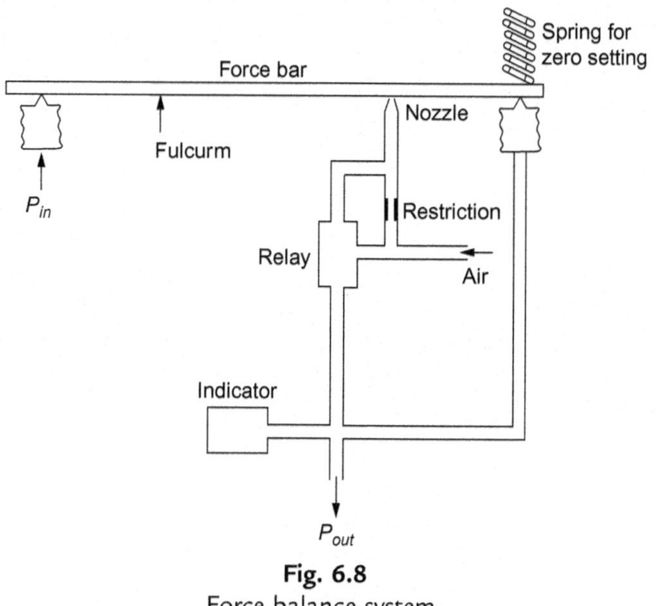

Fig. 6.8
Force balance system.

In Fig. 6.8, an increase in the pressure on the 'in' bellows causes the force-bar to turn clockwise, thus reducing the separation in the flapper—nozzle. This nozzle is fed with air, and the increase in back pressure from the nozzle in turn increases the force developed by the negative feedback bellows producing a counterclockwise movement which restores the flapper–nozzle separation. Balance is achieved when the feedback pressure is proportional to the applied pressure.

In this way, the signal pressure is made to respond to the input pressure and, by adjusting the distance of the bellows from the fulcrum, any range of force may be made to give an output pressure in the range of 120–200 kN/m^2. It is important to note that since the flapper–nozzle unit has a high *gain* (large pressure change for small change in separation), the actual movements are very small, usually within the range of 0.004–0.025 mm.

An instrument in which this feedback arrangement is used is shown in Fig. 6.7. The differential pressure from the meter is applied to the two sides of the diaphragm A giving a force on the bar B. The closure E in the force bar B acts as the fulcrum, so that the flapper–nozzle separation at H responds to the initial difference in pressure. This gives the change in air pressure or signal, which can be transmitted through a considerable distance. A unit of this kind is thus an essential addition to all pressure meters from which a signal in a central control room is required.

6.2.4 Intelligent Electronic Pressure Transmitters

Intelligent pressure transmitters have two major components: (1) a sensor module which comprises the process connections and sensor assembly, and (2) a two-compartment electronics housing with a terminal block and an electronics module that contains signal conditioning

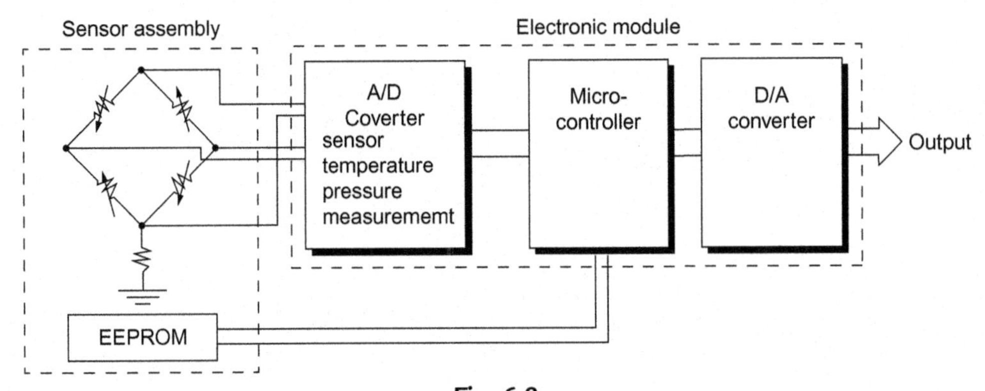

Fig. 6.9
The intelligent transmitter.

circuits and a microprocessor. Fig. 6.9 illustrates how the primary output signal is processed and compensated for errors caused in pressure-sensor temperature. An internal sensor measures the temperature of the pressure sensor. This measurement is fed into the microprocessor where the primary measurement signal is appropriately corrected. This temperature measurement is also transmitted to receivers over the communications network.

The solid state sensor consists of a Wheatstone Bridge circuit shown in Fig. 6.9 which is diffused into a silicon chip, thereby becoming a part of the atomic structure of the silicon. As pressure is applied to the diaphragm (Fig. 6.10), strain is created in the bridge resistors. Piezo-resistive effects created by this strain change resistances in the legs of the bridge, produce a voltage proportional to pressure. Output from the bridge is typically in the range of 75 to 150 mV at full-scale pressure for a bridge excitation of 1.0 mA.

The micro-machined silicon sensor is fabricated in three basic types of pressure sensors. The three types shown in Fig. 6.10, are:

(a) *Gauge pressure* with the sensor referenced to atmosphere.
(b) *Absolute pressure* with the sensor referenced to a full vacuum.

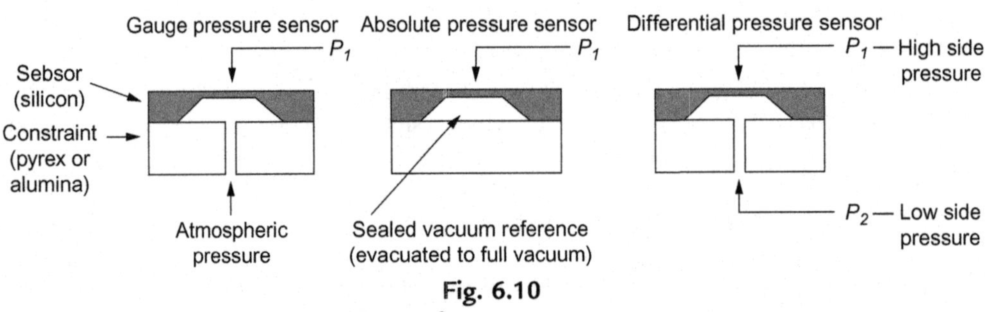

Fig. 6.10
Types of pressure sensors.

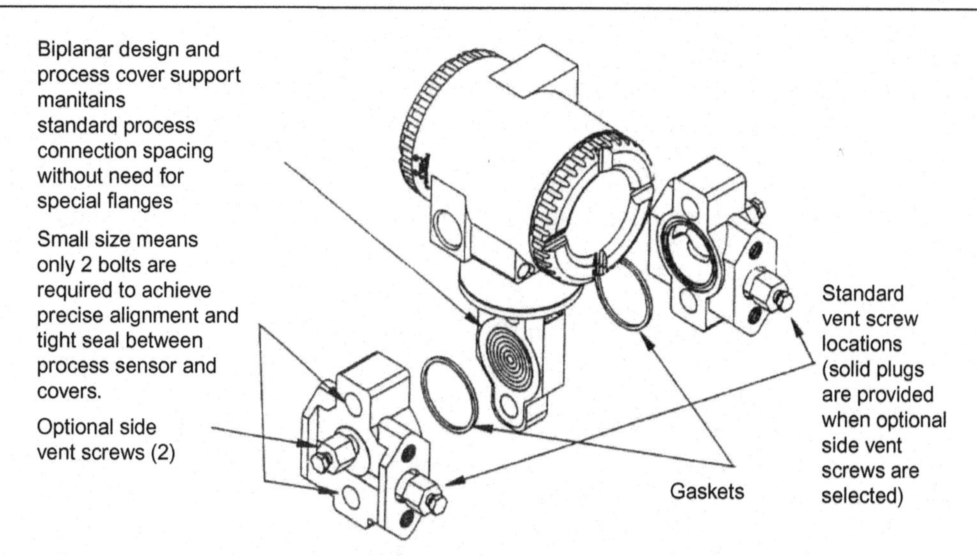
Biplanar design and
process cover support
manitains
standard process
connection spacing
without need for
special flanges
Small size means
only 2 bolts are
required to achieve
precise alignment and
tight seal between
process sensor and
covers.
Optional side
vent screws (2)
Gaskets
Standard
vent screw
locations
(solid plugs
are provided
when optional
side vent
screws are
selected)

Fig. 6.11
Intelligent differential-pressure cell with transmitter.

(c) *Differential pressure* (dp) where the sensor measures the difference between two pressures P_1 and P_2.

Because of the wide range of the sensors, only four different sensor units are needed to cover the entire range of dp spans from 10 kN/m^2 to 20 MN/m^2 (0.5 in water to 3000 lb$_f$/in.^{2}), as shown schematically in Fig. 6.11.

6.2.5 Impact Pressure

The pressure exerted on a plane at right angles to the direction of flow of the fluid consists of two components:

(a) static pressure;
(b) the additional pressure required to bring the fluid to rest at the point.

Consider a fluid flowing between two sections, 1 and 2 (Fig. 6.12), which are sufficiently close for friction losses to be negligible between the two sections; they are a sufficient distance

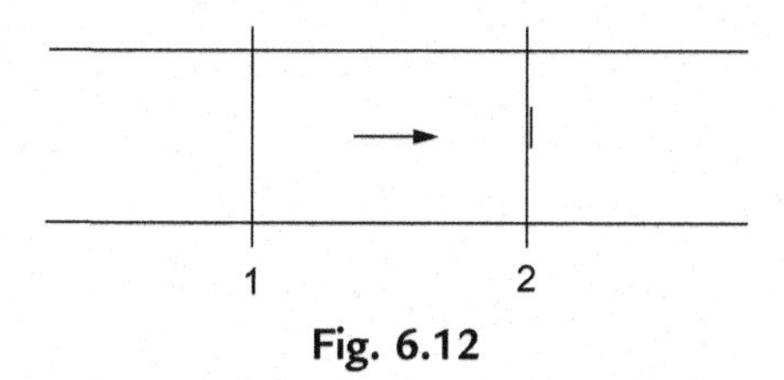
1
2

Fig. 6.12
Impact pressure.

apart, however, for the presence of a small surface at right angles to the direction of flow at section 2 to have negligible effect on the pressure at section 1. These conditions are normally met if the distance between the sections is of the order of one pipe diameter.

Considering a small filament of liquid, which is brought to rest at section 2, and applying the energy balance equation between the two sections, since $g\Delta z$, W_s, and F are all zero:

$$\frac{\dot{u}_1^2}{2} = \frac{\dot{u}_2^2}{2} + \int_{P_1}^{P_2} v\,\mathrm{d}P \quad \text{(from Eq. 2.53)}$$

where $\dot{u}_1$ and $\dot{u}_2$ are velocities at 1 and 2.

If the fluid is incompressible or if the change in the density of the fluid between the sections is negligible, then (since $\dot{u}_2 = 0$):

$$\begin{aligned} \dot{u}_1^2 &= 2v(P_2 - P_1) = 2h_i g \\ \text{or:}\quad \dot{u}_1 &= \sqrt{2v(P_2 - P_1)} = \sqrt{2gh_i} \end{aligned} \tag{6.10}$$

where h_i is the difference between the impact pressure head at section 2 and the static pressure head at section 1.

Little error is introduced if this expression is applied to the flow of a compressible fluid provided that the velocity is not greater than about 60 m/s. When the velocity is high, the equation of state must be used to give the relation between the pressure and the volume of the gas. For nonisothermal flow, $Pv^k = $ a constant, and:

$$\int_{P_1}^{P_2} v\,\mathrm{d}P = \frac{k}{k-1} P_1 v_1 \left[\left(\frac{P_2}{P_1} \right)^{(k-1)/k} - 1 \right] \quad \text{(Eq. 2.73)}$$

so that:

$$\frac{\dot{u}_1^2}{2} = \frac{k}{k-1} P_1 v_1 \left[\left(\frac{P_2}{P_1} \right)^{(k-1)/k} - 1 \right]$$

and:

$$P_2 = P_1 \left(1 + \frac{\dot{u}_1^2}{2} \frac{k-1}{k} \frac{1}{P_1 v_1} \right)^{k/(k-1)} \tag{6.11}$$

For isothermal flow:

$$\frac{\dot{u}_1^2}{2} = P_1 v_1 \ln \frac{P_2}{P_1} \quad \text{(from Eq. 2.69)}$$

$$\therefore \quad P_2 = P_1 \exp \frac{\dot{u}_1^2}{2P_1 v_1}$$
$$\text{or:} \quad P_2 = P_1 \exp \frac{\dot{u}_1^2 M}{2\mathbf{R}T} \qquad (6.12)$$

Eqs (6.11), (6.12) can be used for the calculation of the fluid velocity and the impact pressure in terms of the static pressure a short distance upstream. The two sections are chosen so that they are sufficiently close together for frictional losses to be negligible. Thus P_1 will be approximately equal to the static pressure at both sections and the equations give the relation between the static and impact pressure—and the velocity—at any point in the fluid. Further details regarding pressure measurements, sensors, and actuators can be found in the literature.[6,7]

6.3 Measurement of Fluid Flow

The most important class of flowmeter is that in which the fluid is either accelerated or retarded at the measuring section and the change in kinetic energy is measured by recording the pressure difference produced.

This class includes:

The pitot tube, in which a small element of fluid is brought to rest at an orifice situated at right angles to the direction of flow. The flowrate is then obtained from the difference between the impact and the static pressure. With this instrument the velocity measured is that of a small filament of fluid. Thus, a suitably designed pitot tube can be used to measure local (point) velocity in single phase systems.

The orifice meter, in which the fluid is accelerated at a sudden constriction (the orifice) and the pressure developed is then measured. This is a relatively cheap and reliable instrument, though the overall pressure drop is high because most of the kinetic energy of the fluid at the orifice is dissipated.

The venturi meter, is one in which the fluid is gradually accelerated to a throat and gradually retarded as the flow channel is expanded to the pipe size. A high proportion of the kinetic energy is thus recovered but the instrument is expensive and bulky.

The nozzle, in which the fluid is gradually accelerated up to the throat of the instrument but expansion to pipe diameter is sudden as with an orifice. This instrument is again expensive because of the accuracy required over the inlet section.

The notch or weir, in which the fluid flows over the weir so that its kinetic energy is measured by determining the head of the fluid flowing above the weir. This instrument is used in open-channel flow and extensively in tray towers[8] where the height of the weir is adjusted to provide the necessary liquid depth for a given flow.

Each of these devices will now be considered in more detail together with some less common and special purpose meters. It needs to be emphasised here that this class of devices does disturb the flow and thus is of a invasive kind.

6.3.1 The Pitot Tube

The pitot tube is used to measure the difference between the impact and static pressures in a fluid. It normally consists of two concentric tubes arranged parallel to the direction of flow; the impact pressure is measured on the open end of the inner tube. The end of the outer concentric tube is sealed and a series of orifices on the curved surface give an accurate indication of the static pressure. The position of these orifices must be carefully chosen because there are two disturbances, which may cause an incorrect reading of the static pressure. These are due to:

(1) the head of the instrument;
(2) the portion of the stem which is at right angles to the direction of flow of the fluid.

These two disturbances cause errors in opposite directions, and the static pressure should therefore be measured at the point where the effects are equal and opposite.

If the head and stem are situated at a distance of 14 diameters from each other as on the standard instrument,[9] the two disturbances are equal and opposite at a section 6 diameters from the head and 8 from the stem. This is, therefore, the position at which the static pressure orifices should be located. If the distance between the head and the stem is too high, the instrument will be unwieldy; if it is too short, the magnitude of each of the disturbances will be relatively high, and a small error in the location of the static pressure orifices will appreciably affect the reading.

The two standard instruments are shown in Fig. 6.13A and B; the one with the rounded nose is preferred, since this is less subject to damage.

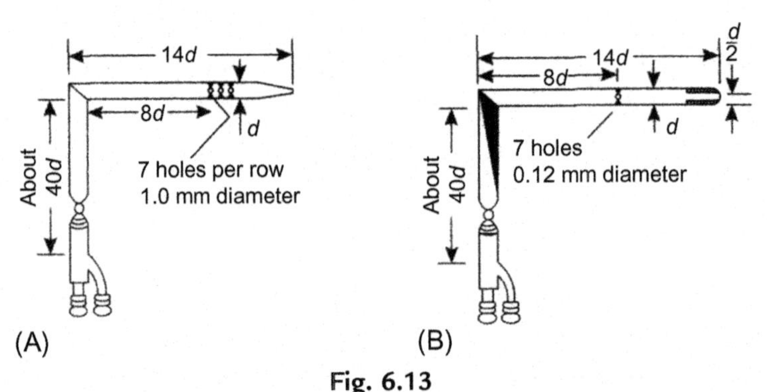

Fig. 6.13
Pitot tubes. (A) Sharp pointed; (B) Rounded.

For Reynolds numbers of 500–300,000, based on the external diameter of the pitot tube, an error of not more than 1% is obtained with this instrument. A Reynolds number of 500 with the standard 7.94 mm pitot tube corresponds to a water velocity of 0.070 m/s or an air velocity of 0.91 m/s. Sinusoidal fluctuations in the flowrate up to 20% do not affect the accuracy by more than 1%, and calibration of the instrument is not necessary.

A very small pressure difference is obtained for low flowrates of gases, and the lower limit of velocity that can be measured is usually set by the minimum difference in pressure that can be measured. This limitation is serious, and various methods have been adopted for increasing the reading of the instrument although they involve the need for calibration. Correct alignment of the instrument with respect to the direction of flow is important; this is attained when the differential reading is a maximum.

For the flow not to be appreciably disturbed, the diameter of the instrument must not exceed about one-fiftieth of the diameter of the pipe; the standard instrument (diameter 7.94 mm) should therefore not be used in pipes of less than 0.4 m diameter. An accurate measurement of the impact pressure can be obtained using a tube of very small diameter with its open end at right angles to the direction of flow; hypodermic tubing is convenient for this purpose. The static pressure is measured using a single piezometer tube or a piezometer ring upstream at a distance approximately equal to the diameter of the pipe: measurement should be made at least 50 diameters from any bend or obstruction.

The pitot tube measures the velocity of only a filament of fluid, and hence it can be used for exploring the velocity distribution across the pipe section. If, however, it is desired to measure the total flow of fluid through the pipe, the velocity must be measured at various distances from the walls and the results integrated. The total flowrate can be calculated from a single reading only if the velocity distribution across the section is already known.

Although a single pitot tube measures the velocity at only one point in a pipe or duct, instruments such as the *averaging pitot tube* or *Annubar*, which employ multiple sampling points over the cross-section, provide information on the complete velocity profile which may then be integrated to give the volumetric flowrate. An instrument of this type has the advantage that it gives rise to a lower pressure drop than most other flow measuring devices, such as the orifice meter described in Section **6.3.2**.

6.3.2 Measurement by Flow Through a Constriction

In measuring devices where the fluid is accelerated by causing it to flow through a constriction, the kinetic energy is thereby increased and the pressure energy therefore decreases. The flowrate is obtained by measuring the pressure difference between the inlet of the meter and a point of reduced pressure, as shown in Fig. 6.14 where the orifice meter, the nozzle, and the venturi meter are illustrated. If the pressure is measured a short distance upstream where

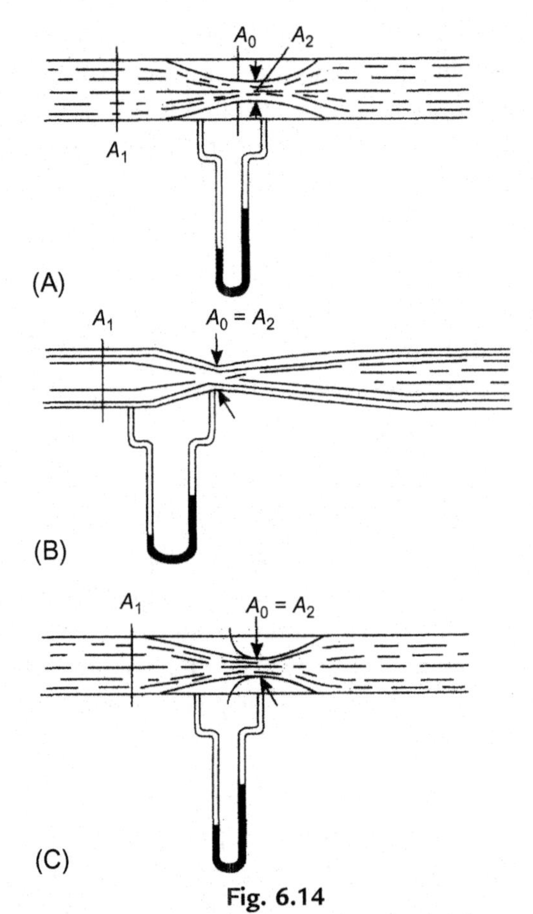

Fig. 6.14
(A) Orifice meter, (B) venturi meter, and (C) nozzle.

the flow is undisturbed (section 1) and at the position where the area of flow is a minimum (section 2), application of the energy and material balance equations gives:

$$\frac{u_2^2}{2\alpha_2} - \frac{u_1^2}{2\alpha_1} + g(z_2 - z_1) + \int_{P_1}^{P_2} v\,dP + W_s + F = 0 \quad \text{(from Eq. 2.55)}$$

and the mass flow,

$$G = \frac{u_1 A_1}{v_1} = \frac{u_2 A_2}{v_2} \quad \text{(from Eq. 2.37)}$$

If the frictional losses are neglected, and the fluid does no work on the surroundings, that is W_s and F are both zero, then:

$$\frac{u_2^2}{2\alpha_2} - \frac{u_1^2}{2\alpha_1} = g(z_1 - z_2) - \int_{P_1}^{P_2} v\,dP \tag{6.13}$$

Inserting the value of u_1 in terms of u_2 in Eq. (6.13) enables u_2 and G to be obtained.

For an incompressible fluid:

$$\int_{P_1}^{P_2} v\mathrm{d}P = v(P_2 - P_1) \quad \text{(Eq. 2.65)}$$

and:

$$u_1 = u_2 \frac{A_2}{A_1}$$

Substituting these values in Eq. (6.13):

$$\frac{u_2^2}{2\alpha_2}\left(1 - \frac{\alpha_2 A_2^2}{\alpha_1 A_1^2}\right) = g(z_1 - z_2) + v(P_1 - P_2)$$

Thus:

$$u_2^2 = \frac{2\alpha_2[g(z_1 - z_2) + v(P_1 - P_2)]}{1 - \dfrac{\alpha_2}{\alpha_1}\left(\dfrac{A_2}{A_1}\right)^2} \tag{6.14}$$

For a horizontal meter $z_1 = z_2$, and:

$$u_2 = \sqrt{\frac{2\alpha_2 v(P_1 - P_2)}{1 - \dfrac{\alpha_2}{\alpha_1}\left(\dfrac{A_2}{A_1}\right)^2}}$$

and

$$G = \frac{u_2 A_2}{v_2} = \frac{A_2}{v}\sqrt{\frac{2\alpha_2 v(P_1 - P_2)}{1 - \dfrac{\alpha_2}{\alpha_1}\left(\dfrac{A_2}{A_1}\right)^2}} \tag{6.15}$$

For an ideal gas in isothermal flow:

$$\int_{P_1}^{P_2} v\mathrm{d}P = P_1 v_1 \ln\frac{P_2}{P_1} \quad \text{(Eq. 2.69)}$$

and

$$u_1 = u_2 \frac{A_2 v_1}{A_1 v_2}$$

And again neglecting terms in z, from Eq. (6.13):

$$\frac{u_2^2}{2\alpha_2}\left[1-\frac{\alpha_2}{\alpha_1}\left(\frac{v_1A_2}{v_2A_1}\right)^2\right]=P_1v_1\ \ln\frac{P_1}{P_2}$$

and:

$$u_2^2=\frac{2\alpha_2P_1v_1\ \ln\dfrac{P_1}{P_2}}{1-\dfrac{\alpha_2}{\alpha_1}\left(\dfrac{v_1A_2}{v_2A_1}\right)^2} \tag{6.16}$$

and the mass flow G is again u_2A_2/v_2.

For an ideal gas in nonisothermal flow. If the pressure and volume are related by $Pv^k=$constant, then a similar analysis gives:

$$\int_{P_1}^{P_2} v\mathrm{d}P=P_1v_1\frac{k}{k-1}\left[\left(\frac{P_2}{P_1}\right)^{(k-1)/k}-1\right] \quad \text{(Eq. 2.73)}$$

and, hence:

$$\frac{u_2^2}{2\alpha_2}\left[1-\frac{\alpha_2}{\alpha_1}\left(\frac{v_1A_2}{v_2A_1}\right)^2\right]=-P_1v_1\frac{k}{k-1}\left[\left(\frac{P_2}{P_1}\right)^{(k-1)/k}-1\right]$$

or:

$$u_2^2=\frac{2\alpha_2P_1v_1\dfrac{k}{k-1}\left[1-\left(\dfrac{P_2}{P_1}\right)^{(k-1)/k}\right]}{1-\dfrac{\alpha_2}{\alpha_1}\left(\dfrac{v_1A_2}{v_2A_1}\right)^2} \tag{6.17}$$

and the mass flow G is again u_2A_2/v_2.

It should be noted that Eqs (6.16), (6.17) apply provided that P_2/P_1 is greater than the critical pressure ratio w_c. This subject is discussed in Chapter 4, where it is shown that when $P_2/P_1<w_c$, the flowrate becomes independent of the downstream pressure P_2 and conditions of *maximum flow* occur.

6.3.3 The Orifice Meter

The most important factors influencing the reading of an orifice meter (Fig. 6.15) are the size of the orifice and the diameter of the pipe in which it is fitted, though a number of other factors do affect the reading to some extent. Thus the exact position and the method of fixing the

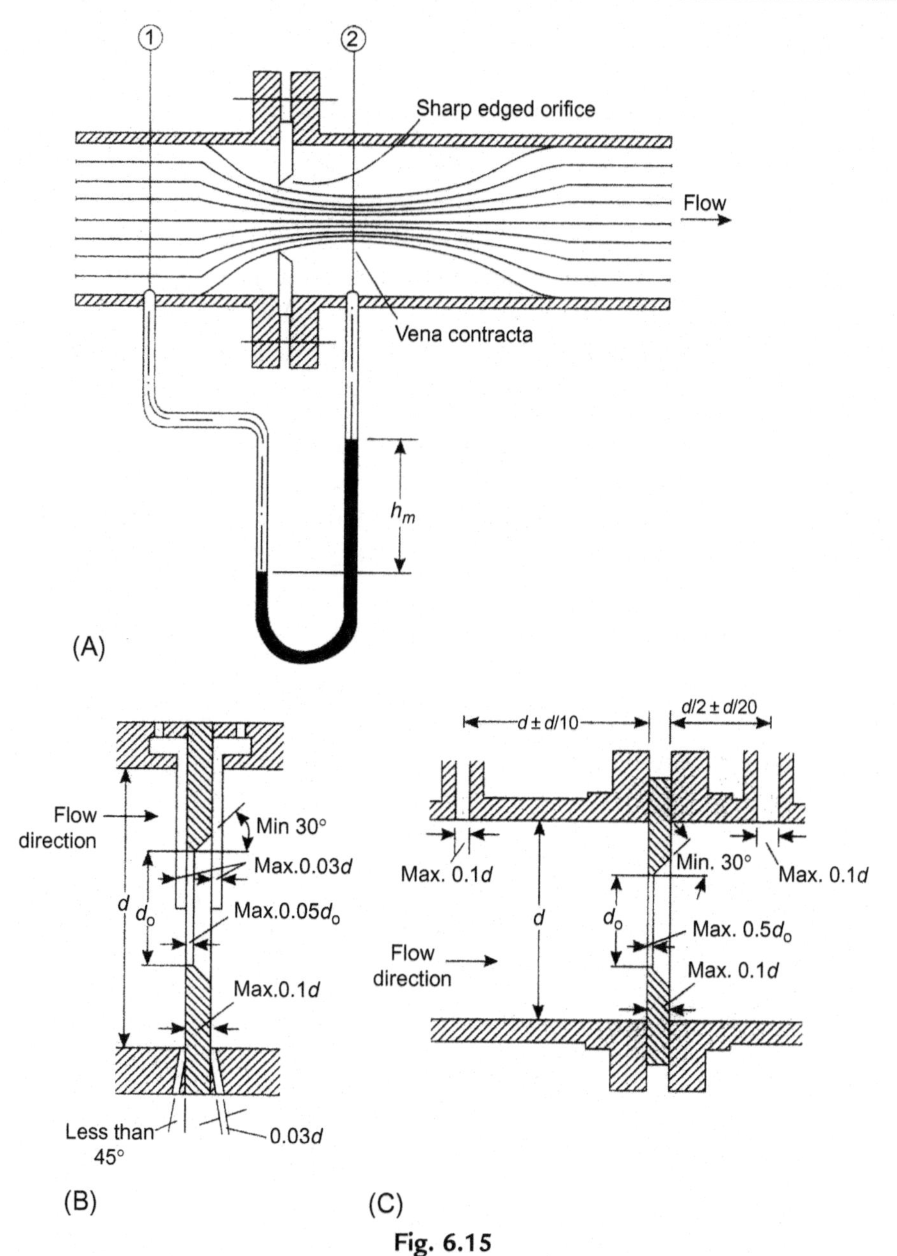

Fig. 6.15

(A) General arrangement. (B) Orifice plate with corner tappings. Upper half shows construction with piezometer ring. Lower half shows construction with tappings into pipe flange. (C) Orifice plate with d and $d/2$ tappings. Nipples must finish flush with wall of pipe without burrs.

pressure tappings are important because the area of flow, and hence the velocity, gradually changes in the region of the orifice. The meter should be located not less than 50 pipe diameters from any pipe fittings. Details of the exact shape of the orifice, the orifice thickness, and other details are given in BS1042[9] and the details must be followed if a standard orifice is to be used without calibration—otherwise the meter must be calibrated. It should be noted that the standard applies only for pipes of at least 150 mm diameter. Further details can be found in standard texts and reference books such as Miller[10] and Baker.[11]

A simple instrument can be made by inserting a drilled orifice plate between two pipe flanges and arranging suitable pressure connections. The orifice must be drilled with sharp edges and is best made from stainless steel, which resists corrosion and abrasion. The size of the orifice should be chosen to give a convenient pressure drop. Although the flowrate is proportional to the square root of the pressure drop, it is difficult to cover a wide range in flow with any given size of orifice. Unlike the pitot tube, the orifice meter gives the average flowrate from a single reading.

The most serious disadvantage of the meter is that most of the pressure drop is not recoverable, that is it is inefficient. The velocity of the fluid is increased at the throat without much loss of energy. The fluid is subsequently retarded as it mixes with the relatively slow-moving fluid downstream from the orifice. A high degree of turbulence is set up and most of the excess kinetic energy is dissipated as heat. Usually only about 5% or 10% of the excess kinetic energy can be recovered as pressure energy. The pressure drop over the orifice meter is therefore high, and this may preclude it from being used in a particular instance.

The area of flow decreases from A_1 at section 1 to A_0 at the orifice and then to A_2 at the *vena contracta* (Fig. 6.15A). The area at the vena contracta can be conveniently related to the area of the orifice by the coefficient of contraction C_c, defined by the relation:

$$C_c = \frac{A_2}{A_0}$$

Inserting the value $A_2 = C_c A_0$ in Eq. (6.15), then for an *incompressible fluid* in a horizontal meter:

$$G = \frac{C_c A_0}{v} \sqrt{\frac{2\alpha_2 v (P_1 - P_2)}{1 - \frac{\alpha_2}{\alpha_1}\left(C_c \frac{A_0}{A_1}\right)^2}} \tag{6.18}$$

Using a coefficient of discharge C_D to take account of the frictional losses in the meter and of the parameters C_c, α_1, and α_2:

$$G = \frac{C_D A_0}{v} \sqrt{\frac{2v (P_1 - P_2)}{1 - \left(\frac{A_0}{A_1}\right)^2}} \tag{6.19}$$

For a meter in which the area of the orifice is small compared with that of the pipe:

$$\sqrt{1 - \left(\frac{A_0}{A_1}\right)^2} \to 1$$

and:

$$G=\frac{C_D A_0}{v}\sqrt{2v(P_1-P_2)}$$
$$=C_D A_0\sqrt{2\rho(P_1-P_2)} \tag{6.20}$$

$$=C_D A_0 \rho\sqrt{2gh_0} \tag{6.21}$$

where h_0 is the difference in head across the orifice expressed in terms of the fluid in question.

This gives a simple working equation for evaluating G, though the coefficient C_D is not a simple function and depends on the values of the Reynolds number in the orifice and the form of the pressure tappings. A value of 0.61 may be taken for the standard meter for Reynolds numbers in excess of 10^4, though the value changes noticeably at lower values of Reynolds number as shown in Fig. 6.16.

For the isothermal flow of an ideal gas, from Eq. (6.16) and using C_D as above:

$$G=\frac{C_D A_0}{v_2}\sqrt{\frac{2P_1 v_1 \ln\left(\frac{P_1}{P_2}\right)}{1-\left(\frac{v_1 A_0}{v_2 A_1}\right)^2}} \tag{6.22}$$

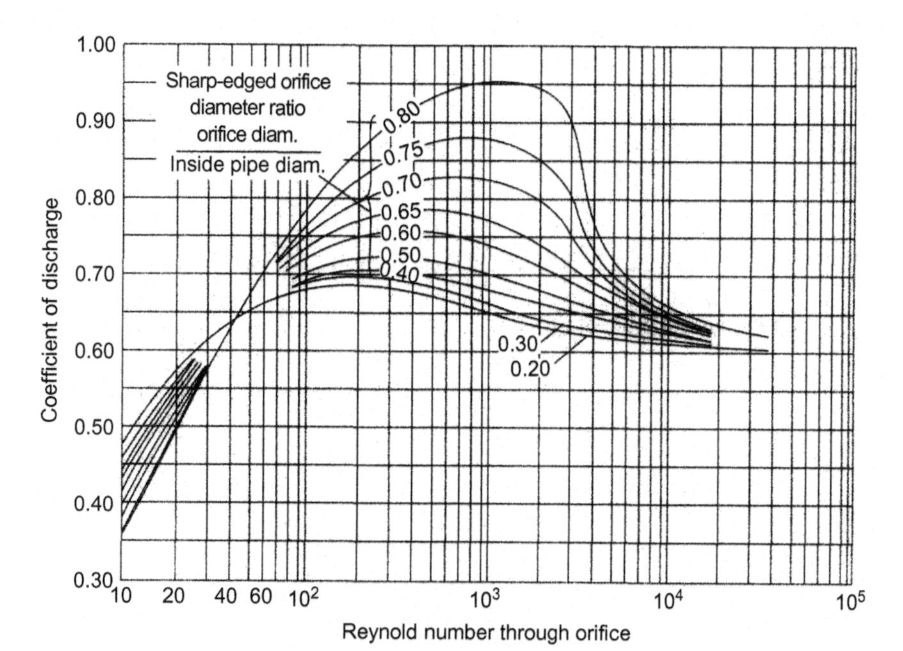

Fig. 6.16
Coefficient for orifice meter.

For a meter in which the area of the orifice is small compared with that of the pipe:

$$G=\frac{C_DA_0}{v_2}\sqrt{2P_1v_1\ \ln\left(\frac{P_1}{P_2}\right)}$$
$$=C_DA_0\sqrt{2\frac{P_2}{v_2}\ \ln\frac{P_1}{P_2}} \tag{6.23}$$

$$=C_DA_0\sqrt{2\frac{P_2}{v_2}\ \ln\left(1-\frac{\Delta P}{P_2}\right)}\quad (\text{where}\ \Delta P=P_2-P_1)$$
$$=C_DA_0\sqrt{2\left(\frac{-\Delta P}{v_2}\right)}\quad \text{if}\ \Delta P\ \text{is small compared with}\ P_2 \tag{6.24}$$

For nonisothermal flow of an ideal gas, from Eq. (6.17):

$$G=\frac{C_DA_0}{v_2}\sqrt{\frac{2P_1v_1\left[\frac{k}{k-1}\right]\left[1-\left(\frac{P_2}{P_1}\right)^{(k-1)/k}\right]}{1-\left(\frac{v_1A_0}{v_2A_1}\right)^2}} \tag{6.25}$$

For a horizontal orifice in which (A_0/A_1) is small:

$$G=\frac{C_DA_0}{v_2}\sqrt{2P_1v_1\frac{k}{k-1}\left[1-\left(\frac{P_2}{P_1}\right)^{(k-1)/k}\right]} \tag{6.26}$$

$$=\frac{C_DA_0}{v_2}\sqrt{2P_1v_1\frac{k}{k-1}\left[1-\left(1+\frac{\Delta P}{P_1}\right)^{(k-1)/k}\right]}$$
$$=C_DA_0\sqrt{2\left(\frac{-\Delta P}{v_2}\right)}\quad \text{if}\ \Delta P\ \text{is small compared with}\ P_2 \tag{6.27}$$

Eqs (6.22), (6.23), (6.25), (6.26) hold provided that P_2/P_1 is greater than the critical pressure ratio w_c.

For $P_2/P_1<w_c$, the flowrate through the orifice is maintained at the maximum rate. From Eq. (4.15), the flowrate is given by Eq. (6.28) and not Eq. (6.23) for isothermal conditions:

$$G_{max}=0.607C_DA_0\sqrt{\frac{P_1}{v_1}} \tag{6.28}$$

Similarly, when the pressure-volume relation is $Pv^k = \text{constant}$, Eq. (4.30) replaces Eq. (6.26):

$$G_{max} = C_D A_0 \sqrt{\frac{kP_1}{v_1}\left(\frac{2}{k+1}\right)^{(k+1)/(k-1)}} \tag{6.29}$$

For isentropic conditions $k = \gamma$ and:

$$G_{max} = C_D A_0 \sqrt{\frac{\gamma P_1}{v_1}\left(\frac{2}{\gamma+1}\right)^{(\gamma+1)/(\gamma-1)}} \tag{6.30}$$

For a diatomic gas, $\gamma = 1.4$ and:

$$G_{max} = 0.685 C_D A_0 \sqrt{\frac{P_1}{v_1}} \tag{6.31}$$

For the flow of steam, a highly nonideal gas, it is necessary to apply a correction to the calculated flowrate, the magnitude of which depends on whether the steam is saturated, wet, or superheated. Correction charts are given by Lyle[12] who also quotes a useful approximation[13]—that a steam meter registers 1% low for every 2% of liquid water in the steam, and 1% high for every 8% of superheat.

Example 6.1

Water flows through an orifice of 25 mm diameter situated in a 75 mm diameter pipe, at a rate of 300 cm^3/s. What will be the difference in level on a water manometer connected across the meter? The viscosity of water is 1 mN s/m^2.

Solution

$$\text{Area of orifice} = \frac{\pi}{4} \times 25^2 = 491\,\text{mm}^2 \text{ or } 4.91 \times 10^{-4}\,\text{m}^2$$

$$\text{Flow of water} = 300\,\text{cm}^3/\text{s or } 3.0 \times 10^{-4}\,\text{m}^3/\text{s}$$

$$\therefore \text{ Velocity of water through the orifice} = \frac{3.0 \times 10^{-4}}{4.91 \times 10^{-4}} = 0.61\,\text{m/s}$$

$$Re \text{ at the orifice} = \frac{25 \times 10^{-3} \times 0.61 \times 1000}{1 \times 10^{-3}} = 15{,}250$$

From Fig. 6.16, the corresponding value of $C_D = 0.61$ (diameter ratio $= 0.33$):

$$\left[1 - \left(\frac{A_0}{A_1}\right)^2\right]^{0.5} = \left[1 - \left(\frac{25^2}{75^2}\right)^2\right]^{0.5} = 0.994 \asymp 1$$

Eq. (6.21) may therefore be applied:

$$G = 3.0 \times 10^{-4} \times 10^3 = 0.30\,\text{kg/s}.$$

$$\therefore \; 0.30 = \left(0.61 \times 4.91 \times 10^{-4} \times 10^3\right)\sqrt{(2 \times 9.81 \times h_0)}.$$

Hence:

$$\sqrt{h_0} = 0.226$$

and:

$$h_0 = 0.051 \text{ m of water}$$
$$= \underline{\underline{51 \text{ mm of water}}}$$

Example 6.2

Sulphuric acid of density 1300 kg/m^3 is flowing through a pipe of 50 mm, internal diameter. A thin-lipped orifice, 10 mm in diameter is fitted in the pipe and the differential pressure shown on a mercury manometer is 0.1 m. Assuming that the leads to the manometer are filled with the acid, calculate (a) the mass flowrate of acid and (b) the approximate drop in pressure caused by the orifice in kN/m^2. The coefficient of discharge of the orifice may be taken as 0.61, the density of mercury as 13,550 kg/m^3, and the density of the water as 1000 kg/m^3.

Solution

(a) The mass flowrate G is given by:

$$G = [(C_D A_0)/v]\sqrt{[(2v(P_1 - P_2))/(1 - (A_0/A_1)^2]} \quad \text{(Eq. 6.19)}$$

where the area of the orifice is small compared with the area of the pipe; that is $\sqrt{\left[1 - (A_0/A_1)^2\right]}$ approximates to 1.0 and:

$$G = C_D A_0 \rho \sqrt{(2gh_0)} \quad \text{(Eq. 6.21)}$$

where h_0 is the difference in head across the orifice expressed in terms of the fluid.

The area of the orifice, is given by:

$$A_0 = (\pi/4)\ (10 \times 10^{-3})^2 = 7.85 \times 10^{-5}\,\text{m}^2$$

and the area of the pipe by:

$$A_1 = (\pi/4)(50 \times 10^{-3})^2 = 196.3 \times 10^{-5}\,\text{m}^2$$

Thus:

$$\left[1 - (A_0/A_1)^2\right] = \left[1 - (7.85/196.3)^2\right] = 0.999 \asymp 1$$

The differential pressure is given by:

$$h_0 = 0.1 \text{ m mercury}$$
$$= 0.1(13{,}550 - 1300)/1300 = 0.94 \text{ m sulphuric acid}$$

and, substituting in Eq. (6.21) gives the mass flowrate as:

$$G = (0.61 \times 7.85 \times 10^{-5})(1300)\sqrt{(2 \times 9.81 \times 0.94)}$$
$$= \underline{\underline{0.268\,\mathrm{kg/s}}}$$

(b) the drop in pressure is then:

$$-\Delta P = \rho g h_0$$
$$= (1300 \times 9.81 \times 0.94) = 11988\,\mathrm{N/m^2} \text{ or } \underline{\underline{12\,\mathrm{kN/m^2}}}$$

Example 6.3

An oil (density = 850 kg/m^3; viscosity = 10 m Pa s) is flowing through a 150 mm diameter pipe at the rate of 0.02 m^3/s and the pressure drop across the orifice plate is measured to be 7.7 kPa. Calculate the diameter of the orifice.

Solution

We can slightly rearrange Eq. (6.19):

$$G = \rho C_D A_0 \sqrt{\frac{(2\Delta P/\rho)}{1-(A_0/A_1)^2}} \quad \text{(Eq. 6.19)}$$

Since the diameter of the orifice is unknown, neither the area term $\{1-(A_0/A_1)^2\}$ in the denominator of this equation nor the value of C_D can be evaluated because the orifice Reynolds number cannot be calculated. We can thus use $C_D = 0.61$ and use the approximation $1-(A_0/A_1)^2 \approx 1$ to begin with.

Substituting values:

$$G = 0.02 \times 850 = 850 \times 0.61 \times A_0\sqrt{2(7.7 \times 10^3/850)}$$

Solving for A_0:

$$A_0 = 7.70 \times 10^{-3}\,\mathrm{m^2} = \frac{\pi d_0^2}{4}$$
$$\text{and } d_0 = 99\,\mathrm{mm}$$

$$\text{Now evaluating the velocity through the orifice} = \frac{0.02}{7.7 \times 10^{-3}}$$
$$= 2.6\,\mathrm{m/s}$$

$$\text{Orifice Reynolds number} = \frac{850 \times 2.6 \times 0.099}{0.01} = 21{,}857$$

$$\frac{d_0}{d} = \frac{99}{150} = 0.66$$

From Fig. 6.16:

$$C_D \cong 0.63$$

$$\text{Checking } 1-\left(\frac{A_0}{A_1}\right)^2=1-\left(\frac{d_0}{d}\right)^4=0.81$$

Thus, using the new values of C_D and $1-\left(\frac{A_0}{A_1}\right)^2$, we can recalculate the value of A_0:

$$0.02\times 850=850\times 0.63\times A_0\sqrt{\frac{\frac{2\times 7.7\times 10^3}{850}}{0.81}}$$

$A_0=6.71\times 10^{-3}\ \text{m}^2$; $d_0=92.5$ mm; $d_0/d=0.616$

Velocity through the orifice $=2.99$ m/s

Orifice Reynolds number $=850\times 2.99\times 0.0925/0.01=23{,}510$

From Fig. 6.16:

$$C_D\cong 0.625;\ 1-(A_0/A_1)^2=0.855$$

While the value of C_D is sufficiently close to the previous value, the value of $1-(A_0/A_1)^2$ is somewhat different. Another iteration yields the following results: $d_0=94$ mm, $C_D\cong 0.62$, and $1-\left(\frac{A_0}{A_1}\right)^2\approx 0.845$ which are all sufficiently close to the previous values.

$$\therefore\ d_0=94\text{mm}$$

While the preceding discussion on orifice plates is limited to the simplest design, the so-called concentric orifice (the centres of the pipe and orifice plate coincide), undoubtedly this is the most common design employed in practice. This configuration is suitable for clean fluids (both gases and liquids) and steam. There are two other designs, eccentric and segmental, shown schematically in Fig. 6.17A and B respectively. The eccentric design is preferred for dilute

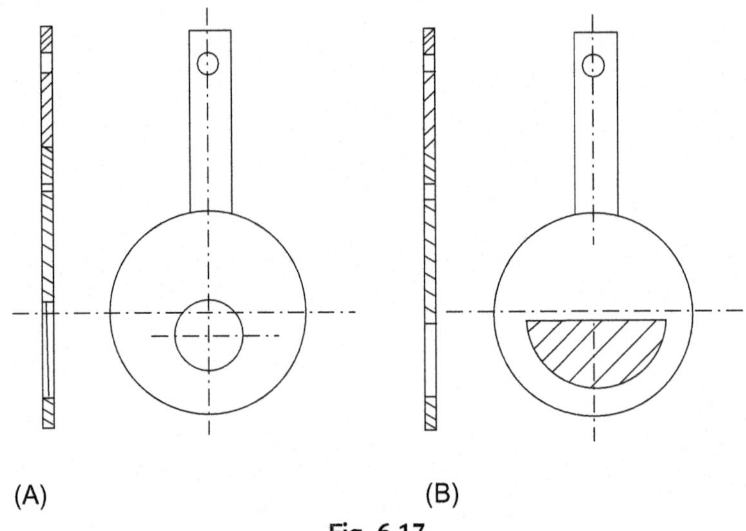

Fig. 6.17
(A) Eccentric orifice plate. (B) Segmented orifice plate.

slurries (when the particles are likely to settle) and for vapours likely to condense and deposit the condensate. Thus, for gases or vapours, it is installed with the eccentric bore top flush with the inner diameter of the pipe and the reverse being the case for the liquid containing solid contaminants. The segmental orifice (Fig. 6.17B) is preferred when there is significant amount of water or air or solid particles present. This design minimises the buildup in front of the orifice. Typically, the orifice hole is situated at the top for liquids and at the bottom for gas streams. For both these designs, some modifications to the relationships derived above are required.

6.3.4 The Nozzle

The nozzle is similar to the orifice meter except that it has a converging tube in place of the orifice plate, as shown in Figs 6.14C and 6.18. The velocity of the fluid is gradually increased and the contours are so designed that almost frictionless flow takes place in the converging portion; the outlet corresponds to the vena contracta of the orifice meter. The nozzle has a constant high coefficient of discharge (ca.0.99) over a wide range of conditions because the coefficient of contraction is unity, though because the simple nozzle is not fitted with a diverging cone, the head lost is very nearly the same as with an orifice. Although much more

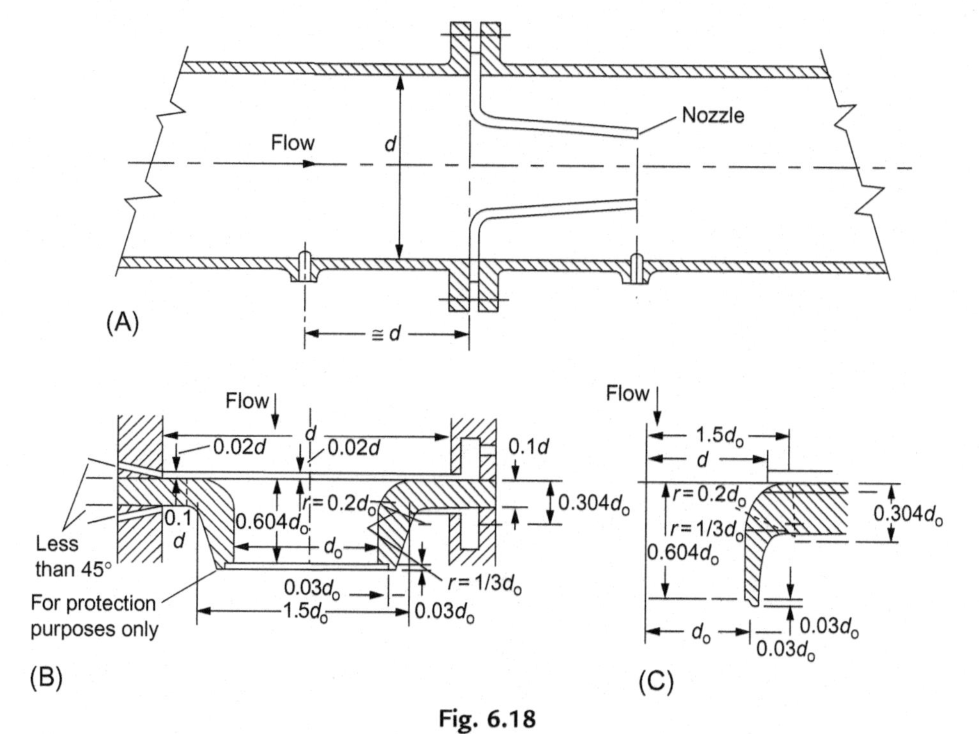

Fig. 6.18
(A) General arrangement. (B) Standard nozzle (A_0/A_1) is less than 0.45. Left half shows construction for corner tappings. Right half shows construction for piezometer ring. (C) Standard nozzle where A_0/A_1 is greater than 0.45.

expensive than the orifice meter, it is extensively used for metering steam. When the ratio of the pressure at the nozzle exit to the upstream pressure is less than the critical pressure ratio w_c, the flowrate is independent of the downstream pressure and can be calculated from the upstream pressure alone.

6.3.5 The Venturi Meter

In this meter, illustrated in Figs 6.14B and 6.19, the fluid is accelerated by its passage through a converging cone of angle 15–20 degrees. The pressure difference between the upstream end of the cone (section 1) and the throat (section 2) is measured and this provides the signal for the rate of flow. The fluid is then retarded in a cone of smaller angle (5–7 degrees) in which a large proportion of the kinetic energy is converted back to pressure energy. Because of the gradual reduction in the area of flow there is no vena contracta and the flow area is a minimum at the throat so that the coefficient of contraction is unity. The main advantage of this meter lies in its high-energy recovery so that it may be used where only a small pressure head is available, though its construction[9–11] is expensive. The flow relationship is given by a similar equation to that for the orifice.

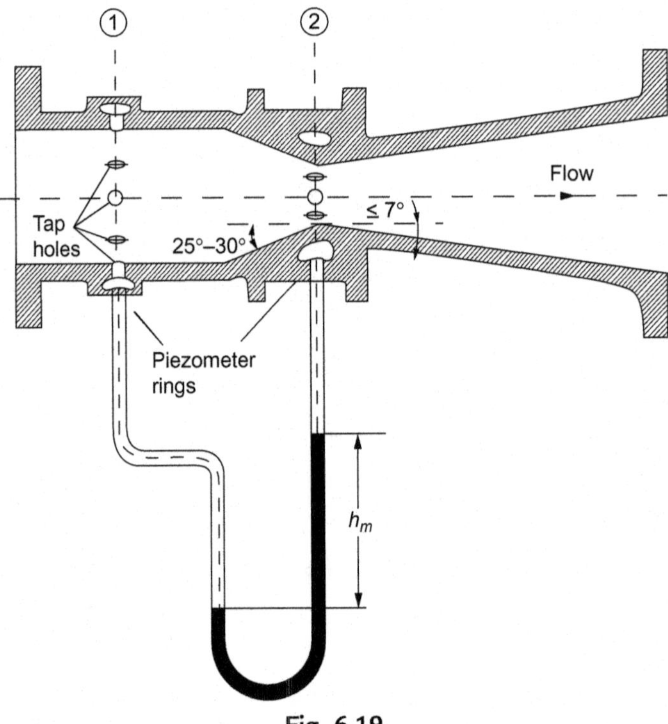

Fig. 6.19
The venturi meter.

For an incompressible fluid in a horizontal meter:

$$G = \frac{C_D A_2}{v}\sqrt{\frac{2v(P_1 - P_2)}{1-(A_2/A_1)^2}} \quad \text{(Eq. 6.19)}$$

$$= C_D \rho \frac{A_1 A_2}{\sqrt{A_1^2 - A_2^2}}\sqrt{2v(P_1 - P_2)} \tag{6.32}$$

$$= C_D \rho C' \sqrt{2gh_v} \tag{6.33}$$

where C' is a constant for the meter and h_v is the loss in head over the converging cone expressed as height of fluid. The coefficient C_D is high, varying from 0.98 to 0.99. The meter is equally suitable for compressible and incompressible fluids.

Example 6.4

The rate of flow of water in a 150 mm diameter pipe is measured with a venturi meter with a 50 mm diameter throat. When the pressure drop over the converging section is 121 mm of water, the flowrate is 2.91 kg/s. What is the coefficient for the converging cone of the meter at this flowrate?

Solution

From Eq. (6.32), the mass rate of flow,

$$G = \frac{C_D \rho A_1 A_2 \sqrt{(2gh_v)}}{\sqrt{(A_1^2 - A_2^2)}}$$

The coefficient for the meter is therefore given by:

$$2.91 = C_D \times 1000\left(\frac{\pi}{4}\right)^2 (150\times10^{-3})^2 (50\times10^{-3})^2$$

$$\times \frac{\sqrt{(2\times 9.81\times 121\times 10^{-3})}}{(\pi/4)\sqrt{\left[(150\times10^{-3})^4 - (50\times10^{-3})^4\right]}}$$

and:

$$C_D = 0.985$$

6.3.6 Pressure Recovery in Orifice-Type Meters

As the discharge coefficient of an orifice meter is about ~0.6, there is a substantial loss of pressure across such a device. This is seen in Fig. 6.20 where the pressure is shown in relation to the position along the tube. There is a steep drop in pressure of about 10% as the vena contracta is reached and the subsequent pressure recovery is poor, amounting to about 50% of the loss. The *Dall Tube* (Fig. 6.21) has been developed to provide a low head loss over the meter while

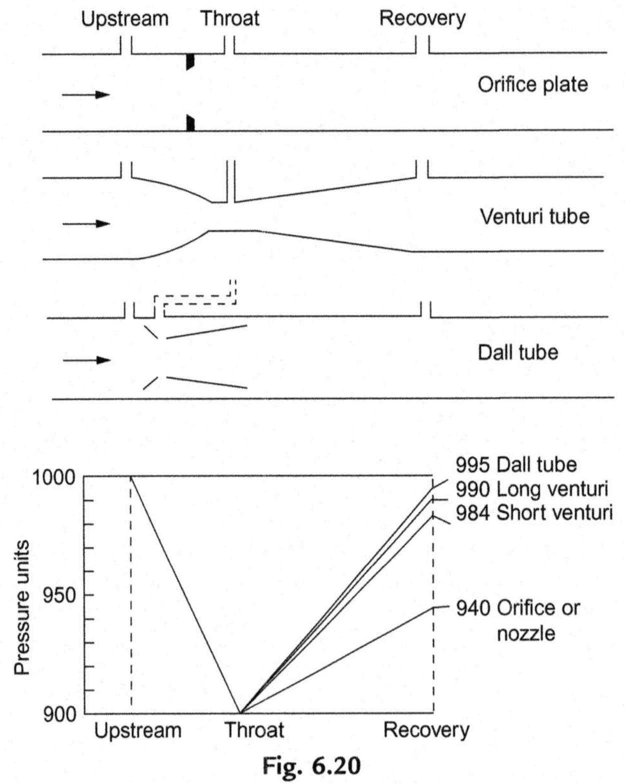

Fig. 6.20
Pressure distribution in orifice plate, venturi meter, and Dall tube. Pressure falls by 10% from upper pressure tapping to throat in each case.

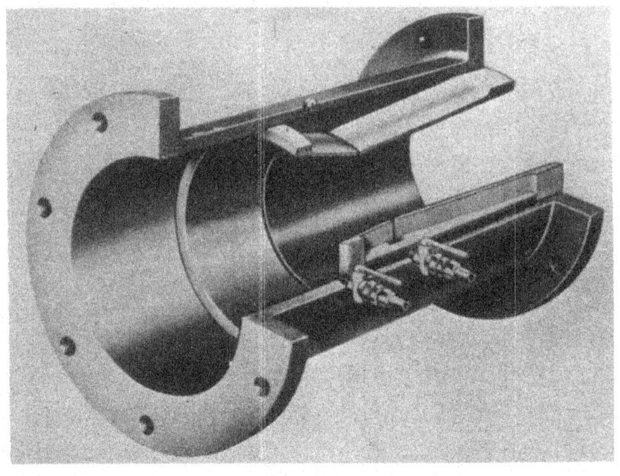

Fig. 6.21
Dall tube.

still giving a reasonably high value of pressure head over the orifice and thus offering great advantages. A short length of parallel lead-in tube is followed by the converging upstream cone and then a diverging downstream cone. This recovery cone is formed by a liner which fits into the meter, the *throat* being formed by a circular slot located between the two cones. One pressure connection is taken to the throat through the annular chamber shown, and the second tapping is on the upstream side. The flow leaves the throat as a diverging jet, which follows the walls of the downstream cone so that eddy losses are almost eliminated. These instruments are made for pipe sizes greater than 150 mm and the pressure loss is only 2%–8% of the differential head. They are cheaper than a venturi, much shorter in length, and correspondingly lighter.

It should be noted that all these restriction type flow meters are primarily intended for pipe sizes greater than 50 mm, and when used on smaller tubes they must be individually calibrated.

Fig. 6.20 shows the pressure changes over the four instruments considered thus far. If the upstream pressure is P_1, the throat pressure P_2, and the final recovery pressure P_3, then the pressure recovery is conveniently expressed as $(P_1 - P_3)/(P_1 - P_2)$, and for these meters, the typical values are given in Table 6.1.

Because of its simplicity the orifice meter is commonly used for process measurements, and this instrument is suitable for providing a signal of the pressure to some comparator as indicated in Fig. 6.1.

6.3.7 Variable Area Meters—Rotameters

In the meters so far described the area of the constriction or orifice is constant and the drop in pressure is dependent on the rate of flow. In the variable area meter, the drop in pressure is constant and the flowrate is a function of the area of the constriction.

A typical meter of this kind, which is commonly known as a *rotameter* (Fig. 6.22), consists of a slightly tapered tube with the smallest diameter at the bottom. The tube contains a freely moving float that rests on a stop at the base of the tube. When the fluid is flowing upward the float rises until its weight is balanced by the upthrust of the fluid, its position then indicating the rate of flow. The pressure difference across the float is equal to its weight divided by its

Table 6.1 Typical meter pressure recovery

Meter Type	Value of C_D	Value of $(P_1 - P_3)/(P_1 - P_2)$ (%)
Orifice	0.6	40–50
Nozzle	0.7	40–50
Venturi	0.9	80–90
Dall	0.9	92–98

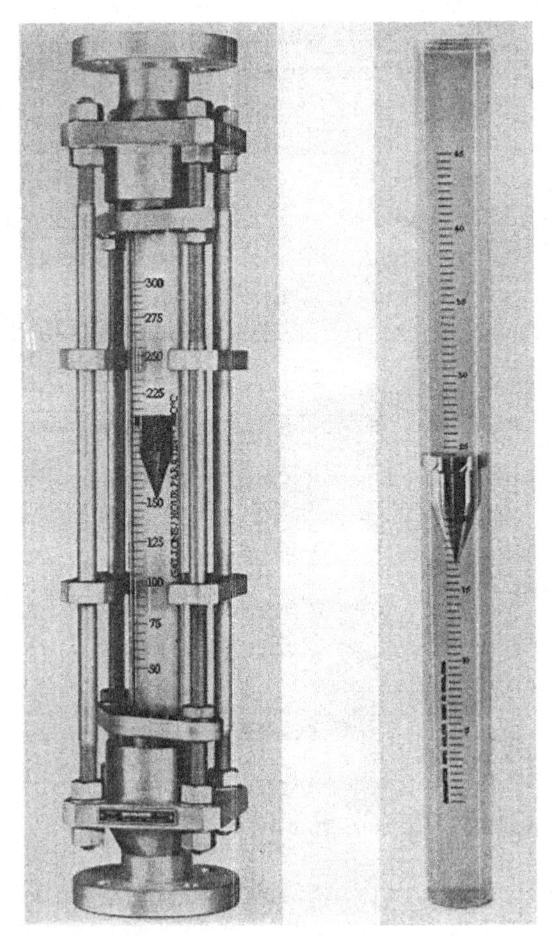

Fig. 6.22
Variable area flowmeters.

maximum cross-sectional area in a horizontal plane. The area for flow is the annulus formed between the float and the wall of the tube.

This meter may thus be considered as an orifice meter with a variable aperture, and the formulae already derived are therefore applicable with only minor changes. Both in the orifice-type meter and in the rotameter the pressure drop arises from the conversion of pressure energy to kinetic energy and from frictional losses, which are accounted for in the coefficient of discharge. The pressure difference over the float $-\Delta P$, is given by:

$$-\Delta P = \frac{V_f(\rho_f - \rho)g}{A_f} \tag{6.34}$$

where V_f is the volume of the float, ρ_f the density of the material of the float, and A_f is the maximum cross-sectional area of the float in a horizontal plane.

If the area of the annulus between the float and tube is A_2 and the cross-sectional area of the tube is A_1, then from Eq. (6.19):

$$G = C_D A_2 \sqrt{\frac{2\rho(-\Delta P)}{1-(A_2/A_1)^2}} \tag{6.35}$$

Substituting for $-\Delta P$ from Eq. (6.34):

$$G = C_D A_2 \sqrt{\frac{2gV_f(\rho_f-\rho)\rho}{A_f\left[1-(A_2/A_1)^2\right]}} \tag{6.36}$$

The coefficient of discharge C_D depends on the shape of the float and the Reynolds number (based on the velocity in the annulus and the mean hydraulic diameter of the annulus) for the flow through the annular space of area A_2. In general, floats, which give the most nearly constant coefficient are of such a shape that they set up eddy currents and give low values of C_D. The variation in C_D largely arises from differences in viscous drag of fluid on the float, and if turbulence is artificially increased, the drag force rises quickly to a limiting but high value. As seen in Fig. 6.23, float *A* does not promote turbulence and the coefficient rises slowly to a high value of 0.98. Float *C* promotes turbulence and C_D rises quickly but only to a low value of 0.60.

The constant coefficient for float *C* arises from turbulence promotion, and for this reason the coefficient is also substantially independent of the fluid viscosity. The meter can be made

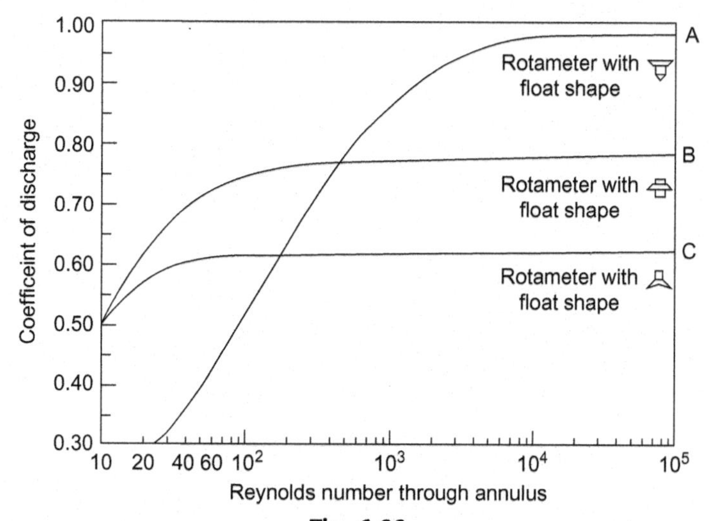

Fig. 6.23
Discharge coefficients for rotameters.

relatively insensitive to changes in the density of the fluid by selection of the density of the float, ρ_f. Thus the flowrate for a given meter will be independent of ρ when $dG/d\rho = 0$.

From Eq. (6.36):

$$\frac{dG}{d\rho} = \frac{C_D A_2}{2}\sqrt{\frac{2gV_f}{A_f\left[1-(A_2/A_1)^2\right]}}\left\{(\rho_f-\rho)^{1/2}\frac{1}{2}\rho^{-1/2} - \rho^{1/2}\frac{1}{2}(\rho_f-\rho)^{-1/2}\right\} \tag{6.37}$$

When:

$$\frac{dG}{d\rho} = 0,$$

i.e.

$$(\rho_f - \rho)\rho^{-1/2} - \rho^{1/2} = 0$$

or,

$$\rho_f = 2\rho \tag{6.38}$$

Thus if the density of the float is twice that of the fluid, then the position of the float for a given flow is independent of the fluid density.

The range of a meter can be increased by the use of floats of different densities, a given float covering a flowrate range of about 10:1. For high-pressure applications, the glass tube is replaced by a metal tube. When a metal tube is used or when the liquid is very dark or dirty an external indicator is required to ascertain the position of the float.

Example 6.5

A rotameter tube is 0.3 m long with an internal diameter of 25 mm at the top and 20 mm at the bottom. The diameter of the float is 20 mm, its density is 4800 kg/m^3, and its volume is 6.0 cm^3. If the coefficient of discharge is 0.7, what is the flowrate of water (density 1000 kg/m^3) when the float is halfway up the tube?

Solution

From Eq. (6.36):

$$G = C_D A_2\sqrt{\frac{2gV_f(\rho_f-\rho)\rho}{A_f\left[1-(A_2/A_1)^2\right]}}$$

$$\text{Cross-sectional area at top of tube} = \frac{\pi}{4}25^2 = 491\,\text{mm}^2 \text{ or } 4.91\times10^{-4}\,\text{m}^2$$

$$\text{Cross-sectional area at bottom of tube} = \frac{\pi}{4}20^2 = 314\,\text{mm}^2 \text{ or } 3.14\times10^{-4}\,\text{m}^2$$

$$\text{Area of float: } A_f = 3.14\times10^{-4}\,\text{m}^3$$

$$\text{Volume of float: } V_f = 6.0\times10^{-6}\,\text{m}^3$$

When the float is halfway up the tube, the area at the height of the float A_1 is given by:

$$A_1 = \frac{\pi}{4}\left(\frac{25+20}{2}\right) \times 10^{-6}\,\mathrm{m}^2$$

or:

$$A_1 = 3.98 \times 10^{-4}\,\mathrm{m}^2$$

The area of the annulus A_2 is given by:

$$A_2 = A_1 - A_f = (3.98 - 3.14) \times 10^{-4} = 0.84 \times 10^{-4}\,\mathrm{m}^2$$

$$A_2/A_1 = \frac{0.84 \times 10^{-4}}{3.98 \times 10^{-4}} = 0.211 \quad \text{and} \quad \left[1-(A_2/A_1)^2\right] = 0.955$$

Substituting into Eq. (6.36):

$$G = (0.70 \times 0.84 \times 10^{-4})\sqrt{\left\{\frac{[2 \times 9.81 \times 6.0 \times 10^{-6}(4800-1000)1000]}{(3.14 \times 10^{-4} \times 0.955)}\right\}}$$

$$= \underline{\underline{0.072\,\mathrm{kg/s}}}$$

6.3.8 The Notch or Weir

The flow of a liquid presenting a free surface can be measured by means of a weir. The pressure energy of the liquid is converted into kinetic energy as it flows over the weir, which may or may not cover the full width of the stream, and a calming screen may be fitted before the weir. Then the height of the weir crest gives a measure of the rate of flow. The velocity with which the liquid leaves depends on its initial depth below the surface. For unit mass of liquid, initially at a depth h below the free surface, discharging through a notch, the energy balance (Eq. 2.67) gives:

$$\Delta\frac{u^2}{2} + v\Delta P = 0 \tag{6.39}$$

for a frictionless flow under turbulent conditions. If the velocity upstream from the notch is small and the flow upstream is assumed to be unaffected by the change of area at the notch:

$$\frac{u_2^2}{2} = -v\Delta P = gh$$
$$\text{and:} \quad u_2 = \sqrt{2gh} \tag{6.40}$$

Rectangular notch

For a rectangular notch (Fig. 6.24), the rate of discharge of fluid at a depth h through an element of cross-section of depth dh will be given by:

$$\mathrm{d}Q = C_D B\,\mathrm{d}h\sqrt{2gh} \tag{6.41}$$

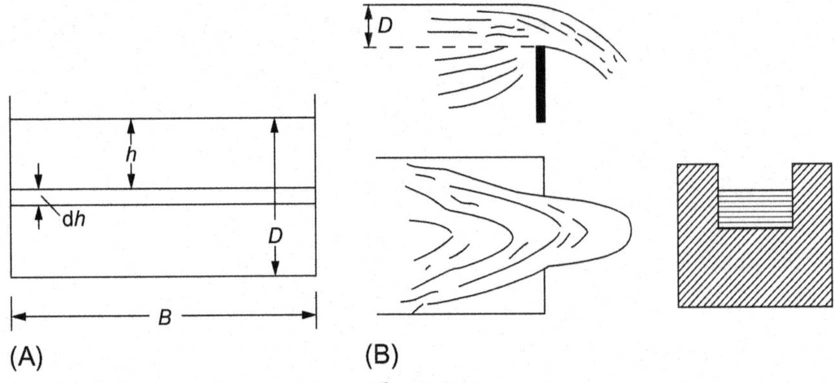

Fig. 6.24
Rectangular notch. (A) Side view; (B) Different types of installation.

where C_D is the coefficient of discharge (usually about 0.6) and B is the width of the notch.

Thus, the total rate of flow,

$$Q = C_D B\sqrt{2g}\left(\frac{2}{3}\right)D^{1.5} \tag{6.42}$$

where D is the total depth of liquid above the bottom of the notch.

The empirical Francis formula:

$$Q = 1.84(B - 0.1nD)D^{1.5} \tag{6.43}$$

gives the flowrate in m^3/s if the dimensions of the notch are expressed in metres with $C_D = 0.62$; $n = 0$ if the notch is the full width of the channel; $n = 1$ if the notch is narrower than the channel but is arranged with one edge coincident with the edge of the channel; $n = 2$ if the notch is narrower than the channel and is situated symmetrically (see Fig. 6.24B): n is known as the number of end contractions.

Example 6.6

Water flows in an open channel across a weir, which occupies the full width of the channel. The length of the weir is 0.5 m and the height of water over the weir is 100 mm. What is the volumetric flowrate of water?

Solution
Use is made of the Francis formula:

$$Q = 1.84(B - 0.1nD)D^{1.5}\left(m^3/s\right) \quad (\text{Eq. } 6.43)$$

where B is the length of the weir (m), n the number of end contractions (in this case $n = 0$), and D the height of liquid above the weir (m).

Thus:

$$Q = 1.84(0.5)0.100^{1.5}$$
$$= \underline{\underline{0.030\,m^3/s}}$$

Example 6.7

An organic liquid flows across a distillation tray and over a weir at the rate of 15 kg/s. The weir is 2 m long and the liquid density is 650 kg/m^3. What is the height of liquid flowing over the weir?

Solution

Use is made of the Francis formula (Eq. 6.43), where, as in the previous example, $n=0$. In the context of this example the height of liquid flowing over the weir is usually designated h_{ow} and the volumetric liquid flow by Q. Rearrangement of Eq. (6.43) gives:

$$h_{ow}=0.666\left(\frac{Q}{L_w}\right)^{0.67}\ (\text{m})$$

where Q is the liquid flowrate (m^3/s) and L_w the weir length (m).

In this case:

$$Q=\left(\frac{15}{650}\right)=0.0230\,\text{m}^3/\text{s and } L_w=2.0\,\text{m}$$

$$h_{ow}=0.666\left(\frac{0.0230}{2}\right)^{0.67}=0.033\,\text{m}$$

$$=\underline{\underline{33.0\,\text{mm}}}$$

Triangular notch

For a triangular notch of angle 2θ, the flow dQ through the thin element of cross-section (Fig. 6.25) is given by:

$$dQ=C_D(D-h)2\tan\theta dh\sqrt{2gh} \tag{6.44}$$

The total rate of flow,

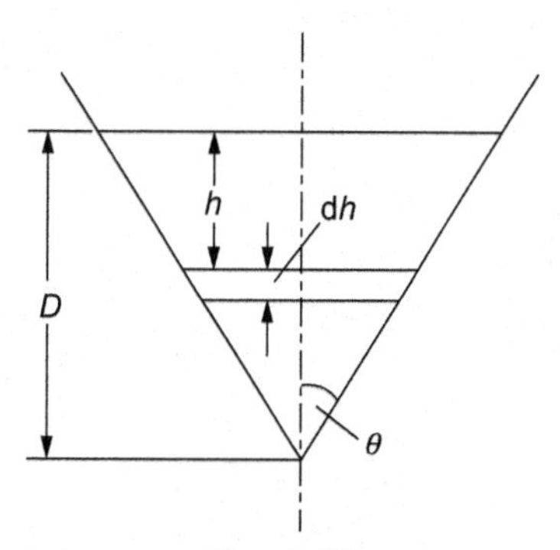

Fig. 6.25
Triangular notch.

$$Q = 2C_D \tan\theta \sqrt{2g} \int_0^D \left(Dh^{1/2} - h^{3/2} \right) dh$$
$$= 2C_D \tan\theta \sqrt{2g} \left(\tfrac{2}{3} D^{5/2} - \tfrac{2}{5} D^{5/2} \right) \tag{6.45}$$
$$= \tfrac{8}{15} C_D \tan\theta \sqrt{2g} D^{5/2}$$

For a 90 degrees notch for which $C_D = 0.6$, and using SI units:

$$Q = 1.42 D^{2.5} \tag{6.46}$$

Thus for a rectangular notch the rate of discharge is proportional to the liquid depth raised to a power of 1.5, and for a triangular notch to a power of 2.5. A triangular notch will therefore handle a wider range of flowrates. In general, the index of D is a function of the contours of the walls of the notch, and any desired relation between flowrate and depth can be obtained by a suitable choice of contour. It is sometimes convenient to employ a notch in which the rate of discharge is directly proportional to the depth of liquid above the bottom of the notch. It can be shown that the notch must have curved walls giving a large width to the bottom of the notch and a comparatively small width towards the top. More detailed discussion of the effect of shape of notches has been given by Chanson.[14]

While open channels are not used frequently for the flow of material other than cooling water between plant units, a weir is frequently installed for controlling the flow within the unit itself, for instance in a distillation column or a reactor.

6.3.9 Other Methods of Measuring Flowrates

The meters which have been described so far depend for their operation on the conversion of some of the kinetic energy of the fluid into pressure energy or vice versa. They form by far the largest class of flowmeters. Other meters are used for special purposes, and brief reference will now be made to a few of these.

For further details reference should be made to Refs. 10,11 which gives a list of the characteristics of flow meters in common use, together with an account of the principles on which they operate and their operational range.

Hot-wire anemometer

If a heated wire is immersed in a fluid, the rate of loss of heat will be a function of the flowrate. In the hot-wire anemometer a fine wire whose electrical resistance has a high temperature coefficient is heated electrically. Under equilibrium conditions, the rate of loss of heat is then proportional to $I^2\Omega$, where Ω is the resistance of the wire and I is the current flowing through the wire.

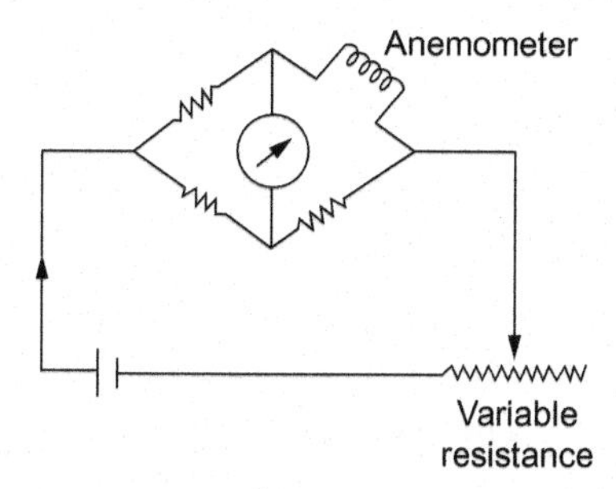

Fig. 6.26
Typical circuit for a hot-wire anemometer.

Either the current or the resistance (and hence the temperature) of the wire is maintained constant. The following is an example of a method in which the resistance is maintained constant. The wire is incorporated as one of the resistances of a Wheatstone network (Fig. 6.26) in which the other three resistances have low temperature coefficients. The circuit is balanced when the wire is immersed in the stationary fluid but, when the fluid is set in motion, the rate of loss of heat increases and the temperature of the wire falls. Its resistance therefore changes and the bridge is thrown out of balance. The balance can be restored by increasing the current so that the temperature and resistance of the wire are brought back to their original values; the other three resistances will be unaffected by the change in current because of their low temperature coefficients. The current flowing in the wire is then measured using either an ammeter or a voltmeter. The rate of loss of heat is found to be approximately proportional to $\sqrt{u\rho+b'}$, so that

$$a'\sqrt{u\rho+b'} = I^2\Omega \text{ under equilibrium conditions} \tag{6.47}$$

where u is the velocity of the fluid (m/s), ρ the density (kg/m^3), and a' and b' are constants for a given meter.

Since the resistance of the wire is maintained constant:

$$u\rho = \frac{I^4\Omega^2}{a'^2} - b' = a''I^4 - b' \tag{6.48}$$

where $a'' = \Omega^2/(a')^2$ remains constant, i.e. the mass rate of flow per unit area is a function of the fourth power of the current, which can be accurately measured.

The hot-wire anemometer is very accurate even for very low rates of flow. It is one of the most convenient instruments for the measurement of the flow of gases at low velocities; accurate readings are obtained for velocities down to about 0.03 m/s. If the ammeter has a high natural frequency, pulsating flows can also be measured. Platinum wire is commonly used.

The magnetic flowmeter

The magnetic flowmeter is a volumetric metering device based on a discovery by Faraday that electrical currents are induced in a conductor that is moved through a magnetic field. Faraday's Law states that the voltage induced in a conductor is proportional to the width of the conductor, the intensity of the magnetic field, and the velocity of the conductor. Thus the output of the flowmeter, a millivoltage, is linearly related to the velocity of the conductor, which is the flowing process fluid. The principle of operation and meter itself is shown in Fig. 6.27.

The induced voltage is in a plane that is mutually perpendicular to both the motion of the conductor and the magnetic field. The process fluid passes through the magnetic field induced by electromagnetic coils or permanent magnets built around the short length of pipe called the metering section. Because of the magnetic field used for the induction, the metering section must be of a nonmagnetic material. Further, if it is metal, it must have a lining that serves as an electrical insulator from the flowing liquid.

These meters have been very useful for handling liquid metal coolants where the conductivity is high though they have now been developed for liquids such as tap water, which have a poor conductivity. Designs are now available for all sizes of tubes, and the instrument is particularly suitable for pharmaceutical and biological purposes where contamination of the fluid must be avoided.

This meter has a number of special features, which should be noted. Thus it gives rise to a negligible drop in pressure for the fluid being metered; fluids containing a high percentage of solids can be metered and the flow can be in either direction. The liquid must have at least a small degree of electrical conductivity. Further details are given by Ginesi.[15]

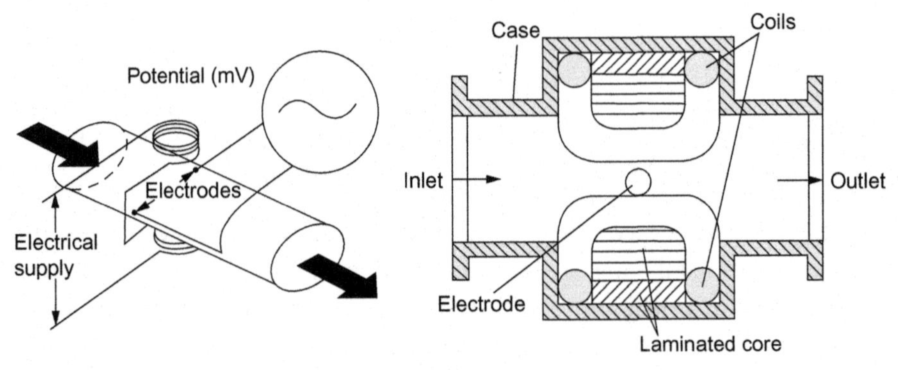

Fig. 6.27
The magnetic flowmeter.

Vortex-shedding flowmeters

When a bluff body is immersed in a fluid stream, the flow is split into two parts. The boundary layer (see Chapter 3 of Vol. 1B), which forms over the surface of the obstruction develops instabilities and vortices are formed and then shed successively from alternate sides of the body, giving rise to what is known as a *von Karman vortex street*. This process sets up regular pressure variations downstream from the obstruction whose frequency is proportional to the fluid velocity, as shown by Strouhal.[16] Vortex flowmeters are very versatile and can be used with almost any fluid—gases, liquids and multiphase fluids. The operation of the vortex meter, illustrated in Fig. 6.28, is described in more detail by Ginesi[15] and in a publication by a commercial manufacturer, Endress and Hauser.[17]

The time-of-flight ultrasonic flowmeter

A high-frequency pressure wave is transmitted at an acute angle to the walls of the pipe and impinges on a receiver on the other side of the pipe. The elapsed time between transmission and reception is a function of the velocity and velocity profile of the fluid, and of the velocity of sound in the fluid and the angle at which the wave is transmitted. These meters which

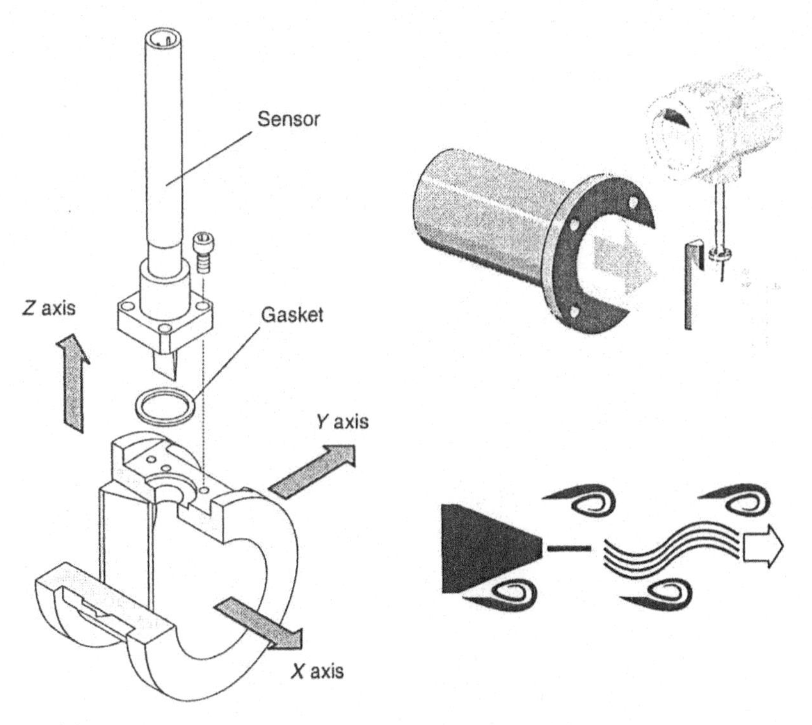

Fig. 6.28
Vortex flow measuring system.

are suitable for the measurement of flowrates of clean liquids only, are described in detail by Lynnworth.[18]

The Doppler ultrasonic flowmeter

The Doppler meter may be used wherever small particulate solids, bubbles or droplets are dispersed in the fluid and are moving at essentially the same velocity as the fluid stream which is to be metered. A continuous ultrasonic wave is transmitted, again at an acute angle to the wall of the duct, and the shift in frequency between the transmitted and scattered waves is measured. This method of measurement of flowrate is frequently used for slurries and dispersions, which present considerable difficulties when other methods are used.

The Coriolis meter

The Coriolis meter (Fig. 6.29) contains a sensor consisting of one or more tubes that are vibrated at their resonant frequency by electromagnetic drivers, and their harmonic vibrations impart an angular motion to the fluid as it passes through the tubes which, in turn, exert Coriolis forces on the tube walls. The magnitude of the Coriolis forces is proportional to the product of the velocity and density of the fluid. A secondary movement of the tubes occurs which is proportional to the mass flowrate and this then becomes superimposed on the primary vibration. A sensor then detects and measures the magnitude of this secondary oscillation. Details of the characteristics of the meters are given by Ginesi,[15] by Plache[19] and by Reizner.[20]

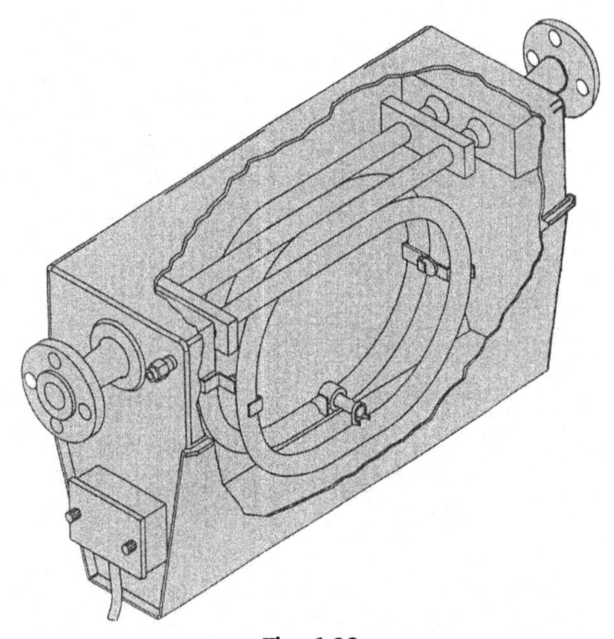

Fig. 6.29
The Coriolis meter.[15]

Quantity meters

The meters, which have been described so far give an indication of the rate of flow of fluid; the total amount passing in a given time must be obtained by integration. Orifice meters are frequently fitted with integrating devices. A number of instruments is available, however, for measuring directly the total quantity of fluid which has passed. An average rate of flow can then be obtained by dividing the quantity by the time of passage.

Gas meters

A simple quantity meter, which is used for the measurement of the flow of gas in an accessible duct is the anemometer (Fig. 6.30). A shaft carrying radial vanes or cups supported on low friction bearings is made to rotate by the passage of the gas; the relative velocity between the gas stream and the surface of the vanes is low because the frictional resistance of the shaft is small. The number of revolutions of the spindle is counted automatically, using a gear train connected to a series of dials. The meter must be calibrated and should be checked at frequent intervals because the friction of the bearings will not necessarily remain constant. The anemometer is useful for gas velocities above about 0.15 m/s.

Quantity meters, suitable for the measurement of the flow of gas through a pipe, include the standard wet and dry meters. In the wet meter, the gas fills a rotating segment and an equal volume of gas is expelled from another segment (Fig. 6.31). The dry gas meter employs a pair of bellows. Gas enters one of the bellows and automatically expels gas from the other; the number of cycles is counted and recorded on a series of dials. Both of these are positive displacement meters and therefore do not need frequent calibration. Gas meters usually appear very bulky for the quantities they are measuring. This is because the linear velocity

Fig. 6.30
Vane anemometer.

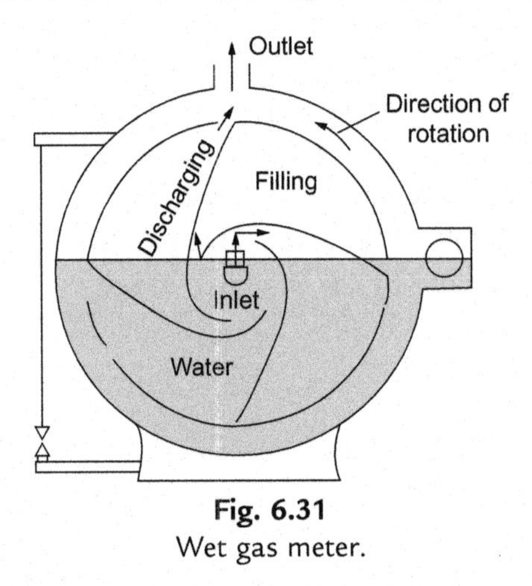

Fig. 6.31
Wet gas meter.

of a gas in a pipe is normally very high compared with that of a liquid, and the large volume is needed so that the speed of the moving parts can be reduced and wear minimised.

Liquid meters

In the oscillating-piston meter, the principle of which is illustrated in Fig. 6.32, the flow of the liquid results in the positive displacement of a rotating element, the cumulative flow being obtained by gearing to a counter.

The body, in the form of a cylindrical chamber, is fitted with a radial partition and a central hub. The circular piston has a slot gap in the circumference to fit over the partition and a peg on its upper face to control movement around the central hub. A rolling seal is thus formed between the piston and hub and between the piston and main chamber. The chamber is thus split into four spaces as shown in Fig. 6.32. The fluid enters the bottom of the chamber through one port and leaves by the other port, the piston forming a movable division between the inlet and outlet.

In an operation cycle, the liquid enters port *A* and fills the spaces 1 and 3, thus forcing the piston to oscillate counterclockwise opening spaces 2 and 4 to port *B*. Because of the partition, the piston moves downwards so that space 3 is cut off from port *A* and becomes space 4. Further movement allows the exit port to be uncovered, and the measured volume between hub and piston is then discharged. The outer space 1 increases until the piston moves upwards over the partition and space 1 becomes space 2 when a second metered volume is discharged by the filling of the inner space 3. Meters of this type will handle flows of between about 0.005 and 15 L/s.

Turbine flow meters are composed of some form of rotary device such as a helical rotor, Pelton wheel or a vane mounted in the flow stream. The fluid passing the rotor causes the rotor to turn at an angular velocity which is proportional to the flow velocity and hence the volumetric flowrate through the meter. The rotary motion of the rotor is sensed by some form of pick-up

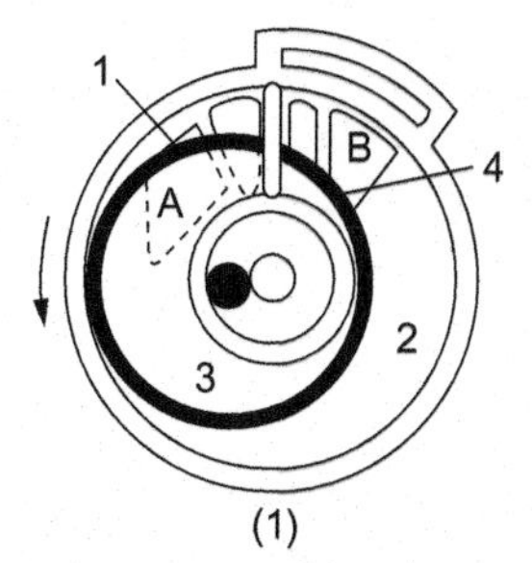

(1)

Spaces 1 and 3 are receiving fluid from the inlet port A, and spaces 2 and 4 are discharging through the outlet port B

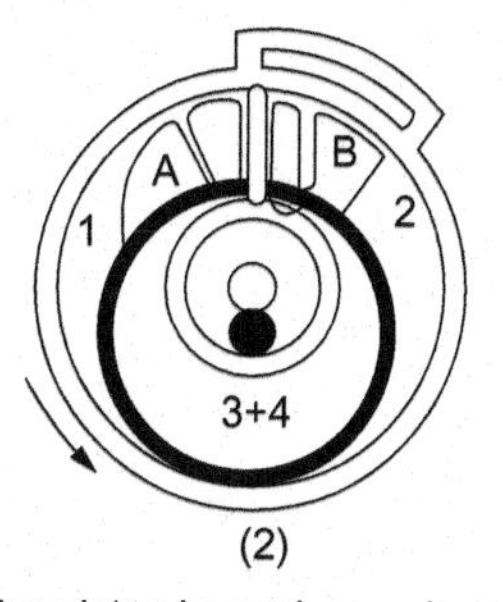

(2)

The piston has advanced and space 1, in connection with the inlet port, has enlarged, and space 2, in connection with the outlet port, has decreased while spaces 3 and 4, which have combined, are about to move into position to discharge through the outlet port

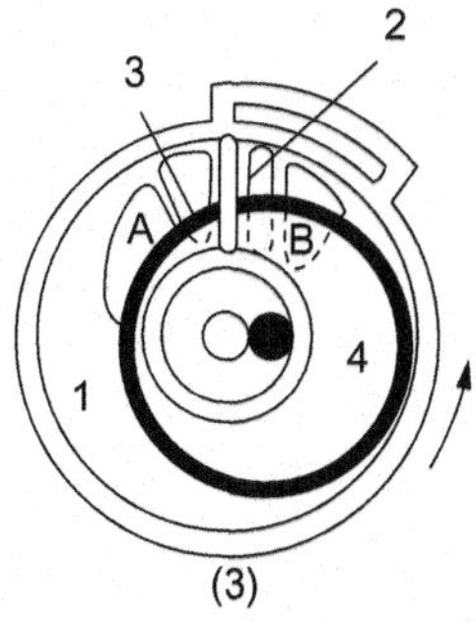

(3)

Space 1 is still admitting fluid from the inlet port and space 3 is just opening up again to the inlet port, while spaces 2 and 4 are discharging through the outlet port

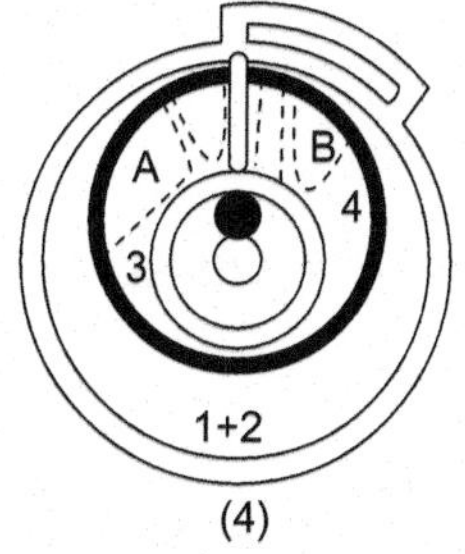

(4)

Fluid is being received into space 3 and discharged from space 4, while spaces 1 and 2 have combined and are about to begin discharging as the piston moves forward again to occupy position as shown in (1)

Fig. 6.32
The oscillating-piston meter.

device that produces an electrical pulse output. The frequency of this signal is proportional to the flowrate and the total count of pulses is proportional to the total volume of liquid passed through the meter.

Turbine flow meters range in size from 5 to 600 mm in diameter, are suitable for temperatures of between 20 and 750 K at pressures of up to 300 bar. A normal range of flows falls between 0.02 and 2000 L/s (2 m^3/s) and a diagram showing a section of an axial turbine flow meter is shown in Fig. 6.33.

Further details of these and other measuring equipment are given by Ower and Pankhurst,[5] Miller,[10] Baker,[11] Considine[21] and in the *Instrument Manual*.[22]

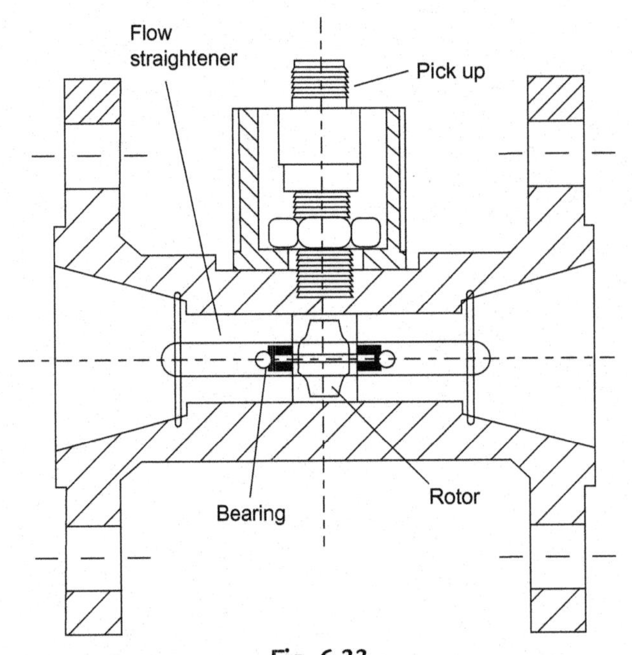

Fig. 6.33
Section through an axial turbine flow meter.

6.4 Nomenclature

		Units in SI system	Dimensions in M, N, L, T, θ, I
A	Area perpendicular to direction of flow	m^2	$\mathbf{L}^2$
A_c	Cross-sectional area of column	m^2	$\mathbf{L}^2$
A_f	Area of rotameter float	m^2	$\mathbf{L}^2$
A_o	Area of orifice	m^2	$\mathbf{L}^2$
A_w	Cross-sectional area of well	m^2	$\mathbf{L}^2$
B	Width of rectangular channel or notch	m	$\mathbf{L}$
C_c	Coefficient of contraction	–	–
C_D	Coefficient of discharge	–	–
C'	Constant for venturi meter	m^2	$\mathbf{L}^2$
D	Depth of liquid or above bottom of notch	m	$\mathbf{L}$
d	Diameter of pipe or pitot tube	m	$\mathbf{L}$
d_o	Diameter of orifice	m	$\mathbf{L}$
e	Error		
F	Energy dissipated per unit mass of fluid	J/kg	$\mathbf{L}^2\mathbf{T}^{-2}$
G	Mass rate of flow	kg/s	$\mathbf{M}\mathbf{T}^{-1}$
g	Acceleration due to gravity	m/s^2	$\mathbf{L}\mathbf{T}^{-2}$

h_i	Difference between impact and static heads on pitot tube	m	**L**
h_m	Reading on manometer	m	**L**
h_v	Fall in head over converging cone of venturi meter	m	**L**
h_w	Decrease in level in well	m	**L**
h_o	Fall in head over orifice meter	m	**L**
I	Electric current	A	**I**
k	Numerical constant used as index for compression	–	–
L	Length of weir or of inclined tube of manometer	m	**L**
M	Molecular weight	kg/kmol	$\mathbf{MN^{-1}}$
n	Number of end contractions	–	–
P	Pressure	N/m^2	$\mathbf{ML^{-1}T^{-2}}$
ΔP	Pressure difference	N/m^2	$\mathbf{ML^{-1}T^{-2}}$
Q	Volumetric rate of flow	m^3/s	$\mathbf{L^3T^{-1}}$
R	Universal gas constant	(8314) J/kmol K	$\mathbf{MN^{-1}L^2T^{-2}}\boldsymbol{\theta}^{-1}$
T	Absolute temperature	K	$\boldsymbol{\theta}$
u	Mean velocity	m/s	$\mathbf{LT^{-1}}$
V_f	Volume of rotameter float	m^3	$\mathbf{L^3}$
v	Volume per unit mass of fluid	m^3/kg	$\mathbf{M^{-1}L^3}$
W_s	Shaft work per unit mass	J/kg	$\mathbf{L^2T^{-2}}$
w	Pressure ratio P_2/P_1	–	–
W_c	Critical pressure ratio	–	–
z	Distance in vertical direction	m	**L**
z_m	Vertical distance between level of manometer liquid and axis of venturi meter	m	**L**
α	Constant in expression for kinetic energy of fluid	–	–
γ	Ratio of specific heats at constant pressure and volume	–	–
Ω	Electrical resistance	ohm	$\mathbf{ML^2T^{-3}I^{-2}}$
ρ	Density of fluid	kg/m^3	$\mathbf{ML^{-3}}$
ρ_f	Density of rotameter float	kg/m^3	$\mathbf{ML^{-3}}$
ρ_m	Density of manometer fluid	kg/m^3	$\mathbf{ML^{-3}}$
θ	Half angle of triangular notch or angle made by inclined manometer tube with horizontal	–	–

References

1. Wightman EJ. *Instrumentation in process control.* London: Butterworths; 1972.
2. Baltikha NE. *The condensed handbook of measurement & control.* 3rd ed. New York: Instrumentation Society of America; 2006.
3. Johnson CD. *Process control instrumentation technology.* 8th ed. New York: Pearson; 2013.
4. Anderson NA. *Instrumentation for process measurement and control.* 3rd ed. Boca Raton, FL: CRC Press; 1997.
5. Ower E, Pankhurst RC. *Measurement of air flow.* 4th ed. Oxford: Pergamon; 1966.
6. Benedict RP. *Fundamentals of temperature, pressure and air flow measurement.* 3rd ed. New York: Wiley; 1984.
7. Bao MH, Middelhoek S. *Micro mechanical transducers, Vol. 8: Pressure sensors, accelerometers and gyroscopes (handbook of sensors & actuators).* Oxford: Elsevier; 2000.
8. Smith BD. *Design of equilibrium stage processes.* New York: McGraw-Hill; 1963.
9. BS 1042: 1943: British Standard 1042 (British Standards Institution, London): Code for flow measurement.
10. Miller RW. *Flow measurement engineering handbook.* 3rd ed. New York: McGraw Hill; 1996.
11. Baker RC. *Flow measurement handbook: industrial designs, operating principles, performance and applications.* 2nd ed. New York: Cambridge University Press; 2016.
12. Lyle O. *The efficient use of fuel.* 2nd ed., 2nd impression. London: HMSO; 1964. p. 237–41.
13. Lyle O. *The efficient use of steam.* London: HMSO; 1963.
14. Chanson H. *The hydraulics of open channel flow: an introduction.* 2nd ed. Oxford: Butterworth-Heinemann; 2004.
15. Ginesi D. A raft of flow meters on tap. *Chem Eng (Albony)* 1991;**5**:147.
16. Strouhal F. Über eine besonderer Art der Tonerregung. *Ann Phys Chem* 1878;**5**:216.
17. Endress and Hauser plc. Vortex flow measuring system, prowirl 70. Technical Information TI 031D/06/e (1997).
18. Lynnworth LC. Ultrasonic flowmeters. In: Mason WP, Thurston RN, editors. *Physical acoustics*, vol. 14. New York: Academic Press; 1979 [chapter 5].
19. Plache KO. Coriolis/gyroscope flow meter. *Mech Eng* 1979;**3**:36.
20. Reizner JR. Exposing coriolis mass flow meters dirty little secrets. *Chem Eng Prog* 2004;**100**(3):24–30.
21. Considine DM. *Process industries and controls handbook.* New York: McGraw-Hill; 1957.
22. Miller JT, editor. *Instrument manual.* 4th ed. London: United Trade Press; 1971.

Further Reading

1. Baker RC. *Measurement handbook: industrial designs, operating principles, performance and applications.* 2nd ed. New York: Cambridge University Press; 2016.
2. Cheremisinoff NP, Cheremisinoff PN. *Flow measurement for engineers and scientists.* New York: Marcel Dekker; 1988.
3. Moore RL. *Basic instrumentation lecture notes and study guide, Vol. 1—Measurement fundamentals, Vol. 2—Process analysers and recorders. Instrument Society of America.* Englewood Cliffs: Prentice Hall; 1982.
4. Scott RWW, editor. *Developments in flow measurement.* London: Applied Science; 1982.
5. Sydenham PH. *Transducers in measurement and control.* London: Adam Hilger; 1984.

CHAPTER 7

Liquid Mixing

7.1 Introduction—Types of Mixing

Mixing is one of the most common operations carried out in the chemical, processing and allied industries. The term "mixing" is applied to the processes used to reduce the degree of nonuniformity, or gradient of a property such as concentration, viscosity, temperature and so on, in a system. Mixing is achieved by moving material from one region to another. It may be of interest simply as a means of achieving a desired degree of homogeneity but it may also be used to promote heat and mass transfer, often where a system is undergoing a chemical reaction.

At the outset it is useful to consider some common examples of problems encountered in industrial mixing operations, since this will not only reveal the ubiquitous nature of the process, but will also provide an appreciation of some of the associated difficulties. Several attempts have been made to classify mixing problems and, for example, Reavell[1] used as a criterion for mixing of powders, the flowability of the final product. Another classification may be based on the mode of operation such as batch, semibatch or continuous mixers, in-line mixers etc. Harnby et al.[2] base their classification on the phases present, that is liquid–liquid, liquid–solid and so on. This is probably the most useful description of mixing as it allows the adoption of a unified approach to the problems encountered in a range of industries. This approach is now followed here.

7.1.1 Single-Phase Liquid Mixing

In many instances, two or more miscible liquids must be mixed to give a product of a desired specification, such as, for example, in the blending of petroleum products of different viscosities. This is the simplest type of mixing as it involves neither heat nor mass transfer, nor indeed a chemical reaction. Even such simple operations can however pose problems when the two liquids have vastly different densities and/or viscosities. Another example is the use of mechanical agitation to enhance the rates of heat and mass transfer between the wall of a vessel, or a coil, and the liquid. Additional complications arise in the case of highly viscous Newtonian and non-Newtonian liquids.

Coulson and Richardson's Chemical Engineering. https://doi.org/10.1016/B978-0-08-101099-0.00007-0

7.1.2 Mixing of Immiscible Liquids

When two immiscible liquids are stirred together, one phase becomes dispersed as tiny droplets in the second liquid which forms a continuous phase. Liquid–liquid extraction, a process that uses successive mixing and settling stages (Chapter 5 of Volume 2B) is one important example of this type of mixing. The liquids are brought into contact with a solvent that will selectively dissolve one of the components present in the mixture. Vigorous agitation causes one phase to disperse into the other and, if the droplet size is small, a high interfacial area is created for interphase mass transfer. When the agitation is stopped, phase separation takes place, but care must be taken to ensure that the droplets are not so small that a diffuse layer appears in the region of the interface; this can remain in a semistable state over a long time and prevent effective separation from occurring. Another important example of dispersion of two immiscible liquids is the production of stable emulsions, such as those encountered in food, brewing and pharmaceutical applications.[3] Because the droplets are very small, the resulting emulsion is usually stable over considerable lengths of time.

7.1.3 Gas–Liquid Mixing

Numerous processing operations involving chemical reactions, such as aerobic fermentation, wastewater treatment, oxidation of hydrocarbons, hydrogenation of oils, and so on; require good contact between a gas and a liquid. The purpose of mixing here is to produce a high interfacial area by dispersing the gas phase in the form of bubbles into the liquid. Also, intense agitation can significantly augment the volumetric mass transfer coefficients in biotechnological applications, hydrogenation and oxidation of oils and hydrocarbons, etc. Generally, gas–liquid mixtures or dispersions are unstable and separate rapidly if agitation is stopped, provided that a foam is not formed. In some cases a stable foam is needed, and this can be formed by injecting gas into a liquid which is rapidly agitated, often in the presence of a surface-active agent. Such foams are encountered in processed foods, cake toppings, personal care products, fire-fighting technologies, for instance.

7.1.4 Liquid–Solids Mixing

Mechanical agitation may be used to suspend particles in a liquid in order to promote mass transfer or a chemical reaction. The liquids involved in such applications are usually of low viscosity, and the particles will settle out when agitation ceases. Additional examples of liquid–solid mixing are found in dissolution and leaching, suspension polymerisation, slurry reactors, crystallisation and precipitation, etc.

At the other extreme, in the formation of composite materials, especially filled polymers, fine particles must be dispersed into a highly viscous Newtonian or non-Newtonian liquid. The incorporation of carbon black powder into rubber is one such operation. Because of the large surface areas involved, surface phenomena play an important role in such applications.

7.1.5 Gas–Liquid–Solids Mixing

In some applications such as catalytic hydrogenation of vegetable oils, slurry reactors, froth flotation, evaporative crystallisation and precipitation, and so on, the success and efficiency of the process is directly influenced by the extent of mixing between the three phases. Despite its high industrial importance, this topic has received only limited attention.

7.1.6 Solids–Solids Mixing

Mixing together of particulate solids, sometimes referred to as blending, is a very complex process in that it is very dependent, not only on the character of the particles—density, size, size distribution, shape and surface properties—but also on the differences of these properties in the components. Mixing of sand, cement and aggregate to form concrete and of the ingredients in gunpowder preparation are longstanding examples of the mixing of solids.

Other industrial sectors employing solids mixing include food, drugs, ceramics, cosmetics and the glass industries, for example. Further examples are found in blending of polymer pellets and granules, breakfast cereal mixes, mixing of agricultural chemicals, etc. All these applications involve only physical contacting, although in recent years, there has been a recognition of the industrial importance of solid–solid reactions, and solid–solid heat exchangers.[4] Unlike liquid mixing, research on solids mixing has not only been limited but also has been carried out only relatively recently. The problems involved in the blending of solids are discussed in Volume 2A, Chapter 1.

7.1.7 Miscellaneous Mixing Applications

Mixing equipment may be designed not only to achieve a predetermined level of homogeneity, but also to improve heat transfer. For example, the rotational speed of an impeller in a mixing vessel is selected so as to achieve a required rate of heat transfer, and the agitation may then be more than sufficient for the mixing duty. Excessive or overmixing should be avoided as it is not only a waste of energy but may be detrimental to product quality. For example, in biological applications, excessively high impeller speeds or energy input may give rise to shear rates which damage the micro-organisms present. In a similar way, where the desirable rheological properties of some polymer solutions and of other structured materials (like foams, emulsions, etc.) may be attributable to structured long-chain molecules, flocs and aggregates, for instance, excessive impeller speeds or agitation over prolonged periods, may damage the structure of the polymer molecules thereby altering their properties. It is therefore important to appreciate that overmixing may often be undesirable because it may result in both excessive energy consumption and impaired product quality. From the examples given of its application, it is abundantly clear that mixing cuts across the boundaries between industries, and indeed it may be required to mix virtually anything with anything else—be it a gas or a liquid or a solid; it is clearly not possible to consider the whole range of mixing problems here. Instead attention will be given primarily to *batch liquid mixing*, and reference will be made to the literature, where appropriate, in this field. This choice has been made largely because it is liquid mixing, which is most commonly encountered in the processing industries. In addition, an extensive literature, mainly dealing with experimental studies, now exists which provides the basis for the design and selection of mixing equipment. It also affords some insight into the nature of the mixing process itself.

In mixing, there are two types of problems to be considered—how to design and select mixing equipment for a given duty, and how to assess whether a given mixer is suitable for a particular application. In both cases, the following aspects of the mixing process should be understood:

i. Mechanisms of mixing.
ii. Scale-up or similarity criteria.
iii. Power consumption.
iv. Flow patterns.
v. Rate of mixing and mixing time.
vi. The range of mixing equipment available and its selection.

Each of these factors is now considered in some detail.

7.2 Mixing Mechanisms

If mixing is to be carried out in order to produce a uniform mixture, it is necessary to understand how liquids move and approach this condition. In liquid mixing devices, it is necessary that two requirements are fulfilled. Firstly, there must be bulk or convective flow so that there are no dead (stagnant) zones. Secondly, there must be a zone of intensive or high-shear mixing in which the inhomogeneities are progressively broken down into smaller entities. Both these processes are energy-consuming and ultimately the mechanical energy is dissipated as heat; the proportion of energy attributable to each varies from one application to another. Depending upon the fluid properties, primarily viscosity, the flow in mixing vessels may be laminar or turbulent, with a substantial transition zone in between the two, and frequently both flow types will occur simultaneously in different parts of the vessel. Laminar and turbulent flow arise from different mechanisms, and it is convenient to consider them separately.

7.2.1 Laminar Mixing

Laminar flow is usually associated with high viscosity liquids (in excess of 10 N s/m^2) which may be either Newtonian or non-Newtonian. The inertial forces therefore tend to die out quickly, and the impeller of the mixer must cover a significant proportion of the cross-section of the vessel to impart sufficient bulk motion to the entire mass of the liquid. Because the velocity gradients close to the rotating impeller are high, the fluid elements in that region deform and stretch. They repeatedly elongate and become thinner each time the fluid elements pass through the high-shear zone. Fig. 7.1 shows the sequence for a fluid element undergoing such a process.

In addition, extensional or elongational flow usually occurs simultaneously. As shown in Fig. 7.2, this is a result of the convergence of the streamlines and consequential increased velocity in the direction of flow. As the volume remains constant, there must, be a thinning or flattening of the fluid elements, as shown in Fig. 7.2. Both these mechanisms (shear and elongation), give rise to stresses in the liquid which then effect a reduction in droplet size and an increase in interfacial area, by which the desired degree of homogeneity is obtained.

In addition, molecular diffusion always tends to reduce inhomogeneities but its effect is not significant until the fluid elements have been reduced in size sufficiently for their specific areas to become large. It must be recognised, however, that the ultimate homogenisation of

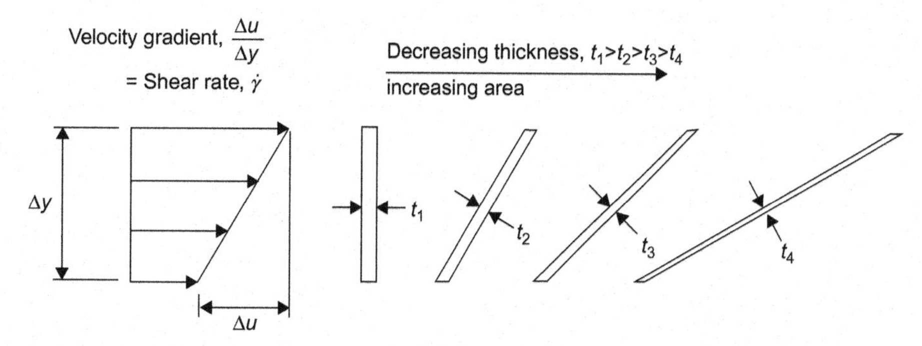

Fig. 7.1
Schematic representation of the thinning of fluid elements due to laminar shear flow.

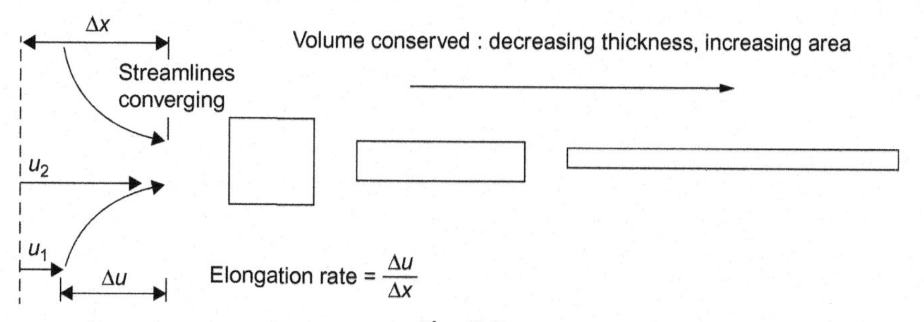

Fig. 7.2
Schematic representation of the thinning of fluid elements due to extensional flow.

miscible liquids, can be only brought about by molecular diffusion. In the case of liquids of high viscosity, this is also a slow process.

In laminar flow, a similar mixing process occurs when the liquid is sheared between two rotating cylinders. During each revolution, the thickness of the fluid element is reduced, and molecular diffusion takes over when the elements are sufficiently thin. This type of mixing is shown schematically in Fig. 7.3 in which the tracer is pictured as being introduced perpendicular to the direction of motion.

Finally, mixing can be induced by physically splicing the fluid into smaller units and redistributing them. In-line or static mixers rely primarily on this mechanism, which is shown in Fig. 7.4.

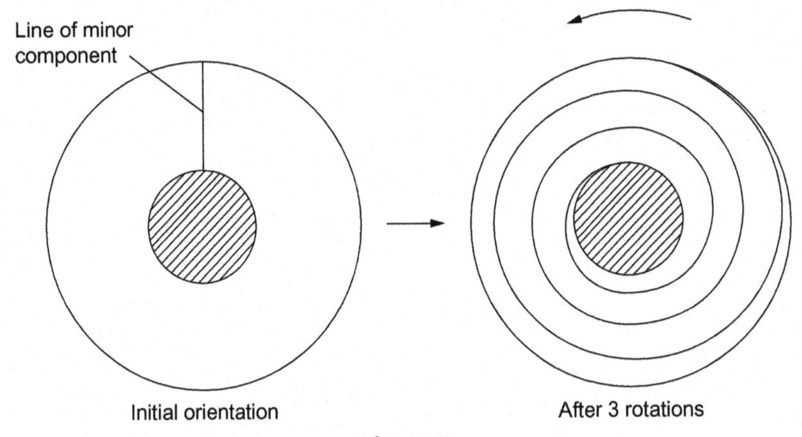

Fig. 7.3
Laminar shear mixing in a coaxial cylinder arrangement.

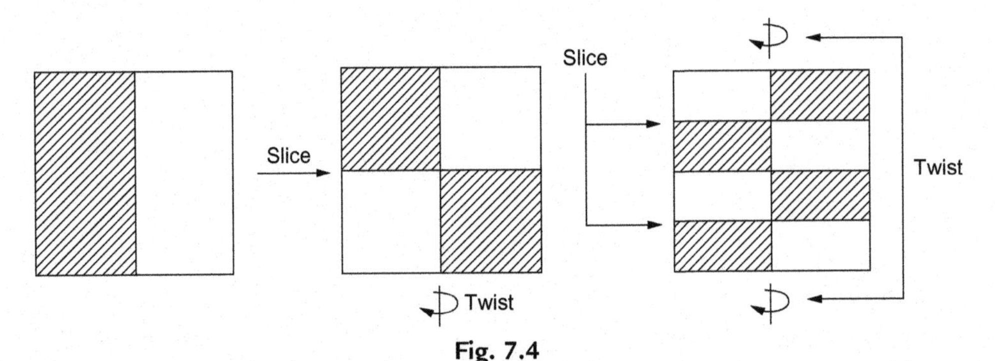

Fig. 7.4
Schematic representation of mixing by cutting and folding of fluid elements.

Thus, mixing in liquids is achieved by several mechanisms, which gradually reduce the size or scale of the fluid elements and then redistribute them in the bulk. If there are initially differences in concentration of a soluble material, uniformity is gradually achieved, and molecular diffusion becomes progressively more important as the element size is reduced. Ottino[5] has illustrated the various stages in mixing by means of a series of coloured photographs.

7.2.2 Turbulent Mixing

For low viscosity liquids (<10 mN s/m^2), the bulk flow pattern in mixing vessels with rotating impellers is generally turbulent. The inertia imparted to the liquid by the rotating impeller is sufficient to cause the liquid to circulate throughout the vessel and return to the impeller. Turbulent eddy diffusion takes place throughout the vessel but is maximum in the vicinity of the impeller. Eddy diffusion is inherently much faster than molecular diffusion and, consequently, turbulent mixing occurs much more rapidly than laminar mixing. Ultimately homogenisation at the molecular level depends on molecular diffusion, which takes place more rapidly in low viscosity liquids. Mixing is most rapid in the region of the impeller because of the high-shear rates due to the presence of trailing vortices, generated by disc-turbine impellers, and associated Reynolds stresses (Chapter 4 of Vol. 1B); furthermore, a high proportion of the energy is dissipated here.

Turbulent flow is inherently complex, and calculation of the flow field prevailing in a mixing vessel is not amenable to rigorous theoretical treatment. If the Reynolds number of the main flow is sufficiently high, some insights into the mixing process can be gained by using the theory of local isotropic turbulence.[6] As is discussed in Chapter 4 of Vol. 1B, turbulent flow may be considered as a spectrum of velocity fluctuations and eddies of different sizes superimposed on an overall time-averaged mean flow. In a mixing vessel, it is reasonable to suppose that the large primary eddies, of a size corresponding approximately to the impeller diameter, would give rise to large velocity fluctuations but would have a low frequency. Such eddies are anisotropic and account for the bulk of the kinetic energy present in the system. Interaction between these primary eddies and slow moving streams produces smaller eddies of higher frequency which undergo further disintegration until finally, dissipate their energy as heat.

During the course of this disintegration process, kinetic energy is transferred from fast moving large eddies to the smaller eddies. The description given here is a gross oversimplification, but it does give a qualitative representation of the salient features of turbulent mixing. This whole process is similar to that for the turbulent flow of a fluid close to a boundary surface described in Chapter 3 of Vol. 1B. Although some quantitative results for the scale size of eddies, have been obtained[3,6] and some workers[7,8] have reported experimental measurements on the structure of turbulence in mixing vessels, it is not at all clear how these data can be used for design purposes.

7.3 Scale-Up of Stirred Vessels

One of the problems confronting the designers of mixing equipment is that of deducing the most satisfactory arrangement for a large unit based on experiments with small units. In order to achieve the same kind of flow pattern in two units, geometrical, kinematic, dynamic similarity and identical boundary conditions must be maintained. This problem of scale-up has been discussed by a number of researchers including Harnby et al.,[2] Paul et al.,[3] Rushton et al.,[9] Kramers et al.,[10] Skelland[11] and Oldshue.[12] It has been found convenient to relate the power consumed by the agitator to the geometrical and mechanical arrangement of the mixer, and thus to obtain a direct indication of the change in power arising from alteration of any of the factors relating to the mixer. A typical (batch) mixer arrangement is shown in Fig. 7.5.

For similarity in two mixing systems, it is important to achieve geometric, kinematic and dynamic similarity.

Geometric similarity prevails between two systems of different sizes if all counterpart length dimensions have a constant ratio. Thus the following ratios must be the same in the two systems:

$$\frac{D_T}{D}; \frac{Z_A}{D}; \frac{W_B}{D}; \frac{W}{D}; \frac{H}{D} \text{ and so on}$$

Kinematic similarity exists in two geometrically similar units when the velocities and acceleration at corresponding points have a constant ratio. Also, the paths of fluid motion (flow patterns) and the boundary conditions must be alike.

Dynamic similarity occurs in two geometrically similar units of different sizes if all corresponding forces at homologous locations have a constant ratio. It is necessary here to distinguish between the various types of forces: inertial, gravitational, viscous and surface tension and other forms, such as normal stresses in the case of viscoelastic and yield-stress for visco-plastic type non-Newtonian liquids. Some or all of these forms may be significant in a

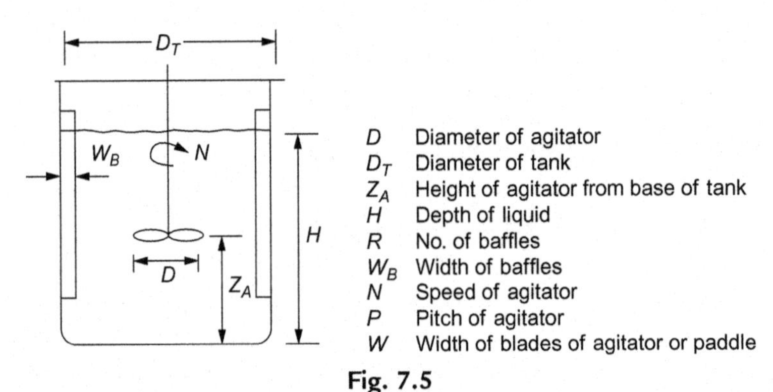

Fig. 7.5
Typical configuration and dimensions of an agitated vessel.

mixing vessel. Considering corresponding positions in systems 1 and 2, which refer to the laboratory and large scale, respectively, when the different types of forces occurring are F_a, F_b, F_c, and so on, the dynamic similarity requires that:

$$\frac{F_{a1}}{F_{a2}}=\frac{F_{b1}}{F_{b2}}=\frac{F_{c1}}{F_{c2}}=\cdots\text{a constant} \tag{7.1}$$

or:

$$\frac{F_{a1}}{F_{b1}}=\frac{F_{a2}}{F_{b2}};\frac{F_{a1}}{F_{c1}}=\frac{F_{a2}}{F_{c2}}\text{etc.} \tag{7.2}$$

Kinematic and dynamic similarities both require geometrical similarity, so the corresponding positions '1' and '2' can be identified in the two systems. Some of the various types of forces that may arise during mixing or agitation will now be formulated.

Inertial force is associated with the reluctance of a body to change its state of rest or motion. Considering a mass m of liquid flowing with a linear velocity u through an area A during the time interval dt; then d$m=\rho uA$ dt where ρ is the fluid density. The inertial force $F_i=$ (mass × acceleration) or:

$$\mathrm{d}F_i=(\rho uA\ \mathrm{d}t)\left(\frac{\mathrm{d}u}{\mathrm{d}t}\right)=\rho uA\ \mathrm{d}u$$

$$\therefore F_i=\int_0^{F_i}\mathrm{d}F_i=\int_0^u\rho uA\ \mathrm{d}u=\rho A\frac{u^2}{2} \tag{7.3}$$

The area for flow is however, $A=$ (constant) L^2, where L is the characteristic linear dimension of the system. In batch-mixing applications, L is usually chosen as the impeller diameter D, and, likewise, the representative velocity u is taken to be the velocity at the tip of the impeller (πDN), where N is the rotational velocity of the impeller in revolutions per unit time. Therefore, the expression for inertial force may be written as:

$$F_i\propto\rho D^4N^2 \tag{7.4}$$

where the constants have been omitted.

The rate change in u due to F_i, du/dt, may be countered by the rate of change in u due to *viscous forces* F_v, which for a Newtonian fluid is given by:

$$F_v=\mu A'\left(\frac{\mathrm{d}u}{\mathrm{d}y}\right) \tag{7.5}$$

where du/dy, the velocity gradient, may be taken to be proportional to u/L, and A' is proportional to L^2. The viscous force is then given by:

$$F_v\propto\mu uL \tag{7.6}$$

which, for an agitated system, becomes $\mu D^2 N$. In a similar way, it is relatively simple to show that the *force due to gravity*, F_g, is given by:

$$F_g = (\text{mass of fluid} \times \text{acceleration due to gravity})$$

or:

$$F_g \propto \rho D^3 g \tag{7.7}$$

and finally the *surface tension* force, $F_s \propto \sigma D$.

Taking F_a, F_b, and so on, in Eq. (7.1) to represent F_i, F_v, F_g, F_s respectively for systems 1 and 2, dynamic similarity of the two systems requires that:

$$\left(\frac{F_i}{F_v}\right)_1 = \left(\frac{F_i}{F_v}\right)_2 \tag{7.8}$$

which upon substitution of the respective expressions for F_i and F_v leads to:

$$\left(\frac{\rho D^2 N}{\mu}\right)_1 = \left(\frac{\rho D^2 N}{\mu}\right)_2 \tag{7.9}$$

This is, of course, the *Reynolds number*, which determines the nature of the flow, as shown in Chapter 3. In a similar fashion, the constancy of the ratios between other forces results in Froude and Weber numbers as follows:

$$\left(\frac{F_i}{F_g}\right)_1 = \left(\frac{F_i}{F_g}\right)_2 \rightarrow \left(\frac{DN^2}{g}\right)_1 = \left(\frac{DN^2}{g}\right)_2 \quad \text{(Froude number)} \tag{7.10}$$

$$\left(\frac{F_i}{F_s}\right)_1 = \left(\frac{F_i}{F_s}\right)_2 \rightarrow \left(\frac{D^3 N^2 \rho}{\sigma}\right)_1 = \left(\frac{D^3 N^2 \rho}{\sigma}\right)_2 \quad \text{(Weber number)} \tag{7.11}$$

Thus, the ratios of the various forces occurring in mixing vessels can be expressed as the above dimensionless groups which, in turn, serve as similarity parameters for scale-up of mixing equipment. It can be shown that the existence of geometric and dynamic similarities also ensures kinematic similarity.

In the case of inelastic, time-independent, non-Newtonian fluids, the fluid viscosity μ, occurring in Reynolds numbers, must be replaced by an apparent viscosity μ_a evaluated at an appropriate value of shear rate. Metzner and coworkers[13,14] have developed a procedure for calculating an average value of shear rate in a mixing vessel as a function of impeller/vessel geometry and speed of rotation; this in turn, allows the value of the apparent viscosity to be obtained either directly from the rheological data or by the use of a rheological model. Though this scheme has proved satisfactory for time-independent non-Newtonian fluids, further complications arise in the scale-up of mixing equipment for viscoelastic liquids (Ulbrecht[15], Astarita[16] and Skelland[17]).

7.4 Power Consumption in Stirred Vessels

From a practical point of view, power consumption is perhaps the most important parameter in the design of stirred vessels. Because of the very different flow patterns and mixing mechanisms involved, it is convenient to consider power consumption in low and high viscosity systems separately.

7.4.1 Low Viscosity Systems

Typical equipment for low viscosity liquids consists of a vertical cylindrical tank, with a height to diameter ratio of 1.5–2, fitted with an agitator. For low viscosity liquids, high-speed propellers of diameter about one-third that of the vessel are suitable, running at 10–25 Hz. Although work on single-phase mixing of low viscosity liquids is of limited value in industrial applications it does, however, serve as a useful starting point for the subsequent treatment of high viscosity liquids.

Considering a stirred vessel in which a Newtonian liquid of viscosity μ, and density ρ is agitated by an impeller of diameter D rotating at a speed N; the tank diameter is D_T, and the other dimensions are as shown in Fig. 7.5, then, the functional dependence of the power input to the liquid **P** on the independent variables (μ, ρ, N, D, D_T, g, other geometric dimensions) may be expressed as:

$$\mathbf{P}=f(\mu,\rho,N,g,D,D_T,\text{other dimensions}) \tag{7.12}$$

In Eq. (7.12), **P** is the impeller power, that is, the energy per unit time dissipated within the liquid. Clearly, the electrical power required to drive the motor will be greater than **P** on account of transmission losses in the gear box, motor, bearings, and so on.

It is readily acknowledged that the functional relationship in Eq. (7.12) cannot be established from first principles. However, by using dimensional analysis, the number of variables can be reduced to give:

$$\frac{\mathbf{P}}{\rho N^3 D^5}=f\left(\frac{\rho N D^2}{\mu},\frac{N^2 D}{g},\frac{D_T}{D},\frac{W}{D},\frac{H}{D},\dots\right) \tag{7.13}$$

where the dimensionless group on the left-hand side is called the *Power number* (N_p); ($\rho ND^2/\mu$) is the *Reynolds number* (Re) and (N^2D/g) is the *Froude number* (Fr). Other dimensionless length ratios, such as (D_T/D), (W/D) and so on, relate to the specific impeller/vessel arrangement. For geometrically similar systems, these ratios must be equal, and the functional relationship between the Power number and the other dimensionless groups reduces to:

$$N_p=f(\text{Re},Fr) \tag{7.14}$$

The simplest form of the function in Eq. (7.14) is a power-law relationship, giving:

$$N_p = K' Re^b Fr^c \tag{7.15}$$

where the values of K', b and c must be determined from experimental measurements, and are dependent upon impeller/vessel configuration and on the flow regime, that is laminar, transition or turbulent, prevailing in the mixing vessel. There are several ways of measuring the power input to the impeller, including the mounting of Prony brakes, using a dynamometer, or a simple calculation from the electrical measurements. Strain gauges and load cells are also used for torque measurements in mixers. Detailed descriptions, along with the advantages and disadvantages of different experimental techniques, have been given by Oldshue,[12] Holland and Chapman,[18] and Paul et al.[3] It is appropriate to note that the uncertainty regarding the actual losses in gear box, bearings, and so on makes the estimation of **P** from electrical measurements rather less accurate than would be desirable.

In Eq. (7.15), the Froude number is usually important only in situations where gross vortexing occurs, and can be neglected if the value of Reynolds number is less than about 300. Thus, in a plot of Power number (N_p) against Reynolds number (Re) with Froude number as parameter, all data fall on a single line for values of $Re < 300$ confirming that in this region Fr has no significant effect on N_p. This behaviour is clearly seen in Fig. 7.6 where the data for a propeller, as reported by Rushton et al.,[9] are plotted. Such a plot is known as a *power curve*.

Thus at $Re < 300$:

$$N_p = K' Re^b \tag{7.16}$$

For values of $Re < 10$, b is found to be -1 (see Fig. 7.6).

Therefore, power **P** is then given by:

$$\mathbf{P} = K' \mu N^2 D^3 \tag{7.17}$$

The value of K' depends on the type of impeller/vessel arrangement, and whether the tank is fitted with baffles. For marine-type three-bladed propellers with pitch equal to diameter, K' has been found to have a value of about 41.

For higher values of the Reynolds number, the Froude number seems to exert some influence on the value of N_p, and separate lines are drawn for various speeds in Fig. 7.6. It will be noted that, on this graph, lines of constant speed of rotation relate to constant values of the Froude number because the diameter of the impeller is constant and equal to 0.3 m. Thus, $Fr = 0.0305\,N^2$. The Reynolds number was varied by using liquids of different viscosities, as well as different rotational speeds, and the slanting lines in Fig. 7.6 represent conditions of constant viscosity. In this region, the effect of Froude number may be minimised, or indeed

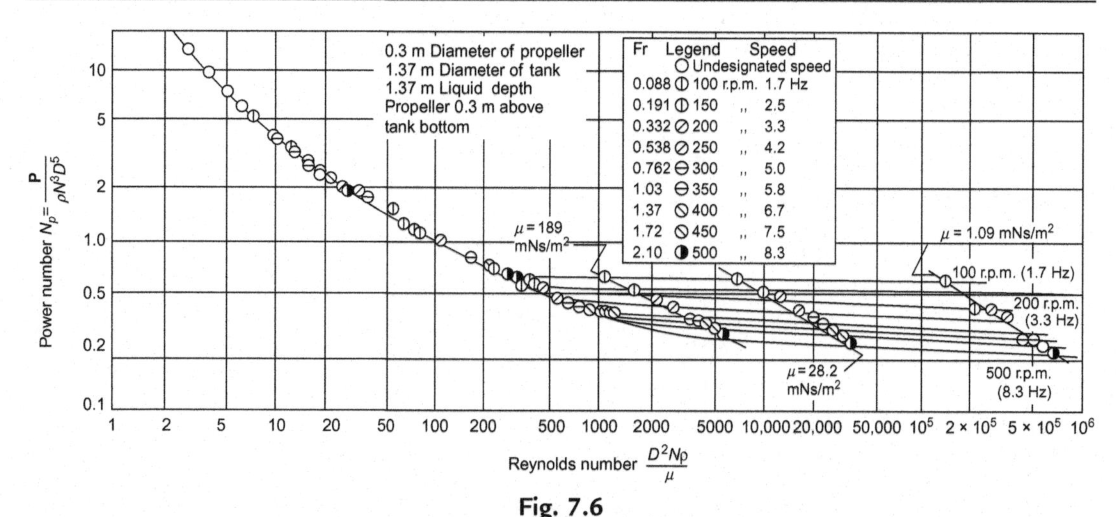

Fig. 7.6
Power number as a function of Reynolds number for a propeller mixer.

eliminated, by the use of baffles or by installing the impeller off-centre. This point is discussed in more detail in Section 7.5.

Example 7.1

On the assumption that the power required for mixing in a stirred tank is a function of the variables given in Eq. (7.12), obtain the dimensionless groups that are important in calculating power requirements for geometrically similar arrangements.

Solution
The variables in this problem, together with their dimensions, are as follows:

$$\begin{array}{ll} \mathbf{P} & \mathbf{ML^2T^{-3}} \\ \mu & \mathbf{ML^{-1}T^{-1}} \\ \rho & \mathbf{ML^{-3}} \\ N & \mathbf{T^{-1}} \\ g & \mathbf{LT^{-2}} \\ D & \mathbf{L} \\ D_T & \mathbf{L} \end{array}$$

This list includes seven variables and there are three fundamental dimensions (**M**, **L**, **T**). By Buckingham's Π theorem, there will be 7–3 = 4 dimensionless groups.

Choosing as the recurring set ρ, N and D, then these three variables themselves cannot be grouped together to give a dimensionless number. **M**, **L** and **T** can now be expressed in terms of combinations of ρ, N and D.

$$\mathbf{L} \equiv D$$
$$\mathbf{T} \equiv N^{-1}$$
$$\mathbf{M} \equiv \rho D^3$$

Dimensionless group 1: $\mathbf{PM^{-1}L^{-2}T^3} = \mathbf{P}(\rho D^3)^{-1}(D)^{-2}(N^{-1})^3 = \mathbf{P}\rho^{-1}D^{-5}N^{-3}$.

Dimensionless group 2: $\mu\mathrm{M}^{-1}\mathbf{LT} = \mu(\rho D^3)^{-1}DN^{-1} = \mu\rho^{-1}D^{-2}N^{-1}$.

Dimensionless group 3: $g\mathbf{L}^{-1}\mathbf{T}^2 = gD^{-1}(N^{-1})^2 = gD^{-1}N^{-2}$.

Dimensionless group 4: $D_T\mathbf{L}^{-1} = D_T(D^{-1}) = D_TD^{-1}$.

Thus, $\mathbf{P}\rho^{-1}D^{-5}N^{-3} = f(\mu\rho^{-1}D^{-2}N^{-1}, gD^{-1}N^{-2}, D_TD^{-1})$.

Rearranging:

$$\frac{\mathbf{P}}{\rho N^3D^5} = f\left(\frac{\rho ND^2}{\mu}, \frac{N^2D}{g}, \frac{D_T}{D}\cdots\right)$$

which corresponds with Eq. (7.13).

Example 7.2

A solution of sodium hydroxide of density 1650 kg/m^3 and viscosity 50 mN s/m^2 is agitated by a propeller mixer of 0.5 m diameter in a tank of 2.28 m diameter, and the liquid depth is 2.28 m. The propeller is situated 0.5 m above the bottom of the tank. What is the power that the propeller must impart to the liquid for a rotational speed of 2 Hz?

Solution

In this problem the geometrical arrangement corresponds with the configuration for which the curves in Fig. 7.6 are applicable.

$$\frac{D_T}{D} = \left(\frac{2.28}{0.5}\right) = 4.56; \quad \frac{H}{D} = \left(\frac{2.28}{0.5}\right) = 4.56; \quad \frac{Z_A}{D} = \left(\frac{0.5}{0.5}\right) = 1$$

$$\mathrm{Re} = \frac{D^2N\rho}{\mu} = \frac{(0.5^2 \times 2 \times 1650)}{50 \times 10^{-3}} = 16{,}500$$

$$Fr = \frac{N^2D}{g} = \frac{2^2 \times 0.5}{9.81} = 0.20$$

From Fig. 7.6:

$$Np = \frac{\mathbf{P}}{\rho N^3D^5} = 0.5$$

$$\text{and} \quad \mathbf{P} = (0.5 \times 1650 \times (2)^3 \times (0.5)^5) = \underline{\underline{206\,\mathrm{W}}}$$

Example 7.3

A reaction is to be carried out in an agitated vessel. Pilot scale tests have been carried out under fully turbulent conditions in a tank 0.6 m in diameter, fitted with baffles and provided with a flat-bladed turbine, and it has been found that satisfactory mixing is obtained at a rotor speed of 4 Hz when the power consumption is 0.15 kW and the Reynolds number 160,000. What should be the rotor speed in order to achieve the same degree of mixing if the linear scale of the equipment is increased by a factor of 6 and what will be the Reynolds number and the power consumption?

Solution

The correlation of power consumption and Reynolds number is given by:

$$\frac{\mathbf{P}}{\rho N^3 D^5} = f\left(\frac{\rho N D^2}{\mu}, \frac{N^2 D}{g}\right) \tag{7.13}$$

in the turbulent regime, at high values of *Re*, the power number *Np* is independent of *Re* and *Fr*, and is a function of geometric configuration only. Thus the power **P** is given by:

$$\mathbf{P} = k' \rho N^3 D^5$$

where k' is a system constant. If the experiments at both scales are carried out using the same fluid, one can incorporate ρ also into the constant k' as is done here.

For the *pilot unit*; taking the impeller diameter as $(D_T/3)$, then:

$$0.15 = k' 4^3 (0.6/3)^5$$

and:

$$k' = 7.32\,\text{kg/m}^3$$

and:

$$\mathbf{P} = 7.32 N^3 D^5 \tag{i}$$

In essence, two criteria may be used to scale-up; constant tip speed and constant power input per unit volume of fluid. These are now considered in turn.

Constant impeller tip speed.

The tip speed is given by $u_t = \pi N D$.

For the pilot unit:

$$u_{t1} = \pi \times 4 \times (0.6/3) = 2.57\,\text{m/s}$$

For the full-scale unit:

$$u_{t2} = 2.57 = \pi N_2 D_2\,\text{m/s}$$

or:

$$2.57 = \pi N_2 (6 \times 0.6/3)$$

and:

$$N_2 = \underline{\underline{0.66\ \text{Hz}}}$$

In Eq. (i):

$$\mathbf{P}_2 = 7.32N_2{}^3D_2{}^5$$
$$= 7.35 \times 0.66^3(6 \times 0.6/3)^5 = \underline{\underline{5.25\ \text{kW}}}$$

For thermal similarity, that is the same temperature in both systems:

$$\mu_1 = \mu_2 \text{ and } \rho_1 = \rho_2$$

and:

$$Re_2/Re_1 = \left(N_2D_2{}^2\right)/\left(N_1D_1{}^2\right)$$

Hence:

$$Re_2/160{,}000 = \left[0.66(6 \times 0.6/3)^2\right]/\left[4 \times (0.6 \times 3)^2\right]$$

and:

$$Re_2 = \underline{\underline{950{,}000}}$$

Constant power input per unit volume.

Assuming the depth of liquid is equal to the tank diameter, then the volume of the pilot scale unit is $[(\pi/4)0.6^2 \times 0.6] = 0.170\ \text{m}^3$ and the power input per unit volume is $(0.150/0.170) = 0.884\ \text{kW/m}^3$.

The volume of the full-scale unit is given by:

$$V_2 = (\pi/4)(6 \times 0.6)^2(6 \times 0.6) = 36.6\text{m}^3$$

and hence the power requirement is:

$$\mathbf{P} = (0.884 \times 36.6) = \underline{\underline{32.4\ \text{kW}}}$$

From Eq. (i):

$$32.4 = 7.32N_2{}^3(6 \times 0.6/3)^5$$

and:

$$N_2 = \underline{\underline{1.21\ \text{Hz}}}$$

The Reynolds number is then given by:

$$\text{Re} = 160{,}000\left[(6 \times 0.6/3)^2 \times 1.21\right]/\left[(0.6/3)^2 \times 4\right]$$
$$= \underline{\underline{1{,}740{,}000}}$$

The choice of scale-up technique depends on the particular system. As a general guide, constant tip speed is used where suspended solids are involved, where heat is transferred to a coil or jacket, and for miscible liquids. Constant power per unit volume is used with immiscible liquids, emulsions, pastes and gas–liquid systems. Constant tip speed seems more appropriate in this case, and hence the rotor speed should be 0.66 Hz. The power consumption will then be 5.25 kW giving a Reynolds number of 950,000.

Fig. 7.7, also taken from the work of Rushton et al.[9] shows N_p versus Re data for a 150 mm diameter turbine with six flat blades. In addition, this figure also shows the effect of introducing baffles in the tank. Bissell et al.[19] have studied the effect of different types of baffles and their configuration on power consumption.

Evidently, for most conditions of practical interest, the Froude number is not a significant variable and the Power number is a unique function of Reynolds number for a fixed impeller/vessel configuration and surroundings. The vast amount of work[2] reported on the mixing of low viscosity liquids suggests that, for a given geometrical design and configuration of impeller and vessel, all single phase experimental data fall on a single power curve. Typically there are three discernible zones in a power curve.

At low values of the Reynolds number, less than about 10, a laminar or viscous zone exists and the slope of the power curve on logarithmic coordinates is -1, which is typical of most viscous flows. This region, which is characterised by slow mixing at both macro- and micro-levels, is where the majority of the highly viscous (Newtonian as well as non-Newtonian) liquids are processed.

At very high values of Reynolds number, greater than about 10^4, the flow is fully turbulent and inertia dominated, resulting in rapid mixing. In this region the Power number is virtually

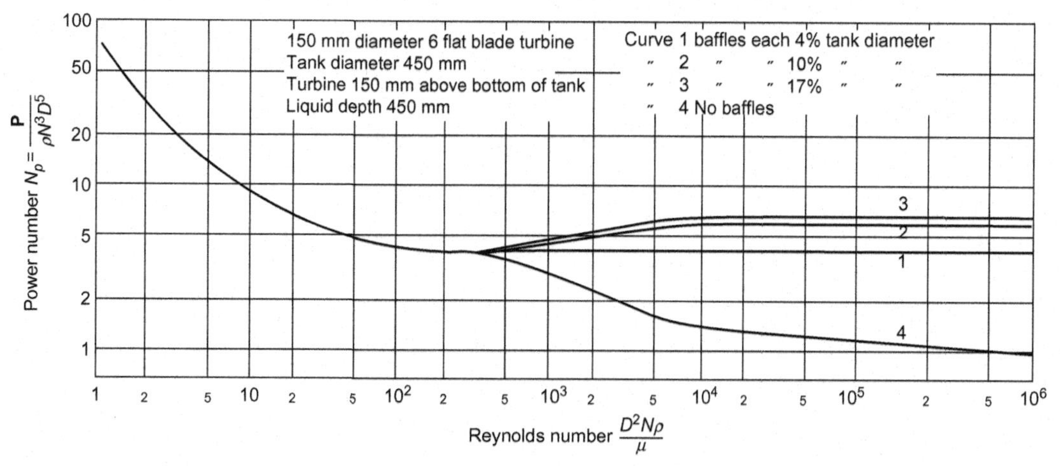

Fig. 7.7
Power number as a function of Reynolds number for a turbine mixer.

constant and independent of Reynolds number, as shown in Figs. 7.6 and 7.7, but depends upon the impeller/vessel configuration. Often gas–liquid, solid–liquid and liquid–liquid contacting operations are carried out in this region. Though the mixing itself is quite rapid, the limiting factor controlling the process may be mass transfer.

Between the laminar and turbulent zones, there exists a substantial transition region in which the viscous and inertial forces are of comparable magnitudes. No simple mathematical relationship exists between N_p and Re in this flow region and, for a given value of Re, N_p must be obtained from the appropriate power curve. In most cases, such curves are based on experimental results, though in recent years there have been significant advances in the use of computational fluid dynamics (CFD) tools to predict the flow patterns, power consumption, etc., in mechanically stirred tanks.[20]

Though it is widely believed that the laminar flow conditions cease to exist at about $Re \sim 10$–50, not only this transition but also the onset of the fully turbulent flow conditions are strongly dependent on the impeller–tank configuration. Grenville et al.[(21)] developed the following relationship between the critical Reynolds number, Re_{cr}, denoting the boundary between the transitional and turbulent flow regimes and the asymptotic value of the power number (N_{pt}) in the fully turbulent regime as:

$$Re_{cr} = \frac{6370}{(N_{pt})^{1/3}} \tag{7.18}$$

Typical values of N_{pt} for a range of conditions are presented in Table 7.1.

Table 7.1 Typical values of N_{pt} (for vessels fitted with four standard baffles)[3]

Type of Impeller	Value of N_{pt}
45° Pitched blade turbine[a]	
– 4 blades	1.27
– 6 blades	1.64
Marine propeller[a]	
– 1.0 pitch	0.34
– 1.5 pitch	0.62
Smith or concave—or hollow[a] blade with 6-blades	4.4
Hollow blade turbine	4.1
Ekato MIG-3 ($D/D_T = 0.7$)	0.55
Ekato Intermig −2 ($D/D_T = 0.7$)	0.61
High-shear disk ($Re = 10^4$)	0.20
Lightnin A 310	0.30
Chemineer HE 3	0.30

[a]For $D/D_T = 1/3$; $(Z_A/D_T) = 1/3$ and $(W/D) = 1/5$.

For the purposes of scale-up, it is generally most satisfactory in the laminar region to maintain a constant speed for the tip of the impeller, and mixing time will generally increase with scale. The most satisfactory basis for scale-up in the turbulent region is to maintain a constant power input per unit volume of fluid.

Power curves for many different impeller geometries, baffle arrangements, and so on are to be found in the literature,[3,11,12,18,22–26] but it must always be remembered that though the power curve is applicable to any single phase Newtonian liquid, at any impeller speed, the curve will be valid for only *one system geometry*.

Sufficient information is now available on low viscosity systems for the estimation of the power requirements for a given duty under most conditions of practical interest.

7.4.2 High Viscosity Systems

As noted previously, mixing in highly viscous liquids is slow both at the molecular scale, on account of the low values of diffusivity, as well as at the macroscopic scale, due to poor bulk flow. Whereas in low viscosity liquids, momentum can be transferred from a rotating impeller through a relatively large body of fluid, in highly viscous liquids only the fluid in the immediate vicinity of the impeller is influenced by the agitator and the flow is normally laminar.

For the mixing of highly viscous and non-Newtonian fluids, it is usually necessary to use specially designed impellers involving close clearances with the vessel walls; these are discussed in a later section. High-speed stirring with small impellers merely wastefully dissipates energy at the central portion of the vessel. The power curve approach is usually applicable and the proportionality constant K' in Eq. (7.17) is a function of the type of rotating member, and the geometrical configuration of the system. Most of the highly viscous fluids of interest in the processing industries exhibit non-Newtonian behaviour, though highly viscous Newtonian fluids include glycerol and many lubricating oils.

A simple relationship has been shown to exist, however, between much of the data on power consumption with time-independent non-Newtonian liquids and Newtonian liquids in the laminar region. This link, which was first established by Metzner and Otto[13] for pseudoplastic liquids, depends on the fact that there appears to be an average angular shear rate $\dot{\gamma}_{ang}$ for a mixer which characterises power consumption, and which is directly proportional to the rotational speed of impeller:

$$\dot{\gamma}_{ang} = k_s N \tag{7.19}$$

where k_s is a function of the type of impeller and the vessel configuration and in some cases of the fluid rheology also.[27,28] If the apparent viscosity corresponding to the average shear rate defined above is used in the equation for a Newtonian liquid, the power consumption for laminar conditions is satisfactorily predicted for the non-Newtonian fluid.

The validity of the linear relationship given in Eq. (7.19) was subsequently confirmed by Metzner and Taylor.[14] The experimental evaluation of k_s for a given geometry proceeds as follows:

i. The Power number (N_p) is determined for a particular value of N.
ii. The corresponding value of Re is obtained from the appropriate power curve as if the liquid were Newtonian.
iii. The equivalent viscosity is computed from the value of Re.
iv. The value of the corresponding shear rate is obtained, either directly from a flow curve obtained by independent experiment, or by use of an appropriate fluid model such as the power-law model.
v. The value of k_s is calculated for a particular impeller configuration.

This procedure can be repeated for different values of N. A compilation of the experimental values of k_s for a variety of impellers, turbines, propellers, paddles, anchors, and so on, has been given by Skelland,[17] and an examination of Table 7.2 suggests that for pseudo-plastic liquids, k_s lies approximately in the range of 10–13 for most configurations of practical interest.[29,30] Skelland[17] has also correlated much of the data on the agitation of purely viscous non-Newtonian fluids, and this is shown in Fig. 7.8.

Table 7.2 Values of k_s for various types of impellers and key to Fig. 7.8[17]

Curve	Impeller	Baffles	D (m)	D_T/D	N (Hz)	k_s ($n < 1$)
A-A	Single turbine with six flat blades	4, $W_B/D_T = 0.1$	0.051–0.20	1.3–5.5	0.05–1.5	11.5 ± 1.5
A-A_1	Single turbine with six flat blades	None	0.051–0.20	1.3–5.5	0.18–0.54	11.5 ± 1.4
B-B	Two turbines, each with six flat blades and $D_T/2$ apart	4, $W_B/D_T = 0.1$	–	3.5	0.14–0.72	11.5 ± 1.4
B-B_1	Two turbines, each with six flat blades and $D_T/2$ apart	4, $W_B/D_T = 0.1$ or none	–	1.023–1.18	0.14–0.72	11.5 ± 1.4
C-C	Fan turbine with six blades at 45°	4, $W_B/D_T = 0.1$ or none	0.10–0.20	1.33–3.0	0.21–0.26	13 ± 2
C-C_1	Fan turbine with six blades at 45°	4, $W_B/D_T = 0.1$ or none	0.10–0.30	1.33–3.0	1.0–1.42	13 ± 2
D-D	Square-pitch marine propellers with	None, (i) shaft vertical at vessel axis,	0.13	2.2–4.8	0.16–0.40	10 ± 0.9

Table 7.2 Values of k_s for various types of impellers and key to —cont'd

Curve	Impeller	Baffles	D (m)	D_T/D	N (Hz)	k_s ($n < 1$)
	three blades (downthrusting)	(ii) shaft 10° from vertical, displaced $r/3$ from centre				
D-D_1	Same as for D-D but upthrusting	None, (i) shaft vertical at vessel axis, (ii) shaft 10° from vertical, displaced $r/3$ from centre	0.13	2.2–4.8	0.16–0.40	10 ± 0.9
D-D_2	Same as for D-D	None, position (ii)	0.30	1.9–2.0	0.16–0.40	10 ± 0.9
D-D_3	Same as for D-D	None, position (i)	0.30	1.9–2.0	0.16–0.40	10 ± 0.9
E-E	Square-pitch marine propeller with three blades	4, $W_B/D_T = 0.1$	0.15	1.67	0.16–0.60	10
F-F	Double-pitch marine propeller with three blades (downthrusting)	None, position (ii)	–	1.4–3.0	0.16–0.40	10 ± 0.9
F-F_1	Double-pitch marine propeller with three blades (downthrusting)	None, position (i)	–	1.4–3.0	0.16–0.40	10 ± 0.9
G-G	Square-pitch marine propeller with four blades	4, $W_B/D_T = 0.1$	0.12	2.13	0.05–0.61	10
G-G_1	Square-pitch marine propeller with four blades	4, $W_B/D_T = 0.1$	0.12	2.13	1.28–1.68	–
H-H	Two-bladed paddle	4, $W_B/D_T = 0.1$	0.09–0.13	2–3	0.16–1.68	10
–	Anchor	None	0.28	1.02	0.34–1.0	11 ± 5
–	Cone impellers	0 or 4, $W_B/D_T = 0.08$	0.10–0.15	1.92–2.88	0.34–1.0	11 ± 5

The prediction of power consumption for the agitation of a given non-Newtonian fluid in a particular mixer, at a desired impeller speed, may be evaluated by the following procedure:

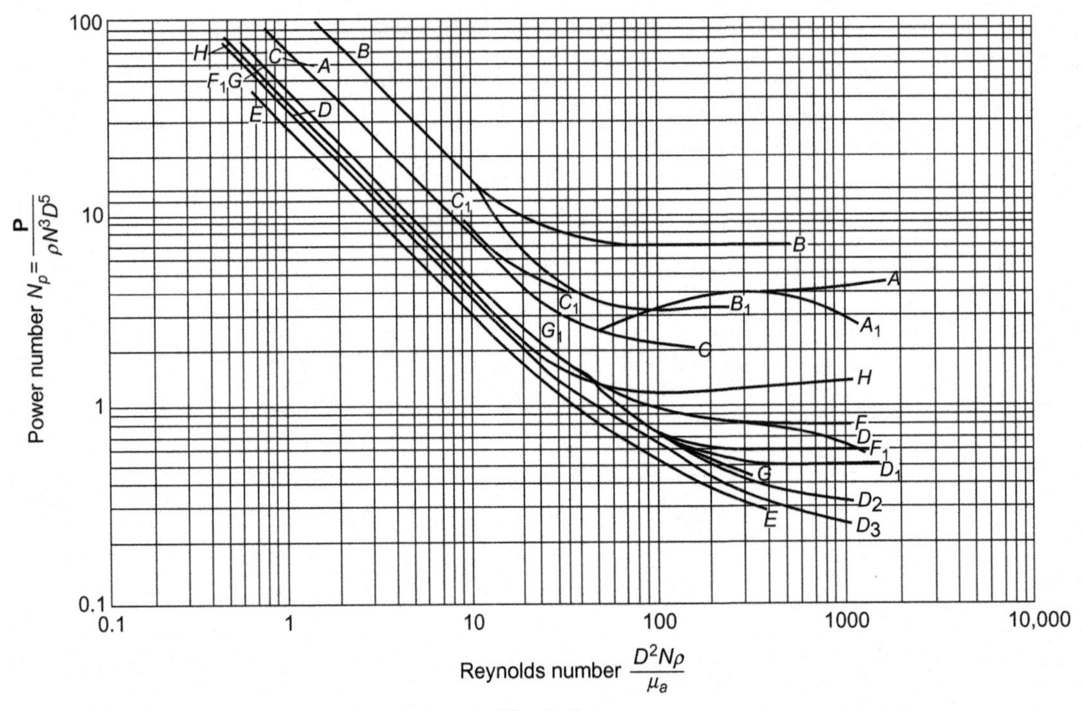

Fig. 7.8
Power curve for pseudoplastic fluids agitated by different types of impeller.

i. Estimate the average shear rate from Eq. (7.19).
ii. Evaluate the corresponding apparent viscosity, either from a flow curve, or by means of the appropriate flow model.
iii. Estimate the value of the Reynolds number as $(\rho N D^2/\mu_a)$ and then the value of the Power number, and hence **P**, from the appropriate curve in Fig. 7.8.

Although this approach of Metzner and Otto,[13] has gained wide acceptance, it has come under some criticism. For instance, Skelland[11] and Mitsuishi and Hirai[31] have argued that this approach does not always yield a unique power curve for a wide range of the flow behaviour index, n. Despite this, it is safe to conclude that this method predicts power consumption with an accuracy of within about 25%–30%. Furthermore, Godfrey[32] has asserted that the constant k_s is independent of equipment size, and thus there are few scale-up problems. It is not yet established, however, how strongly the value of k_s depends upon the rheology. For example, Calderbank and Moo-Young[33] and Beckner and Smith[34] have related k_s to the impeller/vessel configuration and rheological properties (n in the case of power law liquids); the dependence on n, however, is quite weak.

Data for power consumption of Bingham plastic fluids have been reported and correlated by Nagata et al.[35] and of dilatant fluids by Nagata et al.[35] and Metzner et al.[36] Edwards et al.[37] have dealt with the mixing of time-dependent thixotropic materials.

Very little is known about the effect of the viscoelasticity of a fluid on power consumption, but early work[38,39] seems to suggest that it is negligible under the creeping flow conditions. Subsequent work by Oliver et al.[40] and by Prud'homme and Shaqfeh[41] has indicated that the power consumption for Rushton-type turbine impellers, illustrated in Fig. 7.20, may increase or decrease depending upon the value of Reynolds number and the magnitude of elastic effects or the Deborah number. At higher Reynolds numbers, it appears that the elasticity suppresses secondary flows and this results in a reduction in power consumption in comparison with a purely viscous liquid.[42] Many useful review articles have been published on this subject,[2,15,17,43] and theoretical developments relating to the mixing of high viscosity materials have been dealt with by Irving and Saxton.[44]

Finally, it should be noted that the calculation of the power consumption requires a knowledge of the impeller speed which is necessary to blend the contents of a tank in a given time, or of the impeller speed required to achieve a given mass transfer rate in a gas–liquid system. A full understanding of the mass transfer/mixing mechanism is not yet available, and therefore the selection of the optimum operating speed remains primarily a matter of experience. Some practical advice on this as well as on the other aspects of mixing is available in the literature.[45–48] Before concluding this section, it is appropriate to indicate typical power consumptions in kW/m^3 of liquid for various duties, and these are shown in Table 7.3.

7.5 Flow Patterns in Stirred Tanks

A qualitative picture of the flow field created by an impeller in a mixing vessel in a single-phase liquid is useful in establishing whether there are stagnant or dead regions in the vessel, and whether or not particles are likely to be suspended. In addition, the efficiency of mixing equipment, as well as product quality, are influenced by the flow patterns prevailing in the vessel.

Table 7.3 Typical power consumptions

Duty	Power (kW/m^3)
Low power	
Suspending light solids, blending of low viscosity liquids	0.2
Moderate power	
Gas dispersion, liquid–liquid contacting, some heat transfer, etc.	0.6
High power	
Suspending heavy solids, emulsification, gas dispersion, etc.	2
Very high power	
Blending pastes, doughs	4

The flow patterns for single phase, Newtonian and non-Newtonian liquids in tanks agitated by various types of impeller have been reported in the literature.[3,14,34,49,50] The experimental techniques which have been employed include the introduction of tracer liquids, neutrally buoyant particles or hydrogen bubbles, and measurement of the local velocities by means of Pitot tubes, laser-doppler anemometers, particle image velocimetry, and so on. The salient features of the flow patterns encountered with propellers and disc turbines in low viscosity systems are shown in Figs. 7.9 and 7.10.

Basically, the propeller creates an axial flow through the impeller, which may be upwards or downwards depending upon the direction of rotation. The velocities at any point are three-dimensional, and unsteady, but circulation patterns such as those shown in Fig. 7.9 are useful in avoiding dead zones.

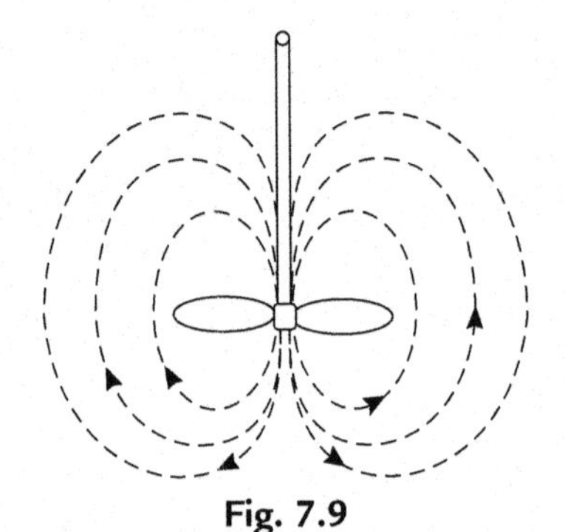

Fig. 7.9
Flow pattern from propeller mixer.

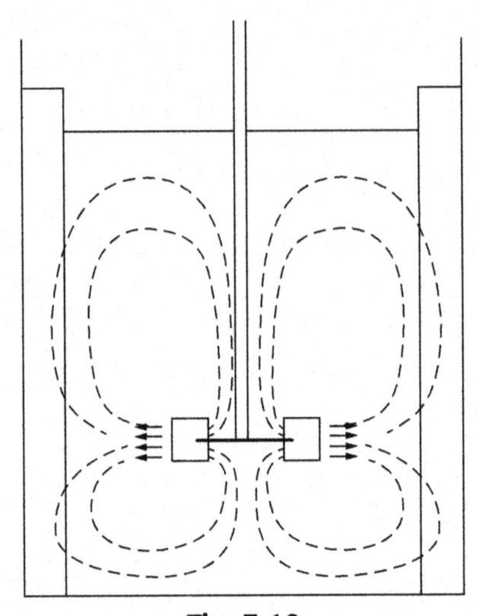

Fig. 7.10
Radial flow pattern from disc turbine.

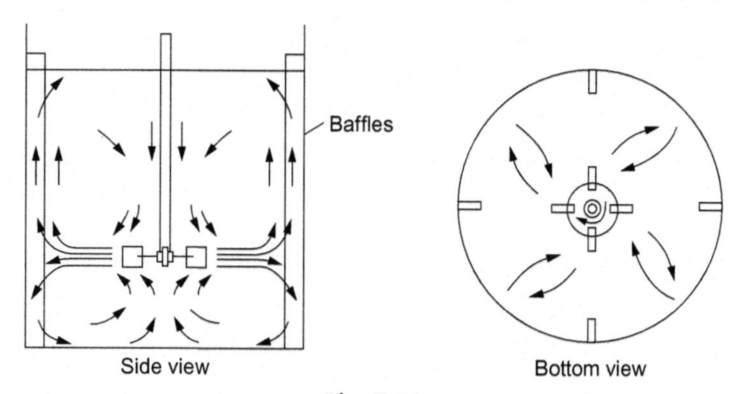

Fig. 7.11
Flow pattern in vessel with vertical baffles.

If a propeller is mounted centrally, and there are no baffles in the tank, there is a tendency for the lighter fluid to be drawn in to form a vortex, and for the degree of agitation to be reduced. To improve the rate of mixing, and to minimise vortex formation, baffles are usually fitted in the tank, and the resulting flow pattern is as shown in Fig. 7.11. The power requirements are however, considerably increased by the incorporation of baffles. Another way of minimising vortex formation is to mount the agitator, off-centre; where the resulting flow pattern is depicted in Fig. 7.12.

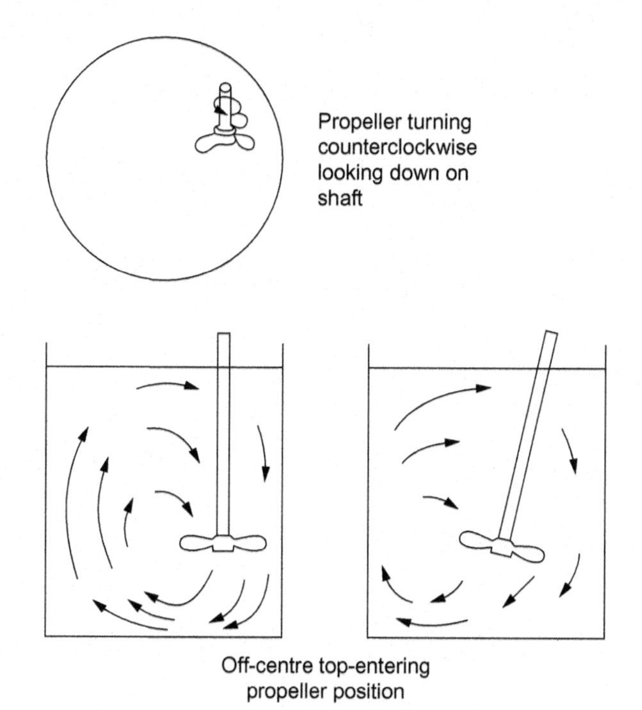

Fig. 7.12
Flow pattern with agitator offset from centre.

The flat-bladed turbine impeller produces a strong radial flow outwards from the impeller (Fig. 7.10), thereby creating circulating zones in the top and bottom of the tank. The flow pattern can be altered by changing the impeller geometry and, for example, if the turbine blades are angled to the vertical, a stronger axial flow component is produced. This can be useful in applications where it is necessary to suspend solids. A flat paddle produces a flow field with large tangential components of velocity, and this does not promote good mixing. Propellers, turbines, and paddles are the principal types of impellers used for low viscosity systems operating in the transition and turbulent regimes.

A brief mention will be made of flow visualisation studies for some of the agitators used for high viscosity liquids; these include anchors, helical ribbons, and screws. Detailed flow pattern studies[51] suggest that both gate and anchor agitators promote fluid motion close to the vessel wall but leave the region in the vicinity of the shaft relatively stagnant. In addition, there is a modest top to bottom turnover and thus vertical concentration gradients may exist, but these may be minimised by means of a helical ribbon or a screw added to the shaft. Such combined impellers would have a ribbon pumping upwards near the wall with the screw twisted in the opposite sense, pumping the fluid downwards near the shaft region. Typical flow patterns for a gate-type anchor are shown in Fig. 7.13. In these systems, the flow patterns change[34] as the stirring speed is increased and the average shear rate cannot be completely described by a linear relationship. Furthermore, any rotational motion induced within a vessel will tend to produce a secondary flow in the vertical direction; the fluid in contact with the bottom of the

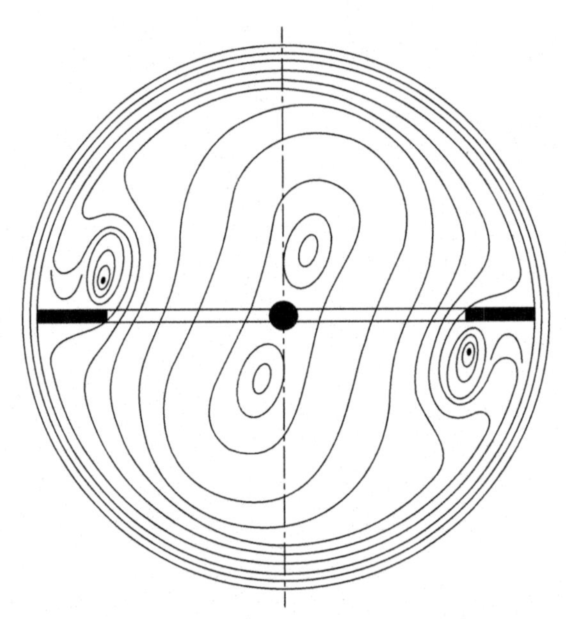

Fig. 7.13
Streamlines in a tank with a gate agitator, drawn relative to the arm of the stirrer.

tank is stationary, while that at higher levels is rotating and will experience centrifugal forces. Consequently, there exist unbalanced pressure forces within the fluid, which lead to the formation of a toroidal vortex. This secondary system may be single-celled or double-celled, as shown in Figs. 7.14 and 7.15, depending upon the viscosity and type of fluid.

Clearly, the flow pattern established in a mixing vessel depends critically upon the vessel/impeller configuration and on the physical properties of the liquid (particularly viscosity). In selecting the appropriate combination of equipment, it must be ensured that the resulting flow pattern is suitable for the required application.

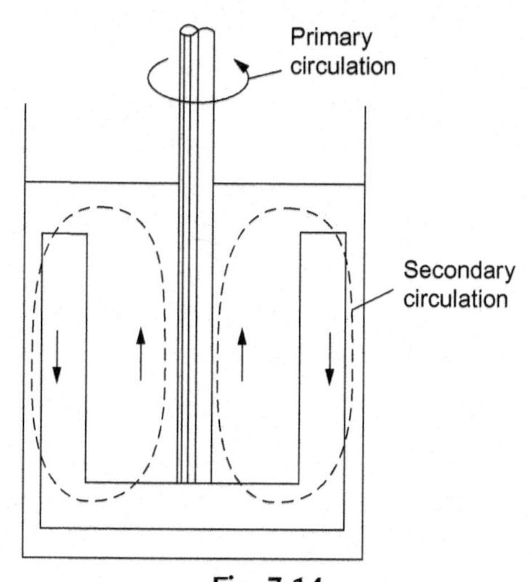

Fig. 7.14
Single-celled secondary flow with an anchor agitator.

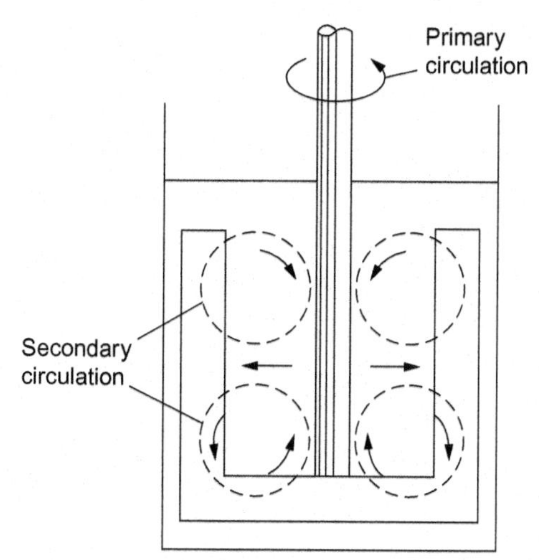

Fig. 7.15
Double-celled secondary flow.

7.6 Rate and Time for Mixing

Before considering rate of mixing and mixing time, it is necessary to have some means of assessing the quality of a mixture, which is the product of a mixing operation. Because of the wide scope and range of mixing problems, several criteria have been developed to assess mixture quality, none of which is universally applicable. One intuitive and convenient, but perhaps unscientific, criterion is whether the product or mixture meets the required specifications. Other ways of judging the quality of mixing have been described by Harnby et al.,[2] Paul et al.[3] and Ottino.[52] Whatever the criteria used, mixing time is defined as the time required to produce a mixture or a product of predetermined quality, and the rate of mixing is the rate at which the mixing progresses towards the final state. For a single-phase liquid in a stirred tank to which a volume of tracer material is added, the mixing time is measured from the instant the tracer is added to the time when the contents of the vessel have reached the required degree of uniformity or mixedness. If the tracer is completely miscible and has the same viscosity and density as the liquid in the tank, the tracer concentration may be measured as a function of time at any point in the vessel by means of a suitable detector, such as a colour meter, or by electrical conductivity. For a given amount of tracer, the equilibrium concentration C_∞ may be calculated; this value will be approached asymptotically at any point as shown in Fig. 7.16.

The mixing time will be that required for the mixture composition to come within a specified deviation from the equilibrium value and this will be dependent upon the way in which the tracer is added and the location of the detector. It may therefore be desirable to record the tracer concentration at several locations, and to define the variance of concentration σ^2 about the equilibrium value as:

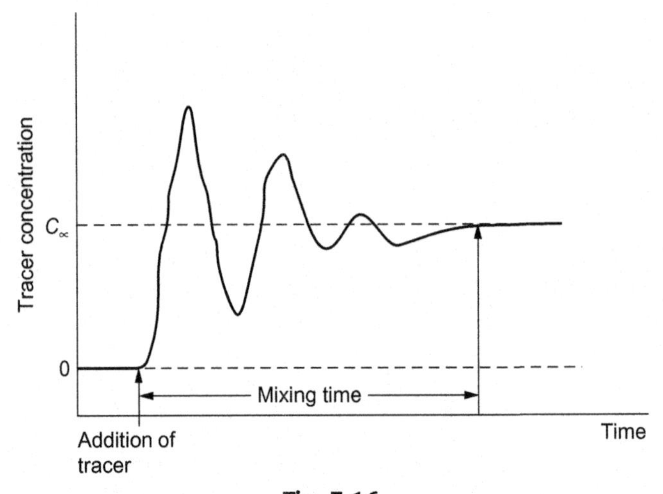

Fig. 7.16
Mixing-time measurement curve.

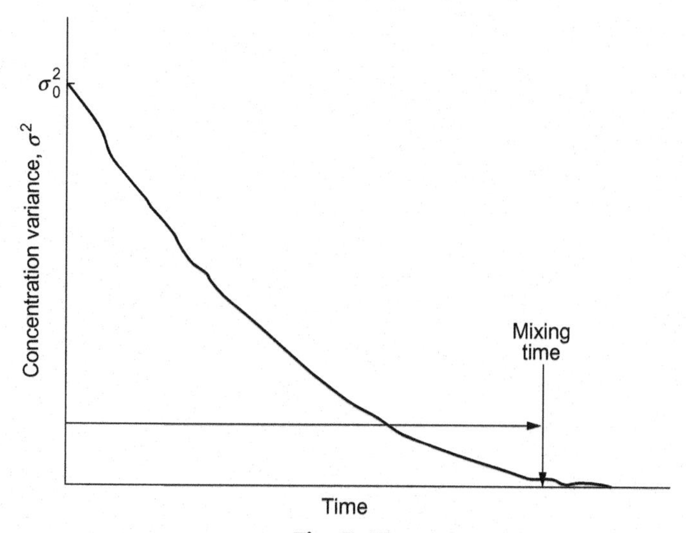

Fig. 7.17
Reduction in variance of concentration with time.

$$\sigma^2 = \frac{1}{\mathbf{n}-1}\sum_{i=0}^{i=n}(C_i - C_\infty)^2 \tag{7.20}$$

where C_i is the tracer concentration at time t recorded by the ith detector. A typical variance curve is shown in Fig. 7.17.

Several experimental techniques may be used, such as acid/base titration, electrical conductivity measurement, temperature measurement, or measurement of optical properties such as refractive index, light absorption, and so on. In each case, it is necessary to specify the manner of tracer addition, the position and number of recording stations, the sample volume of the detection system, and the criteria used in locating the end-point. Each of these factors will influence the measured value of mixing time, and therefore care must be exercised in comparing results from different investigations.[53]

For a given experiment and configuration, the mixing time will depend upon the process and operating variables as follows:

$$t_m = f(\rho, \mu, N, D, g, \text{geometrical dimensions of the system}) \tag{7.21}$$

Using dimensional analysis, the functional relationship may be rearranged as:

$$Nt_m = \theta_m = f\left(\frac{\rho N D^2}{\mu}, \frac{DN^2}{g}, \text{geometrical dimensions as ratios}\right) \tag{7.22}$$

For geometrically similar systems, and assuming that the Froude number DN^2/g is not important:

$$\theta_m = f\left(\frac{\rho N D^2}{\mu}\right) = f(\text{Re}) \tag{7.23}$$

The available experimental data[2] seem to suggest that the dimensionless mixing time (θ_m) is independent of Reynolds number for both laminar and turbulent zones, in each of which it attains a constant value, and changes from the one value to the other in the transition region. $\theta_m - Re$ behaviour is schematically shown in Fig. 7.18, and some typical mixing data obtained by Norwood and Metzner[54] for turbine impellers in baffled vessels using an acid/base titration technique are shown in Fig. 7.19. It is widely reported that θ_m is quite sensitive to impeller/vessel geometry for low viscosity liquids. Bourne and Butler[55] have made some interesting observations which appear to suggest that the rates of mixing and, hence mixing time, are not very sensitive to the fluid properties for both Newtonian and non-Newtonian materials. On the other hand, Godleski and Smith[56] have reported mixing times for pseudoplastic non-Newtonian fluids as being 50 times greater than those expected from the corresponding Newtonian behaviour, and this emphasises the care which must be exercised in applying any generalised conclusions to a particular system.

Although very little is known about the influence of fluid elasticity on rates of mixing and mixing times, the limited work appears to suggest that the role of elasticity strongly depends on the impeller geometry. For instance, Chavan et al.[57] reported a decrease in mixing rate with increasing levels of viscoelasticity for helical ribbons, whereas Hall and Godfrey[58] found the mixing rates to increase with elasticity for sigma blades. Similar conflicting conclusions regarding the influence of viscoelasticity can be drawn from the work of Carreau et al.[50] who found that the direction of impeller rotation had an appreciable effect on mixing rates and times.

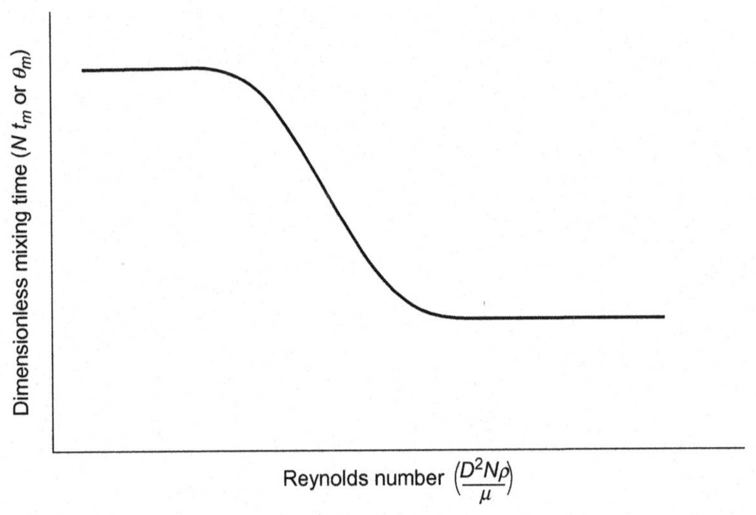

Fig. 7.18
Typical mixing-time behaviour.

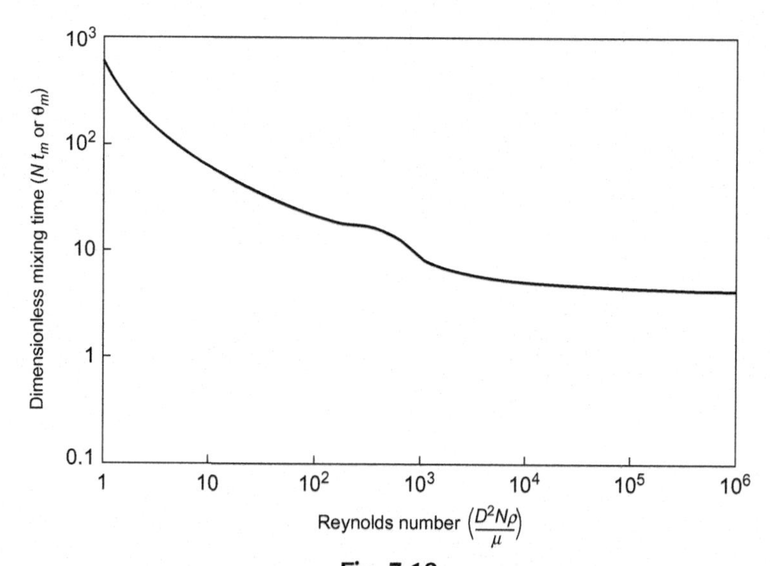

Fig. 7.19
Typical mixing time data for turbines.

Considerable confusion thus seems to exist in the literature regarding the influence of rheological properties on mixing rates and times. Exhaustive reviews are available on this subject.[3,17,43]

7.7 Mixing Equipment

The wide range of mixing equipment available commercially reflects the enormous variety of mixing duties encountered in the processing industries. It is reasonable to expect therefore that no single item of mixing equipment will be able to carry out such a range of duties effectively. This has led to the development of a number of distinct types of mixers over the years. Very little has been done, however, by way of standardisation of equipment and no design codes are available. The choice of a mixer type and its design is therefore primarily governed by experience. In the following sections, the main mechanical features of commonly used types of equipment together with their range of applications are described qualitatively. Detailed description of design and selection of various types of mixers have been presented by Oldshue[12] and Paul et al.[3]

7.7.1 Mechanical Agitation

This is perhaps the most commonly used method of mixing liquids, and essentially there are three elements in such devices.

Vessels

These are often vertically mounted cylindrical tanks, up to 10 m in diameter, which typically are filled to a depth equal to about one diameter, although in some gas–liquid contacting systems tall vessels are used and the liquid depth is up to about three tank diameters; multiple impellers fitted on a single shaft are then frequently used. The base of the tanks may be flat, dished, or conical, or specially contoured, depending upon factors such as ease of emptying, or the need to suspend solids, etc., and so on.

For the batch mixing of viscous pastes and doughs using ribbon impellers and Z-blade mixers, the tanks may be mounted horizontally. In such units, the working volume of pastes and doughs is often relatively small, and the mixing blades are massive in construction.

Baffles

To prevent gross vortexing, which is detrimental to mixing, particularly in low viscosity systems, baffles are often fitted to the walls of the vessel. These take the form of thin strips about one-tenth of the tank diameter in width, and typically four equi-spaced baffles may be used. In some cases, the baffles are mounted flush with the wall, although occasionally a small clearance is left between the wall and the baffle to facilitate fluid motion in the wall region. Baffles are, however, generally not required for high viscosity liquids because the viscous shear is then sufficiently great to damp out the rotary motion. Sometimes, the problem of vortexing is circumvented by mounting impellers off-centre, or from the bottom of the tank, or using side-entering types.

Impellers

Fig. 7.20 shows some of the impellers which are frequently used. Propellers, turbines, paddles, anchors, helical ribbons and screws are usually mounted on a central vertical shaft in a cylindrical tank, and they are selected for a particular duty largely on the basis of liquid viscosity. By and large, it is necessary to move from a propeller to a turbine and then, in order, to a paddle, to an anchor, and then to a helical ribbon and finally to a screw as the viscosity of the fluids to be mixed increases. In so doing the speed of agitation or rotation decreases.

Propellers, turbines, and paddles are generally used with relatively low viscosity systems and operate at high rotational speeds. A typical velocity for the tip of the blades of a turbine is of the order of 3 m/s, with a propeller being a little faster and the paddle a little slower. These are classed as remote-clearance impellers, having diameters in the range (0.13–0.67) × (tank diameter). Furthermore, minor variations within each type are possible. For instance, Fig. 7.20B shows a six-flat bladed Rushton turbine, whereas possible variations are shown in Fig. 7.21. Hence it is possible to have retreating-blade turbines, angled-blade turbines, 4- to 20-bladed turbines, and so on. For dispersion of gases in liquid, turbines are usually employed.

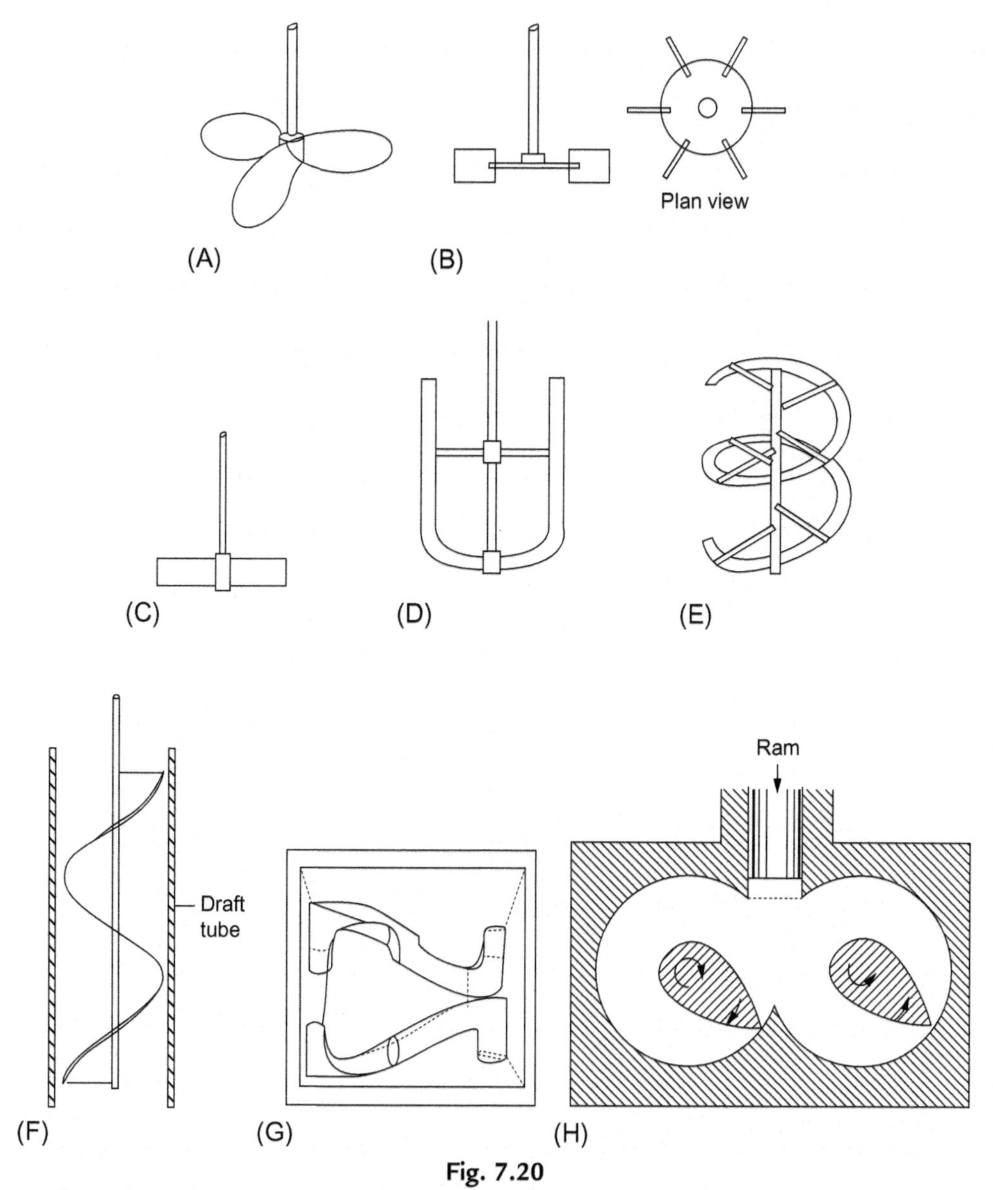

Fig. 7.20
Commonly used impellers. (A) Three-bladed propeller. (B) six-bladed disc turbine (Rushton turbine), (C) simple paddle, (D) anchor impeller, (E) helical ribbon, (F) helical screw with draft tube, (G) *Z*-blade mixer and (H) Banbury mixer.

Propellers are frequently of the three-bladed marine type and are used for in-tank blending operations with low viscosity liquids, and may be arranged as angled side-entry units, as shown in Fig. 7.22. Reavell[1] has shown that the fitting of a cruciform baffle at the bottom of the vessel enables much better dispersion to be obtained, as shown in Fig. 7.23. For large vessels, and

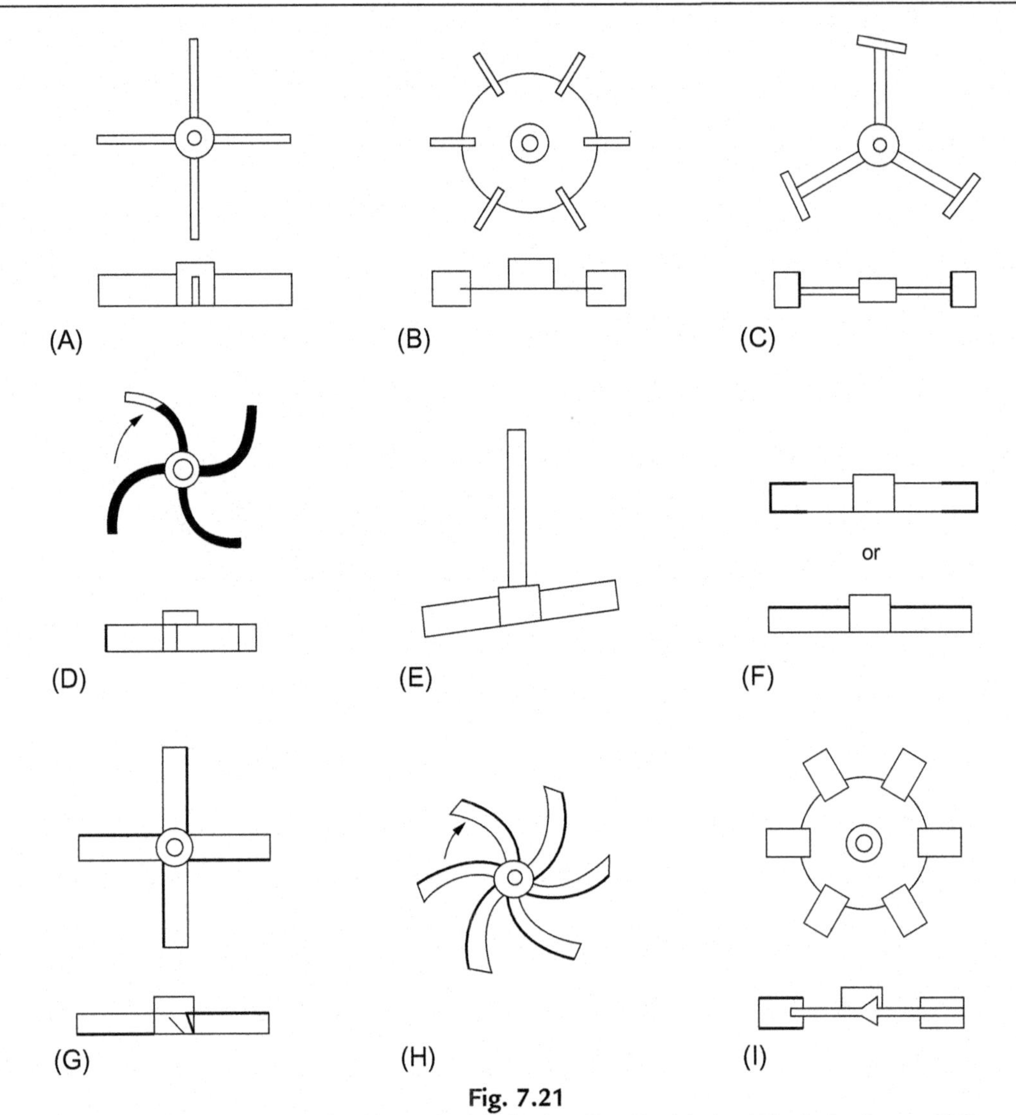

Fig. 7.21
Variation in turbine impeller designs. (A) Flat blade, (B) disc flat blade, (C) pitched vane, (D) curved blade, (E) tilted blade, (F) shrouded, (G) pitched blade, (H) pitched curved blade and (I) arrowhead.

when the liquid depth is large compared with the tank diameter, it is a common practice to mount more than one impeller on the same shaft. With this arrangement the unsupported length of the propeller shaft should not exceed about 2 m. In the case of large vessels, there is some advantage to be gained by using side- or bottom-entry impellers to avoid the large length of unsupported shaft, though a good gland or mechanical seal is needed for such installations or alternatively, a foot bearing is employed. Despite a considerable amount of practical experience, foot bearings can be troublesome owing to the difficulties of lubrication, especially when handling corrosive liquids.

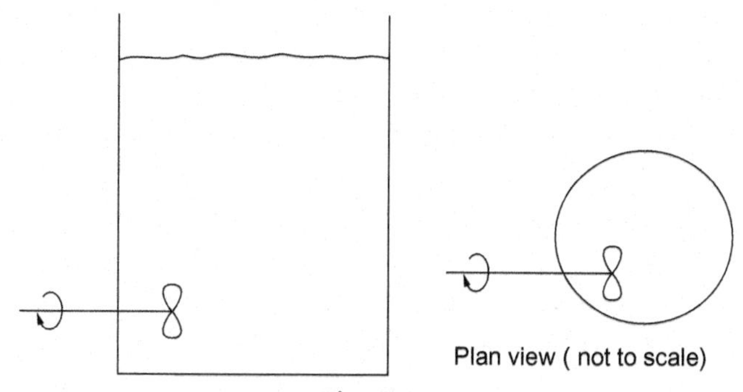

Fig. 7.22
Side entering propeller.

Fig. 7.23
Flow pattern in vessel with cruciform baffle.

In comparing propellers and turbines, the following features may be noted:

Propellers

(a) are self-cleaning in operation,
(b) can be used at a wide range of speeds,
(c) give an excellent shearing effect at high speeds,
(d) do not damage dispersed particles at low speeds,
(e) are reasonably economical in power, provided the pitch is adjusted according to the speed,

(f) by offset mounting, vortex formation is avoided,
(g) if horizontally mounted, a stuffing box is required in the liquid, and they are not effective in viscous liquids.

Shrouded turbines

(a) are excellent for providing circulation,
(b) are normally mounted on a vertical shaft with the stuffing box above the liquid,
(c) are effective in fluids of high viscosity,
(d) are easily fouled or plugged by solid particles,
(e) are expensive to fabricate,
(f) are restricted to a narrow range of speeds, and
(g) do not damage dispersed particles at economical speeds.

Open impellers

(a) are less easily plugged than the shrouded type,
(b) are less expensive, and
(c) give a less well-controlled flow pattern.

Anchors, helical ribbons, and screws, are generally used for high viscosity liquids. The anchor and ribbon are arranged with a close clearance at the vessel wall, whereas the helical screw has a smaller diameter and is often used inside a draft tube to promote fluid motion throughout the vessel. Helical ribbons or interrupted ribbons are often used in horizontally mounted cylindrical vessels.

Kneaders, Z- and sigma-blade, and Banbury mixers as shown in Fig. 7.20, are generally used for the mixing of high viscosity liquids, pastes, rubbers, doughs, and so on. The tanks are usually mounted horizontally and two impellers are often used. The impellers are massive and the clearances between blades, as well as between the wall and blade, are very small thereby ensuring that the entire mass of liquid is sheared.

7.7.2 Portable Mixers

For a wide range of applications, a portable mixer, which can be clamped on the top or side of the vessel is often used. This is commonly fitted with two propeller blades so that the bottom rotor forces the liquid upwards and the top rotor forces the liquid downwards. The power supplied is up to about 2 kW, though the size of the motor becomes too great at higher powers. To avoid excessive strain on the armature, some form of flexible coupling should be fitted between the motor and the main propeller shaft. Units of this kind are usually driven at a fairly high rate (15 Hz), and a reduction gear can be fitted to the unit fairly easily for low-speed operations although this increases the mass of the unit.

7.7.3 Extruders

Mixing duties in the plastics industry are often carried out in either single or twin-screw extruders. The feed to such units usually contains the base polymer in either granular or powder form, together with additives such as stabilisers, pigments, plasticisers, and so on. During processing in the extruder the polymer is melted and the additives mixed. The extrudate is delivered at high pressure and at a controlled rate from the extruder for shaping by means of either a die or a mould. Considerable progress has been made in the design of extruders in recent years, particularly by the application of numerical methods. One of the problems is that a considerable amount of heat is generated and the fluid properties may change by several orders of magnitude as a result of temperature changes. It is therefore always essential to solve the coupled equations of flow and heat transfer.

In the typical single-screw extruder shown in Fig. 7.24, the shearing which occurs in the helical channel between the barrel and the screw is not intense, and therefore this device does not give good mixing. Twin-screw extruders, as shown in Fig. 7.25, may be co- or counter-rotating, and here there are regions where the rotors are in close proximity thereby generating extremely high-shear stresses. Clearly, twin-screw units can yield a product of better mixture quality than

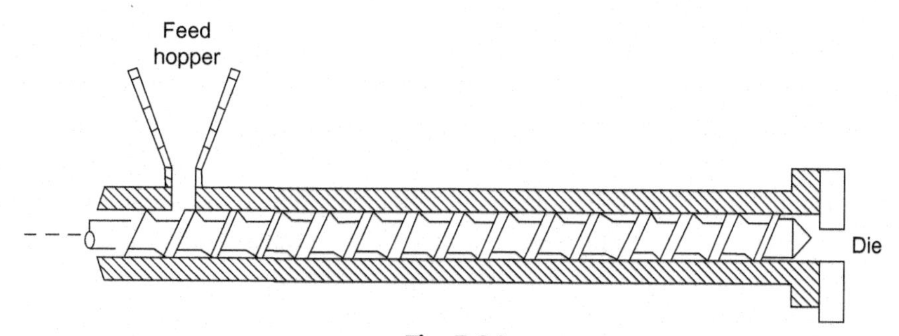

Fig. 7.24
Single-screw extruder.

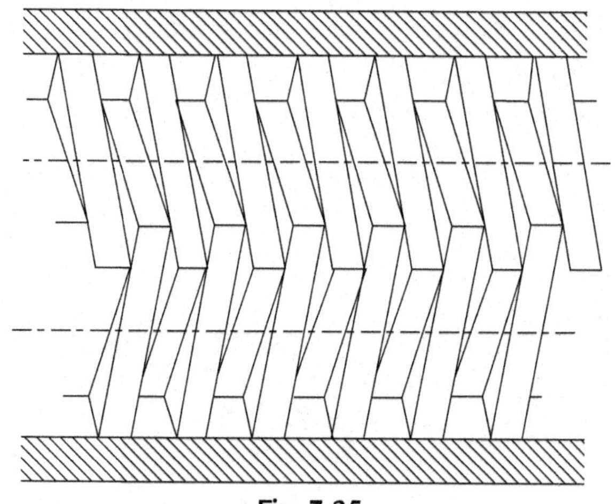

Fig. 7.25
Corotating twin-screw extruder.

a single-screw machine. Detailed accounts of the design and performance of extruders are available in literature.[59–63]

7.7.4 Static Mixers

All the mixers described so far are of the dynamic type in the sense that moving blades are used to impart motion to the fluid and produce the mixing effect. In static mixers,[2] sometimes called 'in-line' or 'motionless' mixers, the fluids to be mixed are pumped through a pipe containing a series of specially shaped stationary blades. Static mixers can be used with liquids of a wide range of viscosities in either the laminar or turbulent regimes, but their special features are perhaps best appreciated in relation to laminar flow mixing. The flow patterns within the mixer are complex, though a numerical simulation of the flow has been carried out by Lang et al.[64]

Fig. 7.26 shows a particular type of static mixer in which a series of stationary helical blades mounted in a circular pipe is used to divide and twist the flowing streams. In laminar flow, (Fig. 7.27), the material divides at the leading edge of each of these elements and follows the channels created by the element shape. At each succeeding element the two channels are further divided, and mixing proceeds by a distributive process similar to the cutting and folding mechanism shown in Fig. 7.4. In principle, if each element divided the streams neatly into two, feeding two dissimilar streams to the mixer would give a striated material in which the thickness of each striation would be of the order $D_T/2^n$ where D_T is the diameter of the tube and n is the number of elements. However, the helical elements shown in Fig. 7.26 also induce further mixing by a laminar shear mechanism (see Figs. 7.1 and 7.3). This, combined with the twisting action of the element, helps to promote radial mixing which is important in eliminating any radial gradients of composition, velocity and possibly temperature that might exist in the material. Fig. 7.28 shows how these mixing processes together produce after only 10–12 elements, a well-blended material. Figs. 7.26–7.28 all refer to static mixers made by Chemineer Ltd., Derby, U.K.

Fig. 7.29 shows a Sulzer type SMX static mixer where the mixing element consists of a lattice of intermeshing and interconnecting bars contained in a pipe 80 mm diameter. It is

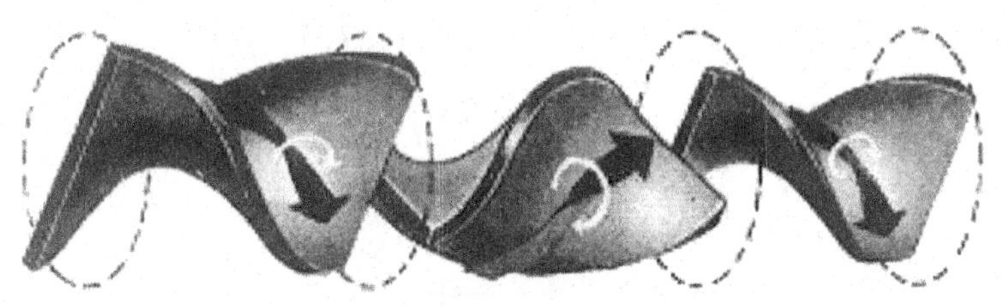

Fig. 7.26
Twisted-blade type of static mixer elements.

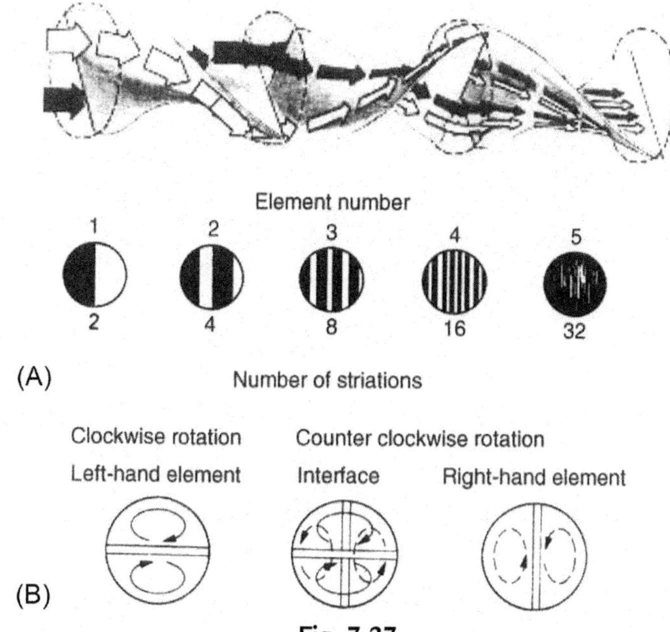

Fig. 7.27
Twisted-blade type of static mixer operating in the laminar flow regime. (A) Distributive mixing mechanism showing, in principle, the reduction in striation thickness produced. (B) Radial mixing contribution from laminar shear mechanism.

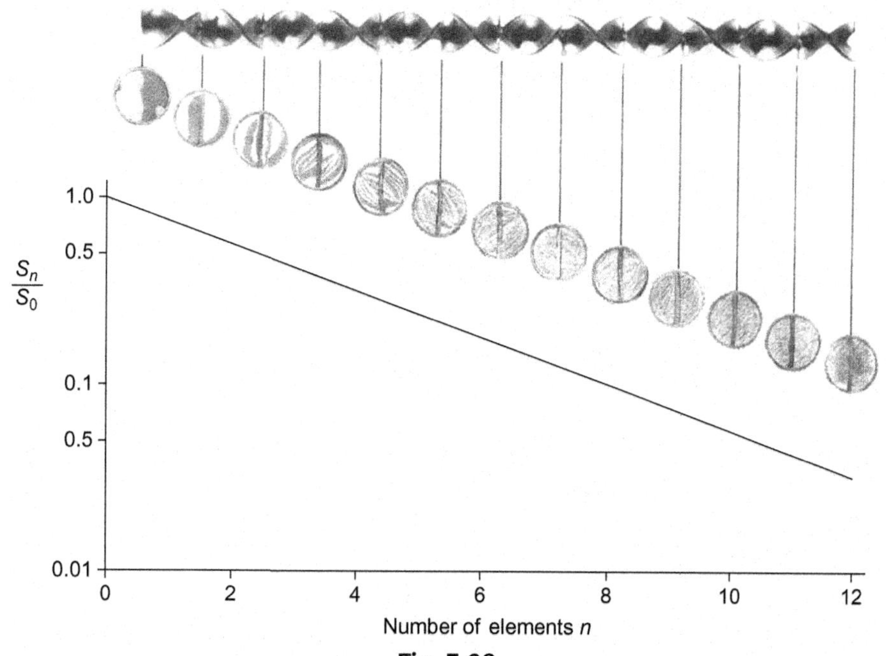

Fig. 7.28
Static mixer in laminar flow: reduction in relative standard deviation of samples indicating improvement in mixture quality with increasing number n of elements traversed.

Fig. 7.29
Static mixer for viscous materials 3.

recommended for viscous materials in laminar flow. The mixer shown is used in food processing, for example mixing fresh cheese with whipped cream.

Quantitatively, a variety of methods[65] has been proposed to describe the degree or quality of mixing produced in a static mixer. One of these measures of mixing quality is the relative standard deviation s/s_0, where s is the standard deviation in composition of a set of samples taken at a certain stage of the mixing operation, and s_0 is the standard deviation for a similar set of samples taken at the mixer inlet. Fig. 7.28 shows schematically how the relative standard deviation falls as the number **n** of elements through which the material has passed increases, a perfectly mixed product having a zero relative standard deviation. One of the problems in using relative standard deviation or a similar index as a measure of mixing is that this depends on sample size, which therefore needs to be taken into account in any assessment.

One of the most important considerations in choosing or comparing static mixers is the power consumed by the mixer to achieve a desired mixture quality. The pressure drop characteristics of a mixer are most conveniently described as the ratio of mixer pressure drop to empty pipe pressure drop for the same flowrate and diameter. Different static mixer designs can be compared[65] on the basis of mixing quality, pressure-drop ratio, initial cost, and convenience of installation.

Static mixers for viscous fluids are widely used in processes producing polymers, fibres, and plastics of all kinds where the materials tend to be viscous, hot, and often at high pressures. However, static mixers have also achieved widespread use for low viscosity fluid mixing for blending, liquid–liquid, and even gas–liquid dispersions. In some cases the designs used for high viscosity liquids have also proved effective in the turbulent mixing regime for low viscosity fluids. In other cases manufacturers have developed special designs for turbulent flow mixing, and a wide variety of static mixer devices is now available.[2,3]

7.7.5 Other Types of Mixer

Only a selection of commercially available mixing equipment has been described here. Indeed, the devices described all exist in a variety of configurations. In addition, there are many items of equipment based on altogether different principles; typical examples include jet mixers, in-line dynamic mixers, mills, valve homogenisers, ultrasonic homogenisers, etc. These, as well as many other types, have been discussed by Harnby et al.,[2] Oldshue,[12] Nagata,[22] Paul et al.[3] and others.[66–68]

Frequently, it is convenient to mix the contents of a tank without the necessity for introducing an agitator and this may present difficulties in the construction of the vessel. It may then be possible to use an external circulating pump (usually centrifugal). If it is desirable not to make special connections to the vessel, it may be possible to connect the suction side of the pump to the bottom outlet of the tank by means of a T-piece and to discharge the liquid into the tank through its open top or through an existing entry point. In such a system, dispersion is effected in the high-shear region in the pump, and the liquid in the tank is maintained in a state of continuous circulation.

Such an arrangement may well be suitable when it is necessary to prevent fine particles from settling out at the bottom of the tank.

7.8 Mixing in Continuous Systems

The mixing problems considered so far have been related to batch systems in which two materials are mixed together and uniformity is maintained by continued operation of the agitator.

Frequently, stirred tanks are used with a continuous flow of material on one side of the tank and with a continuous outflow from the other. A particular application is the use of the tank as a *continuous stirred-tank reactor* (CSTR). Inevitably, there will be a very wide range of residence times for elements of fluid in the tank. Even if the mixing is so rapid that the contents of the tank are always virtually uniform in composition, some elements of fluid will almost immediately flow to the outlet point and others will continue circulating in the tank for a very long period before leaving. The *mean residence time* of fluid in the tank is given by:

$$t_r = \frac{V}{Q} \tag{7.24}$$

where V is the volume of the contents of the tank (assumed constant), and Q is the volumetric throughput.

In a completely mixed system, the composition of the outlet stream will be equal to the composition in the tank.

The variation of time for which fluid elements remain within the tank is expressed as a *residence time distribution* and this can be calculated from a simple material balance if mixing is complete. For incomplete mixing, the calculation presents difficulties.

The problem is of great significance in the design of reactors because a varying residence time will in general, lead to different degrees of chemical conversion of various fluid elements, and this is discussed in some detail in Volume 3A, Chapter 1.

7.9 Nomenclature

		Units in SI System	Dimensions in M, L, T
A	Area of flow	m^2	$\mathbf{L}^2$
A'	Area for shear	m^2	$\mathbf{L}^2$
C	Concentration	–	–
D	Impeller diameter	m	**L**
D_T	Tank *or* tube diameter	m	**L**
F	Force	N	$\mathbf{MLT}^{-2}$
Fr	Froude number (N^2D/g)	–	–
g	Acceleration due to gravity	m/s^2	$\mathbf{LT}^{-2}$
H	Depth of liquid	m	**L**
k_s	Function of impeller speed (Eq. 7.19)	–	–
L	Length	m	**L**
m	Mass of liquid	kg	**M**
N	Speed of rotation (revs/unit time)	Hz	$\mathbf{T}^{-1}$
N_P	Power number ($\mathbf{P}/\rho N^3D^5$)	–	–
n	Power law index	–	–
$\mathbf{n}$	Number of elements	–	–
$\mathbf{P}$	Power	W	$\mathbf{ML}^2\mathbf{T}^{-3}$
P	Pitch of agitator	m	**L**
Q	Volumetric throughput	m^3/s	$\mathbf{L}^3\mathbf{T}^{-1}$
R	Number of baffles	–	–
r	Radius	m	**L**
s, s_0	Standard deviations	–	–
t	Time	s	**T**
t_m	Mixing time	s	**T**
t_r	Residence time	s	**T**
u	Velocity	m/s	$\mathbf{LT}^{-1}$
V	Volume of tank	m^3	$\mathbf{L}^3$
W	Blade width	m	**L**

W_B	Width of baffles	m	**L**
y	Distance	m	**L**
Z_A	Height of agitator from base of tank	m	**L**
$\dot{\gamma}_{ang}$	Shear rate (angular)	s^{-1}	$\mathbf{T}^{-1}$
μ	Viscosity	N s/m^2	$\mathbf{ML}^{-1}\mathbf{T}^{-1}$
μ_a	Apparent viscosity	N s/m^2	$\mathbf{ML}^{-1}\mathbf{T}^{-1}$
ρ	Density	kg/m^3	$\mathbf{ML}^{-3}$
σ	Surface tension	N/m	$\mathbf{MT}^{-2}$
σ^2	Variance	–	–
θ_m	Dimensionless mixing time (Nt_m)	–	–
Re	Reynolds number ($\rho ND^2/\mu$)	–	–

References

1. Reavell BN. Practical aspects of liquid mixing and agitation. *Trans Inst Chem Eng* 1951;**29**:301.
2. Harnby N, Edwards MF, Nienow AW, editors. *Mixing in the process industries*. 2nd ed. London: Butterworth-Heinemann; 1992.
3. Paul EL, Atiemo-Obeng VA, Kresta SM, editors. *Handbook of industrial mixing: science and practice*. New York: Wiley; 2004.
4. Levenspiel O. *Engineering flow and heat exchange*. London: Plenum; 1985.
5. Ottino JM. The mixing of fluids. *Sci Am* 1989;**260**:56.
6. Levich VG. *Physico-chemical hydrodynamics*. London: Prentice-Hall; 1962.
7. van der Molen K, van Maanen HRE. Laser-Doppler measurements of the turbulent flow in stirred vessels to establish scaling rules. *Chem Eng Sci* 1978;**33**:1161.
8. Rao MA, Brodkey RS. Continuous flow stirred tank turbulence parameters in the impeller stream. *Chem Eng Sci* 1972;**27**:137.
9. Rushton JH, Costich EW, Everett HJ. Power characteristics of mixing impellers. Parts I and II. *Chem Eng Prog* 1950;**46**:395, 467.
10. Kramers H, Baars GM, Knoll WH. A comparative study on the rate of mixing in stirred tanks. *Chem Eng Sci* 1953;**2**:35.
11. Skelland AHP. *Non-Newtonian flow and heat transfer*. New York: Wiley; 1967.
12. Oldshue JY. *Fluid mixing technology*. New York: McGraw-Hill; 1983.
13. Metzner AB, Otto RE. Agitation of non-Newtonian fluids. *AICHE J* 1957;**3**:3.
14. Metzner AB, Taylor JS. Flow patterns in agitated vessels. *AICHE J* 1960;**6**:109.
15. Ulbrecht J. Mixing of viscoelastic fluids by mechanical agitation. *Chem Eng (London)* 1974;**286**:347.
16. Astarita G. Scaleup problems arising with non-Newtonian fluids. *J Non-Newtonian Fluid Mech* 1979;**4**:285.
17. Skelland AHP. Mixing and agitation of non-Newtonian fluids. In: Cheremisinoff NP, Gupta R, editors. *Handbook of fluids in motion*. New York: Ann Arbor; 1983.
18. Holland FA, Chapman FS. *Liquid mixing and processing in stirred tanks*. New York: Reinhold; 1966.
19. Bissell ES, Hesse HC, Everett HJ, Rushton JH. Design and utilisation of internal fittings for mixing vessels. *Chem Eng Prog* 1947;**43**:649.
20. Marshall EM, Bakker A. Computational fluid mixing. In: Paul EL, Atiemo-Obeng VA, Kresta SM, editors. *Handbook of industrial mixing: science and practice*. New York: Wiley; 2004 Chapter 5.
21. Grenville, R. K., Ruszkowski, S. and Garret, E. Blending of miscible liquids in the turbulent and transitional regimes. In: Proceedings of the 15th NAMF mixing conference, Banff, Canada (1995).

22. Nagata S. *Mixing: principles and applications.* London: Wiley; 1975.
23. Uhl VW, Gray JB, editors. *Mixing: theory and practice.* **1**:London: Academic Press; 1966.
24. Šterbaček Z, Tausk P. *Mixing in the chemical industry.* Oxford: Pergamon; 1965, Translated by K. Mayer and J.R. Bourne.
25. Tatterson GB. *Scale up and design of industrial mixing processes.* New York: McGraw-Hill; 1994.
26. Ibrahim S, Nienow AW. Power curves and flow patterns for a range of impellers in Newtonian fluids: $40 < Re < 5 \times 10^5$. *Trans Inst Chem Eng* 1995;**73**:485.
27. Chhabra RP, Richardson JF. *Non-Newtonian flow and applied rheology.* 2nd ed. Oxford, UK: Butterworth-Heinemann; 2008.
28. Chhabra RP. Fluid mechanics and heat transfer with non-Newtonian liquids in mechanically agitated vessels. *Adv Heat Tran* 2003;**37**:77.
29. Hoogendoorn CJ, den Hartog AP. Model studies on mixers in the viscous flow region. *Chem Eng Sci* 1967;**22**:1689.
30. Ullrich H, Schreiber H. Rühren in zähen Flüssigkeiten. *Chem Ing Tech* 1967;**39**:218.
31. Mitsuishi N, Hirai NJ. Power requirements in the agitation of non-Newtonian fluids. *J Chem Eng Jpn* 1969;**2**:217.
32. Godfrey JC. Mixing of high viscosity liquids. In: Hamby N, Edwards MF, Nienow AW, editors. *Mixing in the processing industries.* London: Butterworths; 1985. p. 185–201.
33. Calderbank PH, Moo-Young MB. The prediction of power consumption in the agitation of non-Newtonian fluids. *Trans Inst Chem Eng* 1959;**37**:26.
34. Beckner JL, Smith JM. Anchor-agitated systems: power input with Newtonian and pseudo-plastic fluids. *Trans Inst Chem Eng* 1966;**44**:T224.
35. Nagata S, Nishikawa M, Tada H, Hirabayashi H, Gotoh S. Power consumption of mixing impellers in Bingham plastic fluids. *J Chem Eng Jpn* 1970;**3**:237.
36. Metzner AB, Feehs RH, Romos HL, Otto RE, Tuthill JP. Agitation of viscous Newtonian and non-Newtonian fluids. *AIChE J* 1961;**7**:3.
37. Edwards MF, Godfrey JC, Kashani MM. Power requirements for the mixing of thixotropic fluids. *J Non-Newtonian Fluid Mech* 1976;**1**:309.
38. Kelkar JV, Mashelkar RA, Ulbrecht J. On the rotational viscoelastic flows around simple bodies and agitators. *Trans Inst Chem Eng* 1972;**50**:343.
39. Rieger F, Novak V. Power consumption for agitating viscoelastic liquids in the viscous regime. *Trans Inst Chem Eng* 1974;**52**:285.
40. Oliver DR, Nienow AW, Mitson RJ, Terry K. Power consumption in the mixing of Boger fluids. *Chem Eng Res Des* 1984;**62**:123.
41. Prud'homme RK, Shaqfeh E. Effect of elasticity on mixing torque requirements for Rushton turbine impellers. *AIChE J* 1984;**30**:485.
42. Kale DD, Mashelkar RA, Ulbrecht J. High speed agitation of non-Newtonian fluids: influence of elasticity and fluid inertia. *Chem Ing Tech* 1974;**46**:69.
43. Chavan VV, Mashelkar RA. Mixing of viscous Newtonian and non-Newtonian fluids. *Adv Transport Proc* 1980;**1**:210.
44. Irving HF, Saxton RL. Uhl VW, Gray JB, editors. *Mixing theory and practice.* **2**, New York: Academic Press; 1967. pp. 169–224.
45. Hicks RW, Morton JR, Fenic JG. How to design agitators for desired process response. *Chem Eng* 26 Apr. 1976;**83**:102.
46. Gates LE, Hicks RW, Dickey DS. Application guidelines for turbine agitators. *Chem Eng* 1976;**83**(6th Dec.):165.
47. Fasano JB, Bakker A, Penney WR. Advanced impeller geometry boosts liquid agitation. *Chem Eng* 1994;**101** (8):110.
48. Bakker A, Gates LE. Viscous mixing. *Chem Eng Prog* 1995;**91(12)**:25.
49. Gray JB. Uhl VW, Gray JB, editors. *Mixing theory and practice.* **1**, New York: Academic Press; 1966. pp. 179–278.

50. Carreau PJ, Patterson I, Yap CY. Mixing of viscoelastic fluids with helical ribbon agitators. I. Mixing time and flow pattem. *Can J Chem Eng* 1976;**54**:135.
51. Peters DC, Smith JM. Fluid flow in the region of anchor agitator blades. *Trans Inst Chem Eng* 1967;**45**:360.
52. Ottino JM. *The kinematics of mixing: stretching, chaos, and transport.* New York: Cambridge University Press; 1989.
53. Manna L. Comparison between physical and chemical methods for the measurement of mixing times. *Chem Eng J* 1997;**67**:167.
54. Norwood KW, Metzner AB. Flow patterns and mixing rates in agitated vessels. *AICHE J* 1960;**6**:432.
55. Bourne JR, Butler H. On analysis of the flow produced by helical ribbon impellers. *Trans Inst Chem Eng* 1969;**47**:T11.
56. Godleski ES, Smith JC. Power requirements and blend times in the agitation of pseudoplastic fluids. *AICHE J* 1962;**8**:617.
57. Chavan VV, Arumugam M, Ulbrecht J. On the influence of liquid elasticity on mixing in a vessel agitated by a combined ribbon-screw impeller. *AICHE J* 1975;**21**:613.
58. Hall KR, Godfrey JC. The mixing rates of highly viscous Newtonian and non-Newtonian fluids in a laboratory sigma blade mixer. *Trans Inst Chem Eng* 1968;**46**:205.
59. McKelvey JM. *Polymer processing.* New York: Wiley; 1962.
60. Janssen LPBM. *Twin Screw Extrusion.* Amsterdam: Elsevier; 1978.
61. Schenkel G. *Plastics extrusion technology.* London: Cliffe Books; 1966.
62. Bouvier JM, Campanella OH. *Extrusion processing technology: food and non-food biomaterials.* New York: Wiley-Blackwell; 2014.
63. Giles Jr HF, Wagner Jr JR, Mount EM. *Extrusion: the definitive processing guide and handbook.* 2nd ed. Norwich, New York: William Andrew; 2013.
64. Lang E, Drtina P, Streiff F, Fleischli M. Numerical simulation of the fluid flow and mixing process in a static mixer. *Int J Heat Mass Transf* 1995;**38**:2239.
65. Heywood, N. I., Viney, L. J., and Stewart, I. W. Mixing efficiencies and energy requirements of various motionless mixer designs for laminar mixing applications. Inst. Chem. Eng. Symposium Series No. 89, Fluid Mixing II (1984) 147.
66. Zlokarnik M. *Stirring: theory and practice.* New York: Wiley; 2001.
67. Zlokarnik M. *Scale-up in chemical engineering.* 2nd ed. New York: Wiley; 2006.
68. Xuereb C, Poux M, Bertrand J, editors. *Agitation et Melange.* Paris: Dunod; 2006.

Further Reading

1. Harnby N, Edwards MF, Nienow AW, editors. *Mixing in the process industries.* 2nd ed. Oxford: Butterworth-Heinemann; 1992.
2. Holland FA, Chapman FS. *Liquid mixing and processing in stirred tanks.* New York: Reinhold; 1966.
3. McDonough RJ. *Mixing for the process industries.* New York: Reinhold; 1992.
4. Metzner AB. In: Streeter VL, editor. *Handbook of fluid dynamics.* New York: McGraw-Hill; 1961. section 7.
5. Nagata S. *Mixing: principles and applications.* London: Wiley; 1975.
6. Oldshue JY. *Fluid mixing technology.* New York: McGraw-Hill; 1983.
7. Ottino JM. *The kinematics of mixing: stretching, chaos and transport.* New York: Cambridge University Press; 1990.
8. Paul EL, Atiemo-Obeng VA, Kresta SM, editors. *Handbook of industrial mixing: science and practice.* New York: Wiley; 2004.
9. Tatterson GB. *Fluid mixing and gas dispersion in agitated tanks.* New York: McGraw-Hill; 1991.
10. Uhl VW, Gray JB, editors. *Mixing: theory and practice.* vol. 1. London: Academic Press; 1966. vol. 2 (1967); vol. 3 (1986).

CHAPTER 8

Pumping of Fluids

8.1 Introduction

For the pumping of liquids or gases from one vessel to another or through long pipes, some form of mechanical pump is usually employed. The energy required by the pump will depend on the height through which the fluid is raised, the pressure required at delivery point, the length and diameter of the pipe, the rate of flow, together with the physical properties of the fluid, particularly its viscosity and density. The pumping of liquids such as sulphuric acid or petroleum products from bulk store to process buildings, or the pumping of fluids around reaction units and through heat exchangers, are typical illustrations of the use of pumps in the process industries. On the one hand, it may be necessary to inject reactants or catalyst into a reactor at a low, but accurately controlled rate, and on the other to pump cooling water to a power station or refinery at a very high rate. The fluid may be a gas or liquid of low viscosity, or it may be a highly viscous liquid, possibly with non-Newtonian characteristics. It may be clean, or it may contain suspended particles and be very corrosive. All these factors influence the choice of a pump.

Because of the wide variety of requirements, many different types of pumps are in use including centrifugal, piston, gear, screw, and peristaltic pumps, though in the chemical, mineral, and petroleum industries the centrifugal type is by far the most important. The main features considered in this chapter are an understanding of the criteria for pump selection, the determination of size and power requirements, and the positioning of pumps in relation to pipe systems. For greater detail, reference may be made to specialist publications included at the end of this chapter.

Pump design and construction is a specialist field and manufacturers should always be consulted before a final selection is made. In general, pumps used for circulating gases work at higher speeds than those used for liquids, and lighter valves are used. Moreover, the clearances between moving parts are smaller on gas pumps because of the much lower viscosity of gases, giving rise to an increased tendency for leakage to occur. When a pump is used to provide a vacuum, it is even more important to guard against leakage.

The work done by the pump is determined by setting up an energy balance equation. If W_s is the shaft work done by unit mass of fluid on the surroundings, then $-W_s$ is the shaft work done on the fluid by the pump.

Coulson and Richardson's Chemical Engineering. https://doi.org/10.1016/B978-0-08-101099-0.00008-2

$$\text{From Eq.(2.55)}: \quad -W_s = \Delta\frac{u^2}{2\alpha} + g\Delta z + \int_{P_1}^{P_2} v\mathrm{d}P + F \tag{8.1}$$

and

$$\text{from Eq.(2.56)}: \quad -W_s = \Delta\frac{u^2}{2\alpha} + g\Delta z + \Delta H - q \tag{8.2}$$

In any practical system, the pump efficiency must be taken into account, and more energy must be supplied by the motor driving the pump than is given by $-W_s$. If liquids are considered to be incompressible, there is no change in specific volume from the inlet to the delivery side of the pump. The physical properties of gases are, however, considerably influenced by the pressure, and the work done in raising the pressure of a gas is influenced by the rate of heat flow between the gas and the surroundings. Thus, if the process is carried out adiabatically, all the energy added to the system appears in the gas and its temperature rises. If an ideal gas is compressed and then cooled to its initial temperature, its enthalpy will be unchanged and the whole of the energy supplied by the compressor is dissipated to the surroundings. If, however, the compressed gas is allowed to expand it will absorb heat and is therefore capable of doing work at the expense of heat energy from the surroundings. Owing to the underlying differences between the pumping of a liquid and a gas, these are discussed separately.

8.2 Pumping Equipment for Liquids

As already indicated, the liquids used in the chemical industries differ considerably in physical and chemical properties, and it has been necessary to develop a wide variety of pumping equipment. The two main forms are the positive displacement type and centrifugal pumps. In the former, the volume of liquid delivered is directly related to the displacement of the piston and therefore increases directly with speed and is not appreciably influenced by the pressure or by the liquid properties. This group includes the reciprocating piston pump and the rotary gear pump, both of which are commonly used for delivery against high pressures and where nearly constant delivery rates are required. In the centrifugal type, a high kinetic energy is imparted to the liquid, which is then converted as efficiently as possible into pressure energy. The performance of a centrifugal pump is strongly influenced by the viscosity, density, and vapour pressure of the liquid as well as by the type of liquid (Newtonian or non-Newtonian, single phase or multiphase, etc.). For viscous Newtonian and non-Newtonian fluids of moderate to high viscosity, marked increase in pump power, reduction in head developed and some deterioration in pump efficiency occur. Similarly, the performance of a centrifugal pump is strongly influenced by the size, type, and concentration of solids (stones, gravel, mill scale, fine sand, rags, paper sock, etc.) in the slurry. These characteristics not only have an adverse impact on the pump performance but generally also cause significant wear and tear, erosion, and may even lead to operational problems.[1–3] For some applications, such as the handling of liquids, which are particularly corrosive or contain abrasive solids in suspension, compressed air is used as the motive force instead of a mechanical pump. An illustration of the use of this form of equipment is the transfer of the contents of a reaction mixture from one vessel to another.

The following factors influence the choice of pump for a particular operation.

(1) The quantity of liquid to be handled. This primarily affects the size of the pump and determines whether it is desirable to use a number of pumps in parallel.
(2) The head against which the liquid is to be pumped. This will be determined by the difference in pressure, the vertical height of the downstream and upstream reservoirs and by the frictional losses, which occur in the delivery line. The suitability of a centrifugal pump and the number of stages required will largely be determined by this factor.
(3) The nature of the liquid to be pumped. For a given throughput, the viscosity largely determines the frictional losses and hence the power required. The corrosive and abrasive nature will determine the material of construction, both for the pump and the packing. With suspensions, the clearances in the pump must be large compared with the size of the particles.
(4) The nature of the power supply. If the pump is to be driven by an electric motor or internal combustion engine, a high-speed centrifugal or rotary pump will be preferred as it can be coupled directly to the motor. Simple reciprocating pumps can be connected to steam or gas engines.
(5) If the pump is used only intermittently, corrosion problems are more likely than with continuous working.

The cost and mechanical efficiency of the pump must always be considered, and it may be advantageous to select a cheap pump and pay higher replacement or maintenance costs rather than to install a very expensive pump of high efficiency.

8.2.1 Reciprocating Pump

The piston pump

The piston pump consists of a cylinder with a reciprocating piston connected to a rod which passes through a gland at the end of the cylinder as indicated in Fig. 8.1. The liquid enters from the suction line through a suction valve and is discharged through a delivery valve. These pumps may be single-acting, with the liquid admitted only to the portion of the cylinder in front of the piston or double-acting, in which case the feed is admitted to both sides of the piston. The majority of pumps are of the single-acting type typically giving a low flowrate of say 0.02 m^3/s at a high pressure of up to 100 MPa.[4]

The velocity of the piston varies in an approximately sinusoidal manner and the volumetric rate of discharge of the liquid shows corresponding fluctuations. In a single-cylinder pump, the delivery will rise from zero as the piston begins to move forward to a maximum when the piston is fully accelerated at approximately the mid point of its stroke; the delivery will then

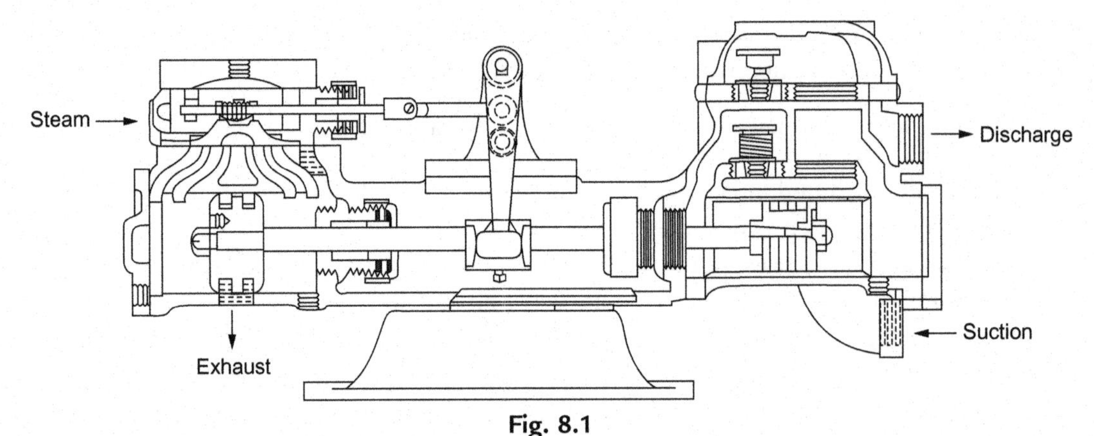

Fig. 8.1
A typical steam-driven piston pump.

gradually fall off to zero. If the pump is single-acting, there will be an interval during the return stroke when the cylinder will fill with liquid and the delivery will remain zero. On the other hand, in a double-acting pump the delivery will be similar in the forward and return strokes. In many cases, however, the cross-sectional area of the piston rod may be significant compared with that of the piston and the volume delivered during the return stroke will therefore be less than that during the forward stroke. A more even delivery is obtained if several cylinders are suitably compounded. If two double-acting cylinders are used there will be a lag between the deliveries of the two cylinders, and the total delivery will then be the sum of the deliveries from the individual cylinders. Typical curves of delivery rate for a single-cylinder (simplex) pump are shown in Fig. 8.2A. The delivery from a two-cylinder (duplex) pump in which both the cylinders are double-acting is shown in Fig. 8.2B; the broken lines indicate the deliveries from the individual cylinders and the continuous line indicates the

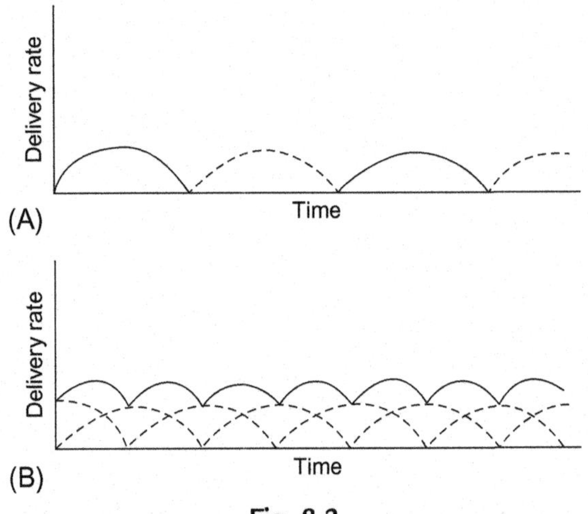

Fig. 8.2
Delivery from (A) simplex and (B) duplex pumps.

total delivery. It will be seen that the delivery is much smoother than that obtained with the simplex pump, the minimum delivery being equal to the maximum obtained from a single cylinder.

The theoretical delivery of a piston pump is equal to the total swept volume of the cylinders. The actual delivery may be less than the theoretical value because of leakage past the piston and the valves or because of inertia of the valves. In some cases, however, the actual discharge is greater than the theoretical value because the momentum of the liquid in the delivery line and sluggishness in the operation of the delivery valve may result in continued delivery during a portion of the suction stroke. The volumetric efficiency, which is defined as the ratio of the actual discharge to the swept volume, is normally $>90\%$.

The size of the suction and delivery valves is determined by the throughput of the pump. Where the rate of flow is high, two or more valves may be used in parallel.

The piston pump can be directly driven by steam, in which case the piston rod is common to both the pump and the steam engine. Alternatively, an electric motor or an internal combustion engine may supply the motive power through a crankshaft; because the load is very uneven, a heavy flywheel should then be fitted and a regulator in the steam supply may often provide a convenient form of speed control.

The pressure at the delivery of the pump is made up of the following components:

(1) The static pressure at the delivery point.
(2) The pressure required to overcome the frictional losses in the delivery pipe.
(3) The pressure for the acceleration of the fluid at the commencement of the delivery stroke.

The liquid in the delivery line is accelerated and retarded in phase with the motion of the piston, and therefore the whole of the liquid must be accelerated at the commencement of the delivery stroke and retarded at the end of it. Every time the fluid is accelerated, work has to be done on it and therefore in a long delivery line, the expenditure of energy is very large since the excess kinetic energy of the fluid is not usefully recovered during the suction stroke. Due to the momentum of the fluid, the pressure at the pump may fall sufficiently low for separation to occur. The pump is then said to *knock*. The flow in the delivery line can be evened out and the energy at the beginning of each stroke reduced, by the introduction of an air vessel at the pump discharge. This consists of a sealed vessel, which contains air at the top and liquid at the bottom. When the delivery stroke commences, liquid is pumped into the air vessel and the air is compressed. When the discharge from the pump decreases towards the end of the stroke, the pressure in the air vessel is sufficiently high for some of the liquid to be expelled into the delivery line. If the air vessel is sufficiently large and is close to the pump, the velocity of the liquid in the delivery line can be maintained approximately constant. The frictional losses are also reduced by the incorporation of an air vessel because the friction loss under turbulent conditions is approximately proportional to the linear velocity in the pipe raised to the power

1.8; i.e. the reduced friction losses during the period of minimum discharge do not compensate for the greatly increased losses when the pump is delivering at maximum rate (see Section 8.6). Further, the maximum stresses set up in the pump are reduced by the use of an air vessel.

Air vessels are also incorporated in the suction line for a similar reason. Here they may be of even greater importance because the pressure drop along the suction line is necessarily limited to rather less than 1 atm if the suction tank is at atmospheric pressure. The flowrate may be limited if part of the pressure drop available must be utilised in accelerating the fluid in the suction line; the air vessel should therefore be sufficiently large for the flowrate to be maintained approximately constant.

The plunger or ram pump

This pump is the same in principle as the piston type but differs in that the gland is at one end of the cylinder making its replacement easier than with the standard piston type. The sealing of the piston and ram pumps has been much improved but, because of the nature of the fluids frequently used, care in selecting and maintaining the seal is very important. The piston or ram pump may be used for injections of small quantities of inhibitors to polymerisation units or of corrosion inhibitors to high-pressure systems, and also for boiler feed water applications.

The diaphragm pump

The diaphragm pump has been developed for handling corrosive liquids and those containing suspensions of abrasive solids. It is in two sections separated by a diaphragm of rubber, leather, or plastics material. In one section, a plunger or piston operates in a cylinder in which a noncorrosive fluid is displaced. The movement of the fluid is transmitted by means of the flexible diaphragm to the liquid to be pumped. The only moving parts of the pump that are in contact with the liquid are the valves, and these can be specially designed to handle the material. In some cases, the movement of the diaphragm is produced by direct mechanical action, or the diaphragm may be air actuated as shown in Fig. 8.3, in which case a particularly simple and inexpensive pump results, capable of operating up to 0.2 MN/m^2.

When pumping non-Newtonian fluids, difficulties are sometimes experienced in initiating the flow of pseudoplastic and viscoplastic materials. Positive displacement pumps can overcome the problem and the diaphragm pump in particular is useful in dealing with agglomerates in suspension. Care must always be taken to ensure that the safe working pressure for the pump is not exceeded. This can be achieved conveniently by using a hydraulic drive for a diaphragm pump, equipped with a pressure limiting relief valve, which ensures that no damage is done to the system, as shown in Fig. 8.4.

By virtue of their construction, diaphragm pumps cannot be used for high-pressure applications. In the Mars pump, there is no need for a diaphragm as the working fluid (oil), of lower density than the liquid to be pumped, forms an interface with it in a vertical chamber.

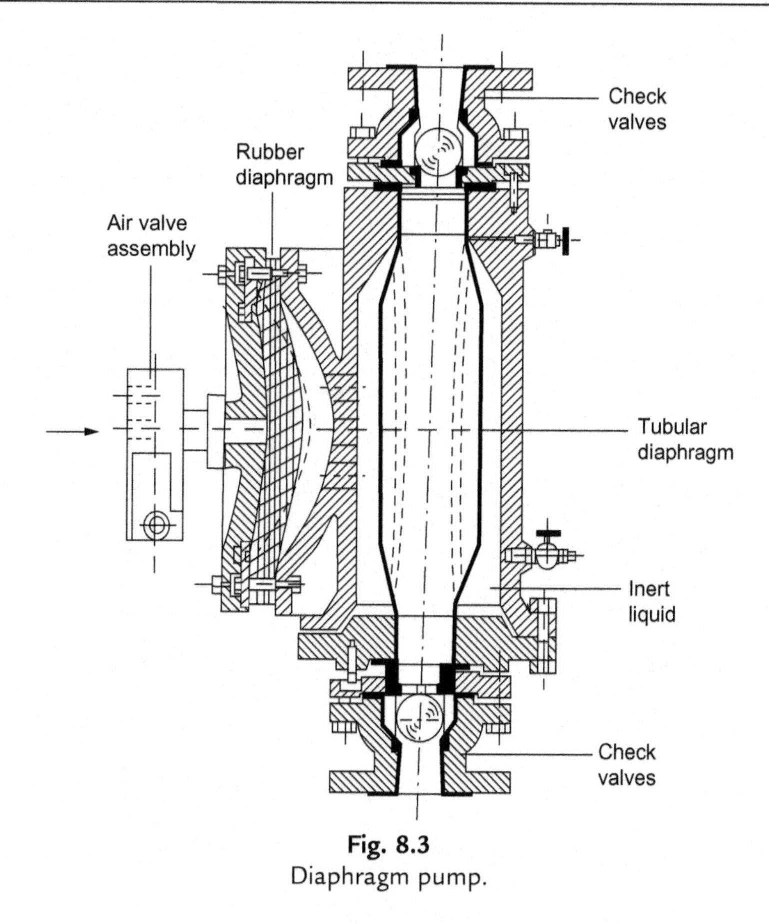

Fig. 8.3
Diaphragm pump.

The pump, which is used extensively for concentrated slurries, is really a development of the old Ferrari's acid pump which was designed for corrosive liquids.

The metering pump

Metering pumps are positive displacement pumps driven by constant speed electric motors. They are used where a constant and accurately controlled rate of delivery of a liquid is required, and they will maintain this constant rate irrespective of changes in the pressure against which they operate. The pumps are usually of the plunger type for low throughput and high-pressure applications; for large volumes and lower pressures a diaphragm is used. In either case, the rate of delivery is controlled by adjusting the stroke of the piston element, and this can be done whilst the pump is in operation. A single-motor driver may operate several individual pumps and in this way give control of the actual flows and of the flow ratio of several streams at the same time. The output may be controlled from zero to maximum flowrate, either manually on the pump or remotely. These pumps may be used for the dosing of works effluents and water supplies, and the feeding of reactants, catalysts, or inhibitors to

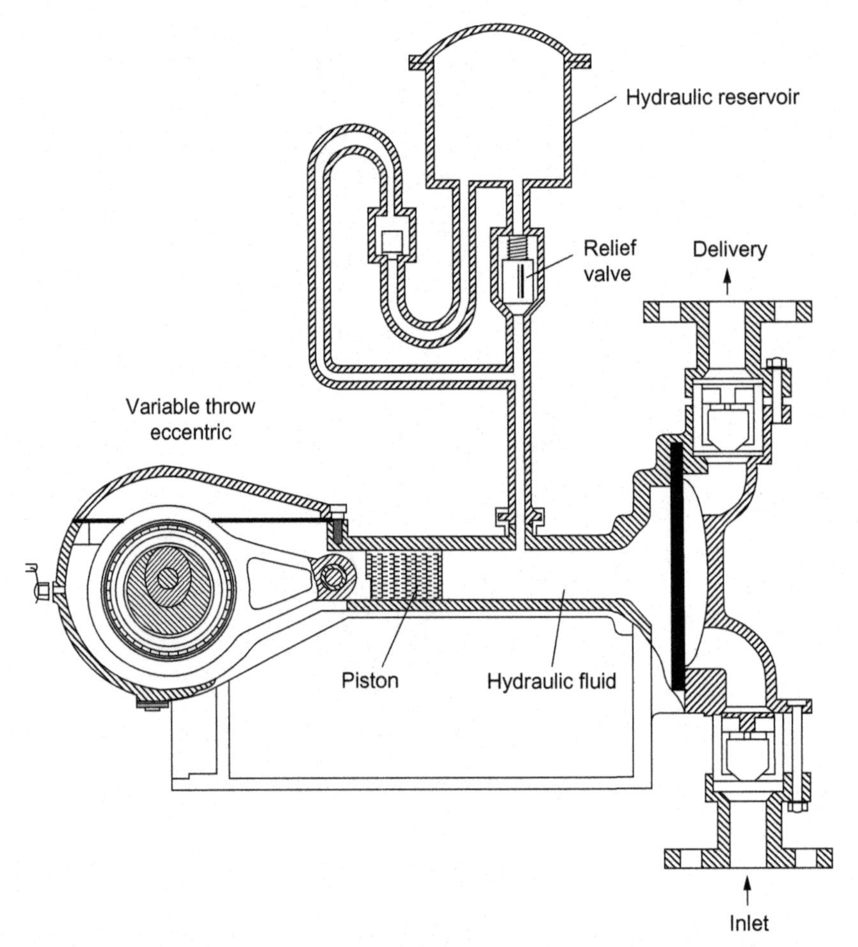

Fig. 8.4
A hydraulic drive to protect a positive displacement pump.

reactors at controlled rates, and although a simple method for controlling flowrate is provided, high precision standards of construction are required.

Example 8.1

A single-acting reciprocating pump has a cylinder diameter of 110 mm and a stroke of 230 mm. The suction line is 6 m long and 50 mm in diameter and the level of water in the suction tank is 3 m below the cylinder of the pump. What is the maximum speed at which the pump can run without an air vessel if separation is not to occur in the suction line? The piston undergoes approximately simple harmonic motion. Atmospheric pressure is equivalent to a head of 10.36 m of water and separation occurs at an absolute pressure corresponding to a head of 1.20 m of water.

Solution

The tendency for separation to occur will be greatest at:

(a) the inlet to the cylinder because here the static pressure is a minimum and the head required to accelerate the fluid in the suction line is a maximum;
(b) the commencement of the suction stroke because the acceleration of the piston is then a maximum.

If the maximum permissible speed of the pump is N Hz:

Angular velocity of the driving mechanism $= 2\pi N$ radians/s.

Acceleration of piston $= 0.5 \times 0.230(2\pi N)^2 \cos(2\pi Nt)$ m/s^2.

Maximum acceleration (when $t = 0$) $= 4.54\ N^2$ m/s^2.

Maximum acceleration of the liquid in the suction pipe

$$= \left(\frac{0.110}{0.05}\right)^2 \times 4.54N^2 = 21.97N^2\,\text{m/s}^2$$

Accelerating force acting on the liquid

$$= 21.97N^2 \frac{\pi}{4}(0.050)^2 \times (6 \times 1000)\text{N}$$

Pressure drop in suction line due to acceleration $= 21.97\ N^2 \times 6 \times 1000$ N/m^2

$$= 1.32 \times 10^5 N^2\,\text{N/m}^2$$

or:

$$\frac{(1.32 \times 10^5 N^2)}{(1000 \times 9.81)} = 13.44N^2\ \text{m water}$$

Pressure head at cylinder when separation is about to occur,

$$1.20 = \left(10.36 - 3.0 - 13.44N^2\right)\ \text{m water}$$

$$\therefore \underline{\underline{N = 0.675\,\text{Hz}}}$$

8.2.2 Positive-Displacement Rotary Pumps

The gear pump and the lobe pump

Gear and lobe pumps operate on the principle of using mechanical means to transfer small elements or "packages" of fluid from the low-pressure (inlet) side to the high-pressure (delivery) side. There is a wide range of designs available for achieving this end. The general characteristics of the pumps are similar to those of reciprocating piston pumps, but the delivery is more even because the fluid stream is broken down into so much smaller elements. The pumps are capable of delivering to a high pressure, and the pumping rate is approximately proportional to the speed of the pump and is not greatly influenced by the pressure against which it is delivering. Again, it is necessary to provide a pressure relief system to ensure that the safe operating pressure is not exceeded. Recent developments in the use of pumps of this type have been described by Harvest.[5]

One of the commonest forms of the pump is the *gear pump* in which one of the gear wheels is driven and the other turns as the teeth engage; two versions are illustrated in Figs. 8.5 and 8.6. The liquid is carried around in the spaces in between the consecutive gear teeth and the outer casing of the pump, and the seal between the high and low-pressure sides of the pump is formed as the gears come into mesh and the elements of fluid are squeezed out. Gear pumps are extensively used for both high-viscosity Newtonian liquids and non-Newtonian fluids. The *lobe-pump* (Figs. 8.7 and 8.8) is similar, but the gear teeth are replaced by two or three lobes and both axles are driven; it is therefore possible to maintain a small clearance between the lobes, and wear is somewhat reduced.

The cam pump

A rotating cam is mounted eccentrically in a cylindrical casing and a very small clearance is maintained between the outer edge of the cam and the casing. As the cam rotates, it expels liquid from the space ahead of it and sucks in liquid behind it. The delivery and suction sides of the pump are separated by a sliding valve which rides on the cam. The characteristics again are similar to those of the gear pump.

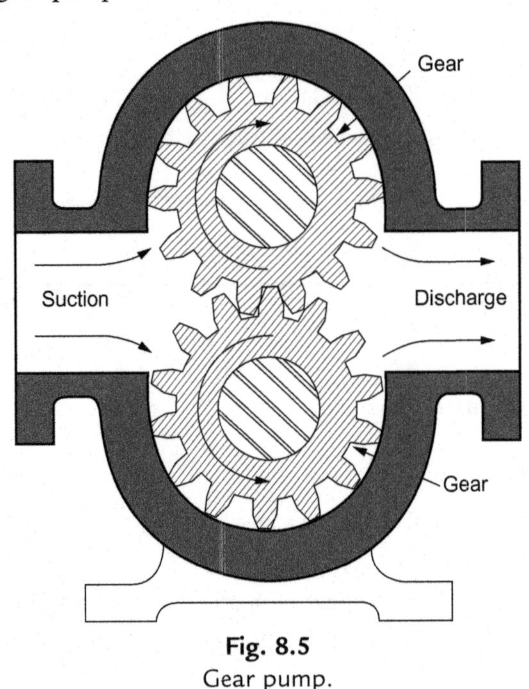

Fig. 8.5
Gear pump.

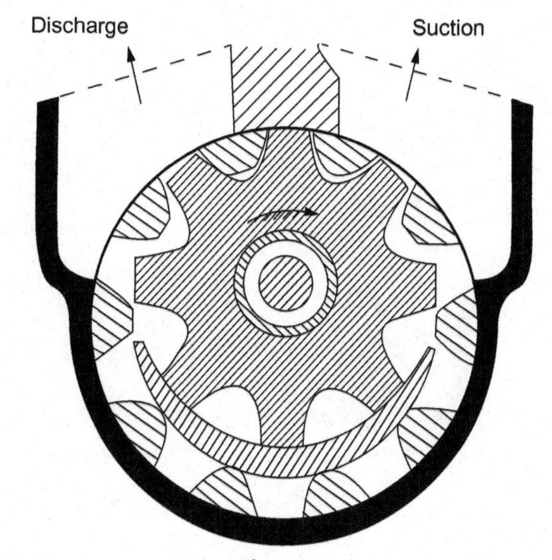

Fig. 8.6
Internal gear pump.

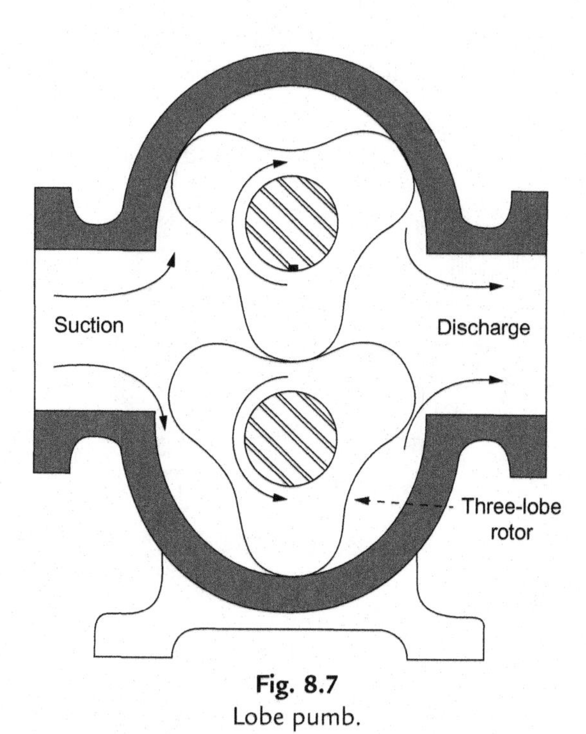

Fig. 8.7
Lobe pumb.

Fig. 8.8
Lobe pump.

The vane pump

The rotor of the vane pump is mounted off centre in a cylindrical casing (Fig. 8.9). It carries rectangular vanes in a series of slots arranged at intervals around the curved surface of the rotor. The vanes are thrown outwards by centrifugal action and the fluid is carried in the spaces

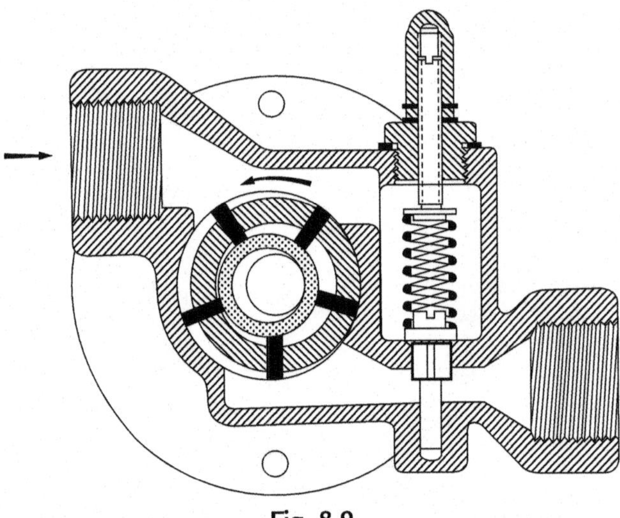

Fig. 8.9
Vane pump.

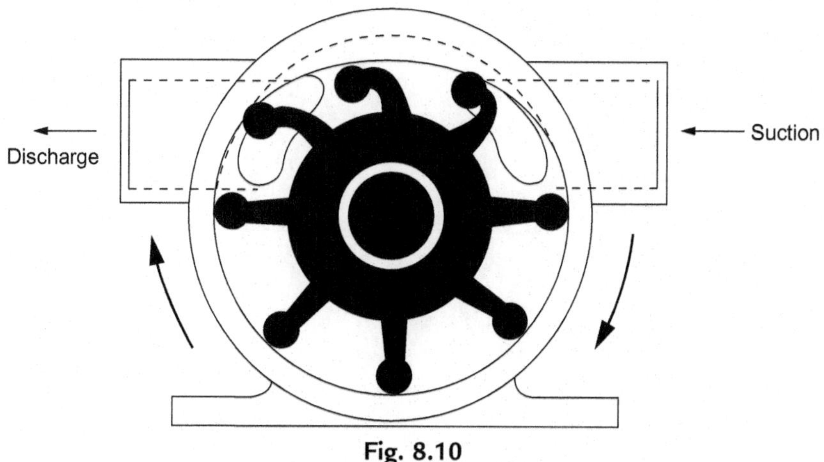

Fig. 8.10
Flexible vane pump.

bounded by adjacent vanes, the rotor, and the casing. Most of the wear is on the vanes and these can readily be replaced.

The flexible vane pump

The pumps described above will not handle liquids containing solid particles in suspension, and the flexible vane pumps have been developed to overcome this disadvantage. In this case, the rotor (Fig. 8.10) is an integral elastomer moulding of a hub with flexible vanes which rotates in a cylindrical casing containing a crescent-shaped block, as in the case of the *internal gear pump*.

The flow inducer or peristaltic pump

This is a special form of pump in which a length of silicone rubber or other elastic tubing, typically of 3 to 25 mm diameter, is compressed in stages by means of a rotor as shown in Fig. 8.11. The tubing is fitted to a curved track mounted concentrically with a rotor carrying three rollers. As the rollers rotate, they flatten the tube against the track at the points of contact. These "flats" move the fluid by positive displacement, and the flow can be precisely controlled by the speed of the motor.

These pumps have been particularly useful for biological fluids where all forms of contact must be avoided. They are being increasingly used and are suitable for pumping emulsions, creams, and similar fluids in laboratories and small plants where the freedom from glands, avoidance of aeration, and corrosion resistance are valuable, if not essential. It is now possible to use[4] thick-wall, reinforced moulded tubes which give a pumping performance of up to 0.02 m^3/s at 1 MN/m^2 and these pumps have been further discussed by Gadsden,[6] Stepanoff[7] and Volk.[8] The control is such that these pumps may conveniently be used as metering pumps for dosage processes.

Fig. 8.11
Flow inducer.

The mono pump

Another example of a positive acting rotary pump is the single screw-extruder pump typified by the Mono pump, illustrated in Fig. 8.12, in which a specially shaped helical metal rotor revolves eccentrically within a double-helix, resilient rubber stator of twice the pitch length of the metal rotor. A continuous forming cavity is created as the rotor turns—the cavity

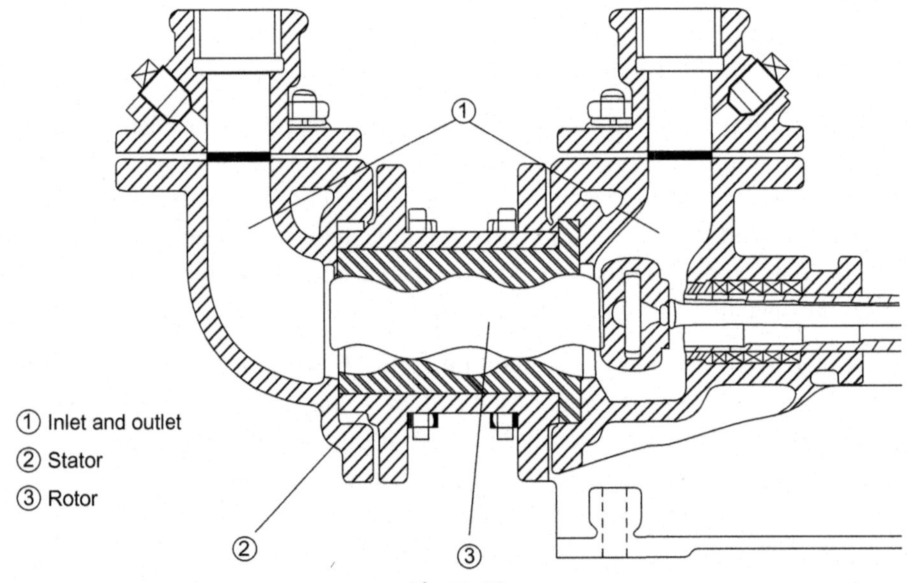

Fig. 8.12
Mono pump.

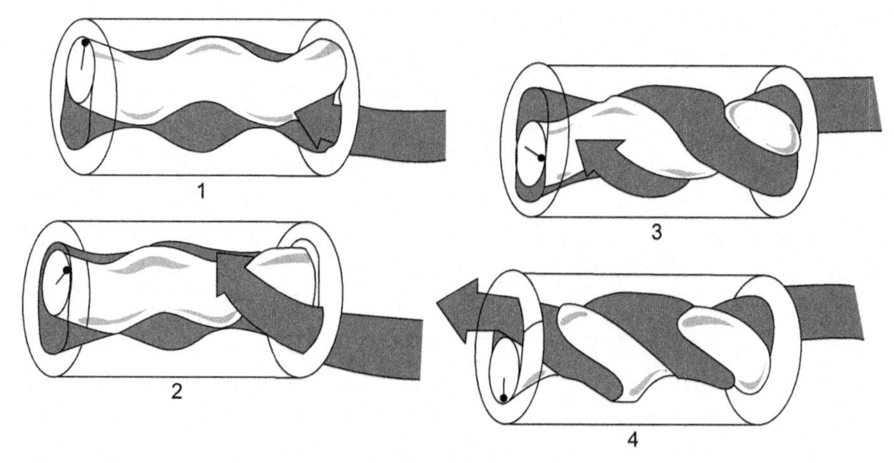

Fig. 8.13
Schematics of the mode of operation of a Mono pump.

progressing towards the discharge, advancing in front of a continuously forming seal line and thus carrying the pumped material with it as shown in Fig. 8.13.

The Mono pump gives a uniform flow and is noiseless in operation. It will pump against high pressures; the higher the required pressure, the longer are the stator and the rotor and the greater the number of turns. The pump can handle corrosive and gritty liquids and is extensively used for feeding slurries to filter presses. It must never be run dry. The Mono Merlin Wide Throat pump is used for highly viscous liquids (Fig. 8.14).

Screw pumps

A most important class of pump for dealing with highly viscous material is represented by the screw extruder used in the polymer industry. Extruders find their main application in the manufacture of simple and complex sections (rods, tubes, beadings, curtain rails, rainwater

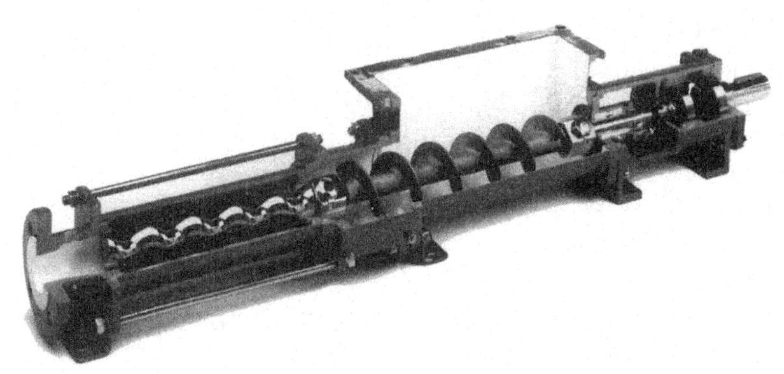

Fig. 8.14
Mono Merlin Wide-Throat pump.

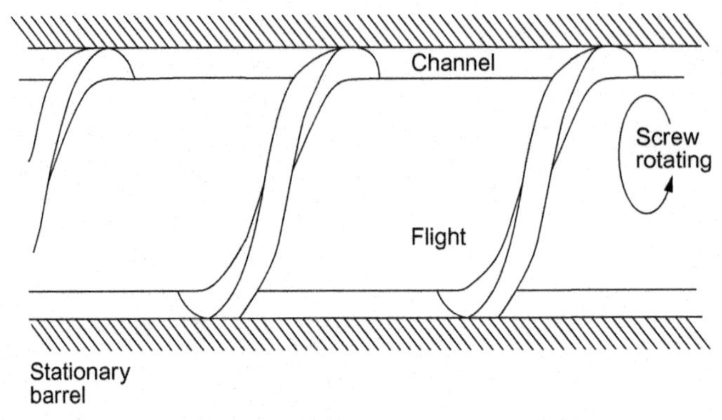

Fig. 8.15
Section of a screw pump.

gutterings and a multitude of other shapes). However, the shape of section produced in a given material is dependent only on the profile of the hole (known as die) through which the fluid is pushed just before it cools and solidifies. The screw pump is of more general application and will be considered first.

The underlying principle is shown in Fig. 8.15. The fluid is sheared in the channel between the screw and the wall of the barrel. The mechanism that generates the pressure can be visualised in terms of a model consisting of an open channel covered by a moving plane surface (Fig. 8.16). This representation of a screw pump takes as the frame of reference a stationary screw with rotating barrel. The planar simplification is not unreasonable, provided that the depth of the screw channel is small with respect to the barrel diameter. It should also be recognised that the distribution of centrifugal forces would be different according to whether the rotating member is the wall or the screw: this distinction would have to be drawn if a detailed force balance were to be attempted, but in any event the centrifugal (inertial) forces are generally far smaller than the viscous forces in most such applications.

If the upper plate moved along in the direction of the channel, then a velocity profile would be established that would approximate to the linear velocity gradient that exists between the planar walls moving parallel to each other. If it moved at right angles to the channel axis, however, a circulation would be developed in the gap, as shown in Fig. 8.17. In fact, the

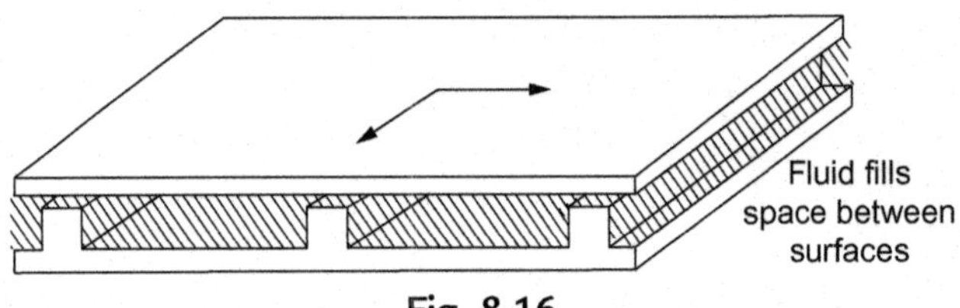

Fig. 8.16
Planar model of part of a screw pump.

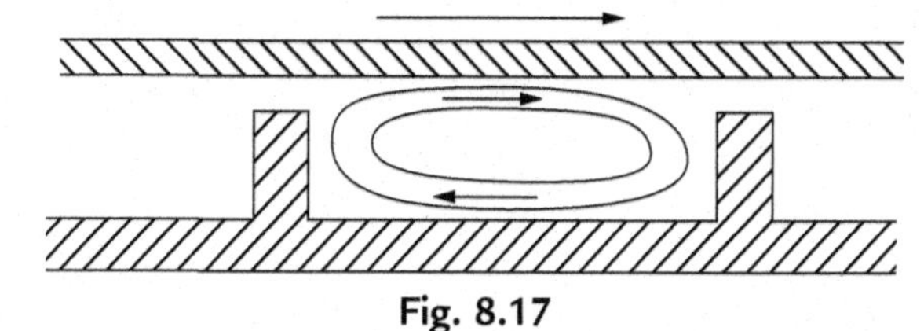

Fig. 8.17
Fluid displacement resulting from movement of plane surface.

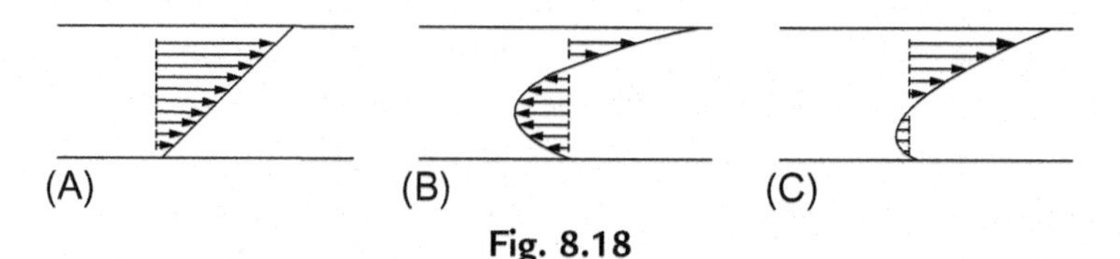

Fig. 8.18
Velocity profile produced between screw pump surfaces (A) with no restriction on fluid flow (B) with no net flow (total restriction) (C) with a partially restricted discharge.

relative movement of the barrel wall is somewhere in between, and is determined by the pitch of the screw. The fluid path in a screw pump is therefore of a complicated helical form within the channel section. The nature of the velocity components along the channel depends on the pressure generated and the amount of resistance at the discharge end. If there is no resistance, the velocity distribution in the channel direction will be the *Couette simple shear* profile shown in Fig. 8.18A. With a totally closed discharge end, the net flow would be zero, but the velocity components at the walls would not be affected. As a result, the flow field necessarily would be of the form shown in Fig. 8.18B.

Viscous forces within the fluid will always prevent a completely unhindered discharge, but in extrusion practice an additional die head resistance is used to generate backflow and mixing, so that a more uniform product is obtained. The flow profile along the channel is then of some intermediate form, such as that shown in Fig. 8.18C.

It must be emphasised here that the flow in a screw pump is produced as a result of viscous forces. Pressures achieved with low viscosity materials are negligible. The screw pump is *not* therefore a modification of the Archimedes screw used in antiquity to raise water—that was essentially a positive displacement device using a deep cut helix mounted at an angle to the horizontal, and not running full. If a detailed analysis of the flow in a screw pump is to be carried out, then it is also necessary to consider the small but finite leakage flow that can occur between the flight and the wall. With the large pressure generation in a polymer extruder, commonly 100 bar (10^7 N/m^2), the flow through this gap, which is typically about 2% of the barrel internal diameter, can be significant. The pressure drop over a single pitch length may be of the order of 10 bar (10^6 N/m^2), and this will force fluid through the gap. Once in this region, the viscous fluid is subject to a high rate of shear (the rotation speed of the screw is often about 2 Hz), and an appreciable part of the total viscous heat generation occurs in this region of an extruder.

More detailed discussions as well as accounts of new developments in the field of positive displacement pumps and their performance are available in the literature.[7–11]

8.2.3 The Centrifugal Pump

The centrifugal pump is by far the most widely used type in the chemical and petroleum industries. It will pump liquids with very wide-ranging properties and suspensions with high solids content including, for example, cement slurries, and may be constructed from a very wide range of corrosion resistant materials. The whole pump casing may be constructed from plastics such as polypropylene or it may be fitted with a corrosion-resistant lining. Because it operates at high speed, it may be directly coupled to an electric motor, or it may be coupled to a variable frequency drive, and it will give a high flowrate for its size.

In this type of pump (Fig. 8.19), the fluid is fed to the centre of a rotating impeller and is thrown outward by centrifugal action. As a result of the high speed of rotation, the liquid acquires a high kinetic energy and the pressure difference between the suction and delivery sides arises from the interconversion of kinetic and pressure energy.

The impeller (Fig. 8.20) consists of a series of curved vanes so shaped that the flow within the pump is as smooth as possible. The greater the number of vanes on the impeller, the greater is the control over the direction of motion of the liquid and hence smaller are the losses due to turbulence and circulation between the vanes. In the open impeller, the vanes are fixed to a central hub, whereas in the closed type the vanes are held between two supporting plates and leakage across the impeller is reduced. As will be seen later, the angle of the tips of the blades very largely determines the operating characteristics of the pump.

Fig. 8.19
Section of centrifugal pump.

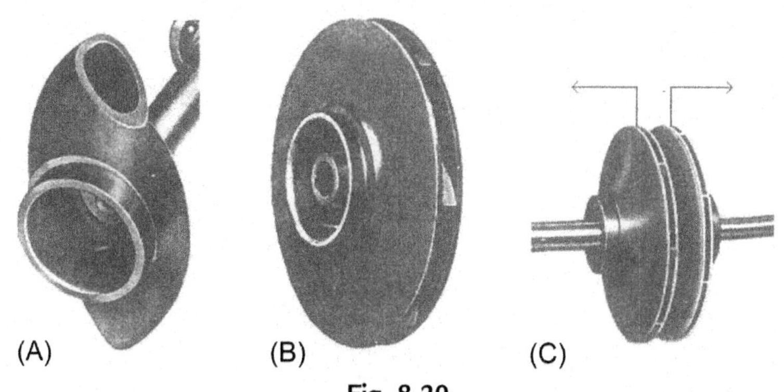

Fig. 8.20
Types of impeller (A) for pumping suspensions (B) standard closed impeller (C) double impeller.

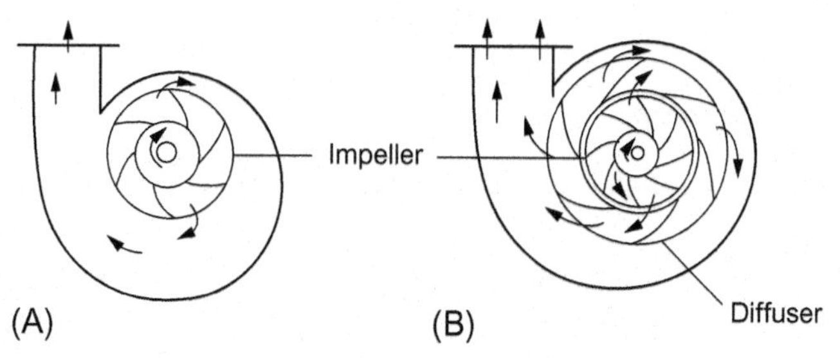

Fig. 8.21
Radial flow pumps (A) with volute (B) with diffuser vanes.

The liquid enters the casing of the pump, normally in an axial direction, and is picked up by the vanes of the impeller. In the simple type of centrifugal pump, the liquid discharges into a volute, a chamber of gradually increasing cross-section with a tangential outlet. A volute type of pump is shown in Fig. 8.21A. In the turbine pump (Fig. 8.21B), the liquid flows from the moving vanes of the impeller through a series of fixed vanes forming a diffusion ring. This gives a more gradual change in direction to the fluid and more efficient conversion of kinetic energy into pressure energy than that is obtained with the volute type. The angle of the leading edge of the fixed vanes should be such that the fluid is received without shock. The liquid flows along the surface of the impeller vane with a certain velocity whilst the tip of the vane is moving relative to the casing of the pump. The direction of motion of the liquid relative to the pump casing—and the required angle of the fixed vanes—is found by compounding these two velocities. In Fig. 8.22, u_v is the velocity of the liquid relative to the vane and u_t is the tangential velocity of the tip of the vane; compounding these two velocities gives the resultant velocity u_2 of the liquid. It is apparent, therefore, that the required vane angle in the diffuser is dependent on the throughput, the speed of rotation, and the angle of the impeller blades. The pump will therefore operate at maximum efficiency only over a narrow range of conditions.

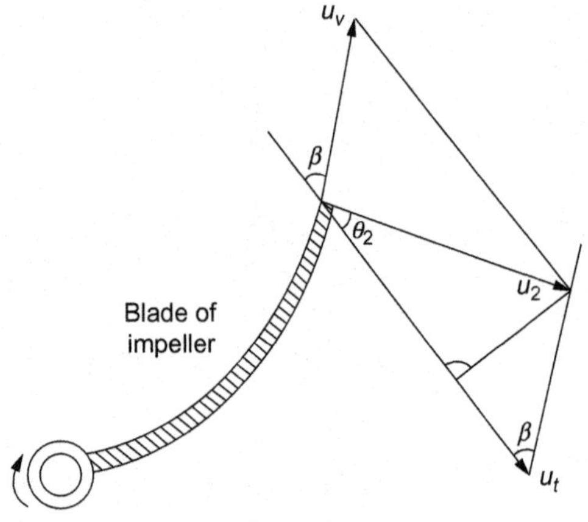

Fig. 8.22
Velocity diagram.

Virtual head of a centrifugal pump

The maximum pressure is developed when the whole of the excess kinetic energy of the fluid is converted into pressure energy. As indicated below, the head is proportional to the square of the radius and to the speed, and is of the order of 60 m for a single-stage centrifugal pump; for higher pressures, multistage pumps must be used. The liquid which is rotating at a distance of between r and $r + dr$ from the centre of the pump (Fig. 8.23) has a mass dM given by $2\pi r\ drb\rho$, where ρ is the density of the fluid and b is the width of the element of fluid.

If the fluid is travelling with a velocity u and at an angle θ to the tangential direction, the angular momentum of this mass of fluid

$$= dM(ur \cos \theta)$$

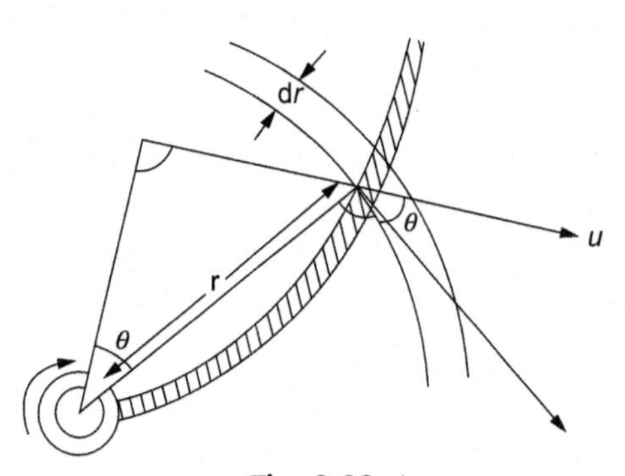

Fig. 8.23
Virtual head.

The torque acting on the fluid $d\tau$ is equal to the rate of change of angular momentum with time, as it goes through the pumps or:

$$\begin{aligned} d\tau &= dM\frac{\partial}{\partial t}(ur\cos\theta) \\ &= 2\pi rb\rho dr\frac{\partial}{\partial t}(ur\cos\theta) \end{aligned} \tag{8.3}$$

The volumetric rate of flow of liquid through the pump:

$$Q = 2\pi rb\frac{\partial r}{\partial t} \tag{8.4}$$

or:

$$d\tau = Q\rho d(ur\cos\theta) \tag{8.5}$$

The total torque acting on the liquid in the pump is therefore obtained by integrating $d\tau$ between the limits denoted by suffix 1 and suffix 2, where suffix 1 refers to the conditions at the inlet to the pump and suffix 2 refers to the conditions at the discharge.

Thus:

$$\tau = Q\rho(u_2r_2\cos\theta_2 - u_1r_1\cos\theta_1) \tag{8.6}$$

The power **P** developed by the pump is equal to the product of the torque and the angular velocity ω:

$$\therefore \mathbf{P} = Q\rho\omega(u_2r_2\cos\theta_2 - u_1r_1\cos\theta_1) \tag{8.7}$$

The power can also be expressed as the product Ghg, where G is the mass rate of flow of liquid through the pump (also given by ρQ, g is the acceleration due to gravity, and h is termed the virtual head developed by the pump.

Thus:

$$Ghg = Q\rho\omega(u_2r_2\cos\theta_2 - u_1r_1\cos\theta_1)$$

and:

$$h = \frac{\omega(u_2r_2\cos\theta_2 - u_1r_1\cos\theta_1)}{g} \tag{8.8}$$

Since u_1 will be approximately zero, the virtual head is then obtained as:

$$h = \frac{\omega u_2r_2\cos\theta_2}{g} \tag{8.9}$$

where g, ω, and r_2 are known in any given instance, and u_2 and θ_2 are to be expressed in terms of the known quantities.

From the geometry of Fig. 8.22

$$u_y \sin \beta = u_2 \sin\theta_2 \tag{8.10}$$

and:

$$u_t = u_y \cos\beta + u_2 \cos\theta_2 \tag{8.11}$$

(where β is the angle between the tip of the blade of the impeller and the tangent to the direction of its motion. If the blade curves backwards, β lies between 0 and $\pi/2$ and if it curves forwards, β lies between $\pi/2$ and π).

The volumetric rate of flow through the pump Q is equal to the product of the area available for flow at the outlet of the impeller and the radial component of the velocity, or

$$Q = 2\pi r_2 b u_2 \sin\theta_2$$

$$= 2\pi r_2 b u_y \sin\beta \quad \text{(from Eq.8.10)} \tag{8.12}$$

$$\therefore\ u_y = \frac{Q}{2\pi r_2 b \sin\beta} \tag{8.13}$$

$$\text{Thus: } h = \frac{\omega r_2 (u_t - u_y \cos\beta)}{g} \quad \text{(from Eqs 8.9 and 8.11)}$$

$$= \frac{\omega}{g} r_2 \left(r_2\omega - \frac{Q}{2\pi r_2 b \tan\beta} \right) \quad (\text{since } u_t = r_2\omega)$$

$$= \frac{r_2^2\omega^2}{g} - \frac{Q\omega}{2\pi b g \tan\beta} \tag{8.14}$$

The virtual head developed by the pump is therefore independent of the density of the fluid, and the pressure will thus be directly proportional to the density. For this reason, a centrifugal pump needs priming. If the pump is initially full of air, the pressure developed is reduced by a factor equal to the ratio of the density of air to that of the liquid, and is insufficient to drive the liquid through the delivery pipe.

For a given speed of rotation, there is a linear relation between the head developed and the rate of flow (Eq. (8.14)). If the tips of the blades of the impeller are inclined backwards, β is less than $\pi/2$, tan β is positive, and therefore the head decreases as the throughput increases. If β is greater than $\pi/2$ (i.e. the tips of the blades are inclined forwards), the head increases as the delivery increases. The angle of the blade tips therefore profoundly affects the performance and characteristics of the pump. For radial blades, the head should be independent of the throughput.

Specific speed

If θ_2 remains approximately constant, u_v, u_t, and u_2 will all be directly proportional to one another, and since $u_t = r_2\omega$, these velocities are proportional to r_2; thus u_v will vary as $r_2\omega$.

The output from a pump is a function of its linear dimensions, the shape, number, and arrangement of the impellers, the speed of rotation, and the head against which it is operating. From Eq. (8.12), for a radial pump with $\beta = \pi/2$ and $\sin\beta = 1$:

$$\begin{aligned} Q &\propto 2\pi r_2 b u_v \\ Q &\propto 2\pi r_2 b r_2 \omega \\ Q &\propto r_2^2 b\omega \end{aligned} \tag{8.15}$$

When $\tan\beta = \tan\pi/2 = \infty$, then from Eq. (8.14):

$$h = \frac{r_2^2\omega^2}{g} \tag{8.16}$$

$$gh = r_2^2\omega^2$$

or:

$$(gh)^{3/4} = r_2^{3/2}\omega^{3/2} \tag{8.17}$$

For a series of geometrically similar pumps, b is proportional to the radius r_2, and thus, from Eq. (8.15):

$$Q \propto r_2^3\omega \tag{8.18}$$

or:

$$Q^{1/2} \propto r_2^{3/2}\omega^{1/2} \tag{8.19}$$

Eliminating r_2 between Eqs (8.17) and (8.19):

$$\frac{Q^{1/2}}{(gh)^{3/4}} = \frac{\omega^{1/2}}{\omega^{3/2}}$$

or:

$$\frac{\omega Q^{1/2}}{(gh)^{3/4}} = \text{constant} = N_s \text{ for geometrically similar pumps} \tag{8.20}$$

Criteria for similarity

The dimensionless quantity $\omega Q^{1/2}/(gh)^{3/4}$ is a characteristic for a particular type of centrifugal pump, and noting that the angular velocity is proportional to the speed N, this group may be rewritten as:

$$N_s = \frac{NQ^{1/2}}{(gh)^{3/4}} \tag{8.21}$$

which is constant for geometrically similar pumps. N_s is defined as the specific speed and is frequently used to classify types of centrifugal pumps. Specific speed may be defined as the speed of the pump which will produce unit flow Q against unit head h under conditions of maximum efficiency.

Eq. (8.21) is dimensionless but specific speed is frequently quoted in the form:

$$N_s = \frac{NQ^{1/2}}{h^{3/4}} \tag{8.22}$$

where the impeller speed N is in rpm, the volumetric flowrate Q in US gpm and the total head developed is in ft. In this form, specific speed has dimensions of $(\mathbf{LT}^{-2})^{3/4}$ and for centrifugal pumps has values between 400 and 10,000 depending on the type of impeller. The use of specific speed in pump selection is discussed more fully in Volume 6.

Operating characteristics

The operating characteristics of a pump are conveniently shown by plotting the head h, power $\mathbf{P}$, and efficiency η against the flow Q as shown in Fig. 8.24. It is important to note that the efficiency reaches a maximum and then falls, whilst the head at first falls slowly with Q but eventually falls off rapidly. The optimum conditions for operation are shown as the duty point, i.e. the point where the head curve cuts the ordinate through the point of maximum efficiency.

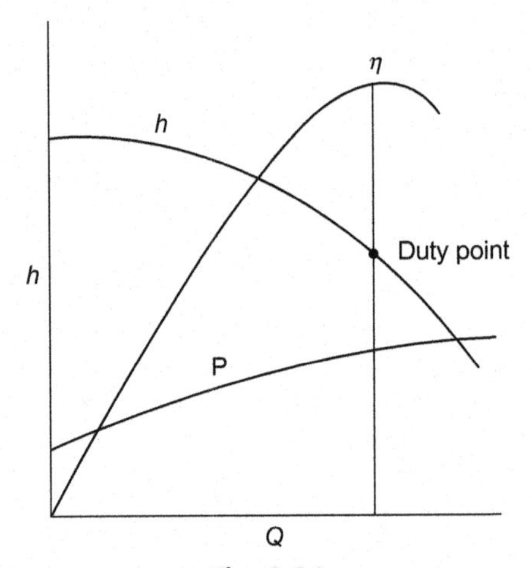

Fig. 8.24
Radial flow pump characteristics.

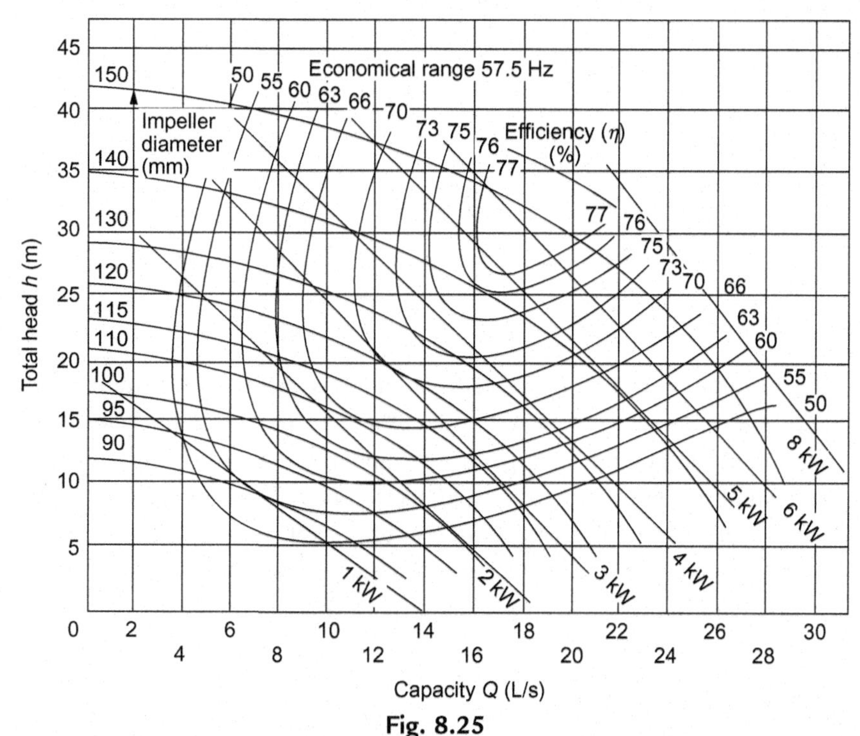

Fig. 8.25
Characteristic curves for a centrifugal pump.

A set of curves for h, η, and $\mathbf{P}$ as a function of Q are shown in Fig. 8.25 from which it is seen that when the pump is operating near optimum conditions, its efficiency remains reasonably constant over a wide range of flowrates. A more general indication of the variation of efficiency with specific speed is shown in Fig. 8.26 for different types of centrifugal pumps. The power developed by a pump is proportional to $Qgh\rho$:

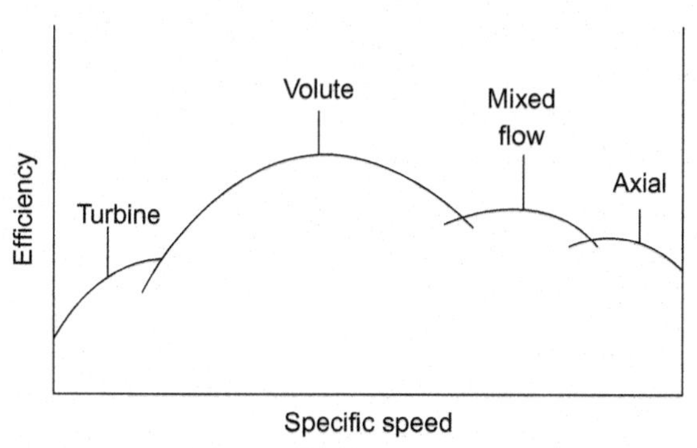

Fig. 8.26
Specific speed and efficiency.

i.e.:

$$\mathbf{P} \propto r_2^2 \omega r_2^2 \omega^2 \rho b$$

or:

$$\mathbf{P} \propto r_2^4 b \omega^3 \rho \tag{8.23}$$

so that $Q \propto \omega$; $h \propto \omega^2$; and $\mathbf{P} \propto \omega^3$, from Eqs (8.15), (8.16), and (8.23).

Cavitation

In designing any installation in which a centrifugal pump is used, careful attention must be paid to check the minimum pressure which will arise at any point. If this pressure is less than the vapour pressure at the pumping temperature, vapourisation will occur and the pump may not be capable of developing the required suction head. Moreover, if the liquid contains gases, these may come out of solution giving rise to pockets of gas. This phenomenon is known as *cavitation* and may result in mechanical damage to the pump as the bubbles collapse. The tendency for cavitation to occur is accentuated by any sudden changes in the magnitude or direction of the velocity of the liquid in the pump. The onset of cavitation is accompanied by a marked increase in noise and vibration as the vapour bubbles collapse, and also a loss of head.

Suction head

Pumps may be arranged so that the inlet is under a suction head or the pump may be fed from a tank. These two systems alter the duty point curves as shown in Fig. 8.27. In developing such curves, the normal range of liquid velocities is 1.5–3 m/s, but lower values are used for pump suction lines. With the arrangement shown in Fig. 8.27A, there can be problems in priming the pump and it may be necessary to use a self-priming centrifugal pump.

For any pump, the manufacturers specify the minimum value of the *net positive suction head* (NPSH), which must exist at the suction point of the pump. The NPSH is the amount by which the pressure at the suction point of the pump, expressed as a head of the liquid to be pumped, must exceed the vapour pressure of the liquid at the operating temperature. For any installation this must be calculated, taking into account the absolute pressure of the liquid, the level of the pump, and the velocity and friction heads in the suction line. The NPSH must allow for the fall in pressure occasioned by the further acceleration of the liquid as it flows on to the impeller and for irregularities in the flow pattern in the pump. If the required value of NPSH is not obtained, partial vapourisation or liberation of dissolved gas is liable to occur, with the result that both suction head and delivery head may be reduced. The loss of suction head is more important because it may cause the pump to be starved of liquid.

In the system shown in Fig. 8.28, the pump is taking liquid from a reservoir at an absolute pressure P_0 in which the liquid level is at a height h_0 above the suction point of the pump. Then,

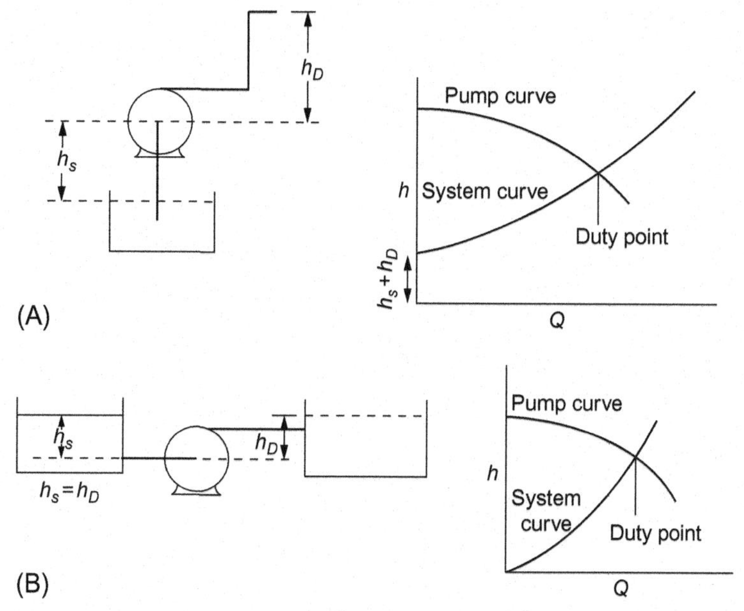

Fig. 8.27
Effect of suction head: (A) systems with suction lift and friction; (B) systems with friction losses only.

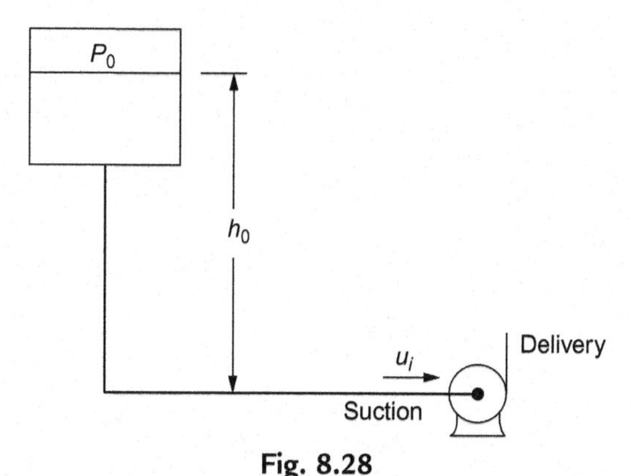

Fig. 8.28
Suction system of centrifugal pump.

if the liquid velocity in the reservoir is negligible, the *absolute* pressure head h_i at the suction point of the pump is obtained by applying the energy or momentum balance:

$$h_i = \frac{P_0}{\rho g} + h_0 - \frac{u_i^2}{2g} - h_f \tag{8.24}$$

where h_f is the head lost in friction, and u_i, is the velocity at the inlet of the pump. If the vapour pressure of the liquid is P_v, the NPSH (= Z) is given by the difference between the *total* head at the suction inlet and the head corresponding to the vapour pressure P_v of the liquid at the pump inlet.

$$Z = \left(h_i + \frac{u_i^2}{2g}\right) - \frac{P_v}{\rho g} \tag{8.25}$$

$$= \frac{P_0}{\rho g} - \frac{P_v}{\rho g} + h_0 - h_f \tag{8.26}$$

In Eq. (8.26), it is implicitly assumed that the kinetic head of the inlet liquid is available for conversion into pressure head. If this is not so, $u_i^2/2g$ must be deducted from the NPSH.

If cavitation and loss of suction head does occur, it can sometimes be overcome by increasing the pressure in the system, either by alteration of the layout to provide a greater hydrostatic pressure or a reduced pressure drop in the suction line. Sometimes, slightly closing the valve on the pump delivery or reducing the pump speed by a small amount may be effective. Generally, a small fast-running pump will require a larger NPSH than a larger slow-running pump.

The efficiency of a centrifugal pump and the head which it is capable of developing are dependent on achieving a good seal between the rotating shaft and the casing of the pump and inefficient operation is frequently due to a problem with the seal or with the gland. When the pump is fitted with the usual type of packed gland, maintenance costs are often very high, especially when organic liquids of low viscosity are being pumped. A considerable reduction in expenditure on maintenance can be effected at the price of a small increase in initial cost by fitting the pump with a mechanical seal, in which the sealing action is achieved as a result of contact between two opposing faces, one stationary and the other rotating. In Fig. 8.29, a mechanical seal is shown in position in a centrifugal pump. The stationary seat *A* is held in position by means of the clamping plate *D*. The seal is made with the rotating face on a carbon ring *B*. The seal between the rotating face and the shaft is made by means of the wedge ring *C*, usually made of polytetrafluoroethylene (PTFE). The drive is through the retainer *E* secured to the shaft, usually by Allen screws. Compression between the fixed and rotating faces is provided by the spiral springs *F*.

It is advantageous to ensure that the seal is fed with liquid which removes any heat generated at the face. In the illustration this is provided by the connection *G*.

Centrifugal pumps must be fitted with good bearings since there is a tendency for an axial thrust to be produced if the suction is taken only on one side of the impeller. This thrust can be balanced by feeding back a small quantity of the high-pressure liquid to a specially designed thrust bearing. By this method, the risk of air leaking into the pump at the gland—and reducing the pressure developed—is minimised. The glandless centrifugal pump, which is used extensively for corrosive liquids, works on a similar principle, and the use of pumps without glands has been increasing both in the nuclear power industry and in the chemical process industry. Such units are totally enclosed and lubrication is provided by the fluid handled. Fig. 8.30 gives the performance characteristics at a particular speed of one group of pumps of this type, and similar data are available for the selection of other types of centrifugal pumps.

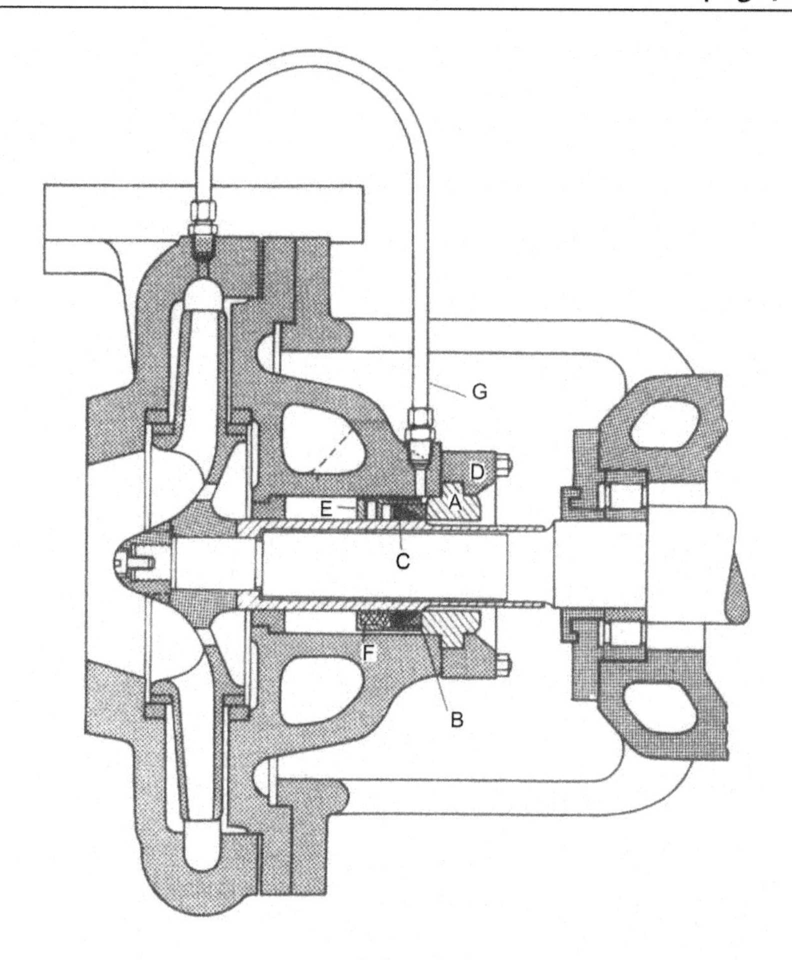

Fig. 8.29
Mechanical seal for centrifugal pump.

Centrifugal pumps are made of a wide range of materials, and in many cases the impeller and the casing are covered with resistant material. Thus stainless steel, nickel, rubber, polypropylene, stoneware, and carbon are all used. When the pump is used with suspensions, the ports and spaces between the vanes must be made sufficiently large to eliminate the risk of blockage. This does mean, however, that the efficiency of the pump is reduced. The *Vacseal pump*, developed by the International Combustion Company for pumping slurries, will handle suspensions containing up to 50% by volume of solids. The whole impeller may be rubber-covered and has three small vanes, as shown in Fig. 8.31. The back plate of the impeller has a second set of vanes of larger diameter. The pressure at the gland is thereby reduced below atmospheric pressure and below the pressure in the suction line; there is, therefore, no risk of the gritty suspension leaking into the gland and bearings. If leakage does occur, air will enter the pump. As mentioned previously, this may reduce the pressure which the pump can deliver, but this is preferable to damaging the bearings by allowing them to become

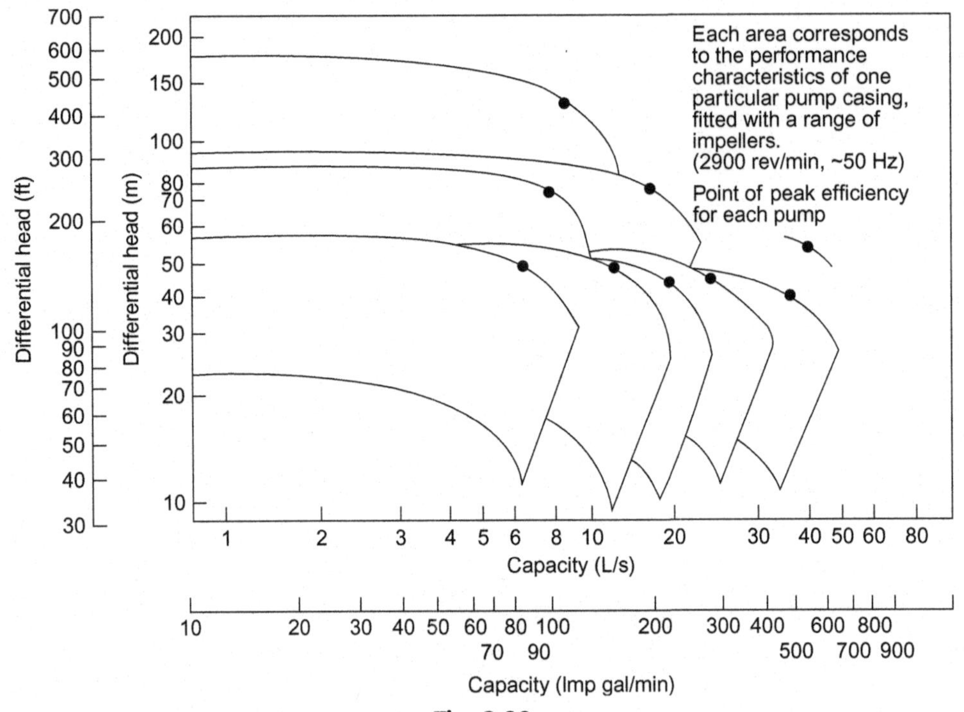

Fig. 8.30
Performance characteristics of a family of glandless centrifugal pumps.

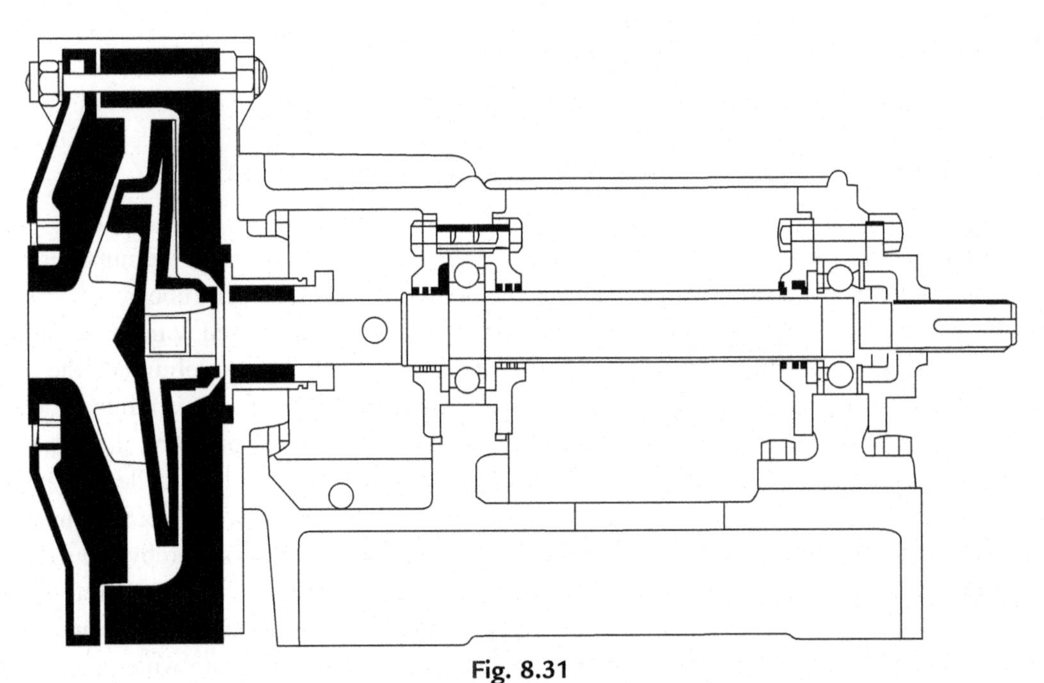

Fig. 8.31
Sectioned arrangement of a 38 mm V-type Vacseal pump with moulded rubber impeller.

contaminated with grit. This is another example for the necessity for tolerating rather low efficiencies in pumps that handle difficult materials. The pumping of slurries has been considered by Steele and Odrowaz-Pieniazek[12] and Wilson et al.[1] who also discuss the many aspects of selection and efficient operation of centrifugal pumps in these demanding circumstances.

The advantages and disadvantages of the centrifugal pump

The main advantages are:

(1) It is simple in construction and can therefore, be made in a wide range of materials.
(2) There is a complete absence of valves.
(3) It operates at high speed (up to 100 Hz) and therefore, can be coupled directly to an electric motor. In general, the higher the speed, the smaller the pump and motor for a given duty.
(4) It gives a steady delivery.
(5) Maintenance costs are lower than for any other type of pump.
(6) No damage is done to the pump if the delivery line becomes blocked, provided it is not run in this condition for a prolonged period.
(7) It is much smaller than other pumps of equal capacity. It can, therefore, be made into a sealed unit with the driving motor, and immersed in the suction tank.
(8) Liquids containing high proportions of suspended solids are readily handled.

The main disadvantages are:

(1) The single-stage pump will not develop a high pressure. Multistage pumps will develop greater heads but they are very much more expensive and cannot readily be made of corrosion-resistant material because of their greater complexity. It is generally better to use very high speeds in order to reduce the number of stages required.
(2) It operates at a high efficiency over only a limited range of conditions: this applies especially to turbine pumps.
(3) It is not usually self-priming.
(4) If a nonreturn valve is not incorporated in the delivery or suction line, the liquid will run back into the suction tank as soon as the pump stops.
(5) Very viscous liquids and slurries cannot be handled efficiently.

Pumping of non-Newtonian fluids

The development of the required pressure at the outlet to a centrifugal pump, depends upon the efficient conversion of kinetic energy into pressure energy. For a pump of this type, the distribution of shear within the pump will vary with position throughput the entire body of the pump. When the discharge valve is completely closed, the highest degree of shearing occurs in the gap between the rotor and the shell, at point B in Fig. 8.32. Between the vanes of the rotor (region A) there will be some circulation as shown in Fig. 8.33, but in the

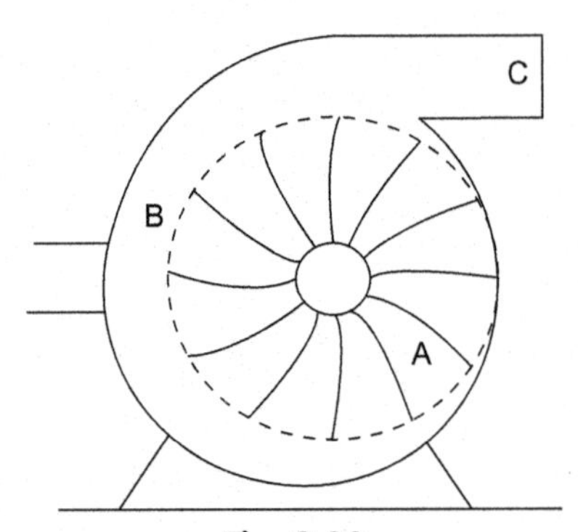

Fig. 8.32
Zones of differing shear in a centrifugal pump.

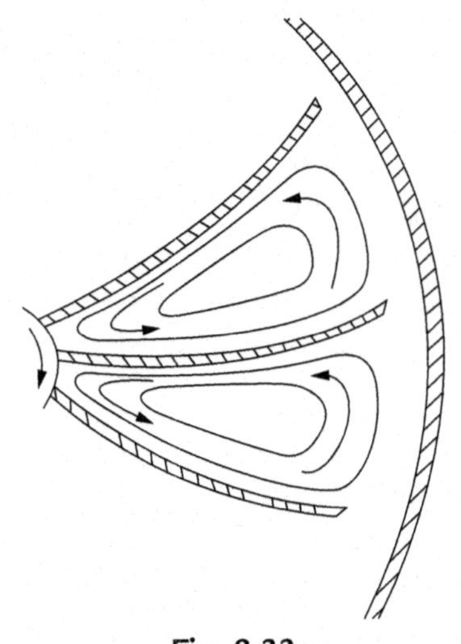

Fig. 8.33
Circulation within a centrifugal pump impeller.

discharge line C, the fluid will be essentially static. When fluid is flowing through the pump, there will still be differences between these shear rates, but they will not be so great. If the fluid has pseudoplastic properties, then the effective viscosity will vary in these different regions, being less at B than at A and C. Under steady-state conditions the pressure developed in the pump may be sufficient to establish a uniform flow. However, there may be problems on startup, when the very high apparent viscosities of the fluid might lead to overloading of the pump motor. The apparent viscosity of the liquid in the delivery line will also be at its maximum value, and the pump may take an inordinately long time to establish the required flowrate. Many pseudoplastic materials are damaged and degraded by prolonged shearing, and such a pump would then be unsuitable. Generally, positive-displacement rotary pumps are more satisfactory with shear-thinning fluids. The question of pump selection for all types of fluids has been subjected to a systems approach by Davidson[13] and Wilson et al.[1]

Example 8.2

A centrifugal pump is required to circulate a liquid of density 800 kg/m^3 and viscosity 0.5×10^{-3} Ns/m^2 from the reboiler of a distillation column through a vapourisor at the rate of 0.0004 m^3/s, and to introduce the superheated vapour above the vapour space in the reboiler which contains a 0.07 m depth of liquid. If smooth-bore 25 mm diameter pipe is to be used, the pressure of vapour in the reboiler is 1 kN/m^2 and the Net Positive Suction Head required by the pump is 2 m of liquid, what is the minimum height required between the liquid level in the reboiler and the pump?

Solution

$$\text{Volumetric flowrate of liquid} = 400 \times 10^{-6} \text{m}^3/\text{s}$$

$$\text{Cross-sectional area of the pipe} = (\pi/4)(0.025)^2 = 0.00049\,\text{m}^2$$

and hence:

$$\text{velocity in the pipe}, u = \left(400 \times 10^{-6}/0.00049\right) = 0.816\,\text{m/s}$$

The Reynolds number is then:

$$\begin{aligned} \text{Re} &= du\rho/\mu \\ &= (0.025 \times 0.816 \times 800)/\left(0.5 \times 10^{-3}\right) = 32{,}700 \end{aligned}$$

From Fig. 3.7, the friction factor for a smooth pipe is:

$$\phi = R/\rho u^2 = 0.0028$$

and the head loss due to friction is given by: (Eq. 3.20)

$$h_f = 4\phi(l/d)\left(u^2/g\right)$$

and:

$$h_f/l = (4 \times 0.0028)(1/0.025)\left(0.816^2/(9.81)\right) = 0.0304\,\text{m/m of pipe}$$

As the liquid is pumped at its boiling point, $(P_0 - P_v)/\rho g = 0$ and Eq. (8.26) becomes:

$$Z = \left(h_0 - h_f\right)$$

and

$$h_0 = \left(Z + h_f\right) = 2.0 + 0.0304l\,\text{m}$$

It should be noted that a slight additional height will be required if the kinetic energy at the pump inlet cannot be utilised.

Thus the height between the liquid level in the reboiler and the pump, h_0, depends on the length of pipe between the reboiler and the pump. If this is say 10 m, the minimum value of h_0 is 2.3 m.

This section is concluded by drawing attention to the fact that the operating envelope (both in terms of capacity and developed head) can be expanded by using several pumps in series and parallel arrangements. If the required head cannot be achieved with one pump, it is possible to nearly double the head by installing two similar pumps in series into a pipeline so that the discharge from the first pump acts as the suction for the second pump. On the other hand, if the flow capacity of one pump is not sufficient, it is not uncommon to use two or more pumps in parallel. The composite pump characteristics curve is obtained by adding the individual flowrates at the same head. Further detailed discussions on these aspects as well as on the mechanical aspects of centrifugal pumps, use of variable frequency drives to control the flow, etc. are available in the literature.[7,8,12–14]

8.3 Pumping Equipment for Gases

Essentially the same types of mechanical equipment are used for handling gases and liquids, though the details of the construction are different in the two cases. Again, there are two basic types, *positive displacement* and *centrifugal*, in which kinetic energy is converted into pressure energy. Over the normal range of operating pressures, the density of a gas is considerably less than that of a liquid with the result that higher speeds of operation can be employed and lighter valves fitted to the delivery and suction lines. Because of the lower viscosity of a gas there is a greater tendency for leakage to occur, and therefore gas compressors are designed with smaller clearances between the moving parts. Further differences in construction are necessitated by the decrease in volume of gas as it is compressed, and this must be allowed for in the design. Since a large proportion of the energy of compression appears as heat in the gas, there will normally be a considerable increase in temperature which may limit the operation of the compressor unless suitable cooling can be effected either within the stages of the compressor or in interstage coolers.

The principal types of compressors for gases will now be described.

8.3.1 Fans and Rotary Compressors

Fans are used for the supply of gases at relatively low pressures (<3.5 kN/m^2), often at very high flowrates. They may be of the *axial flow* type in which the curved blades directly impart an axial motion to the gas, or of the *centrifugal* type. Centrifugal fans, which operate on the same principle as centrifugal pumps for liquids, depend upon the conversion of the kinetic energy of the gas into pressure energy and are capable of developing somewhat higher pressures than the axial type.

Rotary blowers are of the *positive displacement* type, and a typical lobe type of machine is shown in Fig. 8.34. The rotors are driven in opposite directions so that, as each passes the inlet, it takes in gas which is compressed between the impeller and the casing before being expelled. Machines of

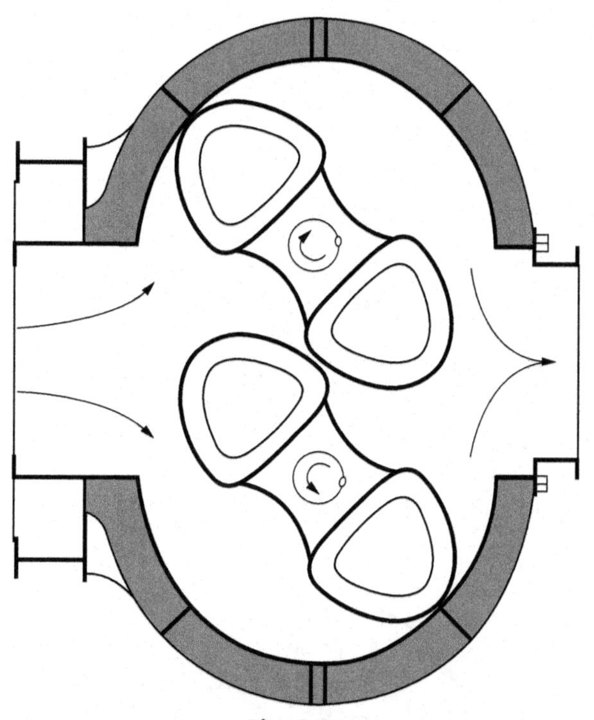

Fig. 8.34
Two lobe-compressor.

this type are capable of developing pressure differentials of up to 100 kN/m^2; they are made in a wide range of sizes, with maximum throughputs of up to 20,000–30,000 m^3/h.

For higher pressures, rotary compressors of the *sliding vane* type will give delivery pressures up to 1 MN/m^2. In a compressor of this type, as illustrated in Fig. 8.35, the compression ratio is achieved by eccentric mounting of the rotor which is slotted to take sliding vanes which subdivide the crescent-shaped space between the rotor and the casing. On rotation, the vanes are thrown out, trapping pockets of gas which are compressed during the rotation and are discharged, as shown, at the delivery port. Sliding vane compressors are also used as vacuum pumps (see Section 8.5).

Liquid-ring pumps, such as the Nash Hytor pump illustrated in Fig. 8.36, are positive-displacement pumps with a specially shaped casing, and a liquid seal which rotates with the impeller. The liquid leaves and re-enters the impeller cells and acts as a piston. The liquid is supplied at a pressure equal to the discharge pressure, and is drawn in automatically to compensate for that discharged from the ports. The energy of compression is converted into heat which is absorbed by the liquid, giving rise to a nearly isothermal process. Downstream, the liquid is separated from the gas, cooled if necessary, and recirculated with make-up liquid.

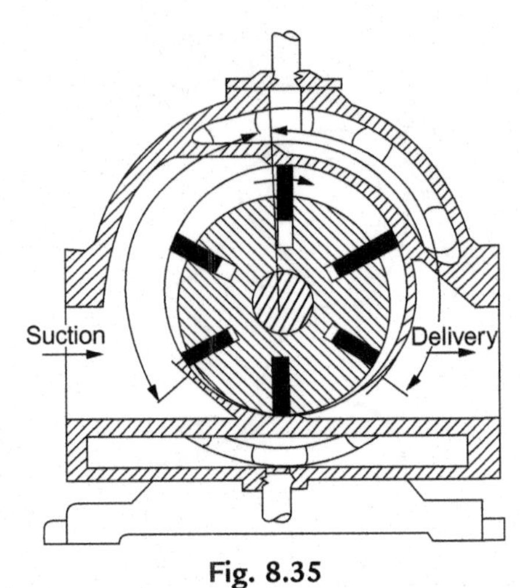

Fig. 8.35
The sliding vane rotary compressor and vacuum pump.

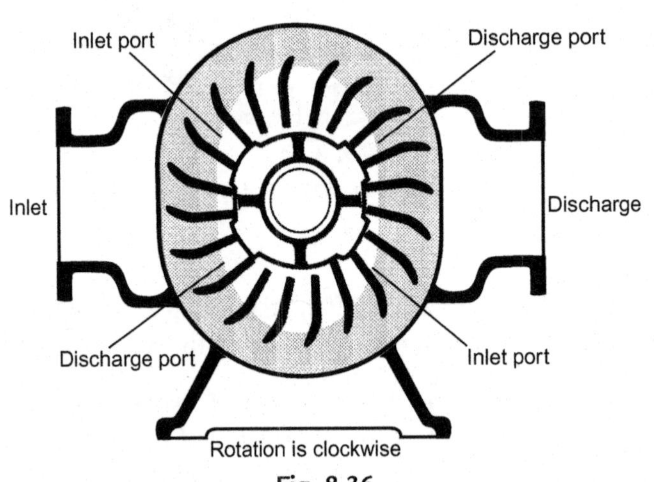

Fig. 8.36
Nash Hytor liquid ring pump.

The shaft and the impeller are the only moving mechanical parts and there is no sliding contact, so no lubricants are required, and the gas under compression does not become contaminated.

8.3.2 Centrifugal and Turbocompressors

These depend on the conversion of kinetic energy into pressure energy. Fans are used for low pressures, and can be made to handle very large quantities of gases. For the higher pressure ratios now in demand, multistage centrifugal compressors are mainly used, particularly for the requirements of high capacity chemical plants. Thus in catalytic reforming, petrochemical

Fig. 8.37
A turbocompressor.

separation plants (ethylene manufacture), ammonia plants with a production rate of 12 kg/s (45 tonne/h), and for the very large capacity needed for natural gas fields, this type of compressor is now supreme. These units now give flowrates up to 140 m^3/s and pressures of up to 5.6 MN/m^2 with the newest range going to 40 MN/m^2. It is important to accept that the very large units offer considerable savings over multiple units and that their reliability is remarkably high. The power required is also very high; thus a typical compressor operating on the process gas stream in the catalytic production of ethylene from naphtha will take 10 MW for a 6.5 kg/s (36 tonne/h) plant. A centrifugal compressor is illustrated in Fig. 8.37.

8.3.3 The Reciprocating Piston Compressor

This type of compressor is the only one capable of developing very high pressures, such as the pressure of 35 MN/m^2 required in the production of polyethylene. Compressors may be either single-stage or multiple-stage where very high pressures are required. A single-stage two-cylinder unit is shown in Fig. 8.38. The cylinders are fitted with jackets through which cooling water is circulated, and interstage coolers are provided on multistage compressors which may consist of anything from 2 to 12 stages. Cooling is essential to avoid the effects of excessively high temperatures on the mechanical operation of the compressor, and in order to reduce the power requirements. The calculation of the power required for compression, and how this is affected by *clearance volume*, is considered in Section 8.3.4. With recent developments in rotary compressors, the use of piston-type compressors is generally restricted to applications where very high pressures are required.

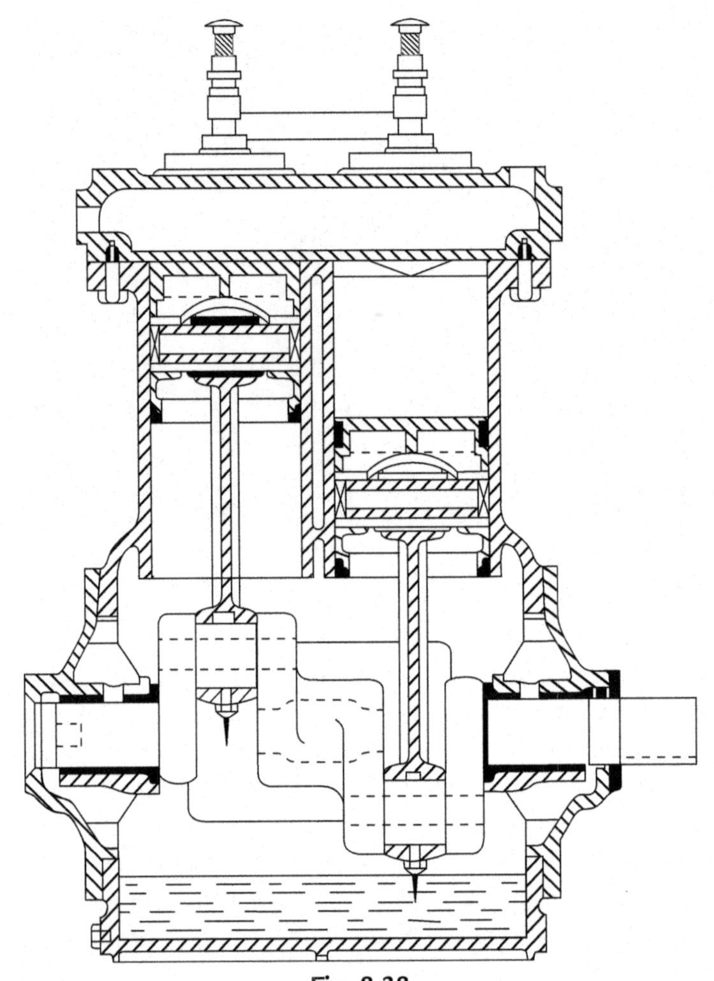

Fig. 8.38
Reciprocating compressor.

8.3.4 Power Required for the Compression of Gases

If during the compression of unit mass of gas, its volume changes by an amount $\mathrm{d}v$ at a pressure P, the net work done on the gas, $-\delta W$, for a reversible change is given by:

$$\begin{aligned} -\delta W &= -P\mathrm{d}v \\ &= v\mathrm{d}P - \mathrm{d}(Pv) \end{aligned} \tag{8.27}$$

and for an irreversible change by:

$$\begin{aligned} -\delta W &= -P\mathrm{d}v + \delta F \quad \text{(from Eq.2.6)} \\ &= v\mathrm{d}P - \mathrm{d}(Pv) + \delta F \end{aligned} \tag{8.28}$$

The work done in a reversible compression will be considered first because this refers to the ideal condition for which the work of compression is a minimum; a reversible compression would have to be carried out at an infinitesimal slow rate and therefore is not relevant in practice. The actual work done will be greater than that calculated, not only because of irreversibility, but also because of frictional loss and leakage in the compressor. These two factors are difficult to separate and will therefore be allowed for in the overall efficiency of the machine.

The total work of compression from a pressure P_1 to a pressure P_2 is found by integrating Eq. (8.27). For an ideal gas undergoing an isothermal compression:

$$-\int_{v_1}^{v_2} P\mathrm{d}v = -W = P_1 v_1 \ln\frac{P_2}{P_1} \quad \text{(from Eq.2.69)}$$

For an isentropic compression of an ideal gas (since $-P\mathrm{d}v = v\mathrm{d}P - \mathrm{d}(Pv)$):

$$-\int_{v_1}^{v_2} P\mathrm{d}v = -W = \left[\frac{\gamma}{\gamma-1}(P_2v_2 - P_1v_1)\right] - (P_2v_2 - P_1v_1) \quad \text{(from Eq.2.71)}$$

$$= \frac{1}{\gamma-1}(P_2v_2 - P_1v_1) \tag{8.29}$$

Thus: (comparing Eqs 2.71 and 8.29)

$$\int_{P_1}^{P_2} v\mathrm{d}P = -\gamma\int_{v_1}^{v_2} P\mathrm{d}v$$

for isentropic conditions and then:

$$-W = \frac{1}{\gamma-1}P_1v_1\left[\left(\frac{P_2}{P_1}\right)^{(\gamma-1)/\gamma} - 1\right] \tag{8.30}$$

Under these conditions, the entire energy of compression appears as heat in the gas.

For unit mass of an ideal gas undergoing an isentropic compression:

$$P_1v_1^\gamma = P_2v_2^\gamma \quad \text{(from Eq.2.30)}$$

and:

$$\frac{P_1v_1}{T_1} = \frac{P_2v_2}{T_2} \quad \text{(from Eq.2.16)}$$

Eliminating v_1 and v_2 gives the ratio of the outlet temperature T_2 to the inlet temperature T_1:

$$\frac{T_2}{T_1} = \left(\frac{P_2}{P_1}\right)^{(\gamma-1)/\gamma} \tag{8.31}$$

In practice, there will be irreversibilities (inefficiencies) associated with the compression process and the additional energy needed will appear as heat, giving rise to an outlet temperature higher than T_2 as given by Eq. (8.31).

For the isentropic compression of a mass m of gas:

$$-Wm = \frac{1}{\gamma - 1} P_1 V_1 \left[\left(\frac{P_2}{P_1} \right)^{(\gamma-1)/\gamma} - 1 \right] \tag{8.32}$$

where V_1 is the volume of a mass m of gas at a pressure P_1.

If the conditions are intermediate between isothermal and isentropic, k must be used in place of γ, where $\gamma > k > 1$.

If the gas deviates appreciably from the ideal gas laws over the range of conditions considered, the work of compression is most conveniently calculated from the change in the thermodynamic properties of the gas.

Thus:

$$dU = TdS - Pdv \quad \text{(from Eq.2.5)}$$

$$\therefore -\delta W = -Pdv = dU - TdS \quad \text{(for a reversible process)}$$

Under isothermal conditions:

$$-W = \Delta U - T\Delta S \tag{8.33}$$

Under isentropic conditions:

$$-W = \Delta U \tag{8.34}$$

These equations give the work done during a simple compression of gas in a cylinder but do not take account of the work done either during the admission of the gas prior to compression or during the expulsion of the compressed gas from the cylinder.

If, after the compression of a volume V_1 of gas at a pressure P_1 to a pressure P_2, the entire gas is expelled at constant pressure P_2 and a fresh charge of gas is admitted at a pressure P_1, then the cycle can be followed in Fig. 8.39, where the pressure P is plotted as ordinate against the volume V as abscissa.

Point 1 represents the initial condition of the gas (P_1 and V_1).

Line 1–2 represents the compression of gas to pressure P_2, volume V_2.

Line 2–3 represents the expulsion of the gas at a constant pressure P_2.

Line 3–4 represents a sudden reduction in the pressure in the cylinder from P_2 to P_1. As the entire gas has been expelled, this can be regarded as taking place instantaneously.

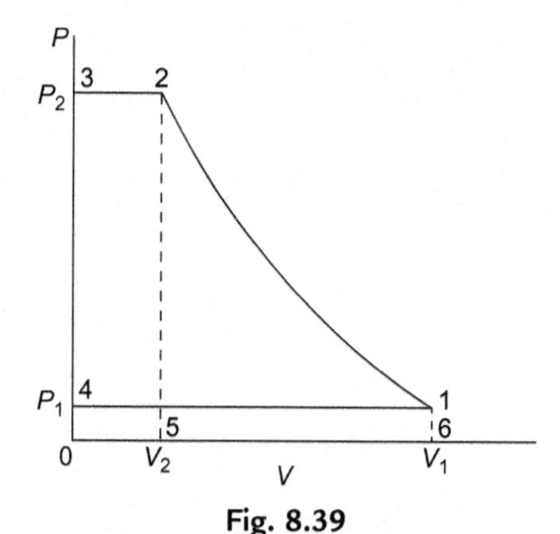

Fig. 8.39
Single-stage compression cycle—no clearance.

Line 4–1 represents the suction stroke of the piston, during which a volume V_1 of gas is admitted at constant pressure, P_1.

It will be noted that the mass of gas in the cylinder varies during the cycle. The work done by the compressor during each phase of the cycle is as follows:

$$\text{Compression} \quad -\int_{V_1}^{V_2} PdV \quad (\text{Area}\,1-2-5-6)$$

$$\text{Expulsion} \quad P_2V_2 \quad (\text{Area}\,2-3-0-5)$$

$$\text{Suction} \quad -P_1V_1 \quad -(\text{Area}\,4-0-6-1)$$

The total work done per cycle

$$\begin{aligned} &= -\int_{V_1}^{V_2} PdV + P_2V_2 - P_1V_1 (\text{Area}\,1-2-3-4) \\ &= \int_{P_1}^{P_2} VdP \end{aligned} \tag{8.35}$$

Note for an ideal gas under isothermal conditions $P_1V_1 = P_2V_2$.

The work of compression for an ideal gas per cycle under isothermal conditions:

$$= P_1V_1 \ln\frac{P_2}{P_1} \tag{8.36}$$

Under isentropic conditions, the work of compression is given as:

$$= P_1V_1\frac{\gamma}{\gamma-1}\left[\left(\frac{P_2}{P_1}\right)^{(\gamma-1)/\gamma} - 1\right] \tag{8.37}$$

Again, working in terms of the thermodynamic properties of the gas:

$$\begin{aligned} v\mathrm{d}P &= \mathrm{d}(Pv) - P\mathrm{d}v \\ &= \mathrm{d}(Pv) + \mathrm{d}U - T\mathrm{d}S \quad \text{(from Eq.2.5)} \\ &= \mathrm{d}H - T\mathrm{d}S \end{aligned} \tag{8.38}$$

For an isothermal process:

$$-mW = m(\Delta H - T\Delta S) \tag{8.39}$$

For an isentropic process:

$$-mW = m\Delta H \tag{8.40}$$

where m is the mass of the gas compressed per cycle.

Clearance volume

In practice, it is not possible to expel the entire gas from the cylinder at the end of the compression; the volume remaining in the cylinder after the forward stroke of the piston is termed the *clearance volume*. The volume displaced by the piston is termed the *swept volume*, and therefore the total volume of the cylinder is made up of the clearance volume plus the swept volume. The *clearance* c is defined as the ratio of the clearance volume to the swept volume.

A typical cycle for a compressor with a finite clearance volume can be followed by reference to Fig. 8.40. A volume V_1 of gas at a pressure P_1 is admitted to the cylinder; its condition is represented by point 1.

Line 1–2 represents the compression of the gas to a pressure P_2 and volume V_2.

Line 2–3 represents the expulsion of gas at constant pressure P_2, so that the volume remaining in the cylinder is V_3.

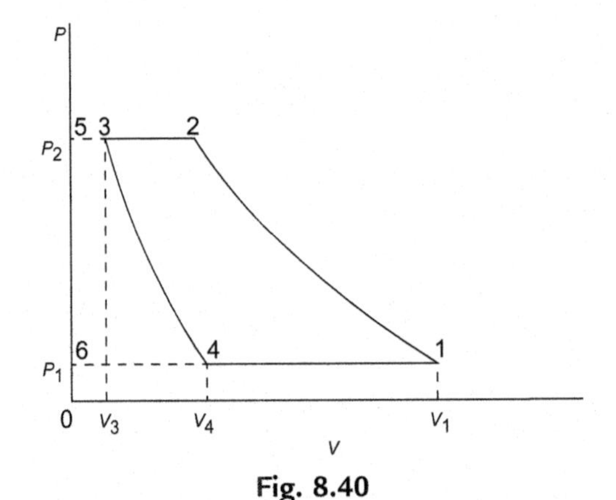

Fig. 8.40
Single-stage compression cycle—with clearance.

Line 3–4 represents an expansion of this residual gas to the lower pressure P_1 and volume V_4 during the return stroke.

Line 4–1 represents the introduction of fresh gas into the cylinder at constant pressure P_1. The work done on the gas during each stage of the cycle is as follows.

Compression

$$-\int_{V_1}^{V_2} PdV$$

Expulsion

$$P_2(V_2 - V_3)$$

Expansion

$$-\int_{V_3}^{V_4} PdV$$

Suction

$$-P_1(V_1 - V_4)$$

The total work done during the cycle is equal to the sum of these four components. It is represented by the area 1–2–3–4, which is equal to area 1–2–5–6 less area 4–3–5–6. If the compression and expansion are taken as isentropic, the work done per cycle is therefore:

$$\begin{aligned} &= P_1V_1\frac{\gamma}{\gamma-1}\left[\left(\frac{P_2}{P_1}\right)^{(\gamma-1)/\gamma} - 1\right] - P_1V_4\frac{\gamma}{\gamma-1}\left[\left(\frac{P_2}{P_1}\right)^{(\gamma-1)/\gamma} - 1\right] \\ &= P_1(V_1 - V_4)\frac{\gamma}{\gamma-1}\left[\left(\frac{P_2}{P_1}\right)^{(\gamma-1)/\gamma} - 1\right] \end{aligned} \tag{8.41}$$

Thus, theoretically, the clearance volume does not affect the work done per unit mass of gas, since $V_1 - V_4$ is the volume admitted per cycle. It does however, influence the quantity of gas admitted and therefore the work done per cycle. In practice however, compression and expansion are not reversible, and losses arise from the compression and expansion of the clearance gases. This effect is particularly serious at high compression ratios.

The value of V_4 is not known explicitly, but can be calculated in terms of V_3, the clearance volume.

For isentropic conditions:

$$V_4 = V_3\left(\frac{P_2}{P_1}\right)^{1/\gamma}$$

and:

$$V_1 - V_4 = (V_1 - V_3) + V_3 - V_3\left(\frac{P_2}{P_1}\right)^{1/\gamma}$$

$$= (V_1 - V_3)\left[1 + \frac{V_3}{V_1 - V_3} - \frac{V_3}{V_1 - V_3}\left(\frac{P_2}{P_1}\right)^{1/\gamma}\right]$$

Now $(V_1 - V_3)$ is the swept volume, V_s, say; and $V_3/(V_1 - V_3)$ is the clearance c.

Thus:

$$V_1 - V_4 = V_s\left[1 + c - c\left(\frac{P_2}{P_1}\right)^{1/\gamma}\right] \tag{8.42}$$

The total work done on the fluid per cycle is therefore, using Eqs (8.41) and (8.42):

$$P_1 V_s \frac{\gamma}{\gamma - 1}\left[\left(\frac{P_2}{P_1}\right)^{(\gamma-1)/\gamma} - 1\right]\left[1 + c - c\left(\frac{P_2}{P_1}\right)^{1/\gamma}\right] \tag{8.43}$$

The factor $[1 + c - c(P_2/P_1)^{1/\gamma}]$ is called the theoretical volumetric efficiency and is a measure of the effect of the clearance on an isentropic compression. The actual volumetric efficiency will be affected, in addition, by the inertia of the valves and leakage past the piston.

The gas is frequently cooled during compression so that the work done per cycle is less than that given by Eq. (8.43), and γ is replaced by some smaller quantity k. The greater the rate of heat removal, the less is the work done. The isothermal compression is usually taken as the condition for least work of compression, but clearly the energy consumption can be reduced below this value if the gas is artificially cooled below its initial temperature as it is compressed. This is not a practicable possibility because of the large amount of energy required to refrigerate the cooling fluid. It can be seen that the theoretical volumetric efficiency decreases as the rate of heat removal is increased since γ is replaced by the smaller quantity k.

In practice the cylinders are usually water-cooled. The work of compression is thereby reduced though the effect is usually small. The reduction in temperature does however, improve the mechanical operation of the compressor and makes lubrication easier.

Multistage compressors

If the required pressure ratio P_2/P_1 is large, it is not practicable to carry out the entire compression in a single cylinder because of the high temperatures which would be set up and the adverse effects of clearance volume on the efficiency. Further, lubrication would be difficult due to carbonisation of the oil, and there would be a risk of causing oil mist explosions in the cylinders when gases containing oxygen are being compressed. The mechanical construction also would be difficult because the single cylinder would have to be strong enough

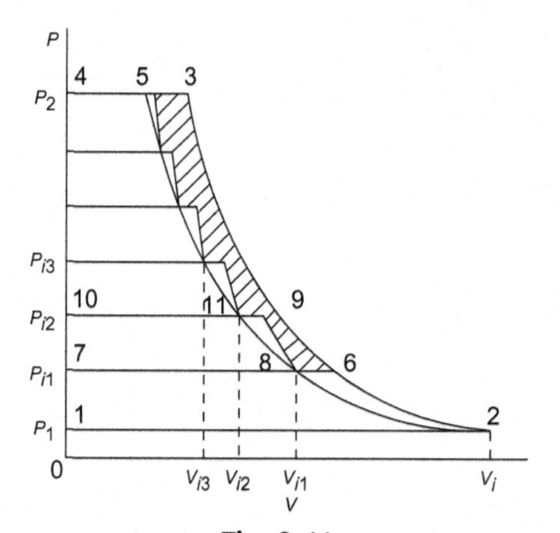

Fig. 8.41
Multistage compression cycle with interstage cooling.

to withstand the final pressure and yet large enough to hold the gas at the initial pressure P_1. In the multistage compressor, the gas passes through a number of cylinders of gradually decreasing volume and can be cooled between the stages. The maximum pressure ratio normally obtained in a single cylinder is 10 but values above 6 are unusual.

The operation of the multistage compressor can conveniently be followed again on a pressure–volume diagram (Fig. 8.41). The effect of clearance volume will be neglected at first. The area 1–2–3–4 represents the work done in compressing isentropically from P_1 to P_2 in a single stage. The area 1–2–5–4 represents the necessary work for an isothermal compression. Now consider a multistage isentropic compression in which the intermediate pressures are P_{i1}, P_{i2}, etc. The gas will be assumed to be cooled to its initial temperature in an interstage cooler before it enters each cylinder.

Line 1–2 represents the suction stroke of the first stage where a volume V_1 of gas is admitted at a pressure P_1.

Line 2–6 represents an isentropic compression to a pressure P_{i1}.

Line 6–7 represents the delivery of the gas from the first stage at a constant pressure P_{i1}.

Line 7–8 represents the suction stroke of the second stage. The volume of the gas has been reduced in the interstage cooler to V_{i1}; that which would have been obtained as a result of an isothermal compression to P_{i1}.

Line 8–9 represents an isentropic compression in the second stage from a pressure P_{i1} to a pressure P_{i2}.

Line 9–10 represents the delivery stroke of the second stage.

Line 10–11 represents the suction stroke of the third stage. Point 11 again lies on the line 2–5, representing an isothermal compression.

It is seen that the overall work done on the gas is intermediate between that for a single stage isothermal compression and that for an isentropic compression. The net saving in energy is shown as the shaded area in Fig. 8.34.

The total work done per cycle W'

$$= P_1 V_1 \frac{\gamma}{\gamma - 1}\left[\left(\frac{P_{i1}}{P_1}\right)^{(\gamma-1)/\gamma} - 1\right] + P_{i1} V_{i1} \frac{\gamma}{\gamma - 1}\left[\left(\frac{P_{i2}}{P_{i1}}\right)^{(\gamma-1)/\gamma} - 1\right] + \cdots$$

for an isentropic compression.

For perfect interstage cooling:

$$P_1 V_1 = P_{i1} V_{i1} = P_{i2} V_{i2} = \cdots$$

$$\therefore\ W' = P_1 V_1 \frac{\gamma}{\gamma - 1}\left[\left(\frac{P_{i1}}{P_1}\right)^{(\gamma-1)/\gamma} + \left(\frac{P_{i2}}{P_{i1}}\right)^{(\gamma-1)/\gamma} + \cdots - n\right]$$

where n is the number of stages.

It is now required to find out how the total work per cycle W' is affected by the choice of the intermediate pressures P_{i1}, P_{i2}, etc. The work will be a minimum when

$$\frac{\partial W'}{\partial P_{i1}} = \frac{\partial W'}{\partial P_{i2}} = \frac{\partial W'}{\partial P_{i3}} = \cdots = 0.$$

When:

$$\frac{\partial W'}{\partial P_{i1}} = 0$$

$$P_1 V_1 \frac{\gamma}{\gamma - 1}\left[\frac{\gamma - 1}{\gamma}\left(\frac{P_{i1}}{P_1}\right)^{(\gamma-1)/\gamma} P_{i1}^{-1} + \frac{1-\gamma}{\gamma}\left(\frac{P_{i2}}{P_{i1}}\right)^{(\gamma-1)/\gamma} P_{i1}^{-1}\right] = 0$$

i.e.:

$$\frac{P_{i1}}{P_1} = \frac{P_{i2}}{P_{i1}} \tag{8.44}$$

The same procedure is then adopted for obtaining the optimum value of P_{i2} and hence:

$$\frac{P_{i2}}{P_{i1}} = \frac{P_{i3}}{P_{i2}} \tag{8.45}$$

Thus the intermediate pressures should be arranged so that the compression ratio is the same in each cylinder; and equal work is then done in each cylinder.

The minimum work of compression in a compressor of n stages is therefore:

$$P_1V_1\frac{\gamma}{\gamma-1}\left[n\left(\frac{P_2}{P_1}\right)^{(\gamma-1)/n\gamma}-n\right]=nP_1V_1\frac{\gamma}{\gamma-1}\left[\left(\frac{P_2}{P_1}\right)^{(\gamma-1)/n\gamma}-1\right] \tag{8.46}$$

The effect of clearance volume may now be taken into account. If the clearances in the successive cylinders are c_1, c_2, c_3, …, the theoretical volumetric efficiency of the first cylinder is

$$=1+c_1-c_1\left(\frac{P_{i1}}{P_1}\right)^{1/\gamma} \quad \text{(from equation 8.42)}$$

Assuming that the same compression ratio is used in each cylinder, then the theoretical volumetric efficiency of the first stage is:

$$1+c_1-c_1\left(\frac{P_2}{P_1}\right)^{1/n\gamma}$$

If the swept volumes of the cylinders are V_{s1}, V_{s2}, …, the volume of the gas admitted to the first cylinder

$$=V_{s1}\left[1+c_1-c_1\left(\frac{P_2}{P_1}\right)^{1/n\gamma}\right] \tag{8.47}$$

The same mass of gas passes through each of the cylinders and therefore, if the interstage coolers are assumed perfectly efficient, the ratio of the volumes of gas admitted to successive cylinders is $(P_1/P_2)^{1/n}$. The volume of gas admitted to the second cylinder is then:

$$V_{s2}\left[1+c_2-c_2\left(\frac{P_2}{P_1}\right)^{1/n\gamma}\right]=V_{s1}\left[1+c_1-c_1\left(\frac{P_2}{P_1}\right)^{1/n\gamma}\right]\left(\frac{P_1}{P_2}\right)^{1/n}$$

$$\therefore \frac{V_{s1}}{V_{s2}}=\frac{1+c_2-c_2(P_2/P_1)^{1/n\gamma}}{1+c_1-c_1(P_2/P_1)^{1/n\gamma}}\left(\frac{P_2}{P_1}\right)^{1/n} \tag{8.48}$$

In this manner, the swept volume of each cylinder can be calculated in terms of V_{s1} and $c_1, c_2, \ldots$ and the cylinder dimensions determined.

When the gas does not behave as an ideal gas, the change in its condition can be followed on a temperature-entropy or an enthalpy-entropy diagram. The intermediate pressures P_{i1}, P_{i2}, …, are then selected so that the enthalpy change (ΔH) is the same in each cylinder.

Several opposing factors will influence the number of stages selected for a given compression. The larger the number of cylinders the greater is the mechanical complexity. Against this must be balanced the higher theoretical efficiency, the smaller mechanical strains set up in the cylinders and the moving parts, and the greater ease of lubrication at the lower temperatures that are experienced. Compressors with as many as nine stages are used for very high pressures.

Compressor efficiencies

The efficiency quoted for a compressor is usually either an isothermal efficiency or an isentropic efficiency. The isothermal efficiency is the ratio of the work required for an ideal isothermal compression to the energy actually expended in the compressor. The isentropic efficiency is defined in a corresponding manner on the assumption that the whole compression is carried out in a single cylinder. Since the energy expended in an isentropic compression is greater than that for an isothermal compression, the isentropic efficiency is always the greater of the two. Clearly the efficiencies will depend on the heat transfer between the gas undergoing compression and the surroundings and on how closely the process approaches a reversible compression.

The efficiency of the compression will also be affected by a number of other factors which are all connected with the mechanical construction of the compressor. Thus the efficiency will be reduced as a result of leakage past the piston and the valves and because of throttling of the gas at the valves. Further, the mechanical friction of the machine will lower the efficiency and the overall efficiency will be affected by the efficiency of the driving motor and transmission.

Example 8.3

A single-acting air compressor supplies 0.1 m^3/s of air measured at, 273 K and 101.3 kN/m^2 which is compressed to 380 kN/m^2 from 101.3 kN/m^2. If the suction temperature is 289 K, the stroke is 0.25 m, and the speed is 4.0 Hz, what is the cylinder diameter? Assuming the cylinder clearance is 4% and compression and reexpansion are isentropic ($\gamma = 1.4$), what are the theoretical power requirements for the compression?

Solution

$$\text{Volume per stroke} = \left(\frac{0.1}{4.0}\right)\left(\frac{289}{273}\right) = 0.0264 \text{ m}^3$$

$$\text{Compression ratio} = \left(\frac{380}{101.3}\right) = 3.75$$

The swept volume is given by Eq. (8.42), using $C = 0.04$:

$$0.0264 = V_s\left[1 + 0.04 - 0.04(3.75)^{1/1.4}\right]$$

$$\therefore V_s = \frac{0.0264}{1.04 - 0.04 \times 2.7} = 0.0283 \text{ m}^3$$

Thus cross-sectional area of cylinder $=\left(\frac{0.0283}{0.25}\right)=0.113\ \text{m}^2$

$$\text{and the cylinder diameter}=\left(\frac{0.113}{\pi/4}\right)^{0.5}=\underline{\underline{0.38\ \text{m}}}$$

From Eq. (8.41), work of compression per cycle

$$=101{,}300\times 0.0264\left[\frac{1.4}{1.4-1.0}\right]\left[(3.75)^{(1.4-1)/1.4}-1\right]$$
$$=9360(1.457-1)=4278\text{J}$$

Theoretical power requirements

$$=(4278\times 4)=17{,}110\text{W or }\underline{\underline{17.1\ \text{kW}}}$$

Example 8.4

Air (molecular weight of 28.8) at 290 K is compressed from 101.3 to 2065 kN/m^2 in a two-stage compressor operating with a mechanical efficiency of 85%. The relation between pressure and volume during the compression stroke and expansion of the clearance gas is $PV^{1.25}=$ constant. The compression ratio in each of the two cylinders is the same, and the interstage cooler may be assumed 100% efficient. If the clearances in the two cylinders are 4% and 5%, respectively, calculate:

a. the work of compression per kg of air compressed;
b. the isothermal efficiency;
c. the isentropic efficiency ($\gamma=1.4$), and
d. the ratio of the swept volumes in the two cylinders.

Solution

Overall compression ratio

$$=\left(\frac{2065}{101.3}\right)=20.4$$

Specific volume of air at 290 K

$$=\left(\frac{22.4}{28.8}\right)\left(\frac{290}{273}\right)=0.826\ \text{m}^3/\text{kg}$$

From Eq. (8.46), work of compression

$$=101{,}300\times 0.826\times 2\left[\frac{1.25}{1.25-1}\right]\left[(20.4)^{(1.25-1)/2\times 1.25}-1\right]$$
$$=836{,}700(1.351-1)$$
$$=293{,}700\text{J/kg}=293.7\text{kJ/kg}$$

Energy supplied to the compressor, which is the work of compression

$$= \left(\frac{293.7}{0.85}\right) = \underline{\underline{345.5\ \text{kJ/kg}}}$$

From Eq. (8.36), the work done in isothermal compression of 1 kg of gas

$$= P_1 V_1 \ln\frac{P_2}{P_1} = (101300 \times 0.826 \ln 20.4)$$
$$= (83700 \times 3.105) = 252{,}300\ \text{J/kg} = 252.3\ \text{kJ/kg}$$

$$\text{Isothermal efficiency} = 100 \times \left(\frac{252.3}{345.5}\right) = \underline{\underline{73\%}}$$

From Eq. (8.37), work done in isentropic compression of 1 kg of gas

$$= P_1 V_1 \left(\frac{\gamma}{\gamma - 1}\right)\left[\left(\frac{P_2}{P_1}\right)^{\frac{\gamma-1}{\gamma}} - 1\right] = 101{,}300 \times 0.826 \times \frac{1.4}{0.4}\left[(20.4)^{0.4/1.4} - 1\right]$$
$$= 292{,}900(2.36 - 1) = 398{,}300\ \text{J/kg} = 398.3\ \text{kJ/kg}$$

$$\text{Isentropic efficiency} = 100 \times \left(\frac{398.3}{345.3}\right) = \underline{\underline{115\%}}$$

From Eq. (8.47), volume swept out in first cylinder in compression of 1 kg of gas is given by:

$$0.826 = V_{s1}\left[1 + 0.04 - 0.04(20.4)^{1/(2\times1.25)}\right]$$

$$= V_{s1} \times 0.906$$

$$\therefore V_{s1} = 0.912\ \text{m}^3/\text{kg}$$

Similarly, the swept volume of the second cylinder is given by:

$$0.826\left(\frac{1}{20.4}\right)^{0.5} = V_{s2}\left[1 + 0.05 - 0.05(20.4)^{1/(2\times1.25)}\right]$$

$$0.183 = 0.883 V_{s2}$$

$$\therefore V_{s2} = 0.207\ \text{m}^3/\text{kg}$$

and:

$$\frac{V_{s1}}{V_{s2}} = \underline{\underline{4.41}}$$

8.4 The Use of Compressed Air for Pumping

Compressed gas is sometimes used for transferring liquid from one position to another in a chemical plant, but more particularly for emptying vessels. It is frequently more convenient to apply pressure by means of compressed gas rather than to install a pump, particularly when the liquid is corrosive or contains solids in suspension. Furthermore, to an increasing extent, chemicals are being delivered in tankers and are discharged by the application of gas under pressure. For instance, phthalic anhydride is now distributed in heated tankers, which are discharged into storage vessels by connecting them to a supply of compressed nitrogen or carbon dioxide.

Several devices have been developed to eliminate the necessity for manual operation of valves, and the automatic acid elevator is an example of equipment incorporating such a device. However, such equipment is becoming less important now since it is possible to construct centrifugal pumps in a wide range of corrosion-resistant materials. The *air-lift pump* makes more efficient use of the compressed air and is used for pumping corrosive liquids. Although it is not used extensively in the chemical industry, it is used for pumping oil from wells, and the principle is incorporated into a number of items of chemical engineering equipment, including the climbing film evaporator.

8.4.1 The Air-Lift Pump

In the air-lift pump (Fig. 8.42), a high efficiency is obtained by allowing compressed air to expand to atmospheric pressure in contact with the liquid. It can be regarded simply as a U-tube in a state of dynamic equilibrium. One limb containing only liquid is relatively short and

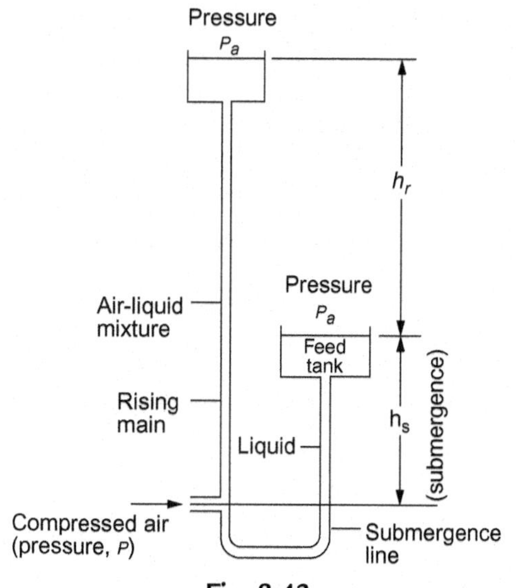

Fig. 8.42
Schematics of an air-lift pump.

this is connected to the feed tank, whilst air is injected near the bottom of the longer limb which therefore contains a mixture of liquid and air that has a lower density. If the air is introduced sufficiently rapidly, liquid will flow from the short to the long limb and be discharged into the delivery tank. The rate of flow will depend on the difference in density and will therefore, rise as the air rate is increased, but will reach a maximum because the frictional resistance increases with the volumetric rate of flow.

The liquid feed line is known as the *submergence limb* and the line carrying the aerated mixture is known as the *rising main*. The ratio of the submergence (h_s) to the total height of rising main above the air injection point ($h_r + h_s$) is known as the *submergence ratio*, $[1 + (h_r/h_s)]^{-1}$.

If a mass G_L of liquid is raised through a net height h_r by a mass G_A of air in unit time, the net rate of energy transfer to the liquid is $G_L g h_r$. This can also be seen as the work done against gravity to lift the mass G_L to height h_r. If the pressure of the entering air is P, the work done by the air in expanding isothermally to atmospheric pressure P_a is given by:

$$P_a v_a G_A \ln\frac{P}{P_a} \tag{8.49}$$

where v_a is the specific volume of air at atmospheric pressure. The expansion will be essentially isothermal because of the intimate contact between the liquid and the air.

The efficiency of the pump η is therefore given by:

$$\eta = \frac{G_L g h_r}{G_A P_a v_a \ln(P/P_a)} \tag{8.50}$$

The mass of air required to pump unit mass of liquid is therefore given by:

$$\frac{G_A}{G_L} = \frac{g h_r}{\eta P_a v_a \ln(P/P_a)} \tag{8.51}$$

If all losses in the operation of the pump were neglected, the pressure at the point of introduction of the compressed air would be equal to atmospheric pressure together with the pressure due to the column of liquid of height h_s, the vertical distance between the liquid level in the suction tank, and the air inlet point. Therefore:

$$P_a = h_a \rho g \text{ (say)} \tag{8.52}$$

and:

$$P = (h_a + h_s)\rho g \tag{8.53}$$

where p is the density of the liquid.

Thus from Eq. (8.51), the mass of air required to pump unit mass of liquid (G_A/G_L) would be equal to:

$$\frac{G_A}{G_L} = \frac{h_r g}{P_a v_a \ln[(h_s + h_a)/h_a]} \tag{8.54}$$

This is the minimum air requirement for the pump if all losses are neglected. It will be seen that (G_A/G_L) decreases as h_s increases; if h_s is zero, G_A/G_L is infinite and therefore the pump will not work. A high submergence h_s is therefore desirable. This can be a considerable practical disadvantage in that it may be necessary to provide a deep "pit" to give the demand submergence.

There are a number of important applications of the air-lift pump in the process industries due to its simplicity. It is particularly useful for handling radioactive materials as there are no mechanical parts in contact with the fluid, and the pump will operate virtually indefinitely without the need for maintenance, which can prove very difficult when handling radioactive liquids.

Example 8.5

An air-lift pump is used for raising 7.5×10^{-4} m^3/s of a liquid of density 1200 kg/m^3 to a height of 20 m. Air is available at a pressure of 450 kN/m^2. Assuming isentropic compression of the air, what is the power requirement of the pump if its efficiency is 30%? ($\gamma = 1.4$). Take the volume of 1 kmol of an ideal gas at 273 K and 101.3 kN/m^2 as 22.4 m^3.

Solution

Mass flow of liquid $= (7.5 \times 10^{-4} \times 1200) = 0.9$ kg/s.

Work per unit time done by the pump $= G_L g h_r = (0.9 \times 9.81 \times 20) = 176.6$ J/s $= 176.6$ W.

Actual work of expansion of air per unit time $= (176.6/0.30) = 588.6$ W.

Taking the molecular weight of air as 28.9 kg/kmol, the specific volume of air at 101.3 kN/m^2 and 273 K, $v_a = (22.4/28.9) = 0.775$ m^3/kg

$$\text{and using Eq.(8.49)}: \; 588.6 = [(101.3 \times 10^3 \times 0.775 G_A)(\ln(450/101.3))]$$

$$\text{from which}: \; G_A = 0.0050 \text{ kg/s}$$

and:

volume flowrate of air, $Q = (0.0050 \times 0.775) = 0.0039$ m^3/s.

From Eq. (8.37):

Power for compression $= [101.3 \times 10^3 \times 0.0039][1.4/(1.4-1)][(450/101.3)^{(1.4-1)/1.4} - 1]$.

$= (395.1 \times 3.5 \times 0.53) = 733$ W.

and:

$$\text{Power required} = \underline{\underline{733 \text{ W}}} \text{ or } \underline{\underline{0.733 \text{ kW}}}$$

Example 8.6

An air-lift pump raises 0.01 m^3/s of water from a well 100 m deep through a 100 mm diameter pipe. The level of water is 40 m below the surface. The air flow is 0.1 m^3/s of free air compressed to 800 kN/m^2. Calculate the efficiency of the pump and the mean velocity of the mixture in the pipe.

Solution

Mass flow of water = (0.01 × 1000) = 10 kg/s.

Work done per unit time = $G_L g h_r$ = (10 × 40 × 9.81) = 3924 W.

Volumetric flowrate of air used = 0.1 m^3/s.

The energy needed to compress 0.1 m^3/s of air is given by:

$$P_1V_1\left[\gamma/(\gamma-1)(P_2/P_1)^{(\gamma-1)/\gamma}-1\right] \text{ (Eq.8.37)}$$

$$=(101300\times 0.1\times 1.4)\left[(800/101.3)^{0.286}-1\right]/0.4=28{,}750\,\text{J}$$

The power required for this compression = 28,750 W.

and hence the efficiency = (3924 × 100)/28, 750 = $\underline{\underline{13.7\%}}$.

The mean velocity depends upon the pressure of air in the pipe. If the air pressure at the bottom of the well is, say, 60 m of water or (60 × 1000 × 9.81)/1000 = 588.6 kN/m^2, and the pressure at the surface is atmospheric, the mean pressure is (101.3 + 588.6)/2 = 345 kN/m^2.

The specific volume v of the air at this pressure and at 273 K is then given by:

$$v=\frac{RT}{MP}=\frac{8314\times 273}{29\times 345{,}000}=0.227\ \text{m}^3/\text{kg}$$

The specific volume v of air at 273 K and 101.3 kN/m^2 is given by:

$$v=\frac{(8314\times 273)}{(29\times 101{,}300)}=0.772\ \text{m}^3/\text{kg}$$

and hence the mass flowrate of the air is:

$$(0.10/0.772)=0.13\ \text{kg/s}$$

Mean volumetric flowrate of air = (0.13 × 0.227) = 0.0295 m^3/s.

Volumetric flowrate of water = 0.01 m^3/s.

Total volumetric flowrate = 0.01 + 0.0295 = 0.0395 m^3/s.

Area of pipe = $(\pi/4) \times 0.1^2$ = 0.00785 m^2.

and hence the mean velocity of the mixture = (0.0395/0.00785) = $\underline{\underline{5.03\ \text{m/s}}}$.

Flow of a vertical column of aerated liquid

The behaviour of a rising column of aerated liquid is important in the way it affects the operation of the air-lift pump. In Chapter 5, the flow of gas–liquid mixtures in pipes has been discussed for conditions where the flowrates of the two fluids are controlled independently. In the air-lift pump, the flowrate of liquid is not normally controlled, but is governed by the air rate and other parameters of the system. For gas–liquid flow, at low degrees of aeration, the gas is distributed in the form of discrete bubbles, which undergo a degree of coalescence. When the proportion of air is increased, '*slug*' flow occurs and the mixture tends to separate into alternate slugs of gas and liquid. The gas slug, however, only occupies the central portion of the cross-section, and is surrounded by liquid, which flows backwards relative to the core of gas, though not necessarily relative to the walls of the pipe. The liquid slug normally has gas bubbles in suspension. At somewhat higher velocities, and with higher proportions of gas, a *churn* flow appears, in which the slugs are broken up and lose their individual identity. At very high ratios of air to water and at high velocities, an annular type of flow is obtained with air passing through a central core in the pipe and dragging upwards a film of liquid at the walls. The boundaries between the various types of flow regimes are not precisely defined and the region of flow may change from the bottom to the top of a tall pipe as a result of the expansion of the gas due to decreasing pressure. In air-lift pumping, *slug* flow and *churn* flow are the types most commonly encountered. Furthermore, although the flow pattern near the air-injection point (at the *foot piece)* will be strongly influenced by the nature of the injector, the flow soon assumes a pattern which is independent of that in the immediate vicinity of the injection point.

Thus, if the gas is injected in the form of small dispersed bubbles in order to reduce the *slip velocity*, coalescence rapidly occurs to give large bubbles and slugs.

The distribution of air and liquid in the pipe, and the proportion of the cross-section occupied by liquid (the *holdup* of the liquid) have an important bearing on the flow of the two fluids, firstly because the hydrostatic pressure is affected by the liquid holdup, and secondly because the nature of the flow affects the frictional pressure drop. Furthermore, the velocity of the air relative to the liquid is also dependent upon its pattern of distribution.

There are several methods whereby the holdup of liquid in a flowing mixture can be determined. One involves a measurement of the attenuation of a beam of γ-rays passed through the mixture; in another, the fluid in a test section is suddenly isolated by rapidly operating valves and the volume of liquid retained in the pipe is measured. In recent years, new measuring tools based on change in capacitance, tomography, etc., have also been developed. The former method gives point values and is therefore useful for determining the distribution throughout the length of the pipe, but the second enables an average value to be obtained from a single reading. Results obtained by the isolating method[15] for the liquid holdup in a 25 mm. diameter pipe are shown graphically in Fig. 8.43 which applies to the regime of slug flow, the limits of which are indicated approximately on the diagram.

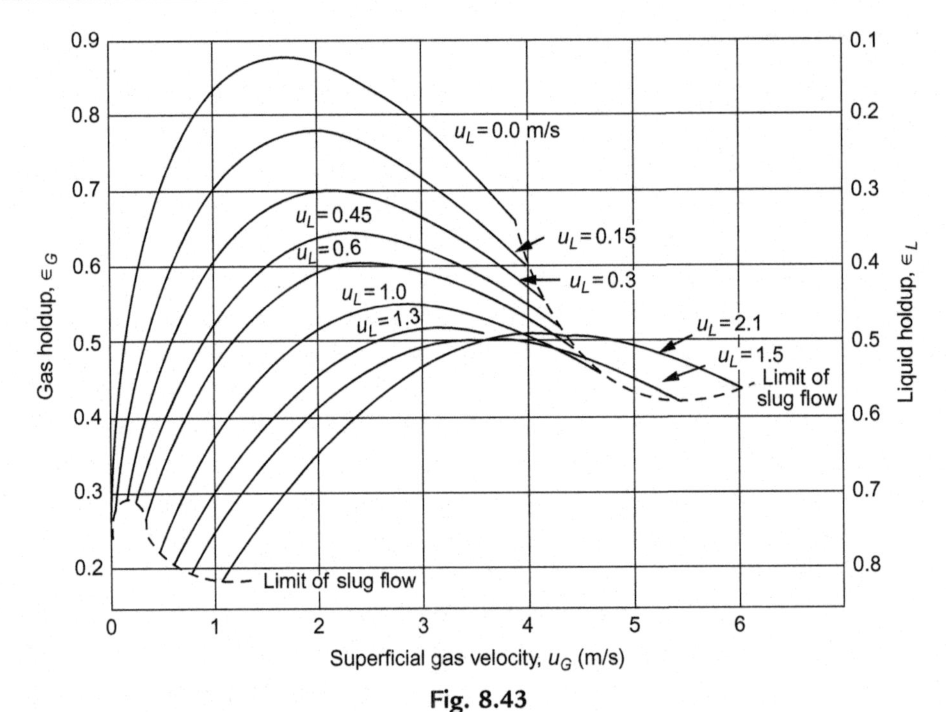

Fig. 8.43

Liquid holdup ϵ_L and gas holdup ϵ_G ($=1-\epsilon_L$) for slug flow in a 25 mm diameter pipe as a function of superficial gas velocity (u_G) and superficial liquid velocity (u_L).

Values of the holdup may be used to estimate the frictional losses in the pipe, since the overall difference in fluid head Δh of the liquid is equal to the friction head less the hold-up of liquid per unit area;

$$\Delta h = \Delta h_F - \frac{V_L}{A} \tag{8.55}$$

$$= \Delta h_F - l\in_L \tag{8.56}$$

where

Δh is the difference in head over a length l of pipe,

Δh_F is the corresponding frictional head,

V_L is the volume of liquid in the pipe of cross-section A, and.

ϵ_L is the mean value of the liquid holdup.

In the usual case h and h_F are falling in the direction of flow and Δh and Δh_F are therefore negative. Values of frictional pressure drop, $-\Delta P_{TPF}$ may conveniently be correlated in terms of the pressure drop $-\Delta P_L$ for liquid flowing alone at the same volumetric rate.

Experimental results obtained for plug flow in a 25 mm. diameter pipe are given as follows by Richardson and Higson[15]:

$$\Phi_G^2 = \frac{-\Delta P_{TPF}}{-\Delta P_L} = \varepsilon_L^{-4/3} \quad (\varepsilon_L > 0.3) \tag{8.57}$$

At high flowrates when chum flow sets in, the pressure drop undergoes a sudden increase and the coefficient in Eq. (8.57) increases:

$$\Phi_G^2 = \frac{-\Delta P_{TPF}}{-\Delta P_L} = 2.25\varepsilon_L^{-4/3} \quad (\varepsilon_L > 0.3) \tag{8.58}$$

Expressions of this type have also been obtained by several other researchers including Armand,[16] Schmidt et al.,[17] Govier et al.,[18,19] Isbin et al.[20] and Moore and Wilde.[21]

Eqs (8.57) and (8.58) are satisfactory except at low liquid rates when the frictional pressure drop is a very small proportion of the total pressure drop. Frictional effects can then even be negative, because the liquid may then flow downwards at the walls, with the gas passing upwards in slugs.

Operation of the air-lift pump

In an experimental study of a small air-lift pump,[15] (25 mm. diameter and 13.8 m overall height), the results were expressed by plotting the efficiency of the pump, defined as the useful work done on the water divided by the energy required for isothermal compression of the air, to a basis of energy input in the air. In each case, the curve was found to rise sharply to a maximum and then to fall off more gradually. Typical results are shown in Fig. 8.44.

The effects of the addition of a surface active agent were investigated because the distribution of air might be affected and the slip velocity reduced. As a result of reducing the surface tension from 0.07 to 0.045 N/m, the maximum efficiency of the pump was increased from 49% to 66%.

A characteristic of the pump was the appearance of a cyclic pattern in the rate of discharge. The resulting fluctuations were responsible for a wastage of energy, not only because of the need continually to reaccelerate the fluid above its equilibrium velocity, but also because frictional losses were thereby increased. The fluctuations were reduced, and the efficiency improved, by a limited throttling of the liquid supply, and by reducing the air capacity on the downstream side of the air control valve. There was a limit, of course, to the improvement in efficiency which would be obtained by throttling the liquid inlet, because frictional losses would be excessive in small throttles. In Fig. 8.44, the continuous curve relates to conditions where the inlet was not throttled (diameter 33 mm), and the broken curve refers to conditions where the water inlet was throttled to 15 mm diameter.

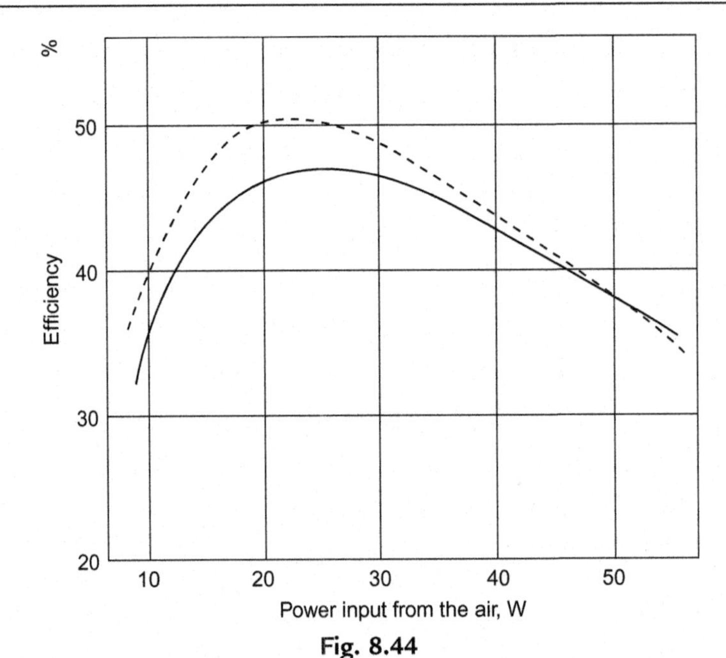

Fig. 8.44
Efficiency of air-lift pump as a function of energy input from air, showing effect of throttling water inlet. Submergence = 50%.

The effect of resistances and capacities in the air line and water feed has been studied theoretically and it has been shown that oscillations will tend to be stable if there is a large air capacity downstream from the control valve, and the inclusion of a resistance in the water supply line should dampen them.

Oscillations occur because the liquid column requires more than the equilibrium quantity of air to produce the initial acceleration. It therefore becomes over-accelerated and excess of liquid enters the limb with the result that it is retarded and then subsequently has to be accelerated again. During the period of retardation some liquid may actually run backwards; this can readily be prevented however, by incorporating a nonreturn valve in the submergence limb.

8.5 Vacuum Pumps

Many chemical plants, particularly distillation columns, operate at low pressures, and pumps are required to create and maintain the vacuum. Vacuum pumps take in gas at a very low pressure and normally discharge at atmospheric pressure, and pressure ratios are therefore high. Rotary pumps are generally preferred to piston pumps.

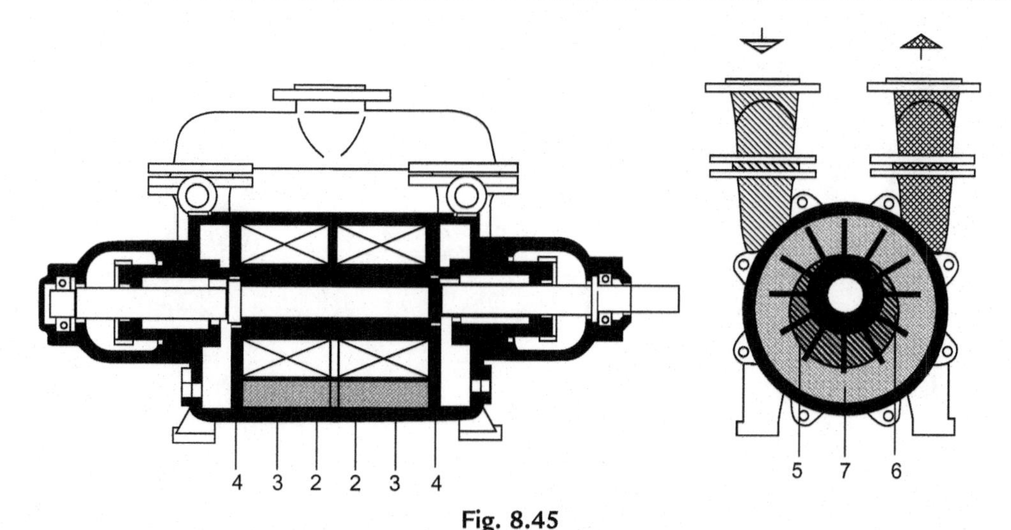

Fig. 8.45
Single-stage, single-action liquid ring gas pump: 1, shaft; 2, impeller; 3, casing; 4, guide plate; 5, suction port; 6, discharge port; 7, liquid ring.

Sliding vane and liquid ring (including the Nash Hytor) types of pumps, already described in Section 8.3.1, are frequently used as vacuum pumps and are capable of achieving pressures down to about 1.3 N/m^2. A pump suitable for vacuum operation is illustrated in Fig. 8.45. Care must be taken to avoid cavitation which can occur if the vapour fraction in the rotating cells exceeds about 50%. When water is used as the sealing liquid, its vapour pressure limits the lower limit of pressure which can be achieved to about 3.3 kN/m^2, and a liquid with a lower vapour pressure must be used if lower pressures are required. Lower pressures may also be obtained by using the liquid-ring pump in series with an additional device, such as the steam-jet ejector, described below.

The *steam-jet ejector*, illustrated in Fig. 8.46, is very commonly used in the process industries since it has no moving parts and will handle large volumes of vapour at low pressures. The operation of the unit is shown in Fig. 8.47. The steam is fed at constant pressure P_1 and expands nearly isentropically along AC; mixing of the steam and vapour which is drawn in, is represented by CE and vapour compression continues to the throat F and to the exit at G. The steam required increases with the compression ratio so that a single-stage unit will operate down to about 17 kN/m^2 corresponding to a compression ratio of 6:1. For lower final pressures multistage units are used, and Fig. 8.48 shows the relationship between the number of stages, steam pressure, and vacuum required. If cooling water is applied between the stages an improved performance will be obtained.

As a guide, a single-stage unit gives a vacuum to 13.5 kN/m^2, a double stage from 3.4 to 13.5 kN/m^2, and a three-stage unit from 0.67 to 2.7 kN/m^2.

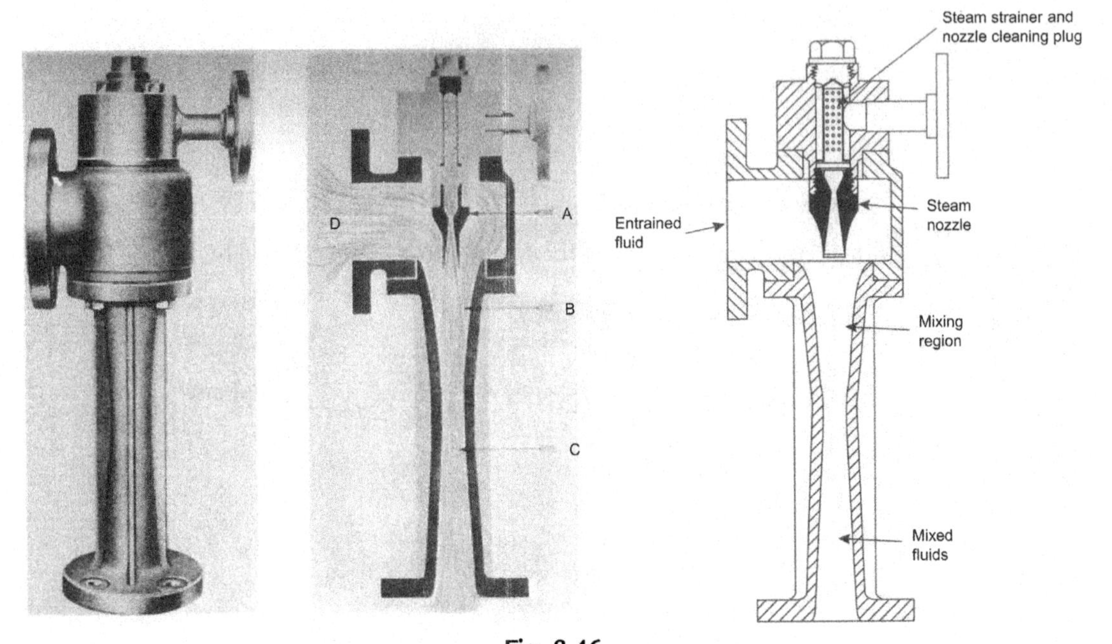

Fig. 8.46
Steam jet ejector. (A) Steam nozzle. (B) Mixing region. (C) Mixed fluids. (D) Entrained fluid.

For very low pressures, a diffusion pump is used with a rotary pump as the first stage. The principle of operation is that the gas diffuses into a stream of oil or mercury and is driven out of the pump by molecular bombardment. Useful practical tips concerning the relative merits and demerits, and selection of vacuum pumps for chemical processing applications has been provided by Collins.[22]

8.6 Power Requirements for Pumping Through Pipelines

A fluid will flow of its own accord as long as its energy per unit mass decreases in the direction of flow. It will flow in the opposite direction only if a pump is used to supply energy, and to increase the pressure at the upstream end of the system.

The energy balance is given by (Eq. 2.55):

$$\Delta \frac{u^2}{2\alpha} + g\Delta z + \int_{P_1}^{P_2} v\,dP + W_s + F = 0$$

The work done on unit mass of fluid is $-W_s$, and the total rate at which energy must be transferred to the fluid is $-GW_s$, for a mass rate of flow G.

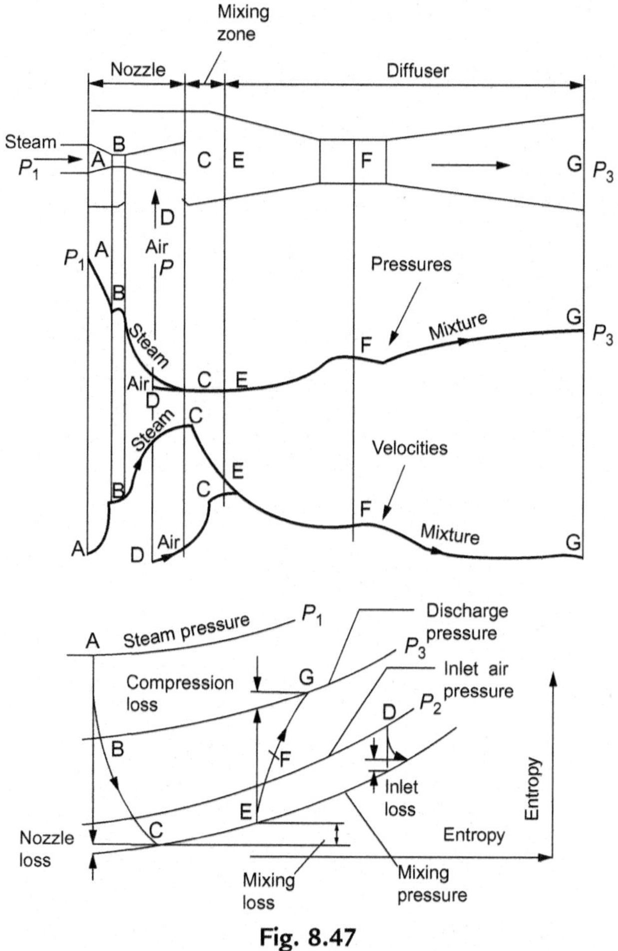

Fig. 8.47
Ejector flow phenomena.

The power requirement **P** is therefore given by:

$$\mathbf{P} = -GW_s = G\left(\Delta\frac{u^2}{2\alpha} + g\Delta z + \int_{P_1}^{P_2} v\mathrm{d}P + F\right) \tag{8.59}$$

8.6.1 Liquids

The term F in Eq. (8.59) may be calculated directly for the flow of liquid through a uniform pipe. If a liquid is pumped through a height Δz from one open tank to another and none of the kinetic energy is recoverable as pressure energy, the fluid pressure is the same at both ends of the system and $\int_{P_1}^{P_2} v\mathrm{d}p$ is zero. The power requirement, is, then:

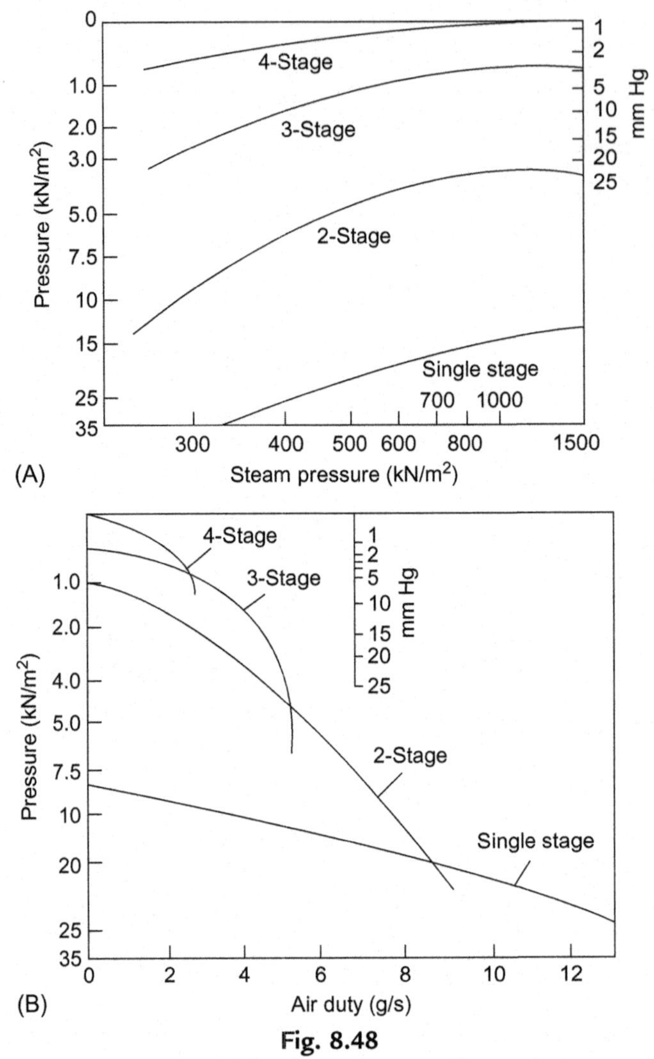

Fig. 8.48

(A) Diagram showing most suitable number of ejector stages for varying vacua and steam pressures. (B) Comparative performance curves of ejectors.

$$\mathbf{P} = G\left(\frac{u^2}{2\alpha} + g\Delta z + F\right) = Ghg \tag{8.60}$$

Taking into account the pump efficiency η, the overall power requirement is obtained from the product of the mass flowrate G, the total head h and the acceleration due to gravity g as:

$$\mathbf{P} = \frac{1}{\eta}Ghg \tag{8.61}$$

The application of this equation is best illustrated by the following examples.

Example 8.7

2.16 m^3/h water at 320 K is pumped through a 40 mm i.d. pipe, through a length of 150 m in a horizontal direction, and up through a vertical height of 10 m. In the pipe there are a control valve, equivalent to 200 pipe diameters, and other pipe fittings equivalent to 60 pipe diameters. Also in the line is a heat exchanger across which the head lost is 2 m water. Assuming the main pipe has a roughness of 0.0002 m, what power must be supplied to the pump if it is 60% efficient?

Solution

$$\text{Area for flow} = (0.040)^2 \times \frac{\pi}{4} = 0.0012\ \text{m}^2$$

$$\text{Flow of water} = \frac{2.16}{3600} = 600 \times 10^{-6}\,\text{m}^3/\text{s}$$

$$\therefore \text{velocity} = \frac{(600 \times 10^{-6})}{0.0012} = 0.50\ \text{m/s}$$

$$\text{At 320 K}, \mu = 0.65\ \text{mNs/m}^2 = 0.65 \times 10^{-3}\text{Ns/m}^2$$

$$\rho = 1000\ \text{kg/m}^3$$

$$\therefore \text{Re} = \frac{(0.040 \times 0.50 \times 1000)}{(0.65 \times 10^{-3})} = 30{,}780$$

$$\therefore \phi = \frac{R}{\rho u^2} = 0.004 \text{ for a relative roughness of } \frac{e}{d} = \left(\frac{0.0002}{0.040}\right) = 0.005 \quad (\text{from Fig.3.7})$$

Equivalent length of pipe = 150 + 10 + (260 × 0.040) = 170.4 m.

$$\therefore h_f = 4\phi \frac{l}{d}\frac{u^2}{g} \quad (\text{from Eq.3.20})$$

$$= 4 \times 0.004 \left(\frac{170.4}{0.040}\right)\left(\frac{0.50^2}{9.81}\right)$$

$$= 1.74\ \text{m}$$

$$\therefore \text{Total head to be developed} = (1.74 + 10 + 2) = 13.74\,\text{m}$$

$$\text{Mass flow of water} = \left(600 \times 10^{-6} \times 1000\right) = 0.60\,\text{kg/s}$$

$$\therefore \text{Power required} = (0.60 \times 13.74 \times 9.81) = 80.9\,\text{W}$$

$$\therefore \text{Power to be supplied} = 80.9 \times \left(\frac{100}{60}\right) = \underline{\underline{135\,\text{W}}}$$

In this solution, the kinetic energy head, $u^2/2g$, has been neglected as this represents only $[0.5^2/(2 \times 9.81)] = 0.013$ m, or 0.1% of the total head.

Example 8.8

It is required to pump cooling water from a storage pond to a condenser in a process plant situated 10 m above the level of the pond. 200 m of 74.2 mm i.d. pipe is available and the pump has the characteristics given below. The head lost in the condenser is equivalent to 16 velocity heads based on the flow in the 74.2 mm pipe.

If the friction factor $\phi = 0.003$, estimate the rate of flow and the power to be supplied to the pump assuming an efficiency of 50%.

Discharge (m^3/s)	0.0028	0.0039	0.0050	0.0056	0.0059
Head developed (m)	23.2	21.3	18.9	15.2	11.0

Solution

The head to be developed,

$$\begin{aligned} h &= 10 + 4\phi \frac{l}{d}\frac{u^2}{g} + 8\frac{u^2}{g} \quad (\text{from } Eq.3.20) \\ &= 10 + 4 \times 0.003\left(\frac{200}{0.0742}\right)\frac{u^2}{g} + \frac{16u^2}{2g} \\ &= 10 + 4.12\, u^2 \text{ m water} \end{aligned}$$

Discharge,

$$Q = (\pi d^2/4)u = (\pi/4) \times 0.0742^2 \times u = 0.0043 u \text{ m}^3/\text{s}$$

$$\therefore\ u = \frac{Q}{0.00434} = 231.3Q \text{ m/s}$$

$$\therefore\ h = 10 + 4.12(231.3Q)^2$$

$$= \left(10 + 2.205 \times 10^5 Q^2\right) \text{m water}$$

Values of Q and h are plotted in Fig. 8.49 and the discharge at the point of intersection between the pump characteristic equation and the line of the above equation is 0.0054 m^3/s.

$$\text{The head developed is thus, } h = 10 + \left(2.205 \times 10^5 \times 0.0054^2\right)$$

$$= 16.43, \text{say } 16 \text{ m water}$$

$$\text{Power required} = \frac{(0.0054 \times 1000 \times 16.43 \times 9.81)}{0.50}$$

$$= 1741 \text{ W or } \underline{\underline{1.74 \text{kW}}}$$

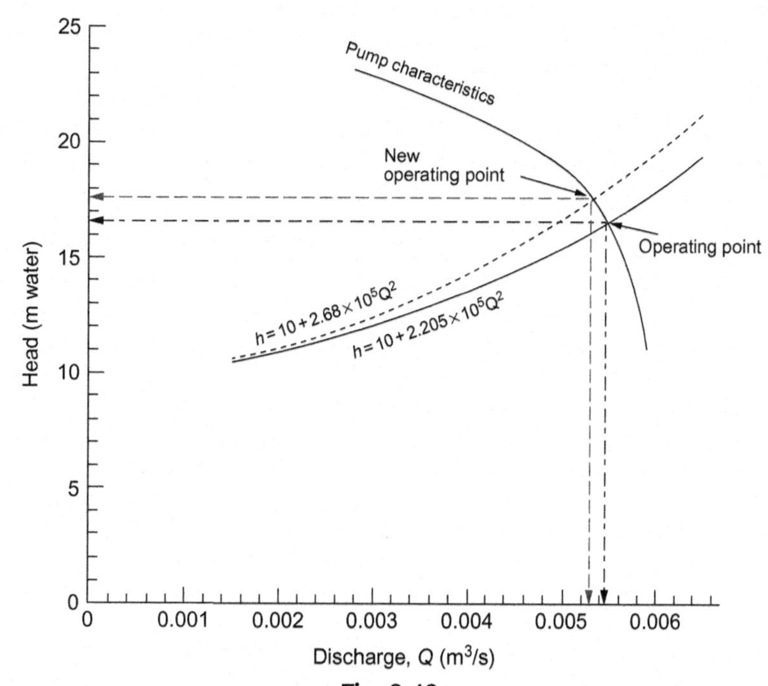

Fig. 8.49
Pump operating point for Examples 8.8 and 8.11.

Selection of pipe diameter

If a fluid is to be pumped between two points, the diameter of the pipeline should be chosen so that the overall cost of operation is a minimum. The smaller the diameter, the lower is the initial cost of the line but the greater is the cost of pumping; an economic balance must therefore, be achieved.

The initial cost of a pipeline, the depreciation, and maintenance costs will be approximately proportional to the diameter raised to a power of between 1.0 and 1.5. The power for pumping an incompressible fluid at a given rate G is made up of two parts:

(1) that necessitated by the difference in static pressure and vertical height at the two ends of the system (this is independent of the diameter of the pipe);
(2) that attributable to the kinetic energy of the fluid and the work done against friction. If the kinetic energy is small, this is equal to:

$$G\left[4\phi\frac{l}{d}u^2\right] \quad \text{(from } Eqs\text{ 3.19 and 8.60)} \tag{8.62}$$

which is proportional to $d^{-4.5 \text{ to } -5}$ for turbulent flow, since $u \propto d^{-2}$ and $\phi \propto u^{-0.25 \text{ to } 0}$, according to the roughness of the pipe.

The power requirement can therefore, be calculated as a function of d and the cost obtained. The total cost *per annum* is then plotted against the diameter of pipe and the optimum conditions are found either graphically or by differentiation as shown in Example 8.9.

Example 8.9

The relation between cost per unit length C of a pipeline installation and its diameter d is given by:

$$C = a + bd$$

where a and b are independent of pipe size. Annual charges are a fraction β of the capital cost. Obtain an expression for the optimum pipe diameter on a minimum cost basis for a fluid of density ρ and viscosity μ flowing at a mass rate of G. Assume that the fluid is in turbulent flow and that the Blasius equation is applicable, that is, the friction factor is proportional to the Reynolds number to the power of minus one quarter. Indicate clearly how the optimum diameter depends on the flowrate and fluid properties.

Solution

The total annual cost of a pipeline consists of a capital charge plus the running costs. The chief element of the running cost will be the power required to overcome the head loss which is given by Eq. (3.20):

$$h_f = 4\phi \frac{l}{d} \frac{u^2}{g}$$

If $\phi = R/\rho u^2 = 0.04/Re^{0.25}$, the head loss per unit length l is given by Eq. (3.20) as:

$$\frac{h_f}{l} = 4 \frac{0.04}{Re^{0.25}} \frac{1}{d} \frac{u^2}{g}$$

$$= 0.16 \frac{u^2}{gd} \left(\frac{\mu}{\rho u d} \right)^{0.25}$$

$$= 0.16 \frac{u^{1.75} \mu^{0.25}}{\rho^{0.25} g d^{1.25}}$$

The velocity,

$$u = G/\rho A = G/\rho(\pi/4)d^2 = 1.27G/\rho d^2$$

$$\therefore \frac{h_f}{l} = \frac{0.16(1.27G/\rho d^2)^{1.75} \mu^{0.25}}{\rho^{0.25} g d^{1.25}}$$

$$= \frac{0.244 G^{1.75} \mu^{0.25}}{\rho^2 g d^{4.75}}$$

The power required for pumping if the pump efficiency η is given as:

$$\mathbf{P} = Gg\frac{0.244G^{1.75}\mu^{.025}}{\rho^2 g d^{4.75}}\frac{1}{\eta}$$

If $\eta = 0.5$ (say), $\mathbf{P} = 0.488G^{2.75}\mu^{0.25}/(\rho^2 d^{4.75})$ (Watt).

If c = power cost/W, the cost of pumping is:

$$\frac{0.488cG^{2.75}\mu^{0.25}}{\rho^2 d^{4.75}}$$

The total annual cost is then

$$C = (\beta a + \beta b d) + \frac{\gamma G^{2.75}\mu^{0.25}}{\rho^2 d^{4.75}}$$

where $\gamma = 0.488c$.

Differentiating the total cost with respect to the diameter gives:

$$\frac{dC}{dd} = \beta b - \frac{4.75\gamma G^{2.75}\mu^{0.25}}{\rho^2 d^{5.75}}$$

For minimum cost, $dC/dd = 0$ and:

$$d^{5.75} = \frac{4.75\gamma G^{2.75}\mu^{0.25}}{\rho^2 \beta b}$$

or:

$$d = \underline{\underline{\frac{KG^{0.48}\mu^{0.043}}{\rho^{0.35}}}}$$

where:

$$K = \left(\frac{4.75\gamma}{\beta b}\right)^{0.174} = \left(\frac{2.32c}{\beta b}\right)^{0.174}$$

Effect of fluctuations in flowrate on power for pumping

The importance of maintaining the flowrate in a pipeline constant may be seen by considering the effect of a sinusoidal variation in flowrate. This corresponds approximately to the discharge conditions in a piston pump during the forward movement of the piston. The flowrate $\dot{Q}$ is given as a function of time t by the relation:

$$\dot{Q} = Q_m \sin \omega t \tag{8.63}$$

The discharge rate rises from zero at $t = 0$ to Q_m when $wt = \pi/2$, i.e. $t = \pi/2\omega$.

The mean discharge rate Q is then given by:

$$Q = \frac{1}{(\pi/2\omega)} \int_0^{\pi/2\omega} Q_m \sin \omega t dt$$
$$= \frac{Q_m \cdot 2\omega}{\pi} \left[-\frac{1}{\omega} \cos \omega t \right]_0^{\pi/2\omega}$$
$$= \frac{2}{\pi} Q_m \tag{8.64}$$

The power required for pumping will be given by the product of the volumetric flowrate and the pressure difference between the pump outlet and the discharge end of the pipeline. Taking note of the fluctuating nature of the flow, it is necessary to consider the energy transferred to the fluid over a small time interval and to integrate over the cycle to obtain the mean value of the power.

$$\text{Thus, the power requirement, } \mathbf{P} = \frac{1}{\pi/2\omega} \int_0^{\pi/2\omega} Q_m \sin \omega t (-\Delta P) dt \tag{8.65}$$

The form of the integral will depend on the relation between $-\Delta P$ and $\dot{Q}$.

Streamline flow

If it is assumed that $-\Delta P = K \dot{Q} = KQ_m \sin \omega t$, then:

$$\mathbf{P} = \frac{2\omega}{\pi} K Q_m^2 \int_0^{\pi/2\omega} \sin^2 \omega t dt$$
$$= \frac{2\omega}{\pi} K \left(Q^2 \frac{\pi^2}{4} \right) \frac{1}{2} \int_0^{\pi/2\omega} (1 - \cos 2\omega t) dt$$
$$= \frac{KQ^2 \pi\omega}{4} \left[t - \frac{1}{2\omega} \sin 2\omega t \right]_0^{\pi/2\omega}$$
$$= \frac{KQ^2 \pi^2}{8}$$

For steady flow:

$$\mathbf{P} = Q(-\Delta P) = KQ^2 \tag{8.66}$$

Thus power is increased by a factor of $\pi^2/8$ (=1.23) as a result of the sinusoidal fluctuations in the flow rate.

Turbulent flow

As a first approximation it may be assumed that pressure drop is proportional to the square of flowrate or:

$$-\Delta P = K^1 Q^2 = K^1 Q_m^2 \sin^2 \omega t \tag{8.67}$$

Then:

$$\begin{aligned}
\mathbf{P} &= \frac{2\omega}{\pi} K^1 Q_m^3 \int_0^{\pi/2\omega} \sin^3 \omega t \, dt \\
&= \frac{2\omega}{\pi} K^1 \left(Q^3 \frac{\pi^3}{8} \right) \frac{1}{4} \int_0^{\pi/2\omega} (3 \sin \omega t - \sin 3\omega t) \, dt \\
&= \frac{K^1 Q^3 \omega \pi^2}{16} \left[-\frac{3}{\omega} \cos \omega t + \frac{1}{3\omega} \cos 3\omega t \right]_0^{\pi/2\omega} \\
&= \frac{K^1 Q^3 \omega \pi^2}{16} \left(\frac{3}{\omega} - \frac{1}{3\omega} \right) \\
&= \frac{K^1 Q^3 \pi^2}{6}
\end{aligned} \tag{8.68}$$

For steady flow:

$$\mathbf{P} = Q(-\Delta P) = K^1 Q^3 \tag{8.69}$$

Thus the power is increased by a factor of $\pi^2/6$ (=1.64) as a result of fluctuations, compared with 1.23 for streamline flow.

In addition to the increased power to overcome friction, it will be necessary to supply the energy required to accelerate the liquid at the beginning of each cycle.

8.6.2 Gases

If a gas is pumped under turbulent flow conditions from a reservoir at a pressure P_1 to a second reservoir at a higher pressure P_2 through a uniform pipe of cross-sectional area A by means of a pump situated at the upstream end, the power required is:

$$\mathbf{P} = G\left(\frac{u^2}{2} + \int_{P_1}^{P_2} v \, dP + F \right) \quad \text{(from Eq.8.59)} \tag{8.70}$$

neglecting pressure changes arising from change in vertical height.

In order to maintain the gas flow, the pump must raise the pressure at the upstream end of the pipe to some value P_3, which is greater than P_2. The required value of P_3 will depend somewhat on the conditions of flow in the pipe. Thus for isothermal conditions, P_3 may be calculated from Eq. (4.55), since the downstream pressure P_2 and the mass rate of flow G are known. For nonisothermal conditions, the appropriate equation, such as (4.66) or (4.77), must be used.

The power requirement is then that for compression of the gas from pressure P_1 to P_3 and for imparting the necessary kinetic energy to it. Under normal conditions however, the kinetic energy term is negligible. Thus for an isothermal efficiency of compression η, the power required is:

$$\mathbf{P} = \frac{1}{\eta} G P_1 v_1 \ln \frac{P_3}{P_1} \tag{8.71}$$

Example 8.10

Hydrogen is pumped from a reservoir at 2 MN/m^2 through a clean horizontal mild steel pipe 50 mm in diameter and 500 m long. The downstream pressure is also 2 MN/m^2 and the pressure of the gas is raised to 2.5 MN/m^2 by a pump at the upstream end of the pipe. The conditions of flow are isothermal and the temperature of the gas is 295 K. What is the flowrate and what is the effective rate of working of the pump if it operates with an efficiency of 60%?

Viscosity of hydrogen = 0.009 mN s/m^2 at 295 K.

Solution

Viscosity of hydrogen = 0.009 mN s/m^2 or 9×10^{-6} N s/m^2.

Density of hydrogen at the mean pressure of 2.25 MN/m^2

$$= \left(\frac{2}{22.4}\right)\left(\frac{2250}{101.3}\right)\left(\frac{273}{295}\right) = 1.83 \text{ kg/m}^3$$

Firstly, an approximate value of G is obtained by neglecting the kinetic energy of the fluid. Taking P_1 and P_2 as the pressures at the upstream and downstream ends of the pipe, 2.5×10^6 and 2.0×10^6 N/m^2, then in Eq. (4.56):

$$P_1 - P_2 = 4\phi \frac{l}{d} \rho_m u_m^2$$

or:

$$0.5 \times 10^6 = 4\phi\left(\frac{500}{0.050}\right) \times 1.83 u_m^2 \text{ N/m}^2$$

$$\therefore \phi u_m^2 = \frac{R}{\rho u^2} u_m^2 = 6.83$$

$$\therefore \phi Re^2 = \frac{R}{\rho u^2} Re^2 = \frac{6.83 \times 0.050^2 \times 1.83^2}{(9 \times 10^{-6})^2}$$
$$= 7.02 \times 10^8$$

Taking the roughness of the pipe surface, e as 0.00005 m:

$$\frac{e}{d}=\frac{0.00005}{0.05}=0.001 \text{ and } \phi Re^2=7.02\times 10^8 \text{ from Fig.3.8}$$

$$\therefore \frac{4G}{\pi\mu d}=5.7\times 10^5$$

$$\therefore G=\left(5.7\times 10^5\times\frac{\pi}{4}\times 9\times 10^{-6}\times 0.050\right)$$

$$=0.201 \text{ kg/s}$$

From Fig. 3.7, $(\phi = R/\rho u^2) = 0.0024$.

Taking the kinetic energy of the fluid into account, Eq. (4.56) may be used:

$$\left(\frac{G}{A}\right)^2 \ln\frac{P_1}{P_2}+(P_2-P_1)\rho_m+4\phi\frac{l}{d}\left(\frac{G}{A}\right)^2=0$$

Using the value of (ϕ) obtained by neglecting the kinetic energy:

$$\left(\frac{G}{A}\right)^2 \ln\frac{2.5}{2.0}-(0.5\times 10^6\times 1.83)+\left[4\times 0.0024\times\left(\frac{500}{0.05}\right)\right]\left(\frac{G}{A}\right)^2=0$$

$$0.223\left(\frac{G}{A}\right)^2-915{,}000+96.0\left(\frac{G}{A}\right)^2=0$$

$$\therefore \frac{G}{A}=97.5$$

$$\therefore G=\left(97.5\times\frac{\pi}{4}\times 0.050^2\right)$$

$$=0.200 \text{ kg/s}$$

This value is very close to that calculated above. Thus, as is commonly the case, when the pressure drop is a relatively small proportion of the total pressure, the change in kinetic energy is negligible compared with the frictional losses. This would not be true had the pressure drop been much greater.

The power requirements for the pump can now be calculated from Eq. (8.71).

$$\text{Power}=\frac{GP_m v_m \ln(P_1/P_2)}{\eta}$$

$$=\frac{\left(0.200\times 2.25\times 10^6\times(1/1.83)\times\ln\left(\frac{2.5\times 10^6}{2\times 10^6}\right)\right)}{0.60}$$

$$=9.14\times 10^4 \text{ W or } \underline{\underline{91.4 \text{ kW}}}$$

8.7 Effect of Minor Losses

With reference to the power calculations for the flow of fluids in pipes, the discussion thus far has been restricted to straight lengths of pipes of uniform size (internal diameter). In practice, however, the pipe sections include numerous obstructions like joining and splitting of fluid streams, (T junctions), change of directions (elbows, bends, etc.), valves to control and regulate the flow (gate, globe, needle, one-way valves, etc.), entrance (storage tank to a pipe), and exit (pipe into a tank) effects, etc. All of these items, shown in Figs. 3.17 and 3.18, add to the overall resistance to flow in the system and therefore, this can influence the operating point on the pump performance curve. While for long pipes, these losses may be insignificant (hence the term minor losses), this may not be so far short pipes used within a plant to transfer fluids from one equipment to another. Therefore, the need to estimate such losses frequently arises in the sizing of a pump or selecting a suitable diameter of the pipe for a specific application. While the additional pressure loss incurred due to a fitting or a valve depends upon the detailed design of a valve, type of fitting (screwed versus flanged), size of fitting, flow pattern, etc., it is customary to introduce the concept of an "equivalent length" (L_e/d) which must be added to the length of the pipe in the calculation of head loss. Another approach expresses this contribution in terms of a constant k_f which multiplied with the kinetic energy term as $k_f.u^2/2\,g$ accounts for the additional pressure drop due to the fitting. Undoubtedly, both the values of the equivalent length (expressed as the number of pipe diameters) and that of k_f depend upon the size of fitting and Reynolds number,[23–25] but the simplest approach treats both these parameters constant. Representative values of the equivalent length and k_f are included in Tables 8.1 and 8.2, similar to that shown in Table 3.3, respectively and more detailed listings can be found in references 23–25, 27.

By convention, the velocity used in the head loss calculation as $k_f u^2/2g$ is that in the pipe.

Table 8.1 Minor losses in valves and fittings[23]

Type of Fitting	Equivalent Length
Gate valve	
25% open	900*d*
50% open	160*d*
75% open	35*d*
Fully open	13*d*
Globe valve (fully open, conventional design)	350–450*d*
90° Elbow (standard)	30*d*
45° Elbow (standard)	16*d*
90° long radius elbow	20*d*
Return bend	50*d*
T-junctions	
Flow in main line	20*d*
Flow in branch	60*d*

Table 8.2 Values of fitting constant k_f^{26}

	Nominal Diameter (mm)								
	Screwed				Flanged				
	12.7	25	50	100	25	50	100	200	500
Valves (fully open):									
Globe	14	8.2	6.9	5.7	13	8.5	6.0	5.8	5.5
Gate	0.30	0.24	0.16	0.11	0.80	0.35	0.16	0.07	0.03
Swing check	5.1	2.9	2.1	2.0	2.0	2.0	2.0	2.0	2.0
Angle	9.0	4.7	2.0	1.0	4.5	2.4	2.0	2.0	2.0
Elbows:									
45° regular	0.39	0.32	0.30	0.29					
45° long radius					0.21	0.20	0.19	0.16	0.14
90° regular	2.0	1.5	0.95	0.64	0.50	0.39	0.30	0.26	0.21
90° long radius	1.0	0.72	0.41	0.23	0.40	0.30	0.19	0.15	0.10
180° regular	2.0	1.5	0.95	0.64	0.41	0.35	0.30	0.25	0.20
180° long radius					0.40	0.30	0.21	0.15	0.10
T-junctions									
Line flow	0.90	0.90	0.90	0.90	0.24	0.19	0.14	0.10	0.07
Branch flow	2.4	1.8	1.4	1.1	1.0	0.80	0.64	0.58	0.41

Example 8.11

Recalculate the rate of flow and pump power for the arrangement considered in Example 8.8 if the piping includes 2 globe (fully opened) and 2 gate (fully open) valves, 10 45° regular elbows. For $d = 74.2$ mm, the values of k_f are:

Globe valve, $k_f = 7$; Gate valve, $k_f = 0.15$ and 45° elbow, $k_f = 0.3$.

Solution

The additional head loss due to these fittings:

$$h_{fittings} = \frac{u^2}{2g}(2 \times 7 + 2 \times 0.15 + 10 \times 0.3)$$

$$= \frac{8.65u^2}{g} = 0.882u^2 \,(\text{m water})$$

$$\therefore \text{Head to be developed now} = 10 + 4.12u^2 + 0.882u^2$$

$$= 10 + 5u^2 \,(\text{m water})$$

In terms of Q:

$$h = \left(10 + 2.68 \times 10^5 Q^2\right) \text{m water}$$

This line is shown as broken line in Fig. 8.49 and the new value of $Q = 0.00528$ m^3/s.

The head developed is thus,

$$h = 10 + 2.68 \times 10^5 \times (0.00528)^2 = 17.47 \text{ m water}$$

$$\text{and power required} = (0.00528 \times 1000 \times 17.47 \times 9.81)/(0.50) = \underline{\underline{1810 \text{ W}}}$$

Example 8.12

For the system shown below, calculate the flowrates in each pipe when the difference between the free surface of water in tanks 1 and 2 is 8 m for the following conditions. For water, $\rho = 1000$ kg. m^{-3}, viscosity $\mu = 10^{-3}$ Pa.s. Neglect the minor losses and the effect of pipe roughness.

Case (i): $d_1 = 600$ mm, $d_2 = d_3 = 300$ mm.

Case (ii): $d_1 = 600$ mm, $d_2 = 400$ mm, $d_3 = 200$ mm.

Case (iii): $d_1 = d_2 = d_3 = 600$ mm.

Assume that the friction factor $(\phi) = 0.0395Re^{-1/4}$.

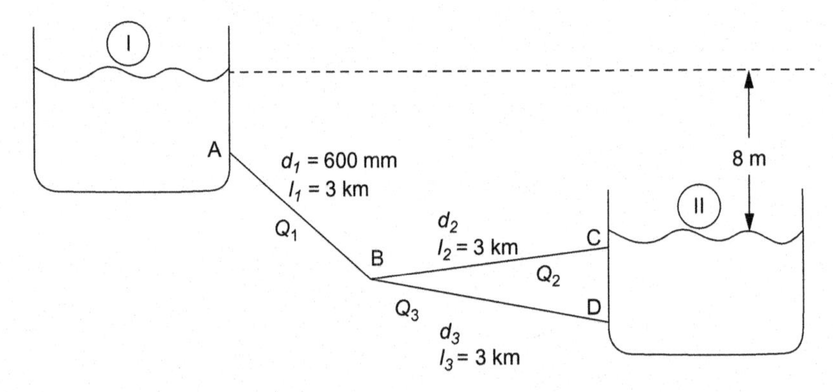

Solution

A fluid element can take the route ABC or ABD to flow from tank I to tank II, but the head available for flow is $z_I - z_{II} = 8$ m in both cases. From the overall continuity:

$$Q_1 = Q_2 + Q_3 \tag{1}$$

$$\Rightarrow \frac{\pi d_1^2}{4} u_1 = \frac{\pi d_2^2}{4} u_2 + \frac{\pi d_3^2}{4} u_3$$

$$\Rightarrow d_1^2 u_1 = d_2^2 u_2 + d_3^2 u_3$$

Now writing mechanical energy balance between points (I) and (II) on the free surface for the two paths:

$$\frac{P_I}{\rho g} + \frac{u_I^2}{2g} + z_I = \frac{P_{II}}{\rho g} + \frac{u_{II}^2}{2g} + z_{II} + h_l$$

$$\therefore h_l = z_I - z_{II} = 8 \text{ m}$$

as $P_I = P_{II} = P_{atm}$ and $u_I = u_{II} \approx 0$.

For path ABC (Neglecting minor losses):

$$h_l = \frac{4\phi_1 l_1 \rho u_1^2}{d_1 \rho g} + \frac{4\phi_2 l_2 \rho u_2^2}{d_2 \rho g} = 8$$

$$\Rightarrow \frac{4\phi_1 l_1 u_1^2}{d_1 g} + \frac{4\phi_2 l_2 u_2^2}{d_2 g} = 8 \quad (2)$$

For path ABD (Neglecting minor losses):

$$h_l = \frac{4\phi_1 l_1 \rho u_1^2}{d_1 \rho g} + \frac{4\phi_3 l_3 \rho u_3^2}{d_3 \rho g} = 8$$

$$\Rightarrow \frac{4\phi_1 l_1 u_1^2}{d_1 g} + \frac{4\phi_3 l_3 u_3^2}{d_3 g} = 8 \quad (3)$$

Using the expression of friction factor

$$\phi = 0.0395 Re^{-1/4}$$

$$\phi_1 = 0.0395 Re_1^{-1/4}$$

$$\Rightarrow \phi_1 = 0.0395 \left(\frac{\rho u_1 d_1}{\mu}\right)^{-1/4}$$

$$\Rightarrow \phi_1 = 0.0395 \left(\frac{\rho\left(Q_1/\left(\pi d_1^2/4\right)\right) d_1}{\mu}\right)^{-1/4}$$

$$\Rightarrow \phi_1 = 0.0395 \left(\frac{4\rho}{\pi\mu}\right)^{-1/4} \cdot \left(\frac{Q_1}{d_1}\right)^{-1/4}$$

$$\therefore \frac{4\phi_1 l_1 u_1^2}{d_1 g} = \frac{(4)(0.0395)\left(\frac{4\rho}{\pi\mu}\right)^{-1/4} \cdot \left(\frac{Q_1}{d_1}\right)^{-1/4} l_1 \left(\frac{4Q_1}{\pi d_1^2}\right)^2}{d_1 g}$$

$$= \frac{(4)(0.0395)}{g}\left(\frac{4\rho}{\pi\mu}\right)^{-1/4}\left(\frac{4}{\pi}\right)^2 \cdot \frac{Q_1^{1.75} l_1}{d_1^{4.75}}$$

$$= A_0 \frac{Q_1^{1.75} l_1}{d_1^{4.75}}$$

here

$$A_0 = \frac{(4)(0.0395)}{g}\left(\frac{4}{\pi}\right)^{1.75}\left(\frac{\rho}{\mu}\right)^{-1/4}$$

$$= 5.1 \times 10^{-4}\left(\frac{4}{\pi}\right)^{1.75} = 7.78 \times 10^{-4}$$

Case (i):

From Eqs (2) and (3):

$$A_0\left(\frac{Q_1^{1.75}l_1}{d_1^{4.75}}+\frac{Q_2^{1.75}l_2}{d_2^{4.75}}\right)=8$$

$$\Rightarrow\frac{7.78\times10^{-4}\times3000}{(0.6)^{4.75}}Q_1^{1.75}+\frac{7.78\times10^{-4}\times3000}{(0.3)^{4.75}}Q_2^{1.75}=8$$

$$\Rightarrow 26.42Q_1^{1.75}+710.92Q_2^{1.75}=8 \quad (4)$$

Similarly,

$$A_0\left(\frac{Q_1^{1.75}l_1}{d_1^{4.75}}+\frac{Q_3^{1.75}l_3}{d_3^{4.75}}\right)=8$$

$$\Rightarrow\frac{7.78\times10^{-4}\times3000}{(0.6)^{4.75}}Q_1^{1.75}+\frac{7.78\times10^{-4}\times3000}{(0.3)^{4.75}}Q_3^{1.75}=8$$

$$\Rightarrow 26.42Q_1^{1.75}+710.9Q_3^{1.75}=8 \quad (5)$$

Subtracting Eqs (4) and (5):

$$Q_2=Q_3$$

From Eq. (1):

$$Q_1=2Q_2 \quad (6)$$

Substituting Eq. (6) in Eq. (4):

$$26.42(2Q_2)^{1.75}+710.9Q_3^{1.75}=8$$

$$\Rightarrow Q_2=Q_3=0.072\,\text{m}^3/\text{s}$$

$$\therefore\ Q_1=0.144\,\text{m}^3/\text{s}$$

Q_1	***Q_2***	***Q_3***	***Re_1***	***Re_2***	***Re_3***
0.144 m^3/s	0.072 m^3/s	0.072 m^3/s	3.18×10^5	3.06×10^5	3.06×10^5

Case (ii):

Following the same procedure as used for case 1:

$$26.42Q_1^{1.75}+181.3Q_2^{1.75}=8 \quad (7)$$

$$26.42Q_1^{1.75}+4878.1Q_3^{1.75}=8 \quad (8)$$

On subtracting Eqs (7) and (8):

$$Q_3=0.153Q_2$$

$$\therefore\ Q_1=Q_2+0.153Q_2=1.153Q_2$$

From Eq. (7):

$$26.42(1.153Q_2)^{1.75} + 181.28Q_2^{1.75} = 8$$

$$\Rightarrow Q_2 = 0.153 \text{ m}^3/\text{s}$$

$$\therefore Q_3 = 0.153 \times Q_2 = 0.023 \text{m}^3/\text{s}$$

and

$$Q_1 = 1.153 \times Q_2 = 0.176 \text{ m}^3/\text{s}$$

Q_1	Q_2	Q_3	Re_1	Re_2	Re_3
0.176 m^3/s	0.153 m^3/s	0.023 m^3/s	3.73×10^5	4.85×10^5	1.47×10^5

Case (iii):

Following the aforementioned procedure:

$$26.42Q_1^{1.75} + 26.42Q_2^{1.75} = 8 \tag{9}$$

$$26.42Q_1^{1.75} + 26.42Q_3^{1.75} = 8 \tag{10}$$

On subtracting Eqs (7) and (8):

$$\Rightarrow Q_2 = Q_3$$

Using Eq. (1):

$$Q_1 = 2Q_2$$

From Eq. (9):

$$26.42(2Q_2)^{1.75} + 26.42Q_2^{1.75} = 8$$

$$Q_2 = 0.218 \text{ m}^3/\text{s}$$

$$\therefore Q_3 = Q_2 = 0.218 \text{ m}^3/\text{s}$$

and

$$Q_1 = 2 \times Q_2 = 0.436 \text{ m}^3/\text{s}$$

Q_1	Q_2	Q_3	Re_1	Re_2	Re_3
0.436 m^3/s	0.218 m^3/s	0.218 m^3/s	9.24×10^5	4.62×10^5	4.62×10^5

Some of the Reynolds number values are beyond the limit of Blasius equation, but the use of more accurate friction factor equations will change the results by 5%–6% only.

8.8 Nomenclature

		Units in SI system	Dimensions in **M, L, T, θ**
A	Cross-sectional area	m^2	L^2
b	Width of pump impeller	m	L
c	Clearance in cylinder, i.e. ratio of clearance volume/swept volume	–	–
d	Diameter	m	**L**
F	Energy degraded due to irreversibility per unit mass of fluid	J/kg	$\mathbf{L^2 T^{-2}}$
G	Mass flowrate	kg/s	$\mathbf{MT^{-1}}$
G_A	Mass flowrate of air	kg/s	$\mathbf{MT^{-1}}$
G_L	Mass flowrate of liquid	kg/s	$\mathbf{MT^{-1}}$
g	Acceleration due to gravity	m/s^2	$\mathbf{LT^{-2}}$
H	Enthalpy per unit mass	J/kg	$\mathbf{L^2 T^{-2}}$
h	Head	m	**L**
h_a	Heat equivalent to atmospheric pressure	m	**L**
h_f	Friction head	m	**L**
h_i	Head at pump inlet	m	**L**
h_0	Height of liquid above pump inlet	m	**L**
h_r	Height through which liquid is raised	m	**L**
h_s	Submergence of air lift pump	m	**L**
K	Coefficient	$(N/m^2)/(m^3/s)$	$\mathbf{ML^{-4} T^{-1}}$
K^1	Coefficient	$(N/m^2)/(m^3/s)^2$	$\mathbf{ML^{-7}}$
l	Length	m	**L**
N	Revolutions per unit time	Hz	$\mathbf{T^{-1}}$
N_s	Specific speed of a pump	–	–
n	Number of stages of compression	–	–
P	Pressure	N/m^2	$\mathbf{ML^{-1} T^{-2}}$
ΔP	Pressure difference	N/m^2	$\mathbf{ML^{-1} T^{-2}}$
P_a	Atmospheric pressure	N/m^2	$\mathbf{ML^{-1} T^{-2}}$
P_i	Intermediate pressure	N/m^2	$\mathbf{ML^{-1} T^{-2}}$
P_v	Vapour pressure of liquid	N/m^2	$\mathbf{ML^{-1} T^{-2}}$
P_0	Pressure at pump suction tank	N/m^2	$\mathbf{ML^{-1} T^{-2}}$
$\mathbf{P}$	Power	W	$\mathbf{ML^{-2} T^{-3}}$
Q	Volumetric rate of flow	m^3/s	$\mathbf{L^3 T^{-1}}$
Q_m	Maximum rate of flow	m^3/s	$\mathbf{L^3 T^{-1}}$
$\dot{Q}$	Rate of flow at time t	m^3/s	$\mathbf{L^3 T^{-1}}$

q	Heat added from surrounding per unit mass of fluid	J/kg	$\mathbf{L}^2\,\mathbf{T}^{-2}$
R	Shear stress at pipe wall	N/m^2	$\mathbf{ML}^{-1}\,\mathbf{T}^{-2}$
r	Radius	m	$\mathbf{L}$
S	Entropy per unit mass	J/kg K	$\mathbf{L}^2\,\mathbf{T}^{-2}\boldsymbol{\theta}^{-1}$
T	Temperature	K	$\boldsymbol{\theta}$
t	Time	s	$\mathbf{T}$
U	Internal energy per unit mass	J/kg	$\mathbf{L}^2\,\mathbf{T}^{-2}$
u	Velocity of fluid	m/s	$\mathbf{LT}^{-1}$
u_G	Superficial gas velocity	m/s	$\mathbf{LT}^{-1}$
u_L	Superficial liquid velocity	m/s	$\mathbf{LT}^{-1}$
u_i	Velocity at pump inlet	m/s	$\mathbf{LT}^{-1}$
u_t	Tangential velocity of tip of vane of compressor or pump	m/s	$\mathbf{LT}^{-1}$
u_v	Velocity of fluid relative to vane of compressor	m/s	$\mathbf{LT}^{-1}$
u_1	Velocity at inlet to centrifugal pump	m/s	$\mathbf{LT}^{-1}$
u_2	Velocity at outlet to centrifugal pump	m/s	$\mathbf{LT}^{-1}$
V	Volume	m^3	$\mathbf{L}^3$
V_s	Swept volume	m^3	$\mathbf{L}^3$
v	Specific volume	m^3/kg	$\mathbf{M}^{-1}\,\mathbf{L}^3$
v_a	Specific volume at atmospheric pressure	m^3/kg	$\mathbf{M}^{-1}\,\mathbf{L}^3$
W	Net work done per unit mass	J/kg	$\mathbf{L}^2\,\mathbf{T}^{-2}$
W_s	Shaft work per unit mass	J/kg	$\mathbf{L}^2\,\mathbf{T}^{-2}$
W'	Work of compressor per cycle	J	$\mathbf{ML}^2\,\mathbf{T}^{-2}$
Z	Net positive suction head	m	$\mathbf{L}$
z	Height	m	$\mathbf{L}$
α	Correction factor for kinetic energy of fluid	–	–
β	Angle between tangential direction and blade of impeller at its tip	–	–
Δ	Finite difference in quantity	–	–
η	Efficiency	–	–
ϵ_G	Mean gas holdup (fraction)	–	–
ϵ_L	Mean liquid holdup (fraction)	–	–
γ	Ratio of specific heat at constant pressure to specific heat at constant volume	–	–
ω	Angular velocity	rad/s	$\mathbf{T}^{-1}$
Π	Dimensionless group	–	–
Φ_G^2	Ratio $-\Delta P_{TPF}/-\Delta P_L$	–	–

ϕ	Friction factor $(R/\rho u^2)$	–	–
ρ	Density	kg/m^3	$\mathbf{ML^{-3}}$
θ	Angle between tangential direction and direction of motion of fluid	–	–
τ	Torque	Nm	$\mathbf{ML^2\,T^{-2}}$

Suffixes

G	Gas
L	Liquid
TPF	Two-phase friction

References

1. Wilson KC, Addie GR, Sellgren A, Clift R. *Slurry transport using centrifugal pumps.* 3rd ed. New York: Springer; 2008.
2. Kalombo JJN, Haldenwang R, Chhabra RP, Fester VG. Centrifugal pump derating for non-Newtonian slurries. *J Fluids Eng* 2014;**136**:031302.
3. Furlan J, Visintainer R, Sellgren A. Centrifugal pump performance when handling highly non-Newtonian clays and tailing slurries. *Hydrotransport* 2014;**19**:117–30.
4. Marshall P. Positive displacement pumps—a brief survey. *Chem Eng (Lond)* 1985;**418**:52.
5. Harvest J. Recent developments in gear pumps. *Chem Eng (Lond)* 1984;**403**:28.
6. Gadsden C. Squeezing the best from peristaltic pumps. *Chem Eng (Lond)* 1984;**404**:42.
7. Stepanoff AJ. *Centrifugal and axial flow pumps.* 2nd ed. Melbourne, FL: Krieger; 1992.
8. Volk M. *Pump characteristics and applications.* 3rd ed. Boca Raton, FL: CRC Press; 2013.
9. AIChE's Equipment Testing Procedures Committee. Assess the performance of positive-displacement pumps. *Chem Eng Prog* 2007;**103**(12):32.
10. AIChE's Equipment Testing Procedure. *Positive displacement pumps: a guide to performance evaluation.* New York: Wiley-Blackwell; 2007.
11. Soares C. *Process engineering equipment handbook.* New York: McGraw Hill; 2002.
12. Kelly JH. Understand the fundamentals of centrifugal pumps. *Chem Eng Prog* 2010;**106**(10):22.
13. Shukla DK, Chaware DK, Swamy RB. Variable frequency drives: an algorithm for selecting VFDs for centrifugal pumps. *Chem Eng* 2010;**117**(2):38.
14. Fernandez K, Pyzdrowski B, Schiller DW, Smith MB. Understand the basics of centrifugal pump operation. *Chem Eng Prog* 2002;**100**:52.
15. Richardson LF, Higson DJ. A study of the energy losses associated with the operation of an air-lift pump. *Trans Inst Chem Eng* 1962;**40**:169.
16. Armand AA. *Hydrodynamics and heat transfer during boiling in high pressure boilers.* Moscow: U.S.S.R. Acad. of Sci; 1995 (USAEC translation (AEC-TR-4490 page 19)).
17. Schmidt E, Behringer P, Schurig W. Wasserumlauf in Dampfkesseln. *Forsch a d Gebiete d Ing Ausgabe B Forsch* 1934;**365**.
18. Govier GW, Radford BA, Dunn JSC. The upwards vertical flow of air-water mixtures 1. Effect of air and water rates on flow pattern, hold-up and pressure drop. *Can J Chem Eng* 1957;**35**:58.
19. Govier GW, Short WL. The upward vertical flow of air-water mixtures II. Effect of tubing diameter on flow pattern, hold-up and pressure drop. *Can J Chem Eng* 1958;**36**:195.
20. Isbin HS, Sher NC, Eddy KC. Void fractions in two-phase steam-water flow. *AIChEJl* 1957;**3**:136.

21. Moore TV, Wilde HD. Experimental measurement of slippage in flow through vertical pipes. *Trans Am Inst Min Met Eng (Pet Div)* 1931;**92**:296.
22. Collins D. Choosing process vacuum pumps. *Chem Eng Prog* 2012;**110**(8):65.
23. Crane Company. *Flow of fluids through valves, fittings and pipes. Technical manual 410.* New York: Crane Co.; 1991. and subsequent revisions.
24. Hooper WB. The two K method predicts head losses in pipe fittings. *Chem Eng* August 24, 1981;**88**:96.
25. Darby R. Correlate pressure drops through fittings. *Chem Eng* 2001;**108**(4):127.
26. Darby R, Chhabra RP. *Chemical engineering fluid mechanics.* 3rd ed. Boca Raton, FL: CRC Press; 2016.
27. White FM. *Fluid mechanics.* 7th ed. New York: McGraw-Hill; 2011.

Further Reading

1. Anderson Jr JD. *Modern compressible flow.* 3rd ed. New York: McGraw-Hill; 2003.
2. Engineering Equipment Users' Association. *Vacuum producing equipment, EEUA handbook no. 11.* London: Constable; 1961.
3. Engineering Equipment Users' Association. *Electrically driven glandless pumps, EEUA handbook no. 26.* London: Constable; 1968.
4. Engineering Equipment Users' Association. *Guide to the selection of rotodynamic pumps, EEUA handbook no. 30.* London: Constable; 1972.
5. Greene W, editor. *The chemical engineering guide to compressors.* New York: McGraw-Hill; 1985.
6. Holland FA, Chapman FS. *Pumping of liquids.* New York: Reinhold; 1966.
7. Karassik IJ, McGuire JT. *Centrifugal pumps.* 2nd ed. New York: Springer; 2012.
8. Karassik IJ, Krutzsch WC, Fraser WH, Messina JP. *Pump handbook.* New York: McGraw-Hill; 1976.
9. Kellogg M. W. Co.. *Design of piping systems.* 2nd ed. Chichester: Wiley; 1964.
10. Lazarkienicz S, Troskolanski AT. *Impeller pumps.* Oxford: Pergamon Press; 1965.
11. McNaughton K, editor. *The chemical engineering guide to pumps.* New York: McGraw-Hill; 1985.
12. Perry RH, Green DW, editors. *Perry's chemical engineers handbook.* 7th ed. New York: McGraw-Hill; 1998.
13. Stepanoff AJ. *Centrifugal and axial flow pumps.* 2nd ed. Melbourne, FL: Krieger; 1992.
14. Streeter VL, Wylie EB. *Fluid mechanics.* New York: McGraw-Hill; 1985.
15. Turton RK. *An introductory guide to pumps and pumping systems.* London: Mechanical Engineering Publications; 1993.
16. Volk M. *Pump characteristics and applications.* 3rd ed. Boca Raton, FL: CRC Press; 2013.
17. Warring RH. *Seals and sealing handbook.* Trade and Technical Press; 1983.
18. Warring RH. *Pumps selection systems and applications.* Trade and Technical Press; 1979.
19. Wilson KC, Addie GR, Sellgren A, Clift R. *Slurry transport using centrifugal pumps.* 3rd ed. New York: Springer; 2008.

Appendix

A.1 Tables of Physical Properties

Table 1 Thermal conductivities of liquids

Liquid	k (W/m K)	(K)	k (Btu/h ft °F)	Liquid	k (W/m K)	(K)	k (Btu/h ft °F)
Acetic acid 100%	0.171	293	0.099	Hexane (*n*-)	0.138	303	0.080
50%	0.35	293	0.20		0.135	333	0.078
Acetone	0.177	303	0.102	Heptyl alcohol (*n*-)	0.163	303	0.094
	0.164	348	0.095		0.157	348	0.091
Allyl alcohol	0.180	298–303	0.104	Hexyl alcohol (*n*-)	0.161	303	0.093
Ammonia	0.50	258–303	0.29		0.156	348	0.090
Ammonia, aqueous	0.45	293	0.261				
	0.50	333	0.29	Kerosene	0.149	293	0.086
Amyl acetate	0.144	283	0.083		0.140	348	0.081
Amyl alcohol (*n*-)	0.163	303	0.094				
	0.154	373	0.089	Mercury	8.36	301	4.83
Amyl alcohol (iso-)	0.152	303	0.088	Methyl alcohol 100%	0.215	293	0.124
	0.151	348	0.087	80%	0.267	293	0.154
Aniline	0.173	273–293	0.100	60%	0.329	293	0.190
				40%	0.405	293	0.234
Benzene	0.159	303	0.092	20%	0.492	293	0.284
	0.151	333	0.087	100%	0.197	323	0.114
Bromobenzene	0.128	303	0.074	Methyl chloride	0.192	258	0.111
	0.121	373	0.070		0.154	303	0.089
Butyl acetate (*n*-)	0.147	298–303	0.085				
Butyl alcohol (*n*-)	0.168	303	0.097	Nitrobenzene	0.164	303	0.095
	0.164	348	0.095		0.152	373	0.088
Butyl alcohol (iso-)	0.157	283	0.091	Nitromethane	0.216	303	0.125
					0.208	333	0.120
Calcium chloride brine 30%	0.55	303	0.32	Nonane (*n*-)	0.145	303	0.084

(Continued)

Table 1 Thermal conductivities of liquids—cont'd

Liquid	k (W/m K)	(K)	k (Btu/h ft °F)	Liquid	k (W/m K)	(K)	k (Btu/h ft °F)
15%	0.59	303	0.34		0.142	333	0.082
Carbon disulphide	0.161	303	0.093	Octane (*n*-)	0.144	303	0.083
	0.152	348	0.088		0.140	333	0.081
Carbon *tetrachloride*	0.185	273	0.107	Oils, petroleum	0.138–0.156	273	0.08–0.09
	0.163	341	0.094	Oil, castor	0.180	293	0.104
Chlorobenzene	0.144	283	0.083		0.173	373	0.100
Chloroform	0.138	303	0.080		0.168	293	0.097
Cymene (*para*)	0.135	303	0.078	Oil, olive	0.164	373	0.095
	0.137	333	0.079				
Decane (*n*-)	0.147	303	0.085	Paraldehyde	0.145	303	0.084
	0.144	333	0.083		0.135	373	0.078
Dichlorodi-fluoromethane	0.099	266	0.057	Pentane (*n*-)	0.135	303	0.078
	0.092	289	0.053		0.128	348	0.074
	0.083	311	0.048	Perchloroethylene	0.159	323	0.092
	0.074	333	0.043	Petroleum ether	0.130	303	0.075
	0.066	355	0.038		0.126	348	0.073
Dichloroethane	0.142	323	0.082	Propyl alcohol (*n*-)	0.171	303	0.099
Dichloromethane	0.192	258	0.111		0.164	348	0.095
	0.166	303	0.096	Propyl alcohol (iso-)	0.157	303	0.091
Ethyl acetate	0.175	293	0.101		0.155	333	0.090
Ethyl alcohol							
100%	0.182	293	0.105	Sodium	0.85	373	49
80%	0.237	293	0.137		0.80	483	46
60%	0.305	293	0.176	Sodium chloride brine 25.0%	0.57	303	0.33
40%	0.388	293	0.224	12.5%	0.59	303	0.34
20%	0.486	293	0.281	Sulphuric acid 90%	0.36	303	0.21
100%	0.151	323	0.087	60%	0.43	303	0.25
Ethyl benzene	0.149	303	0.086	30%	0.52	303	0.30
	0.142	333	0.082	Sulphur dioxide	0.22	258	0.128
Ethyl bromide	0.121	293	0.070		0.192	303	0.111
Ethyl ether	0.138	303	0.080				
	0.135	348	0.078	Toluene	0.149	303	0.086
Ethyl iodide	0.111	313	0.064		0.145	348	0.084
	0.109	348	0.063	β-Trichloroethane	0.133	323	0.077
Ethylene glycol	0.265	273	0.153	Trichloroethylene	0.138	323	0.080
Gasoline	0.135	303	0.078	Turpentine	0.128	288	0.074
Glycerol							
100%	0.284	293	0.164	Vaseline	0.184	288	0.106
80%	0.327	293	0.189				
60%	0.381	293	0.220	Water	0.57	273	0.330
40%	0.448	293	0.259		0.615	303	0.356
20%	0.481	293	0.278		0.658	333	0.381
100%	0.284	373	0.164		0.688	353	0.398
Heptane (*n*-)	0.140	303	0.081	Xylene (*ortho*-)	0.155	293	0.090
	0.137	333	0.079	(*meta*-)	0.155	293	0.090

A linear variation with temperature may be assumed. The extreme values given constitute also the temperature limits over which the data are recommended.

By permission from McAdams WH. Heat transmission, *copyright 1942, McGraw-Hill.*

Table 2 Latent heats of vaporisation

No.	Compound	Range $\theta_c - \theta$ (°F)	θ_c (°F)	Range $\theta_c - \theta$ (K)	θ_c (K)
18	Acetic acid	180–405	610	100–225	594
22	Acetone	216–378	455	120–210	508
29	Ammonia	90–360	271	50–200	406
13	Benzene	18–720	552	10–00	562
16	Butane	162–360	307	90–200	426
21	Carbon dioxide	18–180	88	10–100	304
4	Carbon disulphide	252–495	523	140–275	546
2	Carbon tetrachloride	54–450	541	30–250	556
7	Chloroform	252–495	505	140–275	536
8	Dichloromethane	270–450	421	150–250	489
3	Diphenyl	315–720	981	175–400	800
25	Ethane	45–270	90	25–150	305
26	Ethyl alcohol	36–252	469	20–140	516
28	Ethyl alcohol	252–540	469	140–300	516
17	Ethyl chloride	180–450	369	100–250	460
13	Ethyl ether	18–720	381	10–00	467
2	Freon-11 (CCl_3F)	126–450	389	70–250	471
2	Freon-12 (CCl_2F_2)	72–360	232	40–200	384
5	Freon-21 ($CHCl_2F$)	126–450	354	70–250	451
6	Freon-22 ($CHClF_2$)	90–306	205	50–170	369
1	Freon-113 (CCl_2F-$CClF_2$)	162–450	417	90–250	487
10	Heptane	36–540	512	20–300	540
11	Hexane	90–450	455	50–225	508
15	Isobutane	144–360	273	80–200	407
27	Methanol	72–450	464	40–250	513
20	Methyl chloride	126–450	289	70–250	416
19	Nitrous oxide	45–270	97	25–150	309
9	Octane	54–540	565	30–300	569
12	Pentane	36–360	387	20–200	470
23	Propane	72–360	205	40–200	369
24	Propyl alcohol	36–360	507	20–200	537
14	Sulphur dioxide	162–288	314	90–160	430
30	Water	180–900	705	100–500	647

Example: For water at 373 K, $\theta_c - \theta = (647 - 373) = 274$ K, and the latent heat of vaporisation is 2257 kJ/kg.
By permission from McAdams WH. Heat transmission, *Copyright 1942, McGraw-Hill.*

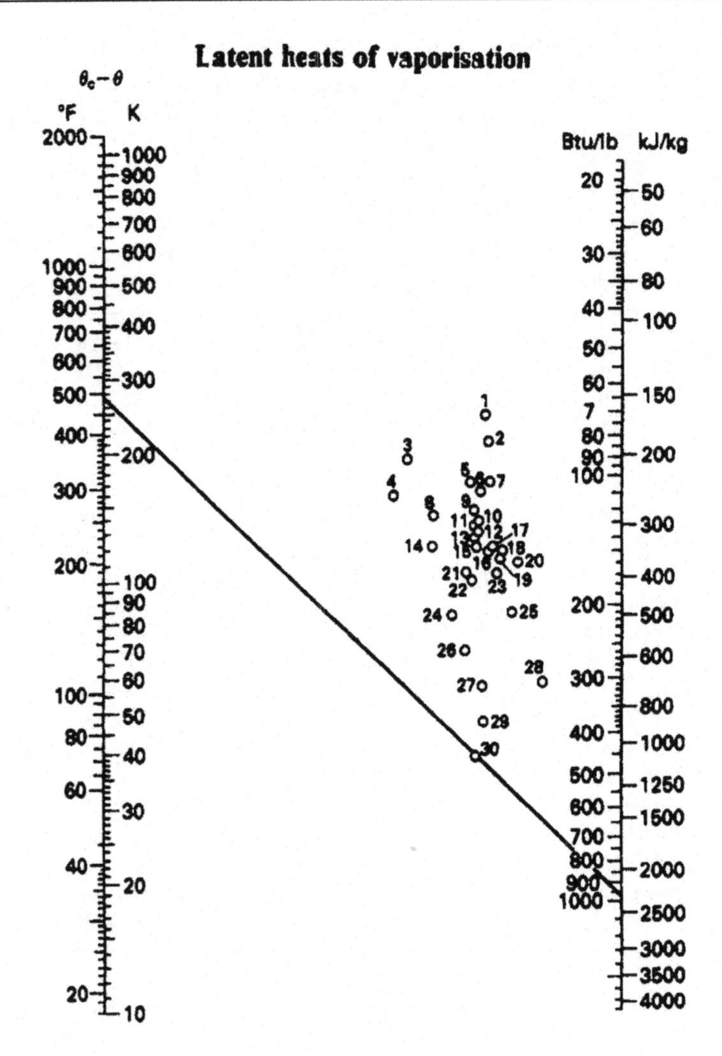

Latent heats of vaporisation

Table 3 Specific heats of liquids

No.	Liquid	Range (K)
29	Acetic acid, 100%	273–353
32	Acetone	293–323
52	Ammonia	203–323
37	Amyl alcohol	223–298
26	Amyl acetate	273–373
30	Aniline	273–403
23	Benzene	283–353
27	Benzyl alcohol	253–303
10	Benzyl chloride	243–303
49	Brine, 25% $CaCl_2$	233–293
51	Brine, 25% NaCl	233–293
44	Butyl alcohol	273–373
2	Carton disulphide	173–298
3	Carbon tetrachloride	283–333
8	Chlorobenzene	273–373
4	Chloroform	273–323
21	Decane	193–298
6A	Dichloroethane	243–333
5	Dichloroethane	233–323
15	Diphenyl	353–393
22	Diphenylmethane	303–373
16	Diphenyl oxide	273–473
16	Dowtherm A	273–473
24	Ethyl acetate	223–298
42	Ethyl alcohol, 100%	303–353
46	Ethyl alcohol, 95%	293–353
50	Ethyl alcohol, 50%	293–353
25	Ethyl benzene	273–373
1	Ethyl bromide	278–298
13	Ethyl chloride	243–313
36	Ethyl ether	173–298
7	Ethyl iodide	273–373
39	Ethylene glycol	233–473
2A	Freon-11 (CCl_3F)	253–343
6	Freon-12 (CCl_2F_2)	233–288
4A	Freon-21 ($CHCl_2F$)	253–343
7A	Freon-22 ($CHClF_2$)	253–333
3A	Freon-113 (CCl_2F-$CClF_2$)	253–343
38	Glycerol	233–293
28	Heptane	273–333
35	Hexane	193–293
48	Hydrochloric acid, 30%	293–373
41	Isoamyl alcohol	283–373
43	Isobutyl alcohol	273–373
47	Isopropyl alcohol	253–323
31	Isopropyl ether	193–293
40	Methyl alcohol	233–293
13A	Methyl chloride	193–293
14	Naphthalene	363–473
12	Nitrobenzene	273–373
34	Nonane	223–298
33	Octane	223–298

(Continued)

Table 3 Specific heats of liquids—cont'd

No.	Liquid	Range (K)
3	Perchloroethylene	243–413
45	Propyl alcohol	253–373
20	Pyridine	223–298
9	Sulphuric acid, 98%	283–318
11	Sulphur dioxide	253–373
23	Toluene	273–333
53	Water	283–473
19	Xylene (*ortho*)	273–373
18	Xylene (*meta*)	273–373
17	Xylene (*para*)	273–373

By permission from McAdams WH. Heat transmission, *copyright 1942, McGraw-Hill.*

Specific heats of liquids

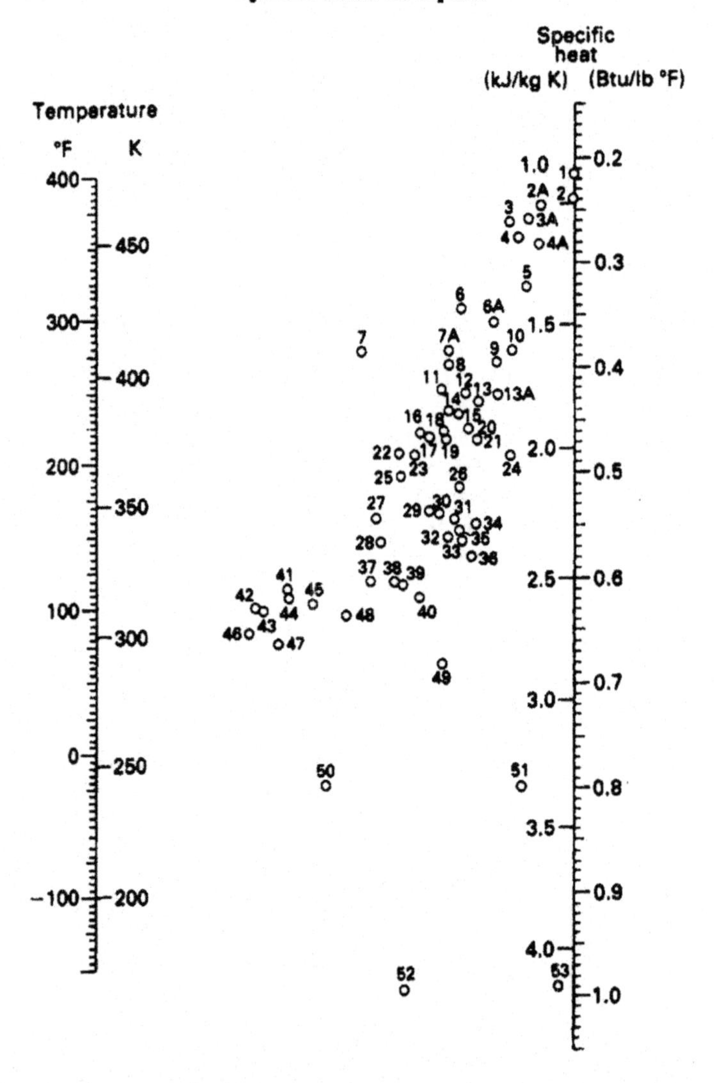

Table 4 Specific heats at constant pressure of gases and vapours at 101.3 kN/m^2

No.	Gas	Range (K)	No.	Gas	Range (K)
10	Acetylene	273–473	1	Hydrogen	273–873
15	Acetylene	473–673	2	Hydrogen	873–1673
16	Acetylene	673–1673	35	Hydrogen bromide	273–1673
27	Air	273–1673	30	Hydrogen chloride	273–1673
12	Ammonia	273–873	20	Hydrogen fluoride	273–1673
14	Ammonia	873–1673	36	Hydrogen iodide	273–1673
18	Carbon dioxide	273–673	19	Hydrogen sulphide	273–973
24	Carbon dioxide	673–1673	21	Hydrogen sulphide	973–1673
26	Carbon monoxide	273–1673	5	Methane	273–573
32	Chlorine	273–473	6	Methane	573–973
34	Chlorine	473–1673	7	Methane	973–1673
3	Ethane	273–473	25	Nitric oxide	273–973
9	Ethane	473–873	28	Nitric oxide	973–1673
8	Ethane	873–1673	26	Nitrogen	273–1673
4	Ethylene	273–473	23	Oxygen	273–773
11	Ethylene	473–873	29	Oxygen	773–1673
13	Ethylene	873–1673	33	Sulphur	573–1673
17B	Freon-11 (CCl_3F)	273–423	22	Sulphur dioxide	273–673
17C	Freon-21 ($CHCl_2F$)	273–423	31	Sulphur dioxide	673–1673
I7A	Freon-22 ($CHClF_2$)	273–423	17	Water	273–1673
17D	Freon-113 (CCl_2F-$CClF_2$)	273–423			

By permission from McAdams WH. Heat transmission, *copyright, 1942, McGraw-Hill.*

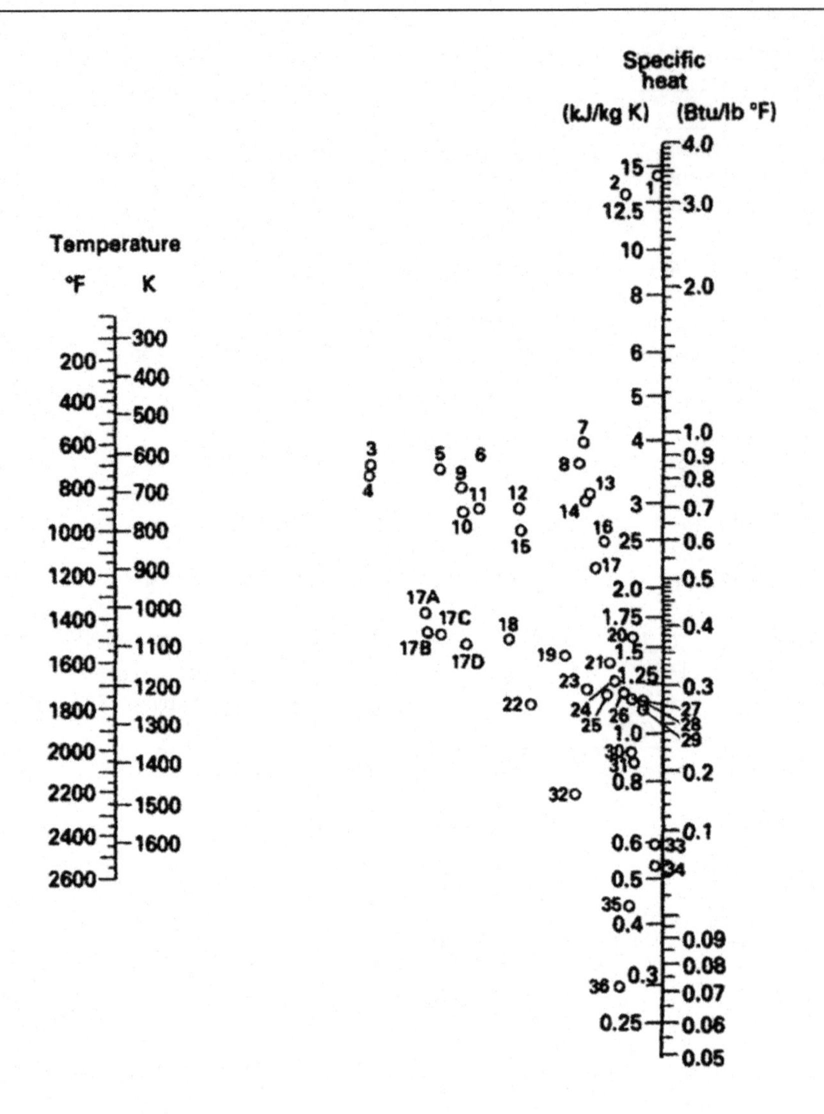
Temperature
°F
K
Specific heat
(kJ/kg K)
(Btu/lb °F)

Table 5 Viscosity of water

Temperature (θ) (K)	Viscosity (μ) (mN s/m^2)	Temperature (θ) (K)	Viscosity (μ) (mN s/m^2)	Temperature (θ) (K)	Viscosity (μ) (mN s/m^2)
273	1.7921	306	0.7523	340	0.4233
274	1.7313	307	0.7371	341	0.4174
275	1.6728	308	0.7225	342	0.4117
276	1.6191	309	0.7085	343	0.4061
277	1.5674	310	0.6947	344	0.4006
278	1.5188	311	0.6814	345	0.3952
279	1.4728	312	0.6685	346	0.3900
280	1.4284	313	0.6560	347	0.3849
281	1.3860	314	0.6439	348	0.3799
282	1.3462	315	0.6321	349	0.3750
283	1.3077	316	0.6207	350	0.3702
284	1.2713	317	0.6097	351	0.3655
285	1.2363	318	0.5988	352	0.3610
286	1.2028	319	0.5883	353	0.3565
287	1.1709	320	0.5782	354	0.3521
288	1.1404	321	0.5683	355	0.3478
289	1.1111	322	0.5588	356	0.3436
290	1.0828	323	0.5494	357	0.3395
291	1.0559	324	0.5404	358	0.3355
292	1.0299	325	0.5315	359	0.3315
293	1.0050	326	0.5229	360	0.3276
293.2	1.0000	327	0.5146	361	0.3239
294	0.9810	328	0.5064	362	0.3202
295	0.9579	329	0.4985	363	0.3165
296	0.9358	330	0.4907	364	0.3130
297	0.9142	331	0.4832	365	0.3095
298	0.8937	332	0.4759	366	0.3060
299	0.8737	333	0.4688	367	0.3027
300	0.8545	334	0.4618	368	0.2994
301	0.8360	335	0.4550	369	0.2962
302	0.8180	336	0.4483	370	0.2930
303	0.8007	337	0.4418	371	0.2899
304	0.7840	338	0.4355	372	0.2868
305	0.7679	339	0.4293	373	0.2838

Calculated by the formula:

$$1/\mu = 21.482\left[(\theta - 281.435) + \sqrt{(8078.4 + (\theta - 281.435)^2}\right] - 1200\,(\mu \text{ in Ns/m}^2)$$

By permission from Bingham EC. Fluidity and plasticity, *Copyright 1922, McGraw-Hill Book Company Inc.*

Table 6 Thermal conductivities of gases and vapours

Substance	k (W/m K)	(K)	k (Btu/h ft °F)	Substance	k (W/m K)	(K)	k (Btu/h ft °F)
Acetone	0.0098	273	0.0057	Chlorine	0.0074	273	0.0043
	0.0128	319	0.0074	Chloroform	0.0066	273	0.0038
	0.0171	373	0.0099		0.0080	319	0.0046
	0.0254	457	0.0147		0.0100	373	0.0058
Acetylene	0.0118	198	0.0068		0.0133	457	0.0077
	0.0187	273	0.0108	Cyclohexane	0.0164	375	0.0095
	0.0242	323	0.0140				
	0.0298	373	0.0172	Dichlorodifluoromethane	0.0083	273	0.0048
Air	0.0164	173	0.0095		0.0111	323	0.0064
	0.0242	273	0.0140		0.0139	373	0.0080
	0.0317	373	0.0183		0.0168	423	0.0097
	0.0391	473	0.0226				
	0.0459	573	0.0265	Ethane	0.0114	203	0.0066
Ammonia	0.0164	213	0.0095		0.0149	239	0.0086
	0.0222	273	0.0128		0.0183	273	0.0106
	0.0272	323	0.0157		0.0303	373	0.0175
	0.0320	373	0.0185	Ethyl acetate	0.0125	319	0.0072
					0.0166	373	0.0096
Benzene	0.0090	273	0.0052		0.0244	457	0.0141
	0.0126	319	0.0073	Alcohol	0.0154	293	0.0089
	0.0178	373	0.0103		0.0215	373	0.0124
	0.0263	457	0.0152	Chloride	0.0095	273	0.0055
	0.0305	485	0.0176		0.0164	373	0.0095
Butane (*n*-)	0.0135	273	0.0078		0.0234	457	0.0135
	0.0234	373	0.0135		0.0263	485	0.0152
(iso-)	0.0138	273	0.0080	Ether	0.0133	273	0.0077
	0.0241	373	0.0139		0.0171	319	0.0099
					0.0227	373	0.0131
Carbon dioxide	0.0118	223	0.0068		0.0327	457	0.0189
	0.0147	273	0.0085		0.0362	485	0.0209
	0.0230	373	0.0133	Ethylene	0.0111	202	0.0064
	0.0313	473	0.0181		0.0175	273	0.0101
	0.0396	573	0.0228		0.0267	323	0.0131
Disulphide	0.0069	273	0.0040		0.0279	373	0.0161
	0.0073	280	0.0042				
Monoxide	0.0071	84	0.0041	Heptane (*n*-)	0.0194	473	0.0112
	0.0080	94	0.0046		0.0178	373	0.0103
	0.0234	213	0.0135	Hexane (*n*-)	0.0125	273	0.0072
Tetrachloride	0.0071	319	0.0041		0.0138	293	0.0080
	0.0090	373	0.0052	Hexene	0.0106	273	0.0061
	0.0112	457	0.0065		0.0109	373	0.0189
Hydrogen	0.0113	173	0.065		0.0225	457	0.0130
	0.0144	223	0.083		0.0256	485	0.0148
	0.0173	273	0.100	Methylene chloride	0.0067	273	0.0039
	0.0199	323	0.115		0.0085	319	0.0049

Table 6 Thermal conductivities of gases and vapours—cont'd

Substance	k (W/m K)	(K)	k (Btu/ h ft °F)	Substance	k (W/m K)	(K)	k (Btu/ h ft °F)
	0.0223	373	0.129		0.0109	373	0.0063
	0.0308	573	0.178		0.0164	485	0.0095
Hydrogen and carbon dioxide		273					
0% H_2	0.0144		0.0083	Nitric oxide	0.0178	203	0.0103
20%	0.0286		0.0165		0.0239	273	0.0138
40%	0.0467		0.0270	Nitrogen	0.0164	173	0.0095
60%	0.0709		0.0410		0.0242	273	0.0140
80%	0.1070		0.0620		0.0277	323	0.0160
100%	0.173		0.10		0.0312	373	0.0180
Hydrogen and nitrogen		273		Nitrous oxide	0.0116	201	0.0067
0% H_2	0.0230		0.0133		0.0157	273	0.0087
20%	0.0367		0.0212		0.0222	373	0.0128
40%	0.0542		0.0313				
60%	0.0758		0.0438	Oxygen	0.0164	173	0.0095
80%	0.1098		0.0635		0.0206	223	0.0119
Hydrogen and nitrous oxide		273			0.0246	273	0.0142
0% H_2	0.0159		0.0092		0.0284	323	0.0164
20%	0.0294		0.0170		0.0321	373	0.0185
40%	0.0467		0.0270				
60%	0.0709		0.0410	Pentane (*n*-)	0.0128	273	0.0074
80%	0.112		0.0650		0.0144	293	0.0083
Hydrogen sulphide	0.0132	273	0.0076	(iso-)	0.0125	273	0.0072
					0.0220	373	0.0127
Mercury	0.0341	473	0.0197	Propane	0.0151	273	0.0087
Methane	0.0173	173	0.0100		0.0261	373	0.0151
	0.0251	223	0.0145				
	0.0302	273	0.0175	Sulphur dioxide	0.0087	273	0.0050
	0.0372	323	0.0215		0.0119	373	0.0069
Methyl alcohol	0.0144	273	0.0083				
	0.0222	373	0.0128	Water vapour	0.0208	319	0.0120
Acetate	0.0102	273	0.0059		0.0237	373	0.0137
	0.0118	293	0.0068		0.0324	473	0.0187
Chloride	0.0092	273	0.0053		0.0429	573	0.0248
	0.0125	319	0.0072		0.0545	673	0.0315
	0.0163	373	0.0094		0.0763	773	0.0441

The extreme temperature values given constitute the experimental range. For extrapolation to other temperatures, it is suggested that the data given be plotted as log *k* versus log *T*, or that use be made of the assumption that the ratio $C_p\mu/k$ is practically independent of temperature (and of pressure, within moderate limits).

By permission from McAdams WH. Heat transmission, *copyright 1942, McGraw-Hill.*

Table 7 Viscosities of gases

No.	Gas	*X*	*Y*
1	Acetic acid	7.7	14.3
2	Acetone	8.9	13.0
3	Acetylene	9.8	14.9
4	Air	11.0	20.0
5	Ammonia	8.4	16.0
6	Argon	10.5	22.4
7	Benzene	8.5	13.2
8	Bromine	8.9	19.2
9	Butene	9.2	13.7
10	Butylene	8.9	13.0
11	Carbon dioxide	9.5	18.7
12	Carbon disulphide	8.0	16.0
13	Carbon monoxide	11.0	20.0
14	Chlorine	9.0	18.4
15	Chloroform	8.9	15.7
16	Cyanogen	9.2	15.2
17	Cyclohexane	9.2	12.0
18	Ethane	9.1	14.5
19	Ethyl acetate	8.5	13.2
20	Ethyl alcohol	9.2	14.2
21	Ethyl chloride	8.5	15.6
22	Ethyl ether	8.9	13.0
23	Ethylene	9.5	15.1
24	Fluorine	7.3	23.8
25	Freon-11 (CCl_8F)	10.6	15.1
26	Freon-12 (CCl_2F_2)	11.1	16.0
27	Freon-21 ($CHCl_2F$)	10.8	15.3
28	Freon-22 ($CHClF_2$)	10.1	17.0
29	Freon-113 (CCl_2F-$CClF_3$)	11.3	14.0
30	Helium	10.9	20.5
31	Hexane	8.6	11.8
32	Hydrogen	11.2	12.4
33	$3H_2 + 1\ N_2$	11.2	17.2
34	Hydrogen bromide	8.8	20.9
35	Hydrogen chloride	8.8	18.7
36	Hydrogen cyanide	9.8	14.9
37	Hydrogen iodide	9.0	21.3
38	Hydrogen sulphide	8.6	18.0
39	Iodine	9.0	18.4
40	Mercury	5.3	22.9
41	Methane	9.9	15.5
42	Methyl alcohol	8.5	15.6
43	Nitric oxide	10.9	20.5
44	Nitrogen	10.6	20.0
45	Nitrosyl chloride	8.0	17.6
46	Nitrous oxide	8.8	19.0
47	Oxygen	11.0	21.3
48	Pentane	7.0	12.8
49	Propane	9.7	12.9

Table 7 Viscosities of gases—cont'd

No.	Gas	*X*	*Y*
50	Propyl alcohol	8.4	13.4
51	Propylene	9.0	13.8
52	Sulphur dioxide	9.6	17.0
53	Toluene	8.6	12.4
54	2,3,3-Trimethylbutane	9.5	10.5
55	Water	8.0	16.0
56	Xenon	9.3	23.0

Co-ordinates for use with graph on facing page.

Viscosities of gases

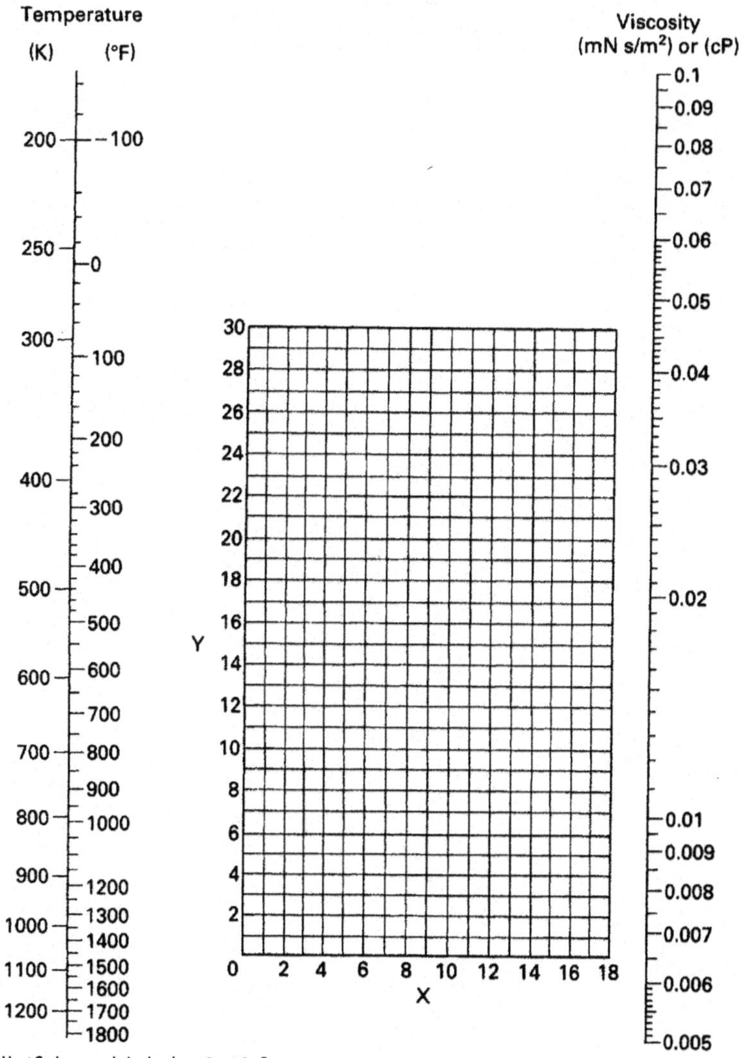

[To convert to lb/ft h multiply by 2.42.]

Table 8 Viscosities and densities of liquids

No.	Liquid	X	Y	Density at 293 K (kg/m^3)	
1	Acetaldehyde	15.2	4.8	783	(291 K)
2	Acetic acid, 100%	12.1	14.2	1049	
3	Acetic acid, 70%	9.5	17.0	1069	
4	Acetic anhydride	12.7	12.8	1083	
5	Acetone, 100%	14.5	7.2	792	
6	Acetone, 35%	7.9	15.0	948	
7	Allyl alcohol	10.2	14.3	854	
8	Ammonia, 100%	12.6	2.0	817	(194 K)
9	Ammonia, 26%	10.1	13.9	904	
10	Amyl acetate	11.8	12.5	879	
11	Amyl alcohol	7.5	18.4	817	
12	Aniline	8.1	18.7	1022	
13	Anisole	12.3	13.5	990	
14	Arsenic trichloride	13.9	14.5	2163	
15	Benzene	12.5	10.9	880	
16	Brine, $CaCl_2$, 25%	6.6	15.9	1228	
17	Brine, NaCl, 25%	10.2	16.6	1186	(298 K)
18	Bromine	14.2	13.2	3119	
19	Bromotoluene	20.0	15.9	1410	
20	Butyl acetate	12.3	11.0	882	
21	Butyl alcohol	8.6	17.2	810	
22	Butyric acid	12.1	15.3	964	
23	Carbon dioxide	11.6	0.3	1101	(236 K)
24	Carbon disulphide	16.1	7.5	1263	
25	Carbon tetrachloride	12.7	13.1	1595	
26	Chlorobenzene	12.3	12.4	1107	
27	Chloroform	14.4	10.2	1489	
28	Chlorosulphonic acid	11.2	18.1	1787	(298 K)
29	Chlorotoluene, *ortho*	13.0	13.3	1082	
30	Chlorotoluene, *meta*	13.3	12.5	1072	
31	Chloroluene, *para*	13.3	12.5	1070	
32	Cresol, *meta*	2.5	20.8	1034	
33	Cyclohexanol	2.9	24.3	962	
34	Dibromoethane	12.7	15.8	2495	
35	Dichloroethane	13.2	12.2	1256	
36	Dichloromethane	14.6	8.9	1336	
37	Diethyl oxalate	11.0	16.4	1079	
38	Dimethyl oxalate	12.3	15.8	1148	(327 K)
39	Diphenyl	12.0	18.3	992	(346 K)
40	Dipropyl oxalate	10.3	17.7	1038	(273 K)
41	Ethyl acetate	13.7	9.1	901	
42	Ethyl alcohol, 100%	10.5	13.8	789	
43	Ethyl alcohol, 95%	9.8	14.3	804	
44	Ethyl alcohol, 40%	6.5	16.6	935	
45	Ethyl benzene	13.2	11.5	867	
46	Ethyl bromide	14.5	8.1	1431	
47	Ethyl chloride	14.8	6.0	917	(279 K)

Table 8 Viscosities and densities of liquids—cont'd

No.	Liquid	X	Y	Density at 293 K (kg/m^3)	
48	Ethyl ether	14.5	5.3	708	(298 K)
49	Ethyl formate	14.2	8.4	923	
50	Ethyl iodide	14.7	10.3	1933	
51	Ethylene glycol	6.0	23.6	1113	
52	Formic acid	10.7	15.8	1220	
53	Freon-11 (CCl_3F)	14.4	9.0	1494	(290 K)
54	Freon-12 (CCl_2F_2)	16.8	5.6	1486	(293 K)
55	Freon-21 ($CHCl_2F$)	15.7	7.5	1426	(273 K)
56	Freon-22 ($CHClF_2$)	17.2	4.7	3870	(273 K)
57	Freon-113 (CCl_2F-$CClF_2$)	12.5	11.4	1576	
58	Glycerol, 100%	2.0	30.0	1261	
59	Glycerol, 50%	6.9	19.6	1126	
60	Heptane	14.1	8.4	684	
61	Hexane	14.7	7.0	659	
62	Hydrochloric acid, 31.5%	13.0	16.6	1157	
63	Isobutyl alcohol	7.1	18.0	779	(299 K)
64	Isobutyric acid	12.2	14.4	949	
65	Isopropyl alcohol	8.2	16.0	789	
66	Kerosene	10.2	16.9	780–820	
67	Linseed oil, raw	7.5	27.2	934±4	(288 K)
68	Mercury	18.4	16.4	13,546	
69	Methanol, 100%	12.4	10.5	792	
70	Methanol, 90%	12.3	11.8	820	
71	Methanol, 40%	7.8	15.5	935	
72	Methyl acetate	14.2	8.2	924	
73	Methyl chloride	15.0	3.8	952	(273 K)
74	Methyl ethyl ketone	13.9	8.6	805	
75	Naphthalene	7.9	18.1	1145	
76	Nitric acid, 95%	12.8	13.8	1493	
77	Nitric acid, 60%	10.8	17.0	1367	
78	Nitrobenzene	10.6	16.2	1205	(291 K)
79	Nitrotoluene	11.0	17.0	1160	
80	Octane	13.7	10.0	703	
81	Octyl alcohol	6.6	21.1	827	
82	Pentachloroethane	10.9	17.3	1671	(298 K)
83	Pentane	14.9	5.2	630	(291 K)
84	Phenol	6.9	20.8	1071	(298 K)
85	Phosphorus tribromide	13.8	16.7	2852	(288 K)
86	Phosphorus trichloride	16.2	10.9	1574	
87	Propionic acid	12.8	13.8	992	
88	Propyl alcohol	9.1	16.5	804	
89	Propyl bromide	14.5	9.6	1353	
90	Propyl chloride	14.4	7.5	890	
91	Propyl iodide	14.1	11.6	1749	
92	Sodium	16.4	13.9	970	
93	Sodium hydroxide, 50%	3.2	25.8	1525	
94	Stannic chloride	13.5	12.8	2226	

(Continued)

Table 8 Viscosities and densities of liquids—cont'd

No.	Liquid	X	Y	Density at 293 K (kg/m^3)	
95	Sulphur dioxide	15.2	7.1	1434	(273 K)
96	Sulphuric acid, 110%	7.2	27.4	1980	
97	Sulphuric acid, 98%	7.0	24.8	1836	
98	Sulphuric acid, 60%	10.2	21.3	1498	
99	Sulphuryl chloride	15.2	12.4	1667	
100	Tetrachloroethane	11.9	15.7	1600	
101	Tetrachloroethylene	14.2	12.7	1624	(288 K)
102	Titanum tetrachloride	14.4	12.3	1726	
103	Toluene	13.7	10.4	866	
104	Trichloroethylene	14.8	10.5	1466	
105	Turpentine	11.5	14.9	861–867	
106	Vinyl acetate	14.0	8.8	932	
107	Water	10.2	13.0	998	
108	Xylene, *ortho*	13.5	12.1	881	
109	Xylene, *meta*	13.9	10.6	867	
110	Xylene, *para*	13.9	10.9	861	

Co-ordinates for graph on following page.

By permission from Perry RH, Green DW, editors. Perry's chemical engineers' handbook, *6th ed. Copyright 1984, McGraw-Hill.*

Viscosities of liquids

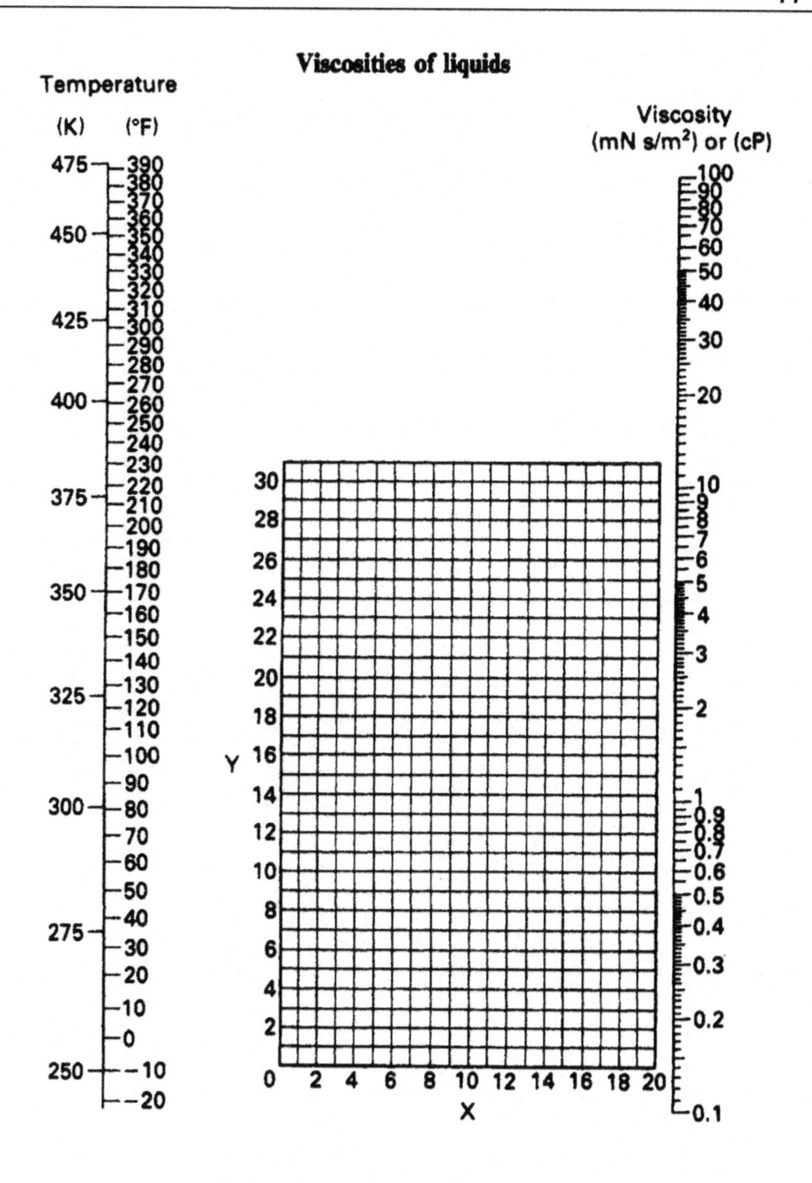

Table 9 Critical constants of gases

	Critical Temperature T_c (K)	Critical Pressure P_c (MN/m^2)	Compressibility Constant in Critical State Z_c
Paraffins			
Methane	191	4.64	0.290
Ethane	306	4.88	0.284
Propane	370	4.25	0.276
n-Butane	425	3.80	0.274
Isobutane	408	3.65	0.282
n-Pentane	470	3.37	0.268
Isopentane	461	3.33	0.268
Neopentane	434	3.20	0.268
n-Hexane	508	3.03	0.264
n-Heptane	540	2.74	0.260
n-Octane	569	2.49	0.258
Mono-olefins			
Ethylene	282	5.07	0.268
Propylene	365	4.62	0.276
1-Butene	420	4.02	0.276
1-Pentene	474	4.05	
Miscellaneous organic compounds			
Acetic acid	595	5.78	0.200
Acetone	509	4.72	0.237
Acetylene	309	6.24	0.274
Benzene	562	4.92	0.274
1,3-Butadiene	425	4.33	0.270
Cyclohexane	553	4.05	0.271
Dichlorodifluoromethane (Freon-12)	385	4.01	0.273
Diethyl ether	467	3.61	0.261
Ethyl alcohol	516	6.38	0.249
Ethylene oxide	468	7.19	0.25
Methyl alcohol	513	7.95	0.220
Methyl chloride	416	6.68	0.276
Methyl ethyl ketone	533	4.00	0.26
Toluene	594	4.21	0.27
Trichlorofluoromethane (Freon-11)	471	4.38	0.277
Trichlorotrifluoroethane (Freon-113)	487	3.41	0.274
Elementary gases			
Bromine	584	10.33	0.307
Chlorine	417	7.71	0.276
Helium	5.3	0.23	0.300
Hydrogen	33.3	1.30	0.304
Neon	44.5	2.72	0.307
Nitrogen	126	3.39	0.291
Oxygen	155	5.08	0.29

Table 9 Critical constants of gases—cont'd

	Critical Temperature T_c (K)	Critical Pressure P_c (MN/m^2)	Compressibility Constant in Critical State Z_c
Miscellaneous inorganic compounds			
Ammonia	406	11.24	0.242
Carbon dioxide	304	7.39	0.276
Carbon monoxide	133	3.50	0.294
Hydrogen chloride	325	8.26	0.266
Hydrogen sulphide	374	9.01	0.284
Nitric oxide (NO)	180	6.48	0.25
Nitrous oxide (N_2O)	310	7.26	0.271
Sulphur	1313	11.75	
Sulphur dioxide	431	7.88	0.268
Sulphur trioxide	491	8.49	0.262
Water	647	22.1	0.23

Selected values from Kobe KA, Lynn RE Jr. Chem Rev *1953;52:117. By permission.*

Table 10 Emissivities of surfaces

Surface	T (K)	Emissivity
A. *Metals and metallic oxides*		
Aluminium		
Highly polished plate	500–850	0.039–0.057
Polished plate	296	0.040
Rough plate	299	0.055
Plate oxidised at 872 K	472–872	0.11–0.19
Aluminium–surfaced roofing	311	0.216
Brass		
Hard–rolled, polished	294	0.038
Polished	311–589	0.096
Rolled–plate, natural surface	295	0.06
Rubbed with coarse emery	295	0.20
Dull plate	322–622	0.22
Oxidised	472–872	0.61–0.59
Chromium—see Nickel alloys		
Copper		
Polished electrolytic	353	0.018
Commercial, emeried and polished	292	0.030
Commercial, scraped shiny	295	0.072
Polished	390	0.023
Plate, covered with thick oxide	498	0.78
Plate heated to 872 K	472–872	0.57–0.57
Cuprous oxide	1072–1372	0.66–0.54

(Continued)

Table 10 Emissivities of surfaces—cont'd

Surface	T (K)	Emissivity
Molten copper	1350–1550	0.16–0.13
Gold		
Pure, highly polished	500–900	0.018–0.35
Iron and steel		
Electrolytic iron, highly polished	450–500	0.052–0.064
Polished iron	700–1300	0.144–0.377
Freshly emeried iron	293	0.242
Polished cast iron	473	0.21
Wrought iron, highly polished	311–522	0.28
Cast iron, newly turned	295	0.435
Steel casting, polished	1044–1311	0.52–0.56
Ground sheet steel	1211–1372	0.55–0.61
Smooth sheet iron	1172–1311	0.55–0.60
Cast iron, turned	1155–1261	0.60–0.70
Oxidised surfaces		
Iron plate, completely rusted	293	0.685
Sheet steel, rolled and oxidised	295	0.657
Iron	373	0.736
Cast iron, oxidised at 872 K	472–872	0.64–0.78
Steel, oxidised at 872 K	472–872	0.79–0.79
Smooth electrolytic iron	500–800	0.78–0.82
Iron oxide	772–1472	0.85–0.89
Ingot iron, rough	1200–1390	0.87–0.95
Sheet steel with rough oxide layer	297	0.80
Cast iron, strongly oxidised	311–522	0.95
Wrought iron, dull oxidised	294–633	0.94
Steel plate, rough	311–644	0.94–0.97
Molten metal		
Cast iron	1572–1672	0.29–0.29
Mild steel	1872–2070	0.28–0.28
Lead		
Pure, unoxidised	400–500	0.057–0.075
Grey, oxidised	297	0.281
Oxidised at 472 K	472	0.03
Mercury	273–373	0.09–0.12
Molybdenum		
Filament	1000–2866	0.096–0.292
Monel		
Metal oxidised at 872 K	472–872	0.41–0.46

Table 10 Emissivities of surfaces—cont'd

Surface	T (K)	Emissivity
Nickel		
Electroplated on polished iron and polished	296	0.045
Techically pure, polished	500–600	0.07–0.087
Electroplated on pickled iron, unpolished	293	0.11
Wire	460–1280	0.096–0.186
Plate, oxidised by heating to 872 K	472–872	0.37–0.48
Nickel oxide	922–1527	0.59–0.86
Nickel alloys		
Chromonickel	325–1308	0.64–0.76
Nickelin, grey oxidised	294	0.262
KA-28 alloy, rough brown, after heating	489–763	0.44–0.36
KA-28 alloy, after heating at 800 K	489–800	0.62–0.73
NCT 3 alloy, oxidised from service	489–800	0.90–0.97
NCT 6 alloy, oxidised from service	544–836	0.89–0.82
Platinum		
Pure, polished plate	500–900	0.054–0.104
Strip	1200–1900	0.12–0.17
Filament	300–1600	0.036–0.192
Wire	500–1600	0.073–0.182
Silver		
Polished, pure	500–900	0.0198–0.0324
Polished	310–644	0.0221–0.0312
Steel—see Iron		
Tantalum		
Filament	1600–3272	0.194–0.31
Tin		
Bright tinned iron sheet	298	0.043 and 0.064
Tungsten		
Filament, aged	300–3588	0.032–0.35
Filament	3588	0.39
Zinc		
Commercially pure, polished	500–600	0.045–0.053
Oxidised by heating to 672 K	672	0.11
Galvanised sheet iron, fairly bright	301	0.228

(Continued)

Table 10 Emissivities of surfaces—cont'd

Surface	T (K)	Emissivity
Galvanised sheet iron, grey oxidised	297	0.276
B. *Refractories, building materials, paints, etc.*		
Asbestos		
Board	297	0.96
Paper	311–644	0.93–0.945
Brick		
Red, rough	294	0.93
Silica, unglazed	1275	0.80
Silica, glazed, rough	1475	0.85
Grog, glazed	1475	0.75
Carbon		
T-carbon	400–900	0.81–0.79
Filament	1311–1677	0.526
Candle soot	372–544	0.952
Lampblack-water-glass coating	372–456	0.957–0.952
Thin layer on iron plate	294	0.927
Thick coat	293	0.967
Lampblack, 0.08 mm or thicker	311–644	0.945
Enamel		
White fused on iron	292	0.897
Glass		
Smooth	295	0.937
Gypsum		
0.5 mm thick on blackened plate	294	0.903
Marble		
Light grey, polished	295	0.931
Oak		
Planed	294	0.895
Oil layers		
On polished nickel		
Polished surface alone		0.045
0.025 mm oil		0.27
0.050 mm oil		0.46
0.125 mm oil		0.72
Thick oil layer		0.82
On aluminium foil		
Aluminium foil alone	373	0.087
1 coat of oil	373	0.561
2 coats of oil	373	0.574

Table 10 Emissivities of surfaces—cont'd

Surface	T (K)	Emissivity
Paints, lacquers, varnishes		
Snow white enamel on rough iron plate	296	0.906
Black, shiny lacquer sprayed on iron	298	0.875
Black, shiny shellac on tinned iron sheet	294	0.821
Black matt shellac	350–420	0.91
Black laquer	311–366	0.80–0.95
Matt black lacquer	311–366	0.96–0.98
White lacquer	311–366	0.80–0.95
Oil paints	373	0.92–0.96
Aluminium paint	373	0.27–0.67
After heating to 600 K	422–622	0.35
Aluminium lacquer	294	0.39
Paper, thin		
Pasted on tinned iron plate	292	0.924
Pasted on rough iron plate	292	0.929
Pasted on black lacquered plate	292	0.944
Roofing	294	0.91
Plaster, lime, rough	283–361	0.91
Porcellain, glazed	295	0.924
Quartz, rough, fused	294	0.932
Refractory materials		
Poor radiators	872–1272	0.65–0.75
Good radiators	872–1272	0.80–0.90
Rubber		
Hard, glossy plate	296	0.945
Soft, grey, rough	298	0.859
Serpentine, polished	296	0.900
Water	273–373	0.95–0.963

From Hottel HC, Sarofim AF. Radiation heat transfer. *New York: McGraw-Hill; 1967.*

A.2 Steam Tables

Tables 11 A, 11B, 11C and 11D are adapted from the *Abridged Callendar Steam Tables* by permission of Messrs Edward Arnold (Publishers) Ltd.

Table 11A Properties of saturated steam (S.I. units)

Absolute Pressure (kN/m²)	Temperature		Enthalpy Per Unit Mass (H_s) (kJ/kg)			Entropy Per Unit Mass (S_s) (kJ/kg K)			Specific Volume (v) (m³/kg)	
	(°C) θ_s	(K) T_s	Water	Latent	Steam	Water	Latent	Steam	Water	Steam
Datum: Triple point of water										
0.611	0.01	273.16	0.0	2501.6	2501.6	0	9.1575	9.1575	0.0010002	206.16
1.0	6.98	280.13	29.3	2485.0	2514.4	0.1060	8.8706	8.9767	0.001000	129.21
2.0	17.51	290.66	73.5	2460.2	2533.6	0.2606	8.4640	8.7246	0.001001	67.01
3.0	24.10	297.25	101.0	2444.6	2545.6	0.3543	8.2242	8.5785	0.001003	45.67
4.0	28.98	302.13	121.4	2433.1	2554.5	0.4225	8.0530	8.4755	0.001004	34.80
5.0	32.90	306.05	137.8	2423.8	2561.6	0.4763	7.9197	8.3960	0.001005	28.19
6.0	36.18	309.33	151.5	2416.0	2567.5	0.5209	7.8103	8.3312	0.001006	23.74
7.0	39.03	312.18	163.4	2409.2	2572.6	0.5591	7.7176	8.2767	0.001007	20.53
8.0	41.54	314.69	173.9	2403.2	2577.1	0.5926	7.6370	8.2295	0.001008	18.10
9.0	43.79	316.94	183.3	2397.9	2581.1	0.6224	7.5657	8.1881	0.001009	16.20
10.0	45.83	318.98	191.8	2392.9	2584.8	0.6493	7.5018	8.1511	0.001010	14.67
12.0	49.45	322.60	206.9	2384.2	2591.2	0.6964	7.3908	8.0872	0.001012	12.36
14.0	52.58	325.73	220.0	2376.7	2596.7	0.7367	7.2966	8.0333	0.001013	10.69
16.0	55.34	328.49	231.6	2370.0	2601.6	0.7721	7.2148	7.9868	0.001015	9.43
18.0	57.83	330.98	242.0	2363.9	2605.9	0.8036	7.1423	7.9459	0.001016	8.45
20.0	60.09	333.24	251.5.	2358.4	2609.9	0.8321	7.0773	7.9094	0.001017	7.65
25.0	64.99	338.14	272.0	2346.4	2618.3	0.8933	6.9390	7.8323	0.001020	6.20
30.0	69.13	342.28	289.3	2336.1	2625.4	0.9441	6.8254	7.7695	0.001022	5.23
35.0	72.71	345.86	304.3	2327.2	2631.5	0.9878	6.7288	7.7166	0.001025	4.53
40.0	75.89	349.04	317.7	2319.2	2636.9	1.0261	6.6448	7.6709	0.001027	3.99
45.0	78.74	351.89	329.6	2312.0	2641.7	1.0603	6.5703	7.6306	0.001028	3.58
50.0	81.35	354.50	340.6	2305.4	2646.0	1.0912	6.5035	7.5947	0.001030	3.24
60.0	85.95	359.10	359.9	2293.6	2653.6	1.1455	6.3872	7.5327	0.001033	2.73
70.0	89.96	363.11	376.8	2283.3	2660.1	1.1921	6.2883	7.4804	0.001036	2.37
80.0	93.51	366.66	391.7	2274.0	2665.8	1.2330	6.2022	7.4352	0.001039	2.09
90.0	96.71	369.86	405.2	2265.6	2670.9	1.2696	6.1258	7.3954	0.001041	1.87
100.0	99.63	372.78	417.5	2257.9	2675.4	1.3027	6.0571	7.3598	0.001043	1.69
101.325	100.00	373.15	419.1	2256.9	2676.0	1.3069	6.0485	7.3554	0.0010437	1.6730
105	101.00	374.15	423.3	2254.3	2677.6	1.3182	6.0252	7.3434	0.001045	1.618
110	102.32	375.47	428.8	2250.8	2679.6	1.3330	5.9947	7.3277	0.001046	1.549

115	103.59	376.74	434.2	2247.4	2681.6	1.3472	5.9655	7.3127	0.001047	1.486
120	104.81	377.96	439.4	2244.1	2683.4	1.3609	5.9375	7.2984	0.001048	1.428
125	105.99	379.14	444.4	2240.9	2685.2	1.3741	5.9106	7.2846	0.001049	1.375
130	107.13	380.28	449.2	2237.8	2687.0	1.3868	5.8847	7.2715	0.001050	1.325
135	108.24	381.39	453.9	2234.8	2688.7	1.3991	5.8597	7.2588	0.001050	1.279
140	109.32	382.47	458.4	2231.9	2690.3	1.4109	5.8356	7.2465	0.001051	1.236
145	110.36	383.51	462.8	2229.0	2691.8	1.4225	5.8123	7.2347	0.001052	1.196
150	111.37	384.52	467.1	2226.2	2693.4	1.4336	5.7897	7.2234	0.001053	1.159
155	112.36	385.51	471.3	2223.5	2694.8	1.4445	5.7679	7.2123	0.001054	1.124
160	113.32	386.47	475.4	2220.9	2696.2	1.4550	5.7467	7.2017	0.001055	1.091
165	114.26	387.41	479.4	2218.3	2697.6	1.4652	5.7261	7.1913	0.001056	1.060
170	115.17	388.32	483.2	2215.7	2699.0	1.4752	5.7061	7.1813	0.001056	1.031
175	116.06	389.21	487.0	2213.3	2700.3	1.4849	5.6867	7.1716	0.001057	1.003
180	116.93	390.08	490.7	2210.8	2701.5	1.4944	5.6677	7.1622	0.001058	0.977
185	117.79	390.94	494.3	2208.5	2702.8	1.5036	5.6493	7.1530	0.001059	0.952
190	118.62	391.77	497.9	2206.1	2704.0	1.5127	5.6313	7.1440	0.001059	0.929
195	119.43	392.58	501.3	2203.8	2705.1	1.5215	5.6138	7.1353	0.001060	0.907
200	120.23	393.38	504.7	2201.6	2706.3	1.5301	5.5967	7.1268	0.001061	0.885
210	121.78	394.93	511.3	2197.2	2708.5	1.5468	5.5637	7.1105	0.001062	0.846
220	123.27	396.42	517.6	2193.0	2710.6	1.5628	5.5321	7.0949	0.001064	0.810
230	124.71	397.86	523.7	2188.9	2712.6	1.5781	5.5018	7.0800	0.001065	0.777
240	126.09	399.24	529.6	2184.9	2714.5	1.5929	5.4728	7.0657	0.001066	0.746
250	127.43	400.58	535.4	2181.0	2716.4	1.6072	5.4448	7.0520	0.001068	0.718
260	128.73	401.88	540.9	2177.3	2718.2	1.6209	5.4179	7.0389	0.001069	0.692
270	129.99	403.14	546.2	2173.6	2719.9	1.6342	5.3920	7.0262	0.001070	0.668
280	131.21	404.36	551.5	2170.1	2721.5	1.6471	5.3669	7.0140	0.001071	0.646
290	132.39	405.54	556.5	2166.6	2723.1	1.6596	5.3427	7.0022	0.001072	0.625
300	133.54	406.69	561.4	2163.2	2724.7	1.6717	5.3192	6.9909	0.001074	0.606
320	135.76	408.91	570.9	2156.7	2727.6	1.6948	5.2744	6.9692	0.001076	0.570
340	137.86	411.01	579.9	2150.4	2730.3	1.7168	5.2321	6.9489	0.001078	0.538
360	139.87	413.02	588.5	2144.4	2732.9	1.7376	5.1921	6.9297	0.001080	0.510
380	141.79	414.94	596.8	2138.6	2735.3	1.7575	5.1541	6.9115	0.001082	0.485
400	143.63	416.78	604.7	2132.9	2737.6	1.7764	5.1179	6.8943	0.001084	0.462
420	145.39	418.54	612.3	2127.5	2739.8	1.7946	5.0833	6.8779	0.001086	0.442
440	147.09	420.24	619.6	2122.3	2741.9	1.8120	5.0503	6.8622	0.001088	0.423
460	148.73	421.88	626.7	2117.2	2743.9	1.8287	5.0186	6.8473	0.001089	0.405
480	150.31	423.46	633.5	2112.2	2745.7	1.8448	4.9881	6.8329	0.001091	0.389
500	151.85	425.00	640.1	2107.4	2747.5	1.8604	4.9588	6.8192	0.001093	0.375

(Continued)

Table 11A Properties of saturated steam (S.I. units)—cont'd

Absolute Pressure (kN/m^2)	Temperature		Enthalpy Per Unit Mass (H_s) (kJ/kg)			Entropy Per Unit Mass (S_s) (kJ/kg K)			Specific Volume (v) (m^3/kg)	
	(°C) θ_s,	(K) T_s	Water	Latent	Steam	Water	Latent	Steam	Water	Steam
			Datum: Triple point of water							
520	153.33	426.48	646.5	2102.7	2749.3	1.8754	4.9305	6.8059	0.001095	0.361
540	154.77	427.92	652.8	2098.1	2750.9	1.8899	4.9033	6.7932	0.001096	0.348
560	156.16	429.31	658.8	2093.7	2752.5	1.9040	4.8769	6.7809	0.001098	0.337
580	157.52	430.67	664.7	2089.3	2754.0	1.9176	4.8514	6.7690	0.001100	0.326
600	158.84	431.99	670.4	2085.0	2755.5	1.9308	4.8267	6.7575	0.001101	0.316
620	160.12	433.27	676.0	2080.8	2756.9	1.9437	4.8027	6.7464	0.001102	0.306
640	161.38	434.53	681.5	2076.7	2758.2	1.9562	4.7794	6.7356	0.001104	0.297
660	162.60	435.75	686.8	2072.7	2759.5	1.9684	4.7568	6.7252	0.001105	0.288
680	163.79	436.94	692.0	2068.8	2760.8	1.9803	4.7348	6.7150	0.001107	0.280
700	164.96	438.11	697.1	2064.9	2762.0	1.9918	4.7134	6.7052	0.001108	0.272
720	166.10	439.25	702.0	2061.1	2763.2	2.0031	4.6925	6.6956	0.001109	0.266
740	167.21	440.36	706.9	2057.4	2764.3	2.0141	4.6721	6.6862	0.001110	0.258
760	168.30	441.45	711.7	2053.7	2765.4	2.0249	4.6522	6.6771	0.001112	0.252
780	169.37	442.52	716.3	2050.1	2766.4	2.0354	4.6328	6.6683	0.001114	0.246
800	170.41	443.56	720.9	2046.5	2767.5	2.0457	4.6139	6.6596	0.001115	0.240
820	171.44	444.59	725.4	2043.0	2768.5	2.0558	4.5953	6.6511	0.001116	0.235
840	172.45	445.60	729.9	2039.6	2769.4	2.0657	4.5772	6.6429	0.001118	0.229
860	173.43	446.58	734.2	2036.2	2770.4	2.0753	4.5595	6.6348	0.001119	0.224
880	174.40	447.55	738.5	2032.8	2771.3	2.0848	4.5421	6.6269	0.001120	0.220
900	175.36	448.51	742.6	2029.5	2772.1	2.0941	4.5251	6.6192	0.001121	0.215
920	176.29	449.44	746.8	2026.2	2773.0	2.1033	4.5084	6.6116	0.001123	0.210
940	177.21	450.36	750.8	2023.0	2773.8	2.1122	4.4920	6.6042	0.001124	0.206
960	178.12	451.27	754.8	2019.8	2774.6	2.1210	4.4759	6.5969	0.001125	0.202
980	179.01	452.16	758.7	2016.7	2775.4	2.1297	4.4602	6.5898	0.001126	0.198
1000	179.88	453.03	762.6	2013.6	2776.2	2.1382	4.4447	6.5828	0.001127	0.194
1100	184.06	457.21	781.1	1998.6	2779.7	2.1786	4.3712	6.5498	0.001133	0.177
1200	187.96	461.11	798.4	1984.3	2782.7	2.2160	4.3034	6.5194	0.001139	0.163
1300	191.60	464.75	814.7	1970.7	2785.4	2.2509	4.2404	6.4913	0.001144	0.151
1400	195.04	468.19	830.1	1957.7	2787.8	2.2836	4.1815	6.4651	0.001149	0.141
1500	198.28	471.43	844.6	1945.3	2789.9	2.3144	4.1262	6.4406	0.001154	0.132
1600	201.37	474.52	858.5	1933.2	2791.7	2.3436	4.0740	6.4176	0.001159	0.124
1700	204.30	477.45	871.8	1921.6	2793.4	2.3712	4.0246	6.3958	0.001163	0.117

1800	207.11	480.26	884.5	1910.3	2794.8	2.3976	3.9776	6.3751	0.001168	0.110
1900	209.79	482.94	896.8	1899.3	2796.1	2.4227	3.9327	6.3555	0.001172	0.105
2000	212.37	485.52	908.6	1888.7	2797.2	2.4468	3.8899	6.3367	0.001177	0.0996
2200	217.24	490.39	930.9	1868.1	2799.1	2.4921	3.8094	6.3015	0.001185	0.0907
2400	221.78	494.93	951.9	1848.5	2800.4	2.5342	3.7348	6.2690	0.001193	0.0832
2600	226.03	499.18	971.7	1829.7	2801.4	2.5736	3.6652	6.2388	0.001201	0.0769
3000	233.84	506.99	1008.3	1794.0	2802.3	2.6455	3.5383	6.1838	0.001216	0.0666
3500	242.54	515.69	1049.7	1752.2	2802.0	2.7252	3.3976	6.1229	0.001235	0.0570
4000	250.33	523.48	1087.4	1712.9	2800.3	2.7965	3.2720	6.0685	0.001252	0.0498
4500	257.41	530.56	1122.1	1675.6	2797.7	2.8612	3.1579	6.0191	0.001269	0.0440
5000	263.92	537.07	1154.5	1639.7	2794.2	2.9207	3.0528	5.9735	0.001286	0.0394
6000	275.56	548.71	1213.7	1571.3	2785.0	3.0274	2.8633	5.8907	0.001319	0.0324
7000	285.80	558.95	1267.5	1506.0	2773.4	3.1220	2.6541	5.8161	0.001351	0.0274
8000	294.98	568.13	1317.2	1442.7	2759.9	3.2077	2.5393	5.7470	0.001384	0.0235
9000	303.31	576.46	1363.8	1380.8	2744.6	3.2867	2.3952	5.6820	0.001418	0.0205
10,000	310.96	584.11	1408.1	1319.7	2727.7	3.3606	2.2592	5.6198	0.001453	0.0180
11,000	318.04	591.19	1450.6	1258.8	2709.3	3.4304	2.1292	5.5596	0.001489	0.0160
12,000	324.64	597.79	1491.7	1197.5	2698.2	3.4971	2.0032	5.5003	0.001527	0.0143
14,000	336.63	609.78	1571.5	1070.9	2642.4	3.6241	1.7564	5.3804	0.0016105	0.01150
16,000	347.32	620.47	1650.4	934.5	2584.9	3.7470	1.5063	5.2533	0.0017102	0.00931
18,000	356.96	630.11	1734.8	779.0	2513.9	3.8766	1.2362	5.1127	0.0018399	0.007497
20,000	365.71	638.86	1826.6	591.6	2418.2	4.0151	0.9259	4.9410	0.0020374	0.005875
22,000	373.68	646.83	2010.3	186.3	2196.6	4.2934	0.2881	4.5814	0.0026675	0.003735
22,120	374.15	647.30	2107.4	0	2107.4	4.4429	0	4.4429	0.0031700	0.003170

Table 11B Properties of saturated steam (Centigrade and Fahrenheit units)

Pressure		Temperature		Enthalpy Per Unit Mass						Entropy (Btu/lb°F)		Specific Volume (ft³/lb)
				Centigrade Units (kcal/kg)			Fahrenheit Units (Btu/lb)					
Absolute (lb/in.²)	Vacuum (in. Hg)	(°C)	(°F)	Water	Latent	Steam	Water	Latent	Steam	Water	Steam	Steam
0.5	28.99	26.42	79.6	**26.45**	582.50	**608.95**	**47.6**	1048.5	**1096.1**	0.0924	2.0367	643.0
0.6	28.79	29.57	85.3	**29.58**	580.76	**610.34**	**53.2**	1045.4	**1098.6**	0.1028	2.0214	540.6
0.7	28.58	32.28	90.1	**32.28**	579.27	**611.55**	**58.1**	1042.7	**1100.8**	0.1117	2.0082	466.6
0.8	28.38	34.67	94.4	**34.66**	577.95	**612.61**	**62.4**	1040.3	**1102.7**	0.1196	1.9970	411.7
0.9	28.17	36.80	98.2	**36.80**	576.74	**613.54**	**66.2**	1038.1	**1104.3**	0.1264	1.9871	368.7
1.0	27.97	38.74	101.7	**38.74**	575.60	**614.34**	**69.7**	1036.1	**1105.8**	0.1326	1.9783	334.0
1.1	27.76	40.52	104.9	**40.52**	574.57	**615.09**	**72.9**	1034.3	**1107.2**	0.1381	1.9702	305.2
1.2	27.56	42.17	107.9	**42.17**	573.63	**615.80**	**75.9**	1032.5	**1108.4**	0.1433	1.9630	281.1
1.3	27.35	43.70	110.7	**43.70**	572.75	**616.45**	**78.7**	1030.9	**1109.6**	0.1484	1.9563	260.5
1.4	27.15	45.14	113.3	**45.12**	571.94	**617.06**	**81.3**	1029.5	**1110.8**	0.1527	1.9501	243.0
1.5	26.95	46.49	115.7	**46.45**	571.16	**617.61**	**83.7**	1028.1	**1111.8**	0.1569	1.9442	228.0
1.6	26.74	47.77	118.0	**47.73**	570.41	**618.14**	**86.0**	1026.8	**1112.8**	0.1609	1.9387	214.3
1.7	26.54	48.98	120.2	**48.94**	569.71	**618.65**	**88.2**	1025.5	**1113.7**	0.1646	1.9336	202.5
1.8	26.33	50.13	122.2	**50.08**	569.06	**619.14**	**90.2**	1024.4	**1114.6**	0.1681	1.9288	191.8
1.9	26.13	51.22	124.2	**51.16**	568.47	**619.63**	**92.1**	1023.3	**1115.4**	0.1715	1.9243	182.3
2.0	25.92	52.27	126.1	**52.22**	567.89	**620.11**	**94.0**	1022.2	**1116.2**	0.1749	1.9200	173.7
3.0	23.88	60.83	141.5	**60.78**	562.89	**623.67**	**109.4**	1013.2	**1122.6**	0.2008	1.8869	118.7
4.0	21.84	67.23	153.0	**67.20**	559.29	**626.49**	**121.0**	1006.7	**1127.7**	0.2199	1.8632	90.63
5.0	19.80	72.38	162.3	**72.36**	556.24	**628.60**	**130.2**	1001.6	**1131.8**	0.2348	1.8449	73.52
6.0	17.76	76.72	170.1	**76.71**	553.62	**630.33**	**138.1**	996.6	**1134.7**	0.2473	1.8299	61.98
7.0	15.71	80.49	176.9	**80.52**	551.20	**631.72**	**144.9**	992.2	**1137.1**	0.2582	1.8176	53.64
8.0	13.67	83.84	182.9	**83.89**	549.16	**633.05**	**151.0**	988.5	**1139.5**	0.2676	1.8065	47.35
9.0	11.63	86.84	188.3	**86.88**	547.42	**634.30**	**156.5**	985.2	**1141.7**	0.2762	1.7968	42.40
10.0	9.59	89.58	193.2	**89.61**	545.82	**635.43**	**161.3**	982.5	**1143.8**	0.2836	1.7884	38.42
11.0	7.55	92.10	197.8	**92.15**	544.26	**636.41**	**165.9**	979.6	**1145.5**	0.2906	1.7807	35.14
12.0	5.50	94.44	202.0	**94.50**	542.75	**637.25**	**170.1**	976.9	**1147.0**	0.2970	1.7735	32.40
13.0	3.46	96.62	205.9	**96.69**	541.34	**638.03**	**173.9**	974.6	**1148.5**	0.3029	1.7672	30.05
14.0	1.42	98.65	209.6	**98.73**	540.06	**638.79**	**177.7**	972.2	**1149.9**	0.3086	1.7613	28.03
14.696	Gauge (lb/in.²)	100.00	212.0	**100.06**	539.22	**639.28**	**180.1**	970.6	**1150.7**	0.3122	1.7574	26.80
15	0.3	100.57	213.0	**100.65**	538.9	**639.5**	**181.2**	970.0	**1151.2**	0.3137	1.7556	26.28

16	1.3	102.40	216.3	**102.51**	537.7	**640.2**	**184.5**	967.9	**1152.4**	0.3187	1.7505	24.74
17	2.3	104.13	219.5	**104.27**	536.5	**640.8**	**187.6**	965.9	**1153.5**	0.3231	1.7456	23.38
18	3.3	105.78	222.4	**105.94**	535.5	**641.4**	**190.6**	964.0	**1154.6**	0.3276	1.7411	22.17
19	4.3	107.36	225.2	**107.53**	534.5	**642.0**	**193.5**	962.2	**1155.7**	0.3319	1.7368	21.07
20	5.3	108.87	228.0	**109.05**	533.6	**642.6**	**196.3**	960.4	**1156.7**	0.3358	1.7327	20.09
21	6.3	110.32	230.6	**110.53**	532.6	**643.1**	**198.9**	958.8	**1157.7**	0.3396	1.7287	19.19
22	7.3	111.71	233.1	**111.94**	531.7	**643.6**	**201.4**	957.2	**1158.6**	0.3433	1.7250	18.38
23	8.3	113.05	235.5	**113 30**	530.8	**644.1**	**203.9**	955.6	**1159.5**	0.3468	1.7215	17.63
24	9.3	114.34	237.8	**114.61**	530.0	**644.6**	**206.3**	954.0	**1160.3**	0.3502	1.7181	16.94
25	10.3	115.59	240.1	**115.87**	529.2	**645.1**	**208.6**	952.5	**1161.1**	0.3534	1.7148	16.30
26	11.3	116.80	242.2	**117.11**	528.4	**645.5**	**210.8**	951.1	**1161.9**	0.3565	1.7118	15.72
27	12.3	117.97	244.4	**118.31**	527.6	**645.9**	**212.9**	949.7	**1162.6**	0.3595	1.7089	15.17
28	13.3	119.11	246.4	**119.47**	526.8	**646.5**	**215.0**	948.3	**1163.3**	0.3625	1.7060	14.67
29	14.3	120.21	248.4	**120.58**	526.1	**646.7**	**217.0**	947.0	**1164.0**	0.3654	1.7032	14.19
30	15.3	121.3	250.3	**121.7**	525.4	**647.1**	**219.0**	945.6	**1164.6**	0.3682	1.7004	13.73
32	17.3	123.3	254.0	**123.8**	524.1	**647.9**	**222.7**	943.1	**1165.8**	0.3735	1.6952	12.93
34	19.3	125.3	257.6	**125.8**	522.8	**648.6**	**226.3**	940.7	**1167.0**	0.3785	1.6905	12.21
36	21.3	127.2	260.9	**127.7**	521.5	**649.2**	**229.7**	938.5	**1168.2**	0.3833	1.6860	11.58
38	23.3	128.9	264.1	**129.5**	520.3	**649.8**	**233.0**	936.4	**1169.4**	0.3879	1.6817	11.02
40	25.3	130.7	267.2	**131.2**	519.2	**650.4**	**236.1**	934.4	**1170.5**	0.3923	1.6776	10.50
42	27.3	132.3	270.3	**132.9**	518.0	**650.9**	**239.1**	932.3	**1171.4**	0.3964	1.6737	10.30
44	29.3	133.9	273.1	**134.5**	516.9	**651.4**	**242.0**	930.3	**1172.3**	0.4003	1.6700	9.600
46	31.3	135.4	275.8	**136.0**	515.9	**651.9**	**244.9**	928.3	**1173.2**	0.4041	1.6664	9.209
48	33.3	136.9	278.5	**137.5**	514.8	**652.3**	**247.6**	926.4	**1174.0**	0.4077	1.6630	8.848
50	35.3	138.3	281.0	**139.0**	513.8	**652.8**	**250.2**	924.6	**1174.8**	0.4112	1.6597	8.516
52	37.3	139.7	283.5	**140.4**	512.8	**653.2**	**252.7**	922.9	**1175.6**	0.4146	1.6566	8.208
54	39.3	141.0	285.9	**141.8**	511.8	**653.6**	**255.2**	921.1	**1176.3**	0.4179	1.6536	7.922
56	41.3	142.3	288.3	**143.1**	510.9	**654.0**	**257.6**	919.4	**1177.0**	0.4211	1.6507	7.656
58	43.3	143.6	290.5	**144.4**	510.0	**654.4**	**259.9**	917.8	**1177.7**	0.4242	1.6478	7.407
60	45.3	144.9	292.7	**145.6**	509.2	**654.8**	**262.2**	916.2	**1178.4**	0.4272	1.6450	7.175
62	47.3	146.1	294.9	**146.8**	508.4	**655.2**	**264.4**	914.6	**1179.0**	0.4302	1.6423	6.957
64	49.3	147.3	296.9	**148.0**	507.6	**655.6**	**266.5**	913.1	**1179.6**	0.4331	1.6398	6.752
66	51.3	148.4	299.0	**149.2**	506.7	**655.9**	**268.6**	911.6	**1180.2**	0.4359	1.6374	6.560
68	53.3	149.5	301.0	**150.3**	505.9	**656.2**	**270.7**	910.1	**1180.8**	0.4386	1.6350	6.378
70	55.3	150.6	302.9	**151.5**	505.0	**656.5**	**272.7**	908.7	**1181.4**	0.4412	1.6327	6.206
72	57.3	151.6	304.8	**152.6**	504.2	**656.8**	**274.6**	907.4	**1182.0**	0.4437	1.6304	6.044
74	59.3	152.6	306.7	**153.6**	503.4	**657.0**	**276.5**	906.0	**1182.5**	0.4462	1.6282	5.890

(Continued)

Table 11B Properties of saturated steam (Centigrade and Fahrenheit units)—cont'd

Pressure		Temperature		Enthalpy Per Unit Mass						Entropy (Btu/lb°F)		Specific Volume (ft³/lb)
Absolute	Vacuum			Centigrade Units (kcal/kg)			Fahrenheit Units (Btu/lb)					
(lb/in.²)	(in. Hg)	(°C)	(°F)	Water	Latent	Steam	Water	Latent	Steam	Water	Steam	Steam
76	61.3	153.6	308.5	**154.7**	502.6	**657.3**	**278.4**	904.6	**1183.0**	0.4486	1.6261	5.743
78	63.3	154.6	310.3	**155.7**	501.8	**657.5**	**280.3**	903.2	**1183.5**	0.4510	1.6240	5.604
80	65.3	155.6	312.0	**156.7**	501.1	**657.8**	**282.1**	901.9	**1184.0**	0.4533	1.6219	5.472
82	67.3	156.5	313.7	**157.7**	500.3	**658.0**	**283.9**	900.6	**1184.5**	0.4556	1.6199	5.346
84	69.3	157.5	315.4	**158.6**	499.6	**658.2**	**285.6**	899.4	**1185.0**	0.4579	1.6180	5.226
86	71.3	158.4	317.1	**159.6**	498.9	**658.5**	**287.3**	898.1	**1185.4**	0.4601	1.6161	5.110
88	73.3	159.4	318.7	**160.5**	498.3	**658.8**	**289.0**	896.8	**1185.8**	0.4622	1.6142	5.000
90	75.3	160.3	320.3	**161.5**	497.6	**659.1**	**290.7**	895.5	**1186.2**	0.4643	1.6124	4.896
92	77.3	161.2	321.9	**162.4**	496.9	**659.3**	**292.3**	894.3	**1186.6**	0.4664	1.6106	4.796
94	79.3	162.0	323.3	**163.3**	496.3	**659.6**	**293.9**	893.1	**1187.0**	0.4684	1.6088	4.699
96	81.3	162.8	324.8	**164.1**	495.7	**659.8**	**295.5**	891.9	**1187.4**	0.4704	1.6071	4.607
98	83.3	163.6	326.6	**165.0**	495.0	**660.0**	**297.0**	890.8	**1187.8**	0.4723	1.6054	4.519
100	85.3	164.4	327.8	**165.8**	494.3	**660.1**	**298.5**	889.7	**1188.2**	0.4742	1.6038	4.434
105	90.3	166.4	331.3	**167.9**	492.7	**660.6**	**302.2**	886.9	**1189.1**	0.4789	1.6000	4.230
110	95.3	168.2	334.8	**169.8**	491.2	**661.0**	**305.7**	884.2	**1189.9**	0.4833	1.5963	4.046
115	100.3	170.0	338.1	**171.7**	489.8	**661.5**	**309.2**	881.5	**1190.7**	0.4876	1.5927	3.880
120	105.3	171.8	341.3	**173.6**	488.3	**661.9**	**312.5**	878.9	**1191.4**	0.4918	1.5891	3.729
125	110.3	173.5	344.4	**175.4**	486.9	**662.3**	**315.7**	876.4	**1192.1**	0.4958	1.5856	3.587
130	115.3	175.2	347.3	**177.1**	485.6	**662.7**	**318.8**	874.0	**1192.8**	0.4997	1.5823	3.456
135	120.3	176.8	350.2	**178.8**	484.2	**663.0**	**321.9**	871.5	**1193.4**	0.5035	1.5792	3.335
140	125.3	178.3	353.0	**180.5**	482.9	**663.4**	**324.9**	869.1	**1194.0**	0.5071	1.5763	3.222
145	130.3	179.8	355.8	**182.1**	481.6	**663.7**	**327.8**	866.8	**1194.6**	0.5106	1.5733	3.116
150	135.3	181.3	358.4	**183.7**	480.3	**664.0**	**330.6**	864.5	**1195.1**	0.5140	1.5705	3.015

Table 11C Enthalpy of superheated steam, H (kJ/kg)

	Saturation		Temperature, θ (°C)	100	200	300	400	500	600	700	800
Pressure P (kN/m^2)	T_s (K)	S_s (kJ/kg)	Temperature, T (K)	373.15	473.15	573.15	673.15	773.15	873.15	973.15	1073.15
100	372.78	2675.4		2676.0	2875.4	3074.6	3278.0	3488.0	3705.0	3928.0	4159.0
200	393.38	2706.3			2870.4	3072.0	3276.4	3487.0	3704.0	3927.0	4158.0
300	406.69	2724.7			2866.0	3069.7	3275.0	3486.0	3703.1	3927.0	4158.0
400	416.78	2737.6			2861.3	3067.0	3273.5	3485.0	3702.0	3926.0	4157.0
500	425.00	2747.5			2856.0	3064.8	3272.1	3484.0	3701.2	3926.0	4156.8
600	431.99	2755.5			2850.7	3062.0	3270.0	3483.0	3701.0	3925.0	4156.2
700	438.11	2762.0			2845.5	3059.5	3269.0	3482.6	3700.2	3924.0	4156.0
800	443.56	2767.5			2839.7	3057.0	3266.8	3480.4	3699.0	3923.8	4155.0
900	448.56	2772.1			2834.0	3055.0	3266.2	3479.5	3698.6	3923.0	4155.0
1000	453.03	2776.2			2828.7	3051.7	3264.3	3478.0	3697.5	3922.8	4154.0
2000	485.59	2797.2				3024.8	3248.0	3467.0	3690.0	3916.0	4150.0
3000	506.98	2802.3				2994.8	3231.7	3456.0	3681.6	3910.4	4145.0
4000	523.49	2800.3				2962.0	3214.8	3445.0	3673.4	3904.0	4139.6
5000	537.09	2794.2				2926.0	3196.9	3433.8	3665.4	3898.0	4135.5
6000	548.71	2785.0				2886.0	3178.0	3421.7	3657.0	3891.8	4130.0
7000	558.95	2773.4				2840.0	3159.1	3410.0	3648.8	3886.0	4124.8
8000	568.13	2759.9				2785.0	3139.5	3398.0	3640.4	3880.8	4121.0
9000	576.46	2744.6					3119.0	3385.5	3632.0	3873.6	4116.0
10,000	584.11	2727.7					3097.7	3373.6	3624.0	3867.2	4110.8
11,000	591.19	2709.3					3075.6	3361.0	3615.5	3862.0	4106.0
12,000	597.79	2698.2					3052.9	3349.0	3607.0	3855.3	4101.2

Table 11D Entropy of superheated steam, S (kJ/kg K)

Pressure P (kN/m^2)	Saturation		Temperature, θ (°C)	100	200	300	400	500	600	700	800
	T_s (K)	H_s (kJ/kg)	Temperature, T (K)	373.15	473.15	573.15	673.15	773.15	873.15	973.15	1073.15
100	372.78	7.3598		7.362	7.834	8.216	8.544	8.834	9.100	9.344	9.565
200	393.38	7.1268			7.507	7.892	8.222	8.513	8.778	9.020	9.246
300	406.69	6.9909			7.312	7.702	8.033	8.325	8.591	8.833	9.057
400	416.78	6.8943			7.172	7.566	7.898	8.191	8.455	8.700	8.925
500	425.00	6.8192			7.060	7.460	7.794	8.087	8.352	8.596	8.820
600	431.99	6.7575			6.968	7.373	7.708	8.002	8.268	8.510	8.738
700	438.11	6.7052			6.888	7.298	7.635	7.930	8.195	8.438	8.665
800	443.56	6.6596			6.817	7.234	7.572	7.867	8.133	8.375	8.602
900	448.56	6.6192			6.753	7.176	7.515	7.812	8.077	8.321	8.550
1000	453.03	6.5828			6.695	7.124	7.465	7.762	8.028	8.272	8.502
2000	485.59	6.3367				6.768	7.128	7.431	7.702	7.950	8.176
3000	506.98	6.1838				6.541	6.922	7.233	7.508	7.756	7.985
4000	523.49	6.0685				6.364	6.770	7.090	7.368	7.620	7.850
5000	537.07	5.9735				6.211	6.647	6.977	7.258	7.510	7.744
6000	548.71	5.8907				6.060	6.542	6.880	7.166	7.422	7.655
7000	558.95	5.8161				5.933	6.450	6.798	7.088	7.345	7.581
8000	568.13	5.7470				5.792	6.365	6.724	7.020	7.280	7.515
9000	576.46	5.6820					6.288	6.659	6.958	7.220	7.457
10,000	584.11	5.6198					6.215	6.598	6.902	7.166	7.405
11,000	591.19	5.5596					6.145	6.540	6.850	7.117	7.357
12,000	597.79	5.5003					6.077	6.488	6.802	7.072	7.312

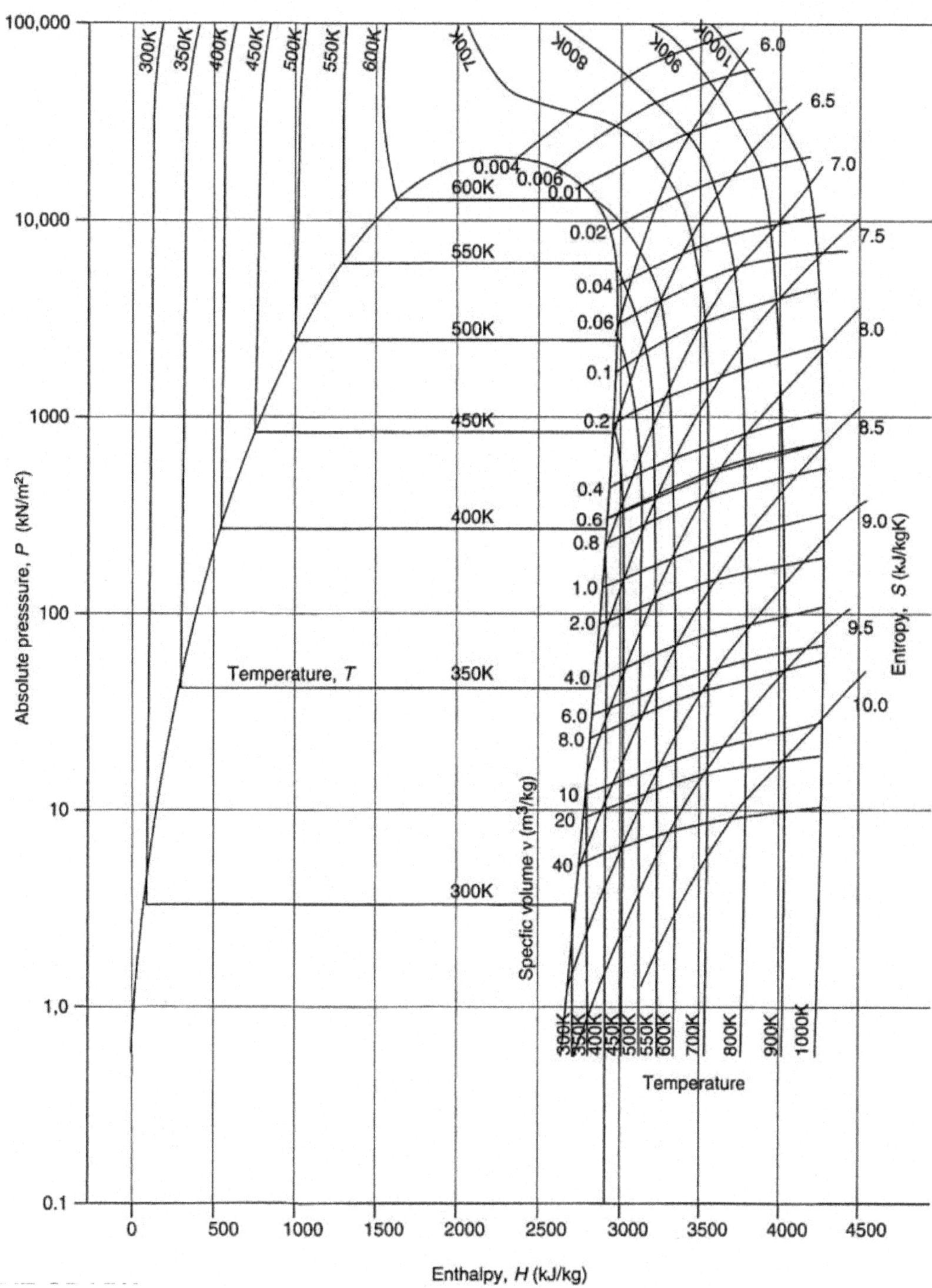

Fig. 11A

Pressure-enthalpy diagram for water and steam.

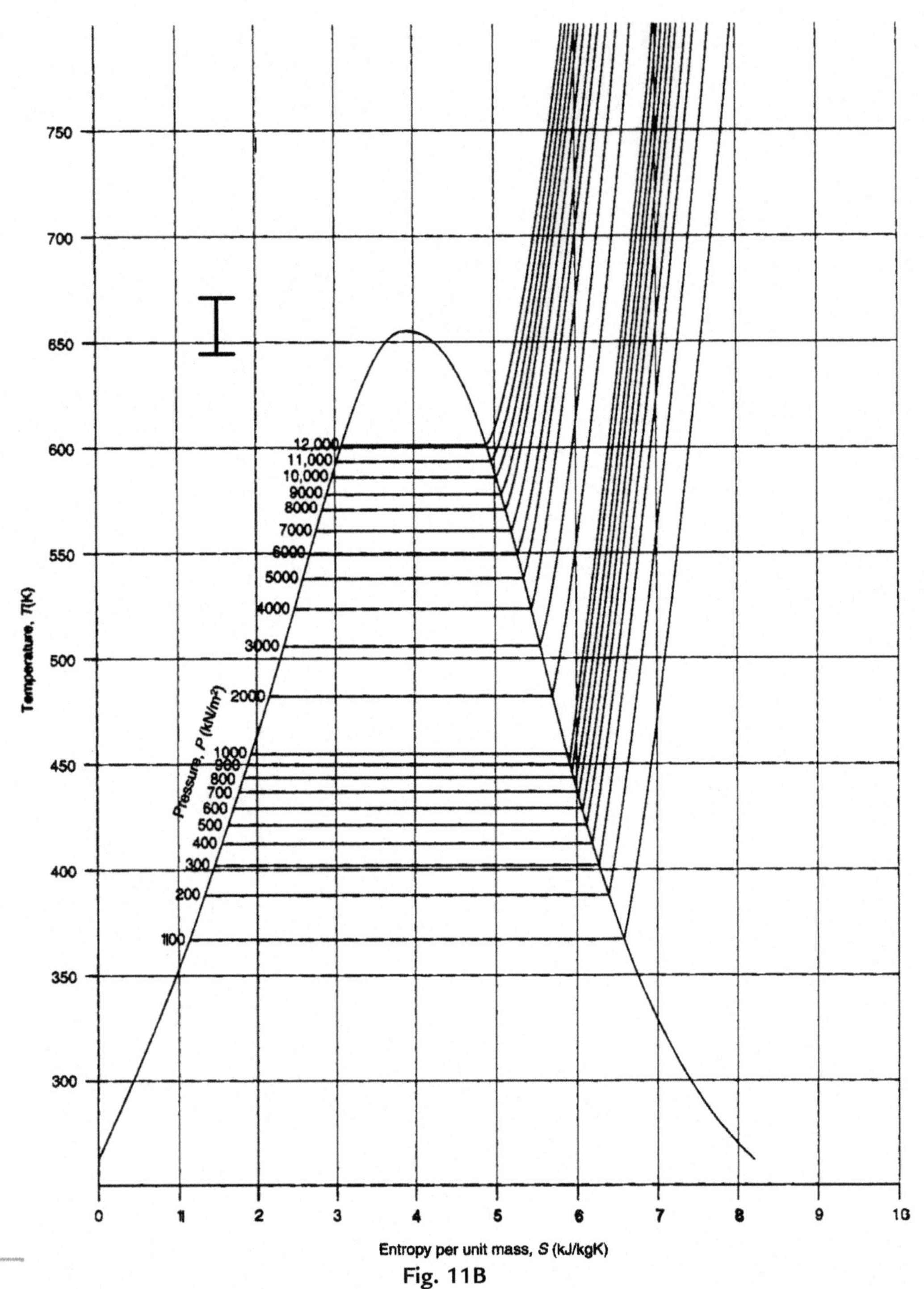

Fig. 11B

Temperature-entropy diagram for water and steam.

A.3 Mathematical Tables

Table 12 Laplace transforms

No.	Transform $\bar{f}(p) = \int_0^\infty e^{-pt} f(t)dt$	Function f(t)
1	$\frac{1}{p}$	1
2	$\frac{1}{p^2}$	t
3	$\frac{1}{p^n} \quad n = 1, 2, 3, \ldots$	$\frac{t^{n-1}}{(n-1)!}$
4	$\frac{1}{\sqrt{p}}$	$\frac{1}{\sqrt{(\pi t)}}$
5	$\frac{1}{p^{3/2}}$	$2\sqrt{\frac{t}{\pi}}$
6	$\frac{1}{p^{n+\frac{1}{2}}} \quad n = 1, 2, 3, \ldots$	$\frac{2^n t^{n-\frac{1}{2}}}{[1.3.5\ldots(2n-1)]\sqrt{\pi}}$
7	$\frac{\Gamma(k)}{p^k} \quad k > 0$	t^{k-1}
8	$\frac{1}{p-a}$	e^{at}
9	$\frac{1}{(p-a)^2}$	te^{at}
10	$\frac{1}{(p-a)^n} \quad n = 1, 2, 3, \ldots$	$\frac{1}{(n-1)!} t^{n-1} e^{at}$
11	$\frac{\Gamma(k)}{(p-a)^k} \quad k > 0$	$t^{k-1} e^{at}$
12	$\frac{1}{(p-a)(p-b)} \quad a \neq b$	$\frac{1}{a-b}(e^{at} - e^{bt})$
13	$\frac{p}{(p-a)(p-b)} \quad a \neq b$	$\frac{1}{a-b}(ae^{at} - be^{bt})$
14	$\frac{1}{(p-a)(p-b)(p-c)} \quad a \neq b \neq c$	$-\frac{(b-c)e^{at} + (c-a)e^{bt} + (a-b)e^{ct}}{(a-b)(b-c)(c-a)}$
15	$\frac{1}{p^2 + a^2}$	$\frac{1}{a} \sin at$
16	$\frac{p}{p^2 + a^2}$	$\cos at$
17	$\frac{1}{p^2 - a^2}$	$\frac{1}{a} \sinh at$
18	$\frac{p}{p^2 - a^2}$	$\cosh at$
19	$\frac{1}{p(p^2 + a^2)}$	$\frac{1}{a^2}(1 - \cos at)$
20	$\frac{1}{p^2(p^2 + a^2)}$	$\frac{1}{a^3}(at - \sin at)$
21	$\frac{1}{(p^2 + a^2)^2}$	$\frac{1}{2a^3}(\sin at - at\cos at)$

(Continued)

Table 12 Laplace transforms—cont'd

No.	Transform $\bar{f}(p) = \int_0^\infty e^{-pt} f(t) dt$	Function f(t)
22	$\frac{p}{(p^2+a^2)^2}$	$\frac{t}{2a}\sin at$
23	$\frac{p^2}{(p^2+a^2)^2}$	$\frac{1}{2a}(\sin at + at\cos at)$
24	$\frac{p^2-a^2}{(p^2+a^2)^2}$	$t\cos at$
25	$\frac{p}{(p^2+a^2)(p^2+b^2)}\quad a^2 \neq b^2$	$\frac{\cos at - \cos bt}{b^2-a^2}$
26	$\frac{1}{(p-a)^2+b^2}$	$\frac{1}{b}e^{at}\sin bt$
27	$\frac{p-a}{(p-a)^2+b^2}$	$e^{at}\cos bt$
28	$\frac{3a^2}{p^3+a^3}$	$e^{-at} - e^{at/2}\left(\cos\frac{at\sqrt{3}}{2} - \sqrt{3}\sin\frac{at\sqrt{3}}{2}\right)$
29	$\frac{4a^3}{p^4+4a^4}$	$\sin at\cosh at - \cos at\sinh at$
30	$\frac{p}{p^4+4a^4}$	$\frac{1}{2a^2}\sin at\sinh at$
31	$\frac{1}{p^4-a^4}$	$\frac{1}{2a^3}(\sinh at - \sin at)$
32	$\frac{p}{p^4-a^4}$	$\frac{1}{2a^2}(\cosh at - \cos at)$
33	$\frac{8a^3p^2}{(p^2+a^2)^3}$	$(1+a^2t^2)\sin at - at\cos at$
34	$\frac{1}{p}\left(\frac{p-1}{p}\right)^n$	$\frac{e^t}{n!}\frac{d^n}{dt^n}(t^n e^{-t})$ = Laguerre polyonomial of degree n
35	$\frac{p}{(p-a)^{3/2}}$	$\frac{1}{\sqrt{(\pi t)}} - e^{at}(1+2at)$
36	$\sqrt{(p-a)} - \sqrt{(p-b)}$	$\frac{1}{2\sqrt{(\pi t^3)}}(e^{bt} - e^{at})$
37	$\frac{1}{\sqrt{p+a}}$	$\frac{1}{\sqrt{(\pi t)}} - ae^{a^2t}\operatorname{erfc}(a\sqrt{t})$
38	$\frac{\sqrt{p}}{p-a^2}$	$\frac{1}{\sqrt{(\pi t)}} + ae^{a^2t}\operatorname{erf}(a\sqrt{t})$
39	$\frac{\sqrt{p}}{p+a^2}$	$\frac{1}{\sqrt{(\pi t)}} - \frac{2ae^{-a^2t}}{\sqrt{\pi}}\int_0^{a\sqrt{t}} e^{\lambda}2d\lambda$
40	$\frac{1}{\sqrt{p(p-a^2)}}$	$\frac{1}{a}e^{a^2t}\operatorname{erf}(a\sqrt{t})$
41	$\frac{1}{\sqrt{p(p+a^2)}}$	$\frac{2e^{-a^2t}}{a\sqrt{\pi}}\int_0^{a\sqrt{t}} e^{\lambda^2}d\lambda$
42	$\frac{b^2-a^2}{(p-a^2)(b+\sqrt{p})}$	$e^{a^2t}[b - a\operatorname{erf}(a\sqrt{t})] - be^{b^2t}\operatorname{erfc}(b\sqrt{t})$

Table 12 Laplace transforms—cont'd

No.	Transform $\bar{f}(p) = \int_0^\infty e^{-pt} f(t) dt$	Function f(t)
43	$\frac{1}{\sqrt{p}(\sqrt{p}+a)}$	$e^{a^2 t} \operatorname{erfc}(a\sqrt{t})$
44	$\frac{1}{(p+a)\sqrt{(p+b)}}$	$\frac{1}{\sqrt{(b-a)}} e^{-at} \operatorname{erf}\left[\sqrt{(b-a)}\sqrt{t}\right]$
45	$\frac{b^2-a^2}{\sqrt{p(p-a^2)}\left(\sqrt{(p+b)}\right)}$	$e^{a^2 t}\left[\frac{b}{a}\operatorname{erf}(a\sqrt{t})-1\right] + e^{b^2 t}\operatorname{erfc}(b\sqrt{t})$
46	$\frac{(1-p)^n}{p^{n+1/2}}$	$\frac{n!}{(2n)!\sqrt{(\pi t)}} H_{2n}(\sqrt{t})$ where $H_n(t) = e^{t^2}\frac{d^n}{dt^n}e^{-t^2}$ is the Hermite polynomial
47	$\frac{(1-p)^n}{p^{n+3/2}}$	$-\frac{n!}{\sqrt{\pi}(2n+1)!} H_{2n+1}(\sqrt{t})$
48	$\frac{\sqrt{(p+2a)}}{\sqrt{p}} - 1$	$a e^{-at}[I_1(at) + I_0(at)]$
49	$\frac{1}{\sqrt{(p+a)}\sqrt{(p+b)}}$	$e^{-\frac{1}{2}(a+b)t} I_0\left(\frac{a-b}{2}t\right)$
50	$\frac{\Gamma(k)}{(p+a)^k(p+b)^k} \quad k>0$	$\sqrt{\pi}\left(\frac{t}{a-b}\right)^{k-\frac{1}{2}} e^{-\frac{1}{2}(a+b)t} I_{k-\frac{1}{2}}\left(\frac{a-b}{2}t\right)$
51	$\frac{1}{\sqrt{(p+a)}(p+b)^{3/2}}$	$t e^{-\frac{1}{2}(a+b)t}\left[I_0\left(\frac{a-b}{2}t\right) + I_1\left(\frac{a-b}{2}t\right)\right]$
52	$\frac{\sqrt{(p+2a)}-\sqrt{p}}{\sqrt{(p+2a)}+\sqrt{p}}$	$\frac{1}{t} e^{-at} I_1(at)$
53	$\frac{(a-b)^k}{\left[\sqrt{(p+a)}+\sqrt{(p+b)}\right]^{2k}} \quad k>0$	$\frac{k}{t} e^{-\frac{1}{2}(a+b)t} I_k\left(\frac{a-b}{2}t\right)$
54	$\frac{\left[\sqrt{(p+a)}+\sqrt{p}\right]^{-2j}}{\sqrt{p}\sqrt{(p+a)}} \quad j>-1$	$\frac{1}{a^j} e^{-\frac{1}{2}at} I_j\left(\frac{1}{2}at\right)$
55	$\frac{1}{\sqrt{(p^2+a^2)}}$	$J_0(at)$
56	$\frac{\left[\sqrt{(p^2+a^2)}-p\right]^j}{\sqrt{(p^2+a^2)}} \quad j>1$	$a^j J_j(at)$
57	$\frac{1}{(p^2+a^2)^k} \quad k>0$	$\frac{\sqrt{\pi}}{\Gamma(k)}\left(\frac{t}{2a}\right)^{k-\frac{1}{2}} J_{k-\frac{1}{2}}(at)$
58	$\left[\sqrt{(p^2+a^2)}-p\right]^k \quad k>0$	$\frac{k a^k}{t} J_k(at)$
59	$\frac{\left[p-\sqrt{(p^2-a^2)}\right]^j}{\sqrt{(p^2-a^2)}} \quad j>-1$	$a^j I_j(at)$
60	$\frac{1}{(p^2-a^2)^k} \quad k>0$	$\frac{\sqrt{\pi}}{\Gamma(k)}\left(\frac{t}{2a}\right)^{k-\frac{1}{2}} I_{k-\frac{1}{2}}(at)$
61	$\frac{e^{-kp}}{p}$	$S_k(t) = \begin{cases} 0 & \text{when } 0<t<k \\ 1 & \text{when } t>k \end{cases}$

(Continued)

Table 12 Laplace transforms—cont'd

No.	Transform $\bar{f}(p)=\int_0^\infty e^{-pt}f(t)dt$	Function f(t)
62	$\frac{e^{-kp}}{p^2}$	$\begin{cases} 0 & \text{when } 0<t<k \\ t-k & \text{when } t>k \end{cases}$
63	$\frac{e^{-kp}}{p^j}\ j>0$	$\begin{cases} 0 & \text{when } 0<t<k \\ \frac{(t-k)^{j-1}}{\Gamma(j)} & \text{when } t>k \end{cases}$
64	$\frac{1-e^{-kp}}{p}$	$\begin{cases} 1 & \text{when } 0<t<k \\ 0 & \text{when } t>k \end{cases}$
65	$\frac{1}{p(1-e^{-kp})}=\frac{1+\coth\frac{1}{2}kp}{2p}$	$S(k,t)=n$ when $(n-1)k<t<nk \quad n=1,2,3,\ldots$
66	$\frac{1}{p(e^{kp}-a)}$	$\begin{cases} 0 \text{ when } 0<t<k \\ 1+a+a^2+\ldots+a^{n-1} \\ \quad \text{when } nk<t<(n+1)k \quad n=1,2,3,\cdots \end{cases}$
67	$\frac{1}{p}\tanh kp$	$M(2k,t)=(-1)^{n-1}$ when $2k(n-1)<t<2kn \quad n=1,2,3,\ldots$
68	$\frac{1}{p(1+e^{-kp})}$	$\frac{1}{2}M(k,t)+\frac{1}{2}=\frac{1-(-1)^n}{2}$ when $(n-1)k<t<nk$
69	$\frac{1}{p^2}\tanh kp$	$H(2k,t)=\begin{cases} t & \text{when } 0<t<2k \\ 4k-t & \text{when } 2k<t<4k \end{cases}$
70	$\frac{1}{p\sinh kp}$	$2S(2k,t+k)-2=2(n-1)$ when $(2n-3)k<t<(2n-1)k \quad t>0$
71	$\frac{1}{p\sinh kp}$	$M(2k,t+3k)+1=1+(-1)^n$ when $(2n-3)k<t<(2n-1)k \quad t>0$
72	$\frac{1}{p}\coth kp$	$2S(2k,t)-1=2n-1$ when $2k(n-1)<t<2kn$
73	$\frac{k}{p^2+k^2}\coth\frac{\pi p}{2k}$	$\|\sin kt\|$
74	$\frac{1}{(p^2+1)(1-e^{-\pi p})}$	$\begin{cases} \sin t & \text{when } (2n-2)\pi<t<(2n-1)\pi \\ 0 & \text{when } (2n-1)\pi<t<2n\pi \end{cases}$
75	$\frac{1}{p}e^{-k/p}$	$J_0[2\sqrt{(kt)}]$
76	$\frac{1}{\sqrt{p}}e^{-k/p}$	$\frac{1}{\sqrt{(\pi t)}}\cos 2\sqrt{(kt)}$
77	$\frac{1}{\sqrt{p}}e^{k/p}$	$\frac{1}{\sqrt{(\pi t)}}\cosh 2\sqrt{(kt)}$
78	$\frac{1}{p^{3/2}}e^{-k/p}$	$\frac{1}{\sqrt{(\pi k)}}\sin 2\sqrt{(kt)}$
79	$\frac{1}{p^{3/2}}e^{k/p}$	$\frac{1}{\sqrt{(\pi k)}}\sinh 2\sqrt{(kt)}$
80	$\frac{1}{p^j}e^{k/p}\ j>0$	$\left(\frac{t}{k}\right)^{(j-1)/2}J_{j-1}[2\sqrt{(kt)}]$

Table 12 Laplace transforms—cont'd

No.	Transform $\bar{f}(p) = \int_0^\infty e^{-pt} f(t)dt$	Function f(t)
81	$\frac{1}{p^j} e^{k/p} \quad j > 0$	$\left(\frac{t}{k}\right)^{(j-1)/2} I_{j-1}\left[2\sqrt{(kt)}\right]$
82	$e^{-k\sqrt{p}} \quad k > 0$	$\frac{k}{2\sqrt{(\pi t^3)}} \exp\left(-\frac{k^2}{4t}\right)$
83	$\frac{1}{p} e^{-k\sqrt{p}} \quad k \geq 0$	$\text{erfc}\left(\frac{k}{2\sqrt{t}}\right)$
84	$\frac{1}{\sqrt{p}} e^{-k\sqrt{p}} \quad k \geq 0$	$\frac{1}{\sqrt{(\pi t)}} \exp\left(-\frac{k^2}{4t}\right)$
85	$p^{-3/2} e^{-k\sqrt{p}} \quad k \geq 0$	$2\sqrt{\frac{t}{\pi}}\left[\exp\left(-\frac{k^2}{4t}\right)\right] - k\,\text{erfc}\left(\frac{k}{2\sqrt{t}}\right)$
86	$\frac{a e^{-k\sqrt{p}}}{p(a+\sqrt{p})} \quad k \geq 0$	$-\exp(ak)\exp(a^2 t)\,\text{erfc}\left(a\sqrt{t} + \frac{k}{2\sqrt{t}}\right) + \text{erfc}\left(\frac{k}{2\sqrt{t}}\right)$
87	$\frac{e^{-k\sqrt{p}}}{\sqrt{p(a+\sqrt{p})}} \quad k \geq 0$	$\exp(ak)\exp(a^2 t)\,\text{erfc}\left(a\sqrt{t} + \frac{k}{2\sqrt{t}}\right)$
88	$\frac{e^{-k\sqrt{[p(p+a)]}}}{\sqrt{[p(p+a)]}}$	$\begin{cases} 0 & \text{when } 0 < t < k \\ \exp(-\frac{1}{2}at) I_0\left[\frac{1}{2}a\sqrt{(t^2 - k^2)}\right] & \text{when } t > k \end{cases}$
89	$\frac{e^{-k\sqrt{(p^2+a^2)}}}{\sqrt{(p^2+a^2)}}$	$\begin{cases} 0 & \text{when } 0 < t < k \\ J_0\left[a\sqrt{(t^2 - k^2)}\right] & \text{when } t > k \end{cases}$
90	$\frac{e^{-k\sqrt{(p^2-a^2)}}}{\sqrt{(p^2-a^2)}}$	$\begin{cases} 0 & \text{when } 0 < t < k \\ I_0\left[a\sqrt{(t^2 - k^2)}\right] & \text{when } t > k \end{cases}$
91	$\frac{e^{-k\left[\sqrt{(p^2+a^2)} - p\right]}}{\sqrt{(p^2+a^2)}} \quad k \geq 0$	$J_0\left[a\sqrt{(t^2 + 2kt)}\right]$
92	$e^{-kp} - e^{-k\sqrt{(p^2+a^2)}}$	$\begin{cases} 0 \text{ when } 0 < t < k \\ \frac{ak}{\sqrt{(t^2 - k^2)}} J_1\left[a\sqrt{(t^2 - k^2)}\right] \\ \text{when } t > k \end{cases}$
93	$e^{-k\sqrt{(p^2-a^2)}} - e^{-kp}$	$\begin{cases} 0 \text{ when } 0 < t < k \\ \frac{ak}{\sqrt{(t^2 - k^2)}} J_1\left[a\sqrt{(t^2 - k^2)}\right] \\ \text{when } t > k \end{cases}$
94	$\frac{a^j e^{-k\sqrt{(p^2+a^2)}}}{\sqrt{(p^2+a^2)}\left[\sqrt{(p^2+a^2)} + p\right]^j}$ $j > -1$	$\begin{cases} 0 \text{ when } 0 < t < k \\ \left(\frac{t-k}{t+k}\right)^{(1/2)j} J_j\left[a\sqrt{(t^2 - k^2)}\right] \\ \text{when } t > k \end{cases}$
95	$\frac{1}{p} \ln p$	$\lambda - \ln t \quad \lambda = -0.5772\ldots$

(Continued)

Table 12 Laplace transforms—cont'd

No.	Transform $\bar{f}(p)=\int_0^\infty e^{-pt}f(t)dt$	Function f(t)
96	$\frac{1}{p^k}\ln p \quad k>0$	$t^{k-1}\left\{\frac{\lambda}{[\Gamma(k)]^2}-\frac{\ln t}{\Gamma(k)}\right\}$
97[a]	$\frac{\ln p}{p-a} \quad a>0$	$(\exp at)[\ln a - Ei(-at)]$
98[b]	$\frac{\ln p}{p^2+1}$	$\cos t\mathrm{Si}(t)-\sin t\mathrm{Ci}(t)$
99[b]	$\frac{p\ln p}{p^2+1}$	$-\sin t\mathrm{Si}(t)-\cos t\mathrm{Ci}(t)$
100[a]	$\frac{1}{p}\ln(1+kp) \quad k>0$	$-Ei\left(-\frac{1}{k}\right)$
101	$\ln\frac{p-a}{p-b}$	$\frac{1}{t}(e^{bt}-e^{at})$
102[b]	$\frac{1}{p}\ln(1+k^2p^2)$	$-2Ci\left(\frac{t}{k}\right)$
103[b]	$\frac{1}{p}\ln(p^2+a^2) \quad a>0$	$2\ln a - 2\,\mathrm{Ci}(at)$
104[b]	$\frac{1}{p^2}\ln(p^2+a^2) \quad a>0$	$\frac{2}{a}[at\ln a+\sin at - atCi(at)]$
105	$\ln\frac{p^2+a^2}{p^2}$	$\frac{2}{t}(1-\cos at)$
106	$\ln\frac{p^2-a^2}{p^2}$	$\frac{2}{t}(1-\cosh at)$
107	$\tan^{-1}\frac{k}{p}$	$\frac{1}{t}\sin kt$
108[b]	$\frac{1}{p}\tan^{-1}\frac{k}{p}$	$\mathrm{Si}\,(kt)$
109	$\exp(k^2p^2)\,\mathrm{erfc}(kp) \quad k>0$	$\frac{1}{k\sqrt{\pi}}\exp\left(-\frac{t^2}{4k^2}\right)$
110	$\frac{1}{p}\exp(k^2p^2)\,\mathrm{erfc}(kp) \quad k>0$	$\mathrm{erf}\left(\frac{t}{2k}\right)$
111	$\exp(kp)\,\mathrm{erfc}\left[\sqrt{(kp)}\right] \quad k>0$	$\frac{\sqrt{k}}{\pi\sqrt{t(t+k)}}$
112	$\frac{1}{\sqrt{p}}\mathrm{erfc}\left[\sqrt{(kp)}\right]$	$\begin{cases}0 & \text{when } 0<t<k\\ (\pi t)^{-\frac{1}{2}} & \text{when } t>k\end{cases}$
113	$\frac{1}{\sqrt{p}}\exp(kp)\,\mathrm{erfc}\left[\sqrt{(kp)}\right] \quad k>0$	$\frac{1}{\sqrt{[\pi(t+k)]}}$
114	$\mathrm{erf}\left(\frac{k}{\sqrt{p}}\right)$	$\frac{1}{\pi t}\sin(2k\sqrt{t})$
115	$\frac{1}{\sqrt{p}}\exp\left(\frac{k^2}{p}\right)\mathrm{erfc}\left(\frac{k}{\sqrt{p}}\right)$	$\frac{1}{\sqrt{(\pi t)}}\exp(-2k\sqrt{t})$
116[c]	$K_0(kp)$	$\begin{cases}0 & \text{when } 0<t<k\\ (t^2-k^2)^{-\frac{1}{2}} & \text{when } t>k\end{cases}$

Table 12 Laplace transforms—cont'd

No.	Transform $\bar{f}(p)=\int_0^\infty e^{-pt}f(t)dt$	Function f(t)
117[c]	$K_0(k\sqrt{p})$	$\frac{1}{2t}\exp\left(-\frac{k^2}{4t}\right)$
118[c]	$\frac{1}{p}\exp(kp)K_1(kp)$	$\frac{1}{k}\sqrt{[t(t+2k)]}$
119[c]	$\frac{1}{\sqrt{p}}K_1(k\sqrt{p})$	$\frac{1}{k}\exp\left(-\frac{k^2}{4t}\right)$
120[c]	$\frac{1}{\sqrt{p}}\exp\left(\frac{k}{p}\right)K_0\left(\frac{k}{p}\right)$	$\frac{2}{\sqrt{(\pi t)}}K_0\left[2\sqrt{(2kt)}\right]$
121[d]	$\pi\exp(-kp)I_0(kp)$	$\begin{cases}[t(2k-t)]^{-\frac{1}{2}} & \text{when } 0<t<2k\\ 0 & \text{when } t>2k\end{cases}$
122[d]	$\exp(-kp)I_1(kp)$	$\begin{cases}\frac{k-t}{\pi k\sqrt{[t(2k-t)]}} & \text{when } 0<t<2k\\ 0 & \text{when } t>2k\end{cases}$
123	Unity	Unit impulse

[a] $Ei(-t)=-\int_t^\infty \frac{e^{-x}}{x}dx$ (f or $t>0$) = exponential integral function.

[b] $Si(t)=\int_0^t \frac{\sin x}{x}dx$ = sine integral function and $Ci(t)=-\int_t^\infty \frac{\cos x}{x}dx$ = cosine integral function.

[c] $K_n(x)$ denotes the Bessel function of the second kind for the imaginary argument.

[d] $I_n(x)$ denotes the Bessel function of the first kind for the imaginary argument.

These functions are tabulated in Abramowitz M, Stegun IA. *Handbook of mathematical functions*. New York: Dover Publications; 1965.

By permission from Churchill RV. Operational mathematics. *McGraw-Hill; 1958.*

Table 13 Error function and its derivative

x	$\frac{2}{\sqrt{\pi}}e^{-x^2}$	erfx	x	$\frac{2}{\sqrt{\pi}}e^{-x^2}$	erfx
0.00	1.12837	0.00000	0.50	0.87878	0.52049
0.01	1.12826	0.01128	0.51	0.86995	0.52924
0.02	1.12792	0.02256	0.52	0.86103	0.53789
0.03	1.12736	0.03384	0.53	0.85204	0.54646
0.04	1.12657	0.04511	0.54	0.84297	0.55493
0.05	1.12556	0.05637	0.55	0.83383	0.56332
0.06	1.12432	0.06762	0.56	0.82463	0.57161
0.07	1.12286	0.07885	0.57	0.81536	0.57981
0.08	1.12118	0.09007	0.58	0.80604	0.58792
0.09	1.11927	0.10128	0.59	0.79666	0.59593
0.10	1.11715	0.11246	0.60	0.78724	0.60385

(Continued)

Table 13 Error function and its derivative—cont'd

x	$\frac{2}{\sqrt{\pi}}e^{-x^2}$	erfx	x	$\frac{2}{\sqrt{\pi}}e^{-x^2}$	erfx
0.11	1.11480	0.12362	0.61	0.77777	0.61168
0.12	1.11224	0.13475	0.62	0.76826	0.61941
0.13	1.10946	0.14586	0.63	0.75872	0.62704
0.14	1.10647	0.15694	0.64	0.74914	0.63458
0.15	1.10327	0.16799	0.65	0.73954	0.64202
0.16	1.09985	0.17901	0.66	0.72992	0.64937
0.17	1.09623	0.18999	0.67	0.72027	0.65662
0.18	1.09240	0.20093	0.68	0.71061	0.66378
0.19	1.08837	0.21183	0.69	0.70095	0.67084
0.20	1.08413	0.22270	0.70	0.69127	0.67780
0.21	1.07969	0.23352	0.71	0.68159	0.68466
0.22	1.07506	0.24429	0.72	0.67191	0.69143
0.23	1.07023	0.25502	0.73	0.66224	0.69810
0.24	1.06522	0.26570	0.74	0.65258	0.70467
0.25	1.06001	0.27632	0.75	0.64293	0.71115
0.26	1.05462	0.28689	0.76	0.63329	0.71753
0.27	1.04904	0.29741	0.77	0.62368	0.72382
0.28	1.04329	0.30788	0.78	0.61408	0.73001
0.29	1.03736	0.31828	0.79	0.60452	0.73610
0.30	1.03126	0.32862	0.80	0.59498	0.74210
0.31	1.02498	0.33890	0.81	0.58548	0.74800
0.32	1.01855	0.34912	0.82	0.57601	0.75381
0.33	1.01195	0.35927	0.83	0.56659	0.75952
0.34	1.00519	0.36936	0.84	0.55720	0.76514
0.35	0.99828	0.37938	0.85	0.54786	0.77066
0.36	0.99122	0.38932	0.86	0.53858	0.77610
0.37	0.98401	0.39920	0.87	0.52934	0.78143
0.38	0.97665	0.40900	0.88	0.52016	0.78668
0.39	0.96916	0.41873	0.89	0.51103	0.79184
0.40	0.96154	0.42839	0.90	0.50196	0.79690
0.41	0.95378	0.43796	0.91	0.49296	0.80188
0.42	0.94590	0.44746	0.92	0.48402	0.80676
0.43	0.93789	0.45688	0.93	0.47515	0.81156
0.44	0.92977	0.46622	0.94	0.46635	0.81627
0.45	0.92153	0.47548	0.95	0.45761	0.82089
0.46	0.91318	0.48465	0.96	0.44896	0.82542
0.47	0.90473	0.49374	0.97	0.44037	0.82987
0.48	0.89617	0.50274	0.98	0.43187	0.83423
0.49	0.88752	0.51166	0.99	0.42345	0.83850
1.00	0.41510	0.84270	1.50	0.11893	0.96610
1.01	0.40684	0.84681	1.51	0.11540	0.96727
1.02	0.39867	0.85083	1.52	0.11195	0.96841
1.03	0.39058	0.85478	1.53	0.10859	0.96951
1.04	0.38257	0.85864	1.54	0.10531	0.97058
1.05	0.37466	0.86243	1.55	0.10210	0.97162

Table 13 Error function and its derivative—cont'd

x	$\frac{2}{\sqrt{\pi}}e^{-x^2}$	erfx	x	$\frac{2}{\sqrt{\pi}}e^{-x^2}$	erfx
1.06	0.36684	0.86614	1.56	0.09898	0.97262
1.07	0.35911	0.86977	1.57	0.09593	0.97360
1.08	0.35147	0.87332	1.58	0.09295	0.97454
1.09	0.34392	0.87680	1.59	0.09005	0.97546
1.10	0.33647	0.88020	1.60	0.08722	0.97634
1.11	0.32912	0.88353	1.61	0.08447	0.97720
1.12	0.32186	0.88678	1.62	0.08178	0.97803
1.13	0.31470	0.88997	1.63	0.07917	0.97884
1.14	0.30764	0.89308	1.64	0.07662	0.97962
1.15	0.30067	0.89612	1.65	0.07414	0.98037
1.16	0.29381	0.89909	1.66	0.07173	0.98110
1.17	0.28704	0.90200	1.67	0.06938	0.98181
1.18	0.28037	0.90483	1.68	0.06709	0.98249
1.19	0.27381	0.90760	1.69	0.06487	0.98315
1.20	0.26734	0.91031	1.70	0.06271	0.98379
1.21	0.26097	0.91295	1.71	0.06060	0.98440
1.22	0.25471	0.91553	1.72	0.05856	0.98500
1.23	0.24854	0.91805	1.73	0.05657	0.98557
1.24	0.24248	0.92050	1.74	0.05464	0.98613
1.25	0.23652	0.92290	1.75	0.05277	0.98667
1.26	0.23065	0.92523	1.76	0.05095	0.98719
1.27	0.22489	0.92751	1.77	0.04918	0.98769
1.28	0.21923	0.92973	1.78	0.04747	0.98817
1.29	0.21367	0.93189	1.79	0.04580	0.98864
1.30	0.20820	0.93400	1.80	0.04419	0.98909
1.31	0.20284	0.93606	1.81	0.04262	0.98952
1.32	0.19757	0.93806	1.82	0.04110	0.98994
1.33	0.19241	0.94001	1.83	0.03963	0.99034
1.34	0.18734	0.94191	1.84	0.03820	0.99073
1.35	0.18236	0.94376	1.85	0.03681	0.99111
1.36	0.17749	0.94556	1.86	0.03547	0.99147
1.37	0.17271	0.94731	1.87	0.03417	0.99182
1.38	0.16802	0.94901	1.88	0.03292	0.99215
1.39	0.16343	0.95067	1.89	0.03170	0.99247
1.40	0.15894	0.95228	1.90	0.03052	0.99279
1.41	0.15453	0.95385	1.91	0.02938	0.99308
1.42	0.15022	0.95537	1.92	0.02827	0.99337
1.43	0.14600	0.95685	1.93	0.02721	0.99365
1.44	0.14187	0.95829	1.94	0.02617	0.99392
1.45	0.13783	0.95969	1.95	0.02517	0.99417
1.46	0.13387	0.96105	1.96	0.02421	0.99442
1.47	0.13001	0.96237	1.97	0.02328	0.99466
1.48	0.12623	0.96365	1.98	0.02237	0.99489
1.49	0.12254	0.96489	1.99	0.02150	0.99511
			2.00	0.02066	0.99532

Problems

A solutions manual is available for the problems in Volume 1 of *Chemical Engineering* from booksellers and from

Heinemann Customer Services
Halley Court
Jordan Hill
Oxford 0X2 8YW
The United Kingdom

Tel: 01865 888180
E-mail: bhuk.orders@repp.co.uk

1.1. 98% sulphuric acid of viscosity 0.025 Ns/m^2 and density 1840 kg/m^3 is pumped at 685 cm^3/s through a 25 mm line. Calculate the value of the Reynolds number.

1.2. Compare the costs of electricity at 1 p/kWh and gas at 15 p/therm.

1.3. A boiler plant raises 5.2 kg/s of steam at 1825 kN/m^2 pressure, using coal of calorific value 27.2 MJ/kg. If the boiler efficiency is 75%, how much coal is consumed per day? If the steam is used to generate electricity, what is the power generation in kilowatts, assuming a 20% conversion efficiency of the turbines and generators?

1.4. The power required by an agitator in a tank is a function of the following four variables:

(a) Diameter of impeller
(b) Number of rotations of impeller per unit time
(c) Viscosity of liquid
(d) Density of liquid

From a dimensional analysis, obtain a relation between the power and the four variables. The power consumption is found, experimentally, to be proportional to the square of the speed of rotation. By what factor would the power be expected to increase if the impeller diameter were doubled?

1.5. It is found experimentally that the terminal settling velocity u_0 of a spherical particle in a fluid is a function of the following quantities:

Particle diameter d
Buoyant weight of particle (weight of particle-weight of displaced fluid) W
Fluid density ρ
Fluid viscosity μ

Obtain a relationship for u_0 using dimensional analysis.

Stokes established, from theoretical considerations, that for small particles that settle at very low velocities, the settling velocity is independent of the density of the fluid except in so far as this affects the buoyancy. Show that the settling velocity *must* then be inversely proportional to the viscosity of the fluid.

1.6. A drop of liquid spreads over a horizontal surface. What are the factors that will influence

(a) the rate at which the liquid spreads,
(b) the final shape of the drop?

Obtain dimensionless groups involving the physical variables in the two cases.

1.7. Liquid is flowing at a volumetric flowrate Q per unit width down a vertical surface. Obtain from dimensional analysis the form of the relationship between flowrate and film thickness. If the flow is streamline, show that the volumetric flowrate is directly proportional to the density of the liquid.

1.8. Obtain, by dimensional analysis, a functional relationship for the heat-transfer coefficient for forced convection at the inner wall of an annulus through which a cooling liquid is flowing.

1.9. Obtain by dimensional analysis a functional relationship for the wall heat-transfer coefficient for a fluid flowing through a straight pipe of circular cross-section. Assume that the effects of natural convection can be neglected in comparison with those of forced convection.

It is found by experiment that, when the flow is turbulent, increasing the flowrate by a factor of 2 always results in a 50% increase in the coefficient. How would a 50% increase in density of the fluid be expected to affect the coefficient, all other variables remaining constant?

1.10. A stream of droplets of liquid is formed rapidly at an orifice submerged in a second, immiscible liquid. What physical properties would be expected to influence the mean size of droplet formed? Using dimensional analysis, obtain a functional relation between the variables.

1.11. Liquid flows under steady-state conditions along an open channel of fixed inclination to the horizontal. On what factors will the depth of liquid in the channel depend? Obtain a relationship between the variables using dimensional analysis.

1.12. Liquid flows down an inclined surface as a film. On what variables will the thickness of the liquid film depend? Obtain the relevant dimensionless groups. It may be assumed that the surface is sufficiently wide for edge effects to be negligible.

1.13. A glass particle settles under the action of gravity in a liquid. Upon which variables would you expect the terminal velocity of the particle to depend? Obtain a relevant dimensionless grouping of the variables. The falling velocity is found to be proportional to the square of the particle diameter when other variables are kept constant. What will be the effect of doubling the viscosity of the liquid? What does this suggest about the nature of the flow?

1.14. Heat is transferred from condensing steam to a vertical surface, and the resistance to heat transfer is attributable to the thermal resistance of the condensate layer on the surface. What variables will be expected to affect the film thickness at a point? Obtain the relevant dimensionless groups.

For streamline flow, it is found that the film thickness is proportional to the one-third power of the volumetric flowrate per unit width. Show that the heat-transfer coefficient would be expected to be inversely proportional to the one-third power of viscosity.

1.15. A spherical particle settles in a liquid contained in a narrow vessel. Upon what variables would you expect the falling velocity of the particle to depend? Obtain the relevant dimensionless groups.

For particles of a given density settling in a vessel of large diameter, the settling velocity is found to be inversely proportional to the viscosity of the liquid. How would you expect it to depend on particle size?

1.16. A liquid is in steady-state flow in an open trough of rectangular cross-section inclined at an angle θ to the horizontal. On what variables would you expect the mass flow per unit time to depend? Obtain the dimensionless groups that are applicable to this problem.

1.17. The resistance force on a spherical particle settling in a fluid is given by Stokes' law. Obtain an expression for the terminal falling velocity of the particle. It is convenient to express the results of experiments in the form of a dimensionless group that may be plotted against a Reynolds group with respect to the particle. Suggest a suitable form for this dimensionless group.

Force on particle from Stokes' law $= 3\pi\mu du$, where μ is the fluid viscosity, d is the particle diameter, and u is the velocity of the particle relative to the fluid.

What will be the terminal falling velocity of a particle of diameter 10 μm and of density 1600 kg/m^3 settling in a liquid of density 1000 kg/m^3 and of viscosity 0.001 Ns/m^2?

If Stokes' law applies for particle Reynolds numbers up to 0.2, what is the diameter of the largest particle whose behaviour is governed by Stokes' law for this solid and liquid?

1.18. A sphere, initially at a constant temperature, is immersed in a liquid whose temperature is maintained constant. The time t taken for the temperature of the centre of the sphere to reach a given temperature θ_c is a function of the following variables:

Diameter of sphere	d
Thermal conductivity of sphere	k
Density of sphere	ρ
Specific heat capacity of sphere	C_p
Temperature of fluid in which it is immersed	θ_s

Obtain relevant dimensionless groups for this problem.

1.19. Upon what variables would you expect the rate of filtration of a suspension of fine solid particles to depend? Consider the flow through unit area of filter medium and express the variables in the form of dimensionless groups.

It is found that the filtration rate is doubled if the pressure difference is doubled. What effect would you expect from raising the temperature of filtration from 293 to 313 K?

The viscosity of the liquid is given by

$$\mu = \mu_0(1 - 0.015(T - 273))$$

where μ is the viscosity at a temperature T (K) and μ_0 is the viscosity at 273 K.

2.1. Calculate the ideal available energy produced by the discharge to atmosphere through a nozzle of air stored in a cylinder of capacity 0.1 m^3 at a pressure of 5 MN/m^2. The initial temperature of the air is 290 K, and the ratio of the specific heats is 1.4.

2.2. Obtain expressions for the variation of

(a) internal energy with change of volume,

(b) internal energy with change of pressure,

(c) enthalpy with change of pressure.

All at constant temperature, for a gas whose equation of state, is given by van der Waals' law.

2.3. Calculate the energy stored in 1000 cm^3 of gas at 80 MN/m^2 and 290 K using a datum of STP.

2.4. Compressed gas is distributed from works in cylinders that are filled to a pressure P by connecting them to a large reservoir of gas that remains at a steady pressure P and temperature T. If the small cylinders are initially at a temperature T and pressure P_0, what is the final temperature of the gas in the cylinders if heat losses can be neglected and if the compression can be regarded as reversible? Assume that the ideal gas laws are applicable.

3.1. Calculate the hydraulic mean diameter of the annular space between a 40 and a 50 mm tube.

3.2. 0.015 m^3/s of acetic acid is pumped through a 75 mm diameter horizontal pipe 70 m long. What is the pressure drop in the pipe?

$$\text{Viscosity of acid} = 2.5\ \text{mN s/m}^2$$

$$\text{Density of acid} = 1060\ \text{kg/m}^3$$

$$\text{Roughness of pipe surface} = 6 \times 10^{-5}\ \text{m}$$

3.3. A cylindrical tank, 5 m in diameter, discharges through a mild steel pipe 90 m long and 230 mm in diameter connected to the base of the tank. Find the time taken for the water level in the tank to drop from 3 to 1 m above the bottom. Take the viscosity of water as 1 mN s/m^2.

3.4. Two storage tanks A and B containing a petroleum product discharge through pipes each 0.3 m in diameter and 1.5 km long to a junction at D. From D, the product is carried by a 0.5 m diameter pipe to a third storage tank C, 0.8 km away. The surface of the liquid in A is initially 10 m above that in C, and the liquid level in B is 7 m higher than that in A. Calculate the initial rate of discharge of the liquid if the pipes are of mild steel. Take the density of the petroleum product as 870 kg/m^3 and the viscosity as 0.7 mN s/m^2.

3.5. Find the drop in pressure due to friction in a glazed porcelain pipe 300 m long and 150 mm in diameter when water is flowing at the rate of 0.05 m^3/s.

3.6. Two tanks, the bases of which are at the same level, are connected with one another by a horizontal pipe 75 mm in diameter and 300 m long. The pipe is bellmouthed at each end so that losses on entry and exit are negligible. One tank is 7 m in diameter and contains water to a depth of 7 m. The other tank is 5 m in diameter and contains water to a depth of 3 m. If the tanks are connected to each other by means of the pipe, how long will it take before the water level in the larger tank has fallen to 6 m? Assume the pipe to be of aged mild steel.

3.7. Two immiscible fluids A and B, of viscosities μ_A and μ_B, flow under streamline conditions between two horizontal parallel planes of width b, situated a distance $2a$ apart (where a is much less than b), as two distinct parallel layers one above the other, each of depth a. Show that the volumetric rate of flow of A is

$$\left(\frac{-\Delta P a^3 b}{12\mu_A l}\right) \times \left(\frac{7\mu_A + \mu_B}{\mu_A + \mu_B}\right),$$

where $-\Delta P$ is the pressure drop over a length l in the direction of flow.

3.8. Glycerol is pumped from storage tanks to rail cars through a single 50 mm diameter main 10 m long, which must be used for all grades of glycerol. After the line has been used for commercial material, how much pure glycerol must be pumped before the issuing liquid contains not more than 1% of the commercial material? The flow in the pipeline is streamline, and the two grades of glycerol have identical densities and viscosities.

3.9. A viscous fluid flows through a pipe with slightly porous walls so that there is a leakage of kP m^3/m^2 s, where P is the local pressure measured above the discharge pressure and k is a constant. After a length L, the liquid is discharged into a tank. If the internal diameter of the pipe is D m and the volumetric rate of flow at the inlet is Q m^3/s, show that the pressure drop in the pipe is given by

$$-\Delta P = \left(\frac{Q}{\pi k D}\right) a \tanh aL,$$

where

$$a = \left(\frac{128 k \mu}{D^3}\right)^{0.5}$$

Assume a fully developed flow with $(R/\rho\mu^2) = 8Re^{-1}$.

3.10. A petroleum product of viscosity 0.5 mN s/m^2 and specific gravity 0.7 is pumped through a pipe of 0.15 m diameter to storage tanks situated 100 m away. The pressure drop along the pipe is 70 kN/m^2. The pipeline has to be repaired, and it is necessary to pump the liquid by an alternative route consisting of 70 m of 20 cm pipe followed by 50 m of 10 cm pipe. If the existing pump is capable of developing a pressure of 300 kN/m^2, will it be suitable for use during the period required for the repairs? Take the roughness of the pipe surface as 0.00005 m.

3.11. Explain the phenomenon of hydraulic jump that occurs during the flow of a liquid in an open channel.

A liquid discharges from a tank into an open channel under a gate so that the liquid is initially travelling at a velocity of 1.5 m/s and a depth of 75 mm. Calculate, from first principles, the corresponding velocity and depth after the jump.

3.12. What is a non-Newtonian fluid? Describe the principal types of behaviour exhibited by these fluids. The viscosity of a non-Newtonian fluid changes with the rate of shear according to the approximate relationship:

$$\mu_a = k\left(-\frac{du_x}{dr}\right)^{-0.5}$$

where μ_a is the apparent viscosity and du_x/dr is the velocity gradient normal to the direction of motion.

Show that the volumetric rate of streamline flow through a horizontal tube of radius a is

$$\frac{\pi}{5} a^5 \left(\frac{-\Delta P}{2kl}\right)^2$$

where $-\Delta P$ is the pressure drop over length l of the tube.

3.13. Calculate the pressure drop when 3 kg/s of sulphuric acid flows through 60 m of 25 mm pipe ($\rho = 1840$ kg/m^3 and $\mu = 0.025$ Ns/m^2).

3.14. The relation between cost per unit length C of a pipeline installation and its diameter d is given by

$$C = a + bd,$$

where a and b are independent of pipe size. Annual charges are a fraction β of the capital cost. Obtain an expression for the optimum pipe diameter on a minimum cost basis for a fluid of density ρ and viscosity μ flowing at a mass rate of G. Assume that the fluid is in turbulent flow and that the Blasius equation is applicable, i.e. the friction factor is proportional to the Reynolds number to the power of minus one-quarter. Indicate clearly how the optimum diameter depends on flowrate and fluid properties.

3.15. A heat exchanger is to consist of a number of tubes each 25 mm in diameter and 5 m long arranged in parallel. The exchanger is to be used as a cooler with a rating of 4 MW, and the temperature rise in the water feed to the tubes is to be 20 K.

If the pressure drop over the tubes is not to exceed 2 kN/m^2, calculate the minimum number of tubes that are required. Assume that the tube walls are smooth and that entrance and exit effects can be neglected. Viscosity of water $= 1$ mN s/m^2.

3.16. Sulphuric acid is pumped at 3 kg/s through a 60 m length of smooth 25 mm pipe. Calculate the drop in pressure. If the pressure drop falls by one-half, what will be the new flowrate?

$$\text{Density of acid} = 1840\,\text{kg/m}^3; \quad \text{Viscosity of acid} = 25\,\text{mN s/m}^2$$

3.17. A Bingham plastic material is flowing under streamline conditions in a pipe of circular cross-section. What are the conditions for one-half of the total flow to be within the central core across which the velocity profile is flat? The shear stress acting within the fluid R_y varies with velocity gradient du_x/dy according to the relation

$$R_y - R_c = -k\frac{du_x}{dy},$$

where R_c and k are constants for the material.

3.18. Oil of viscosity 10 mN s/m^2 and density 950 kg/m^3 is pumped 8 km from an oil refinery to a distribution depot through a 75 mm diameter pipeline and is then despatched to customers at a rate of 500 tonne/day. Allowance must be made for periods of maintenance that may interrupt the supply from the refinery for up to 72 h. If the maximum permissible pressure drop over the pipeline is 3450 kN/m^2, what is the shortest time in which the storage tanks can be completely recharged after a 72 h shutdown? Take the roughness of the pipe surface as 0.05 mm.

3.19. Water is pumped at 1.4 m^3/s from a tank at a treatment plant to a tank at a local works through two parallel pipes, 0.3 and 0.6 m in diameter, respectively. What is the velocity in each pipe, and if a single pipe is used, what diameter will be needed if this flow of water is to be transported, the pressure drop being the same? Assume turbulent flow, with the friction factor inversely proportional to the one-quarter power of the Reynolds number.

3.20. Oil of viscosity 10 mN s/m^2 and specific gravity 0.90 flows through 60 m of 100 mm diameter pipe, and the pressure drop is 13.8 kN/m^2. What will be the pressure drop for a second oil of viscosity 30 mN s/m^2 and specific gravity 0.95 flowing at the same rate through the pipe? Assume the pipe wall to be smooth.

3.21. Crude oil is pumped from a terminal to a refinery through a foot diameter pipeline. As a result of frictional heating, the temperature of the oil is 20 K higher at the refinery end than at the terminal end of the pipe, and the viscosity has fallen to one-half its original value. What is the ratio of the pressure gradient in the pipeline at the refinery end to that at the terminal end? Viscosity of oil at terminal = 90 mN s/m^2; Density of oil (approximately constant) = 960 kg/m^3; Flowrate of oil = 20,000 tonne/day.
Outline a method for calculating the temperature of the oil as a function of distance from the inlet for a given value of the heat-transfer coefficient between the pipeline and the surroundings.

3.22. Oil with a viscosity of 10 mN s/m^2 and density 900 kg/m^3 is flowing through a 500 mm diameter pipe 10 km long. The pressure difference between the two ends of the pipe is 10^6 N/m^2. What will the pressure drop be at the same flowrate if it is necessary to replace the pipe by one only 300 mm in diameter? Assume the pipe surface to be smooth.

3.23. Oil of density 950 kg/m^3 and viscosity 10^{-2} Ns/m^2 is to be pumped 10 km through a pipeline, and the pressure drop must not exceed 2×10^5 N/m^2. What is the minimum diameter of pipe that will be suitable if a flowrate of 50 tonne/h is to be maintained? Assume the pipe wall to be smooth. Use either the pipe friction chart *or* the Blasius equation ($R/\rho u^2 = 0.0396 Re^{-1/4}$).

3.24. On the assumption that the velocity profile in a fluid in turbulent flow is given by the Prandtl one-seventh power law, calculate the radius at which the flow between it and the centre is equal to that between it and the wall, for a pipe 100 mm in diameter.

3.25. A pipeline 0.5 m in diameter and 1200 m long is used for transporting an oil of density 950 kg/m^3 and of viscosity 0.01 Ns/m^2 at 0.4 m^3/s. If the roughness of the pipe surface is 0.5 mm, what is the pressure drop? With the same pressure drop, what will be the flowrate of a second oil of density 980 kg/m^3 and of viscosity 0.02 Ns/m^2?

3.26. Water (density of 1000 kg/m^3 and viscosity of 1 mN s/m^2) is pumped through a 50 mm diameter pipeline at 4 kg/s, and the pressure drop is 1 MN/m^2. What will be the pressure drop for a solution of glycerol in water (density of 1050 kg/m^3 and viscosity of 10 mN s/m^2) when pumped at the same rate? Assume the pipe to be smooth.

3.27. A liquid is pumped in streamline flow through a pipe of diameter d. At what distance from the centre of the pipe will the fluid be flowing at the average velocity?

3.28. Cooling water is supplied to a heat exchanger and flows through 25 mm diameter tubes each 5 m long arranged in parallel. If the pressure drop over the heat exchanger is not to exceed 8000 N/m^2, how many tubes must be included for a total flowrate of water of 110 tonne/h? Density of water = 1000 kg/m^3; Viscosity of water = 1 mN s/m^2; Assume pipes to be smooth-walled.

If 10% of the tubes became blocked, what would the new pressure drop be?

3.29. The effective viscosity of a non-Newtonian fluid may be expressed by the relationship:

$$\mu_a = k''\left(-\frac{du_x}{dr}\right)$$

where k'' is constant.

Show that the volumetric flowrate of this fluid in a horizontal pipe of radius a under isothermal laminar flow conditions with a pressure gradient $-\Delta P/l$ per unit length is

$$Q = \frac{2\pi}{7}a^{7/2}\left(\frac{-\Delta P}{2k''l}\right)$$

3.30. Determine the yield stress of a Bingham fluid of density 2000 kg/m^3 that will just flow out of an open-ended vertical tube of diameter 300 mm under the influence of its own weight.

3.31. A fluid of density 1.2×10^3 kg/m^3 flows down an inclined plane at 15 degrees to the horizontal. If the viscous behaviour is described by the relationship

$$R_{yx} = -k\left(\frac{du_x}{dy}\right)^n$$

where $k = 4.0$ Ns$^{0.4}$/m^2 and $n = 0.4$, calculate the volumetric flowrate per unit width if the fluid film is 10 mm thick.

3.32. A fluid with a finite yield stress is sheared between two concentric cylinders, 50 mm long. The inner cylinder is 30 mm in diameter, and the gap is 20 mm. The outer cylinder is held stationary, whilst a torque is applied to the inner. The moment required just to produce motion was 0.01 N m. Calculate the torque needed to ensure all the fluid is flowing under shear if the plastic viscosity is 0.1 Ns/m^2.

3.33. Experiments, carried out with a capillary viscometer of length 100 mm and diameter 2 mm on a fluid, gave the following results:

Applied Pressure Difference $-\Delta P$ (N/m^2)	Volumetric Flowrate Q (m^3/s)
1×10^3	1×10^{-7}
2×10^3	2.8×10^{-7}
5×10^3	1.1×10^{-7}
1×10^4	3×10^{-6}
2×10^4	9×10^{-6}
5×10^4	3.5×10^{-5}
1×10^5	1×10^{-4}

Suggest a suitable model to describe the fluid properties.

3.34. Data obtained with a cone and plate viscometer (cone half-angle 89 degrees and cone radius 50 mm) were the following:

Cone Speed (Hz)	Measured Torque (Nm)
0.1	4.6×10^{-1}
0.5	7×10^{-1}
1	1.0
5	3.4
10	6.4
50	30.0

Suggest a suitable model to describe the fluid properties.

3.35. Tomato purée of density 1300 kg/m^3 is pumped through a 50 mm diameter factory pipeline at a flowrate of 0.00028 m^3/s. It is suggested that in order to double production,

(a) a similar line with pump should be put in parallel to the existing one,

(b) a large pump should force the material through the present line,

(c) a large pump should supply the liquid through a line of twice the cross-sectional area.

Given that the flow properties of the purée can be described by the Casson equation,

$$(-R_y)^{1/2} = (-R_Y)^{1/2} + \left(-\mu_c \frac{du_x}{dy}\right)^{1/2}$$

where

R_Y is a yield stress, here 20 N/m^2;

μ_c is a characteristic Casson plastic viscosity, 5 Ns/m^2;

$\frac{du_x}{dy}$ is the velocity gradient.

Evaluate the relative pressure drops of the three suggestions, assuming laminar flow throughout.

3.36. The rheological properties of a particular suspension can be approximated reasonably well by either a 'power-law' or a 'Bingham plastic' model over the shear rate range of 10–50 s^{-1}. If the consistency k is 10 Ns^n/m^2 and the flow behaviour index n is 0.2 in the power-law model, what will be the approximate values of the yield stress and of the plastic viscosity in the Bingham plastic model?

What will be the pressure drop, when the suspension is flowing under laminar conditions in a pipe 200 m long and 40 mm in diameter and when the centreline velocity is 1 m/s, according to the power-law model? Calculate the centreline velocity for this pressure drop for the Bingham plastic model and comment on the result.

3.37. Show how, by suitable selection of the index n, the power law may be used to describe the behaviour of both shear-thinning and shear-thickening non-Newtonian fluids over a limited range of shear rates. What are the main objections to the use of the power law? Give some examples of different types of shear-thinning fluids.

A power-law fluid is flowing under laminar conditions through a pipe of circular cross-section. At what radial position is the fluid velocity equal to the mean velocity in the pipe? Where does this occur for a fluid with an n-value of 0.2?

3.38. A liquid whose rheology can be represented by the 'power-law' model is flowing under streamline conditions through a pipe of 5 mm diameter. If the mean velocity of flow in 1 m/s and the velocity at the pipe axis is 1.2 m/s, what is the value of the power-law index n?

Water, of viscosity 1 mN s/m^2 flowing through the pipe at the same mean velocity, gives rise to a pressure drop of 10^4 N/m^2 compared with 10^5 N/m^2 for the non-Newtonian fluid. What is the consistency ('k' value) of the non-Newtonian fluid?

3.39. Two liquids of equal densities, the one Newtonian and the other a non-Newtonian 'power-law' fluid, flow at equal volumetric rates down two wide vertical surfaces of the same widths. The non-Newtonian fluid has a power-law index of 0.5 and has the same apparent viscosity in SI units as the Newtonian fluid when its shear rate is 0.01 s^{-1}. Show that, for equal surface velocities of the two fluids, the film thickness for the Newtonian fluid is 1.125 times that of the non-Newtonian fluid.

3.40. A fluid that exhibits non-Newtonian behaviour is flowing in a pipe of diameter 70 mm, and the pressure drop over a 2 m length of pipe is 4×10^4 N/m^2. (When the flowrate is doubled, the pressure drop increases by a factor of 1.5.) A pitot tube is used to measure the velocity profile over the cross-section. Confirm that the information given below is consistent with the laminar flow of a *power-law* fluid.

Any equations used must be derived from the basic relation between shear stress R and shear rate $\dot{\gamma}$:

$$R = k(\dot{\gamma})^n$$

Radial Distance (s mm) From the Centre of Pipe	Velocity (m/s)
0	0.80
10	0.77
20	0.62
30	0.27

3.41. A Bingham plastic fluid (yield stress of 14.35 N/m^2 and plastic viscosity of 0.150 Ns/m^2) is flowing through a pipe of diameter 40 mm and length 200 m. Starting with the rheological equation, show that the relation between pressure gradient $-\Delta P/l$ and volumetric flowrate Q is

$$Q = \frac{\pi(-\Delta P)r^4}{8l\mu_p}\left[1 - \frac{4}{3}X + \frac{1}{3}X^4\right]$$

where

r is the pipe radius,

μ_p is the plastic viscosity, and

X is the ratio of the yield stress to the shear stress at the pipe wall.

Calculate the flowrate for this pipeline when the pressure drop is 600 kN/m^2. It may be assumed that the flow is laminar.

4.1. A gas, having a molecular weight of 13 kg/kmol and a kinematic viscosity of 0.25 cm^2/s, is flowing through a pipe 0.25 m in internal diameter and 5 km long at the rate of 0.4 m^3/s and is delivered at atmospheric pressure. Calculate the pressure required to maintain this rate of flow under isothermal conditions. The volume occupied by 1 kmol at 273 K and 101.3 kN/m^2 is 22.4 m^3.

What would be the effect on the required pressure if the gas were to be delivered at a height of 150 m (i) above and (ii) below its point of entry into the pipe?

4.2. Nitrogen at 12 MN/m^2 is fed through a 25 mm diameter mild steel pipe to a synthetic ammonia plant at the rate of 1.25 kg/s. What will be the drop in pressure over a 30 m length of pipe for isothermal flow of the gas at 298 K? Absolute roughness of the pipe surface = 0.005 mm; Kilogram molecular volume = 22.4 m^3; Viscosity of nitrogen = 0.02 mN s/m^2.

4.3. Hydrogen is pumped from a reservoir at 2 MN/m^2 pressure through a clean horizontal mild steel pipe 50 mm in diameter and 500 m long. The downstream pressure is also 2 MN/m^2, and the pressure of this gas is raised to 2.6 MN/m^2 by a pump at the upstream end of the pipe. The conditions of flow are isothermal, and the temperature of the gas is 293 K. What is the flowrate and the effective rate of working of the pump?

Viscosity of hydrogen = 0.009 mN s/m^{2}at 293 K.

4.4. In a synthetic ammonia plant, the hydrogen is fed through a 50 mm steel pipe to the converters. The pressure drop over the 30 m length of pipe is 500 kN/m^2, the pressure at the downstream end being 7.5 MN/m^2. What power is required in order to overcome friction losses in the pipe? Assume isothermal expansion of the gas at 298 K. What error is introduced by assuming the gas to be an incompressible fluid of density equal to that at the mean pressure in the pipe? μ = 0.02 mN s/m^2.

4.5. A vacuum distillation plant operating at 7 kN/m^2 at the top has a boil-up rate of 0.125 kg/s of xylene. Calculate the pressure drop along a 150 mm bore vapour pipe used to connect the column to the condenser. The pipe length may be taken as equivalent to 6 m, e/d = 0.002 and μ = 0.01 mN s/m^2.

4.6. Nitrogen at 12 MN/m^2 pressure is fed through a 25 mm diameter mild steel pipe to a synthetic ammonia plant at the rate of 0.4 kg/s. What will be the drop in pressure over a

30 m length of pipe assuming isothermal expansion of the gas at 300 K? What is the average quantity of heat per unit area of pipe surface that must pass through the walls in order to maintain isothermal conditions? What would be the pressure drop in the pipe if it were perfectly lagged? ($\mu = 0.02$ mN s/m^2)

4.7. Air, at a pressure of 10 MN/m^2 and a temperature of 290 K, flows from a reservoir through a mild steel pipe 10 mm in diameter and 30 m long to a second reservoir at a pressure P_2. Plot the mass rate of flow of the air as a function of the pressure P_2. Neglect any effects attributable to differences in level and assume an adiabatic expansion of the air. $\mu = 0.018$ mN s/m^2 and $\gamma = 1.36$.

4.8. Over a 30 m length of 150 mm vacuum line carrying air at 293 K, the pressure falls from 1 to 0.1 kN/m^2. If the relative roughness e/d is 0.002, what is the approximate flowrate?

4.9. A vacuum system is required to handle 10 g/s of vapour (molecular weight of 56 kg/kmol) so as to maintain a pressure of 1.5 kN/m^2 in a vessel situated 30 m from the vacuum pump. If the pump is able to maintain a pressure of 0.15 kN/m^2 at its suction point, what diameter pipe is required? The temperature is 290 K, and isothermal conditions may be assumed in the pipe, whose surface can be taken as smooth. The ideal gas law is followed. Gas viscosity $= 0.01$ mN s/m^2.

4.10. In a vacuum system, air is flowing isothermally at 290 K through a 150 mm diameter pipeline 30 m long. If the relative roughness of the pipe wall e/d is 0.002 and the downstream pressure is 130 N/m^2, what will the upstream pressure be if the flowrate of air is 0.025 kg/s?

Assume that the ideal gas law applies and that the viscosity of air is constant at 0.018 mN s/m^2.

What error would be introduced if the change in kinetic energy of the gas as a result of expansion were neglected?

4.11. Air is flowing at the rate of 30 kg/m^{2}s through a smooth pipe 50 mm in diameter and 300 m long. If the upstream pressure is 800 kN/m^2, what will the downstream pressure be if the flow is isothermal at 273 K? Take the viscosity of air as 0.015 mN s/m^2 and the kilogram molecular volume as 22.4 m^3. What is the significance of the change in kinetic energy of the fluid?

4.12. If temperature does not change with height, estimate the boiling point of water at a height of 3000 m above sea level. The barometer reading at sea level is 98.4 kN/m^2, and the temperature is 288.7 K. The vapour pressure of water at 288.7 K is 1.77 kN/m^2. The effective molecular weight of air is 29 kg/kmol.

4.13. A 150 mm gas main is used for transferring a gas (molecular weight of 13 kg/kmol and kinematic viscosity of 0.25 cm^2/s) at 295 K from a plant to a storage station 100 m away, at a rate of 1 m^3/s. Calculate the pressure drop, if the pipe can be considered to be smooth.

If the maximum permissible pressure drop is 10 kN/m^2, is it possible to increase the flowrate by 25%?

5.1. It is required to transport sand of particle size 1.25 mm and density 2600 kg/m^3 at the rate of 1 kg/s through a horizontal pipe, 200 m long. Estimate the air flowrate required, the pipe diameter, and the pressure drop in the pipeline.

5.2. Sand of mean diameter 0.2 mm is to be conveyed by water flowing at 0.5 kg/s in a 25 mm ID horizontal pipe, 100 m long. What is the maximum amount of sand that may be transported in this way if the head developed by the pump is limited to 300 kN/m^2? Assume fully suspended heterogeneous flow.

5.3. Explain the various mechanisms by which particles may be maintained in suspension during hydraulic transport in a horizontal pipeline and indicate when each is likely to be important.

A highly concentrated suspension of flocculated kaolin in water behaves as a pseudohomogeneous fluid with shear-thinning characteristics that can be represented approximately by the Ostwald-de Waele power law, with an index of 0.15. It is found that, if air is injected into the suspension when in laminar flow, the pressure gradient may be reduced even though the flowrate of suspension is kept constant. Explain how this is possible in 'slug' flow and estimate the possible reduction in pressure gradient for equal volumetric flowrates of suspension and air.

6.1. Sulphuric acid of density 1300 kg/m^3 is flowing through a pipe of 50 mm internal diameter. A thin-lipped orifice, 10 mm in diameter, is fitted in the pipe, and the differential pressure shown by a mercury manometer is 10 cm. Assuming that the leads to the manometer are filled with the acid, calculate (a) the mass of acid flowing per second and (b) the approximate loss of pressure (in kN/m^2) caused by the orifice.

The coefficient of discharge of the orifice may be taken as 0.61, the density of mercury as 13,550 kg/m^3, and the density of water as 1000 kg/m^3.

6.2. The rate of discharge of water from a tank is measured by means of a notch for which the flowrate is directly proportional to the height of liquid above the bottom of the notch. Calculate and plot the profile of the notch if the flowrate is 0.1 m^3/s when the liquid level is 150 mm above the bottom of the notch.

6.3. Water flows at between 3000 and 4000 L/s through a 50 mm pipe and is metered by means of an orifice. Suggest a suitable size of orifice if the pressure difference is to be measured with a simple water manometer. What is the approximate pressure difference recorded at the maximum flowrate?

6.4. The rate of flow of water in a 150 mm diameter pipe is measured by means of a venturi meter with a 50 mm diameter throat. When the drop in head over the converging section is 100 mm of water, the flowrate is 2.7 kg/s. What is the coefficient for the converging cone of the meter at that flowrate, and what is the head lost due to friction? If the total loss of head over the meter is 15 mm water, what is the coefficient for the diverging cone?

6.5. A venturi meter with a 50 mm throat is used to measure a flow of slightly salty water in a pipe of inside diameter 100 mm. The meter is checked by adding 20 cm^3/s of normal sodium chloride solution above the meter and analysing a sample of water downstream from the meter. Before addition of the salt, 1000 cm^3 of water requires 10 cm^3 of 0.1 M silver nitrate solution in a titration. 1000 cm^3 of the downstream sample required 23.5 cm^3 of 0.1 M silver nitrate. If a mercury-under-water manometer connected to the meter gives a reading of 221 mm, what is the discharge coefficient of the meter? Assume that the density of the liquid is not appreciably affected by the salt.

6.6. A gas cylinder containing 30 m^3 of air at 6 MN/m^2 pressure discharges to the atmosphere through a valve that may be taken as equivalent to a sharp-edged orifice of 6 mm diameter (coefficient of discharge = 0.6). Plot the rate of discharge against the pressure in the cylinder. How long will it take for the pressure in the cylinder to fall to (a) 1 MN/m^2 and (b) 150 kN/m^2?

Assume an adiabatic expansion of the gas through the valve and that the contents of the cylinder remain constant at 273 K.

6.7. Air at 1500 kN/m^2 and 370 K flows through an orifice of 30 mm^2 to atmospheric pressure. If the coefficient of discharge is 0.65, the critical pressure ratio is 0.527, and the ratio of the specific heats is 1.4, calculate the mass flowrate.

6.8. Water flows through an orifice of 25 mm diameter situated in a 75 mm pipe at the rate of 300 cm^3/s. What will be the difference in level on a water manometer connected across the meter? Take the viscosity of water as 1 mN s/m^2.

6.9. Water flowing at 1500 cm^3/s in a 50 mm diameter pipe is metered by means of a simple orifice of diameter 25 mm. If the coefficient of discharge of the meter is 0.62, what will be the reading on a mercury-under-water manometer connected to the meter?

What is the Reynolds number for the flow in the pipe?

$$\text{Density of water} = 1000\,\text{kg/m}^3$$

$$\text{Viscosity of water} = 1\,\text{mN s/m}^2$$

6.10. What size of orifice would give a pressure difference of 0.3 m water gauge for the flow of a petroleum product of specific gravity 0.9 at 0.05 m^3/s in a 150 mm diameter pipe?

6.11. The flow of water through a 50 mm pipe is measured by means of an orifice meter with a 40 mm aperture. The pressure drop recorded is 150 mm on a mercury-under-water manometer, and the coefficient of discharge of the meter is 0.6. What is the Reynolds number in the pipe, and what would you expect the pressure drop over a 30 m length of the pipe to be?

$$\text{Friction factor, } \phi = \frac{R}{\rho u^2} = 0.0025$$

$$\text{Specific gravity of mercury} = 13.6$$

$$\text{Viscosity of water} = 1\,\text{mN s/m}^2.$$

What type of pump would you use, how would you drive it, and what material of construction would be suitable?

6.12. A rotameter has a tube 0.3 m long that has an internal diameter of 25 mm at the top and 20 mm at the bottom. The diameter of the float is 20 mm, its effective specific gravity is

4.80, and its volume is 6.6 cm^3. If the coefficient of discharge is 0.72, at what height will the float be when metering water at 100 cm^3/s?

6.13. Explain why there is a critical pressure ratio across a nozzle at which, for a given upstream pressure, the flowrate is a maximum.

Obtain an expression for the maximum flow for a given upstream pressure for isentropic flow through a horizontal nozzle. Show that for air (ratio of specific heats $\gamma = 1.4$), the critical pressure ratio is 0.53 and calculate the maximum flow through an orifice of area 30 mm^2 and coefficient of discharge 0.65 when the upstream pressure is 1.5 MN/m^2 and the upstream temperature 293 K:

$$\text{Kilogram molecular volume} = 22.4\,m^3.$$

6.14. A gas cylinder containing air discharges to atmosphere through a valve whose characteristics may be considered similar to those of a sharp-edged orifice. If the pressure in the cylinder is initially 350 kN/m^2, by how much will the pressure have fallen when the flowrate has decreased to one-quarter of its initial value?

The flow through the valve may be taken as isentropic and the expansion in the cylinder as isothermal. The ratio of the specific heats at constant pressure and constant volume is 1.4.

6.15. Water discharges from the bottom outlet of an open tank 1.5 m $\times$ 1 m in cross-section. The outlet is equivalent to an orifice of 40 mm diameter with a coefficient of discharge of 0.6. The water level in the tank is regulated by a float valve on the feed supply that shuts off completely when the height of water above the bottom of the tank is 1 m and gives a flowrate that is directly proportional to the distance of the water surface below this maximum level. When the depth of water in the tank is 0.5 m, the inflow and outflow are directly balanced.

As a result of a short interruption in the supply, the water level in the tank falls to 0.25 m above the bottom but is then restored again. How long will it take the level to rise to 0.45 m above the bottom?

6.16. The flowrate of air at 298 K in a 0.3 m diameter duct is measured with a pitot tube that is used to traverse the cross-section. Readings of the differential pressure recorded on a water manometer are taken with the pitot tube at 10 different positions in the cross-section. These positions are so chosen as to be the midpoints of 10 concentric annuli each of the same cross-sectional area. The readings are as follows:

Position	1	2	3	4	5
Manometer reading (mm water)	18.5	18.0	17.5	16.8	15.7
Position	6	7	8	9	10
Manometer reading (mm water)	14.7	13.7	12.7	11.4	10.2

The flow is also metered using a 15 cm orifice plate across which the pressure differential is 50 mm on a mercury-under-water manometer. What is the coefficient of discharge of the orifice meter?

6.17. Explain the principle of operation of the pitot tube and indicate how it can be used in order to measure the total flowrate of fluid in a duct.

If a pitot tube is inserted in a circular cross-section pipe in which a fluid is in streamline flow, calculate at what point in the cross-section it should be situated so as to give a direct reading representative of the mean velocity of flow of the fluid.

6.18. The flowrate of a fluid in a pipe is measured using a pitot tube that gives a pressure differential equivalent to 40 mm of water when situated at the centreline of the pipe and 22.5 mm of water when midway between the axis and the wall. Show that these readings are consistent with streamline flow in the pipe.

6.19. Derive a relationship between the pressure difference recorded between the two orifices of a pitot tube and the velocity of flow of an incompressible fluid. A pitot tube is to be situated in a large circular duct in which fluid is in turbulent flow so that it gives a direct reading of the mean velocity in the duct. At what radius in the duct should it be located, if the radius of the duct is r?

The point velocity in the duct can be assumed to be proportional to the one-seventh power of the distance from the wall.

6.20. A gas of molecular weight 44 kg/kmol, temperature 373 K, and pressure 202.6 kN/m^2 is flowing in a duct. A pitot tube is located at the centre of the duct and is connected to a differential manometer containing water. If the differential reading is 38.1 mm water, what is the velocity at the centre of the duct?

The volume occupied by 1 kmol at 273 K and 101.3 kN/m^2 is 22.4 m^3.

6.21. Glycerol, of density 1260 kg/m^3 and viscosity 50 mN s/m^2, is flowing through a 50 mm pipe, and the flowrate is measured using an orifice meter with a 38 mm orifice. The pressure differential is 150 mm as indicated on a manometer filled with a liquid of the same density as the glycerol. There is reason to suppose that the orifice meter may have become partially blocked and that the meter is giving an erroneous reading. A check is therefore made by inserting a pitot tube at the centre of the pipe. It gives a reading of 100 mm on a water manometer. What does this suggest?

6.22. The flowrate of air in a 305 mm diameter duct is measured with a pitot tube that is used to traverse the cross-section. Readings of the differential pressure recorded on a water manometer are taken with the pitot tube at 10 different positions in the cross-section. These positions are so chosen as to be the midpoints of 10 concentric annuli each of the same cross-sectional area. The readings are as follows:

Position	1	2	3	4	5
Manometer reading (mm water)	18.5	18.0	17.5	16.8	15.8
Position	6	7	8	9	10
Manometer reading	14.7	13.7	12.7	11.4	10.2

The flow is also metered using a 50 mm orifice plate across which the pressure differential is 150 mm on a mercury-under-water manometer. What is the coefficient of discharge of the orifice meter?

6.23. The flow of liquid in a 25 mm diameter pipe is metered with an orifice meter in which the orifice has a diameter of 19 mm. The aperture becomes partially blocked with dirt from the liquid. What fraction of the area can become blocked before the error in flowrate at a given pressure differential exceeds 15%? Assume that the coefficient of discharge of the meter remains constant when calculated on the basis of the actual free area of the orifice.

6.24. Water is flowing through a 100 mm diameter pipe, and its flowrate is metered by means of a 50 mm diameter orifice across which the pressure drop is 13.8 kN/m^2. A second stream, flowing through a 75 mm diameter pipe, is also metered using a 50 mm diameter orifice across which the pressure differential is 150 mm measured on a mercury-under-water manometer. The two streams join and flow through a 150 mm diameter pipe. What would you expect the reading to be on a mercury-under-water manometer connected across a 75 mm diameter orifice plate inserted in this pipe?

The coefficients of discharge for all the orifice meters are equal.

$$\text{Density of mercury} = 13{,}600\,\text{kg/m}^3$$

6.25. Water is flowing through a 150 mm diameter pipe, and its flowrate is measured by means of a 50 mm diameter orifice, across which the pressure differential is 2.27×10^4 N/m^2. The coefficient of discharge of the orifice meter is independently checked by means of a pitot tube that, when situated at the axis of the pipe, gave a reading of 15.6 mm on a mercury-under-water manometer. On the assumption that the flow in the pipe is turbulent and that the velocity distribution over the cross-section is given by the Prandtl one-seventh power law, calculate the coefficient of discharge of the orifice meter.

6.26. Air at 323 K and 152 kN/m^2 flows through a duct of circular cross-section, diameter of 0.5 m. In order to measure the flowrate of air, the velocity profile across a diameter of the duct is measured using a pitot-static tube connected to a water manometer inclined at an angle of $\cos^{-1} 0.1$ to the vertical. The following results are obtained:

Distance From Duct Centreline (m)	Manometer Reading h_m (mm)
0	104
0.05	100
0.10	96
0.15	86
0.175	79
0.20	68
0.225	50

Calculate the mass flowrate of air through the duct, the average velocity, the ratio of the average to the maximum velocity, and the Reynolds number. Comment on these results.

Discuss the application of this method of measuring gas flowrates with particular emphasis on the best distribution of experimental points across the duct and on the accuracy of the results.

(Take the viscosity of air as 1.9×10^{-2} mN s/m^2 and the molecular weight of air as 29 kg/kmol.)

7.1. A reaction is to be carried out in an agitated vessel. Pilot-plant experiments were performed under fully turbulent conditions in a tank 0.6 m in diameter, fitted with baffles and provided with a flat-bladed turbine. It was found that the satisfactory mixing was obtained at a rotor speed of 4 Hz, when the power consumption was 0.15 kW and the Reynolds number 160,000. What should be the rotor speed in order to retain the same mixing performance if the linear scale of the equipment is increased six times? What will be the power consumption and the Reynolds number?

7.2. A three-bladed propeller is used to mix a fluid in the laminar region. The stirrer is 0.3 m in diameter and is rotated at 1.5 Hz. Due to corrosion, the propeller has to be replaced by a flat two-bladed paddle, 0.75 m in diameter. If the same motor is used, at what speed should the paddle rotate?

7.3. Compare the capital and operating costs of a three-bladed propeller with those of a constant speed six-bladed turbine, both constructed from mild steel. The impeller diameters are 0.3 and 0.45 m, and both stirrers are driven by a 1 kW motor. What is the recommended speed of rotation in each case? Assume operation for 8000 h/year, power at £0.01/kWh, and interest and depreciation at 15% per year.

7.4. In a leaching operation, the rate at which solute goes into solution is given by an equation of the form

$$\frac{dM}{dt} = k(c_s - c) \text{ kg/s}$$

where M (kg) is the amount of solute dissolving in t (s), k (m^3/s) is a constant, and c_s and c are the saturation and bulk concentrations of the solute, respectively, in kg/m^3. In a pilot test on a vessel 1 m^3 in volume, 75% saturation was attained in 10 s. If 300 kg of a solid containing 28% by mass of a water soluble solid is agitated with 100 m^3 of water, how long will it take for all the solute to dissolve assuming conditions are the same as in the pilot unit? Water is saturated with the solute at a concentration of 2.5 kg/m^3.

7.5. For producing an oil-water emulsion, two portable three-bladed propeller mixers are available, a 0.5 m diameter impeller rotating at 1 Hz and a 0.35 m impeller rotating at 2 Hz. Assuming that turbulent conditions prevail, which unit will have the lower power consumption?

7.6. Tests on a small-scale tank 0.3 m in diameter (Rushton impeller, diameter of 0.1 m) have shown that a blending process between two miscible liquids (aqueous solutions, properties approximately the same as water, i.e., viscosity of 1 mN s/m^2 and density of 1000 kg/m^3) is satisfactorily completed after 1 min using an impeller speed of 250 rev/min. It is decided to scale up the process to a tank of 2.5 m diameter using the criterion of constant tip speed:

(a) What speed should be chosen for the larger impeller?

(b) What power will be required?

(c) What will be the blend time in the large tank?

7.7. An agitated tank with a standard Rushton impeller is required to disperse gas in a solution of properties similar to those of water. The tank will be 3 m in diameter (1 m diameter impeller). A power level of 0.8 kW/m^3 is chosen. Assuming fully turbulent conditions, the presence of the gas does not significantly affect the relation between the power and Reynolds numbers:

(a) What power will be required by the impeller?

(b) At what speed should the impeller be driven?

(c) If a small pilot-scale tank 0.3 m in diameter is to be constructed to test the process, at what speed should the impeller be driven?

8.1. A three-stage compressor is required to compress air from 140 kN/m^2 and 283 K to 4000 kN/m^2. Calculate the ideal intermediate pressures, the work required per kilogram of gas, and the isothermal efficiency of the process. Assume the compression to be adiabatic and the interstage cooling to cool the air to the initial temperature. Show qualitatively, by means of temperature-entropy diagrams, the effect of unequal work distribution and imperfect intercooling, on the performance of the compressor.

8.2. A twin-cylinder, single-acting compressor, working at 5 Hz, delivers air at 515 kN/m^2 pressure, at the rate of 0.2 m^3/s. If the diameter of the cylinder is 20 cm, the cylinder clearance ratio 5%, and the temperature of the inlet air 283 K, calculate the length of stroke of the piston and the delivery temperature.

8.3. A single-stage double-acting compressor running at 3 Hz is used to compress air from 110 kN/m^2 and 282 K to 1150 kN/m^2. If the internal diameter of the cylinder is 20 cm, the length of stroke 25 cm, and the piston clearance 5%, calculate (a) the maximum capacity of the machine, referred to air at the initial temperature and pressure, and (b) the theoretical power requirements under isentropic conditions.

8.4. Methane is to be compressed from atmospheric pressure to 30 MN/m^2 in four stages.

Calculate the ideal intermediate pressures and the work required per kilogram of gas. Assume compression to be isentropic and the gas to behave as an ideal gas. Indicate on a temperature-entropy diagram the effect of imperfect intercooling on the work done at each stage.

8.5. An airlift raises 0.01 m^3/s of water from a well of 100 m deep through a 100 mm diameter pipe. The level of the water is 40 m below the surface. The air consumed is 0.1 m^3/s of free air compressed to 800 kN/m^2.

Calculate the efficiency of the pump and the mean velocity of the mixture in the pipe.

8.6. In a single-stage compressor,

$$\text{Suction pressure} = 101.3\,kN/m^2$$

$$\text{Suction temperature} = 283\,K$$

$$\text{Final pressure} = 380\,kN/m^2$$

If each new charge is heated 18 K by contact with the clearance gases, calculate the maximum temperature attained in the cylinder, assuming adiabatic compression.

8.7. A single-acting reciprocating pump has a cylinder diameter of 115 mm and a stroke of 230 mm. The suction line is 6 m long and 50 mm in diameter, and the level of the water in the suction tank is 3 m below the cylinder of the pump. What is the maximum speed at which the pump can run without an air vessel if separation is not to occur in the suction line? The piston undergoes approximately simple harmonic motion. Atmospheric pressure is equivalent to a head of 10.4 m of water, and separation occurs at a pressure corresponding to a head of 1.22 m of water.

8.8. An airlift pump is used for raising 0.8 l/s of a liquid of specific gravity 1.2 to a height of 20 m. Air is available at 450 kN/m^2. If the efficiency of the pump is 30%, calculate the power requirement, assuming isentropic compression of the air ($\gamma = 1.4$).

8.9. A single-acting air compressor supplies 0.1 m^3/s of air (at STP) compressed to 380 kN/m^2 from 101.3 kN/m^2 pressure. If the suction temperature is 288.5 K, the stroke is 250 mm, and the speed is 4 Hz, find the cylinder diameter. Assume the cylinder clearance is 4%, and compression and reexpansion are isentropic ($\gamma = 1.4$). What is the theoretical power required for the compression?

8.10. Air at 290 K is compressed from 101.3 to 2000 kN/m^2 pressure in a two-stage compressor operating with a mechanical efficiency of 85%. The relation between pressure and volume during the compression stroke and expansion of the clearance gas is $PV^{1.25} = \text{constant}$. The compression ratio in each of the two cylinders is the same, and the interstage cooler may be taken as perfectly efficient. If the clearances in the two cylinders are 4% and 5%, respectively, calculate

(a) the work of compression per unit mass of gas compressed,
(b) the isothermal efficiency,
(c) the isentropic efficiency ($\gamma = 1.4$),
(d) the ratio of the swept volumes in the two cylinders.

8.11. Explain briefly the significance of the 'specific speed' of a centrifugal- or axial-flow pump.

A pump is designed to be driven at 10 Hz and to operate at a maximum efficiency when delivering 0.4 m^3/s of water against a head of 20 m. Calculate the specific speed. What type of pump does this value suggest?

A pump, built for these operating conditions, has a measured maximum overall efficiency of 70%. The same pump is now required to deliver water at 30 m head. At what speed should the pump be driven if it is to operate at maximum efficiency? What will be the new rate of delivery and the power required?

8.12. A centrifugal pump is to be used to extract water from a condenser in which the vacuum is 640 mm of mercury. At the rated discharge, the net positive suction head must be at least 3 m above the cavitation vapour pressure of 710 mm mercury vacuum. If losses in the suction pipe account for a head of 1.5 m, what must be the least height of the liquid level in the condenser above the pump inlet?

8.13. What is meant by the net positive suction head (NPSH) required by a pump? Explain why it exists, and how it can be made as low as possible. What happens if the necessary NPSH is not provided?

A centrifugal pump is to be used to circulate liquid, of density 800 kg/m^3 and viscosity 0.5 mN s/m^2, from the reboiler of a distillation column through a vaporiser at the rate of 400 cm^3/s and to introduce the superheated liquid above the vapour space in the reboiler that contains liquid to a depth of 0.7 m. Suggest a suitable layout if a smooth bore 25 mm pipe is to be used. The pressure of the vapour in the reboiler is 1 kN/m^2, and the NPSH required by the pump is 2 m of liquid.

8.14. 1250 cm^3/s of water is to be pumped through a steel pipe, 25 mm in diameter and 30 m long, to a tank 12 m higher than its reservoir. Calculate the approximate power required. What type of pump would you install for the purpose, and what power motor (in kW) would you provide?

$$\text{Viscosity of water} = 1.30\,\text{mN s/m}^2$$
$$\text{Density of water} = 1000\,\text{kg/m}^3$$

8.15. Calculate the pressure drop in and the power required to operate a condenser consisting of 400 tubes 4.5 m long and 10 mm in internal diameter. The coefficient of contraction at the entrance of the tubes is 0.6, and 0.04 m^3/s of water is to be pumped through the condenser.

8.16. 75% sulphuric acid, of density 1650 kg/m^3 and viscosity 8.6 mN s/m^2, is to be pumped for 0.8 km along a 50 mm internal diameter pipe at the rate of 3.0 kg/s and then raised vertically 15 m by the pump. If the pump is electrically driven and has an efficiency of 50%, what power will be required? What type of pump would you use, and of what material would you construct the pump and pipe?

8.17. 60% sulphuric acid is to be pumped at the rate of 4000 cm^3/s through a lead pipe 25 mm in diameter and raised to a height of 25 m. The pipe is 30 m long and includes two right-angled bends. Calculate the theoretical power required.

The kinematic viscosity of the acid is 4.25×10^{-5} m^2/s, and its density is 1531 kg/m^3. The density of water may be taken as 1000 kg/m^3.

8.18. 1.3 kg/s of 98% sulphuric acid is to be pumped through a 25 mm diameter pipe, 30 m long, to a tank 12 m higher than its reservoir. Calculate the power required and indicate the type of pump and material of construction of the line that you would choose:

$$\text{Viscosity of acid} = 0.025\ \text{N s/m}^2$$
$$\text{Density} = 1840\ \text{kg/m}^3$$

8.19. A petroleum fraction is pumped 2 km from a distillation plant to storage tanks through a mild steel pipeline, 150 mm in diameter, at the rate of 0.04 m^3/s. What is the pressure drop along the pipe and the power supplied to the pumping unit if it has an efficiency of 50%?

The pump impeller is eroded, and the pressure at its delivery falls to one-half. By how much is the flowrate reduced?

$$\text{Density of the liquid} = 705\ \text{kg/m}^3$$
$$\text{Viscosity of the liquid} = 0.5\ \text{mN s/m}^2$$
$$\text{Roughness of pipe surface} = 0.004\ \text{mm}$$

8.20. Calculate the power required to pump oil of density 850 kg/m^3 and viscosity 3 mN s/m^2 at 4000 cm^3/s through a 50 mm pipeline 100 m long, the outlet of which is 15 m higher than the inlet. The efficiency of the pump is 50%. What effect does the nature of the surface of the pipe have on the resistance?

8.21. 600 cm^3/s of water at 320 K is pumped in a 40 mm i.d. pipe through a length of 150 m in a horizontal direction and up through a vertical height of 10 m. In the pipe, there is a control valve that may be taken as equivalent to 200 pipe diameters, and other pipe fittings are equivalent to 60 pipe diameters. Also in the line, there is a heat exchanger across which there is a loss in head of 1.5 m of water. If the main pipe has a roughness of 0.0002 m, what power must be delivered to the pump if the unit is 60% efficient?

8.22. A pump developing a pressure of 800 kN/m^2 is used to pump water through a 150 mm pipe 300 m long to a reservoir 60 m higher. With the valves fully open, the flowrate obtained is 0.05 m^3/s. As a result of corrosion and scaling, the effective absolute roughness of the pipe surface increases by a factor of 10. By what percentage is the flowrate reduced?

$$\text{Viscosity of water} = 1\ \text{mN s/m}^2.$$

Index

Note: Page numbers followed by "*f*" indicate figures, and "*t*" indicate tables.

A

Abridged callendar steam tables
  pressure-enthalpy diagram, 493*f*
  saturated steam
    Centigrade and Fahrenheit units, 488–490*t*
    S.I. units, 484–487*t*
  superheated steam, 491–492*t*
  temperature-entropy diagram, 494*f*
Absolute pressure, 294, 294*f*
Aerofoil effect, 262, 262*f*
Air compressor. *See* Gas compressors
Air-lift pumping
  aerated liquid, vertical column in, 433–435, 434*f*
  atmospheric pressure, 430
  efficiency, 432
  operation of, 435–436, 436*f*
  power requirement, 431
  pump unit mass of liquid, 430–431
  schematic diagram, 429–430, 429*f*

B

Bingham-plastic fluid, 160
Blasius equation, 100
Boundary layer, 70–71
British Engineering System, 2
Buckingham's π theorem
  dimensionless groups, 20–23, 20*t*
  drop of liquid spreads, 22
  m fundamental dimensions, 18
  n variables, 18
  recurring set, 18–19
  scaling laws, 23–25

C

Centigrade unit, 488–490*t*
Centimetre-Gram-Second (cgs) system, 4
Centrifugal pump
  advantages, 409
  cavitation, 404
  disadvantages, 409
  impeller types, 396, 397*f*
  kinetic and pressure energy, 396
  non-Newtonian fluids, 409–412, 410*f*
  operating characteristics, 402–404, 402–403*f*
  radial flow pumps, 397, 397*f*
  similarity, 401–402
  specific speed, 401
  suction head (*see* Suction system)
  velocity diagram, 397, 398*f*
  virtual head, 398–400, 398*f*
Chezy equation, 116
Coarse solids
  horizontal flow
    additional hydraulic gradient, 247
    γ-ray absorption system, 250
    bed deposit, 246–247
    dense phase flow, 243
    drag coefficient, 248
    excess hydraulic gradient, 249*f*, 250
    excess pressure gradient, 247
    heavy and shear-thinning media transportation, 254–256
    heterogeneous suspension, 248
    hold-up and slip velocity, 243–246
    hydraulic gradient-velocity, 246, 246*f*
    low density transportation, 256
    pressure gradient, 247
    principal types, 243
    Reynolds numbers, 247–248
    two-layer model (*see* Two-layer model)
  vertical flow
    free-falling velocity, 258
    frictional pressure drop, 258
    γ-ray technique, 257
    hydraulic gradient, 256–257
    hydrostatic pressure gradient, 257
    nonsettling suspensions, 258
    terminal falling velocity, 258–259
Coarse suspensions, 240
Compressible fluids flow
  converging-diverging nozzles, 185*f*
    flow in pipe (*see* Pipeline flow)
    maximum flow and critical pressure ratio, 185–186

Compressible fluids flow (*Continued*)
pressure and flow area, 186–188
shock waves (*see* Shock wave)
nozzle/orifice (*see* Nozzle/orifice gas flow)
pressure and density, 171
velocity of propagation, 182–184
Continuous stirred-tank reactor (CSTR), 373
Coriolis meter, 326, 326*f*
Couette simple shear profile, 394–395, 395*f*

D

Dall tube, 313–315, 314*f*
Deborah number, 142
Derived units, 7–8
Diaphragm pump, 384–385, 385*f*
Differential pressure cell system, 292–293, 292–293*f*
Dimensional analysis
application of, 14
dimensionless groups, 13–14
dimensionless numbers, 13–14
influence of linear size, 17
mass, length, and time, 15
Reynolds number, 15–16
simultaneous equation, 16
Dimensionless groups, 20–23, 20*t*
Doppler ultrasonic flowmeter, 326
Drag ratio, 227, 228–229*f*

E

Electrical units, 10
Emulsion rheology, 232–233
Energy balance equation, 286
Eye and/or on photographic methods, 233

F

Fahrenheit unit, 488–490*t*
Faraday's Law, 324
Fine suspensions, 240
Flapper–nozzle separation, 293
Fluid flows
constriction
energy and material balance equations, 299–300
horizontal meter, 301
incompressible fluid, 301
isothermal flow, 301
nonisothermal flow, 302
Coriolis meter, 326, 326*f*
Doppler ultrasonic flowmeter, 326
forced vortex, 60–63
free vortex, 63–64
gas meters, 327–328, 327–328*f*
hot-wire anemometer, 322–323
internal energy
entropy, 32
heat units, 32
infinitesimal change, 32
irreversible process, 32, 34
mechanical energy, 32, 34
physical state of, 31–32
stationary system, 33
liquid meters, 328–329, 329–330*f*
magnetic flowmeter, 324, 324*f*
motion
constant flow per unit area, 55
continuity, 47–48
flow patterns, 46, 46–47*f*
momentum changes, 48–51
pressure and fluid head, 54–55
reaction force, 50
separation, 55
time-average flowrate, 46
total fluid energy, 51–54
water hammer, 51
notch/weir, 297, 319–322, 320*f*
nozzle, 297, 311–312, 311*f*
orifice meter
definition, 297
diameter calculation, 309–310
disadvantage, 304
drilled orifice plate, 304
eccentric orifice plate, 310–311, 310*f*
flow of steam, 307–310
frictional losses, 304
horizontal orifice, 306
incompressible fluid, 304
influencing factors, 302–304, 303*f*
isothermal flow, 305
mass flowrate, 308–309
nonisothermal flow, 306
pressure drop, 309
pressure recovery, 313–315, 314*f*, 315*t*
Reynolds numbers, 305
segmented orifice plate, 310–311
vena contracta, 304
viscosity of water, 307
pitot tube, 297–299, 298*f*
pressure-volume relationship
compressible fluids, 56–58
incompressible fluids *vs.* independent of pressure, 56
quantity meters, 327
static pressure, 286–287
time-of-flight ultrasonic flowmeter, 325–326
types
externally applied pressure, 35
ideal gas (*see* Ideal gas fluid)
incompressible fluid, 36
nonideal gas (*see* Nonideal gas fluids)
shear stress, 35
stress distribution, 35–36
volume changes, 35
variable area meters (*see* Rotameter)
venturi meter, 297, 312–313, 312*f*
vortex-shedding flowmeters, 325, 325*f*
Fluid pressure
Bourdon gauge, 290, 291*f*
differential pressure cell, 292–293, 292–293*f*
impact pressure, 295–297, 295*f*
inclined manometer, 289, 289*f*
intelligent pressure transmitters, 293–295, 294–295*f*
inverted manometer, 289, 290*f*
simple manometer, 287
two-liquid manometer, 289, 290*f*
well-type manometer, 287–288
Foot–Pound–Second (fps) system, 5
Force balance system, 292–293, 293*f*

Forced vortex, 60–63
Francis formula, 320–321
Free vortex, 63–64
Froude number, 119

G

γ-ray absorption technique, 253
Gas compressors
air-lift pumping
aerated liquid, vertical column in, 433–435, 434f
atmospheric pressure, 430
efficiency, 432
operation of, 435–436, 436f
power requirement, 431
pump unit mass of liquid, 430–431
schematic diagram, 429–430, 429f
centrifugal and turbocompressors, 414–415, 415f
clearance volume, 420–422, 420f
efficiency, 426–428
fans and rotary compressors
axial flow type, 412
Nash Hytor liquid ring pump, 413–414, 414f
positive displacement type, 412–413, 413f
sliding vane type, 413, 414f
irreversible change, 416
isentropic compression, 417–418, 420
isothermal compression, 417, 419–420
multistage compressors
interstage cooling, 423–424, 423f
net saving, 424
temperature-entropy, 425
theoretical volumetric efficiency, 425
outlet and inlet temperatures, 417–418
reciprocating piston compressor, 415, 416f
reversible change, 416
single-stage compression, 418, 419f
thermodynamic properties, 420
Gas meters
vane anemometer, 327, 327f
wet gas meter, 327–328, 328f
Gas-solids mixtures flow
dense phase transport, 260–261
horizontal transport
cross-correlation methods, 265
dense phase conveying, 271
dilute phase conveying, 263–264
dilute phase/lean phase transport, 260–261
electrostatic charging, 270–271
flow patterns, 261–262
mechanism of suspension, 262–263
solid velocities, 264–265
hydraulic conveying, 260
pneumatic conveying, 260
pressure drops and solid velocities, 266
solid velocities (*see* Pressure drops)
vertical transport
additional pressure gradient, 271–272
dense-phase plug-flow, 272
design of pneumatic conveying, 273–275
fluidised bed reactor systems, 271
free-falling velocity, 274
frictional pressure drop, 272
mean air velocity, 275–277, 275f
minimum fluidising velocity, 271
over-riding factor, 272–273
pressure gradient and flowrates of gas, 272
75 μm catalyst particles, 272
solids to gas ratio, 273
stepped pipeline parameters, 275–277, 276f
Gas viscosity, 472–473t
Gas whose equation, 43
Gear pump, 388, 388–389f
cam pump, 388–389
flexible vane pump, 391, 391f
flow inducer/peristaltic pump, 391, 392f
mono pump, 392–393, 392–393f
screw pump, 393–396, 394–395f
vane pump, 390–391, 390f
Generalised Bingham equation, 133

H

Herschel–Bulkley equation, 133
Herschel–Bulkley fluid, 160
Hooke's law, 125, 137
Hot-wire anemometer, 322–323
Hydraulic mean diameter, 104

I

Ideal gas fluid, 36–40
fluid pressure, 36
internal energy of, 38
isentropic process, 39–40
isothermal process, 38–39
molecular weight, 36–37
unit mass of, 36–37
Impact pressure
energy balance equation, 296
isothermal flow, 296–297
nonisothermal flow, 296
static pressure, 295
Incompressible fluid, 36
Intelligent pressure transmitters
absolute pressure, 294, 294f
differential pressure, 295, 295f
Gauge pressure, 294, 294f
piezo-resistive effects, 294
sensor module, 293–294
two-compartment electronics, 293–294
Wheatstone Bridge circuit, 294, 294f
Internal energy, fluids flow, 52
entropy, 32
heat units, 32
infinitesimal change, 32
irreversible process, 32, 34
mechanical energy, 32, 34
physical state of, 31–32
stationary system, 33

Isentropic compression, 417–418, 420
Isothermal compression, 417, 419–420

J

Joule–Thomson effect, 45–46

K

Kinetic energy, 52–54

L

Laminar mixing
  coaxial cylinder arrangement, 338, 338*f*
  cutting and folding, 338, 338*f*
  extensional flow, 336–337, 337*f*
  shear flow, 336–337, 337*f*
Laminar sublayer, 71
Laplace transforms, 495–501*t*
Laval nozzles, 184
Liquid–liquid mixtures flow
  average holdup, 237–238
  degree of subjectivity and arbitrariness, 233
  dispersed flow, 233
  emulsion rheology, 232–233
  Eotvös number, 235–237
  eye and/or on photographic methods, 233
  flow patterns, 233
  gas–liquid systems, 233–235
  material of construction, 232
  oil–water systems, 233–235
  pressure gradient, 238–239
  prevailing flow pattern, 232
  pseudo-homogeneous single phase fluids, 232–233
Liquid meters
  oscillating-piston meter, 328, 329*f*
  turbine flow meters, 328–329, 330*f*
Liquid mixing
  continuous systems, 373–374
  extruders, 368–370, 369*f*
  gas–liquid mixing, 334
  gas–liquid–solids mixing, 335
  homogeneity, 335–336
  immiscible liquids, 334
  Laminar mixing
    coaxial cylinder arrangement, 338, 338*f*
    cutting and folding, 338, 338*f*
    extensional flow, 336–337, 337*f*
    shear flow, 336–337, 337*f*
  liquid–solids mixing, 334
  mechanical agitation
    anchors, helical ribbons, and screws, 365*f*, 368
    baffles, 364
    Banbury mixer, 365*f*, 368
    kneaders, Z-and sigma-blade, 365*f*, 368
    propellers, turbines, and paddles, 364, 365–367*f*, 367
    shrouded turbines, 368
    vessels, 364
  portable mixers, 368
  rate and time
    acid/base titration technique, 362, 363*f*
    behaviour, 362, 362*f*
    definition, 360
    dimensional analysis, 361
    experiment and configuration, 361
    measurement curve, 360, 360*f*
    tracer concentration, 360–361, 361*f*
    viscoelasticity, 362
  single-phase liquid mixing, 333
  solids–solids mixing, 335
  static mixers, 370–372, 370–372*f*
  stirred vessels, scale-up of (*see* Stirred vessels)
  turbulent mixing, 339
  types of, 373
Liquids density, 474–476*t*
Liquids flow
  gas–liquid mixtures, 68
  ideal gas law, 67
  nature of
    fluctuating velocity, 70
    laminar/streamline, 68–69
    laminar sublayer, 68–69
    over a surface, 70–71
    pipe, flow in, 71
    Reynolds' method, 68–69, 69*f*
    turbulent flow, 68–69
  Newtonian behaviour, 67
  Newtonian fluid (*see* Newtonian fluids)
  non-Newtonian behaviour, 68
  non-Newtonian fluid (*see* Non-Newtonian fluids)
  power requirements, 67
  volume changes, 67
Liquids viscosity, 474–476*t*
Lobe pump, 388, 389–390*f*. *See also* Gear pump
Lockhart–Martinelli approach, 238–239

M

Magnetic flowmeter, 324, 324*f*
Magnus principle, 263, 263*f*
Mathematical tables
  error function and derivative, 501–503*t*
  Laplace transforms, 495–501*t*
Mechanical agitation
  anchors, helical ribbons, and screws, 365*f*, 368
  baffles, 364
  Banbury mixer, 365*f*, 368
  kneaders, Z-and sigma-blade, 365*f*, 368
  propellers, turbines, and paddles, 364, 365–367*f*, 367
  shrouded turbines, 368
  vessels, 364
Metre–Kilogram–Second (mks) system, 4–5
Metzner and Reed correlation, 162, 163*f*
Molar units, 9
Momentum balance, 223
Mono Merlin Wide Throat pump, 393, 393*f*
Mono pump, 392–393, 392–393*f*
Motion, fluids flow
  constant flow per unit area, 55
  continuity, 47–48
  flow patterns, 46, 46–47*f*
  momentum changes, 48–51

pressure and fluid head, 54–55
reaction force, 50
separation, 55
time-average flowrate, 46
total fluid energy, 51–54
water hammer, 51
Multiphase mixtures flow
electrostatic charging, 216
flow systems, 215
gas–solids mixtures (*see* Gas-solids mixtures flow)
liquid–liquid mixtures
aforementioned flow patterns, 233
average holdup, 237–238
degree of subjectivity and arbitrariness, 233
dispersed flow, 233
emulsion rheology, 232–233
Eotvös number, 235–237
eye and/or on photographic methods, 233
gas–liquid systems, 233–235
material of construction, 232
oil–water systems, 233–235
pressure gradient, 238–239
prevailing flow pattern, 232
pseudo-homogeneous single phase fluids, 232–233
Newtonian liquids, 215–216
non-Newtonian fluids, 215–216
phase density, 216
pressure drop, 215–216
slip velocity, 215–216
solids–liquid mixtures
coarse particles (*see* Coarse solids)
coarse suspensions, 240
fine suspensions, 240
homogeneous nonsettling suspensions, 240–243
hydraulic transport, 239–240
variables, 240
two-phase gas (vapour)-liquid flow
continuous flow of liquid, 217
critical flow, 225–227
erosion, 231–232
evaluating pressure drop, 224–225
Flow Pattern Map, 217
hold-up, 221–223
horizontal flow, 218–220, 218*f*, 219*t*, 220*f*
non-Newtonian flow, 227–231
separated flow model, 223
streamline/turbulent flow, 217
vertical flow, 218*f*, 220–221, 220*f*

N

Net positive suction head (NPSH), 404
Newtonian fluids
banks of tubes, 112
free surface flow
Chezy equation, 116
energy of liquid, 117–119
hydraulic jump, 121–124
laminar flow, 112–114, 113*f*
turbulent flow, 114
uniform flow, 114–117
velocity of transmission, 119–121
friction losses
heat exchanger, 110
initial rate of discharge, 111
pipe fittings, 108–111, 109*t*
sudden contraction, 107–108
sudden enlargement, 105–107
laminar to turbulent flow transition, 98–99
pressure drop (*see* Newtonian liquids flow)
Reynolds number and shear stress, 89–90
shearing characteristics, 71–73, 71*f*
stable streamline flow, 104
velocity distributions and volumetric flowrates
annulus flow, 96–98, 96*f*
average velocity, 92–93
circular cross-section, pipe of, 90–95
kinetic energy, 93–95, 101–103
mean velocity, 100–101, 101*f*
noncircular ducts, 103–104
Prandtl one-seventh power law, 102–103
two parallel plates flow, 95–98
volumetric rate of flow, 92–93
Newtonian liquids flow
allowable water velocity, 82
effect of roughness, 80–89, 80*t*
flow rate, 87
hydraulic gradient *vs.* velocity, 73*f*
kinetic energy, 84–87
liquid flowing, 78–79, 79*f*
pipe diameter, 87
resistance to flow, 75–77
Reynolds number, 74, 81
shear stress, 74–75
turbulent flow, 82–84
Newton's law, 2
Noncoherent system, 6–7
Nonideal gas fluids
atmospheric pressure, 41–45
generalised compressibility-factor chart, 41, 41*f*
internal energy and temperature, 43
Joule–Thomson effect, 45–46
temperature and pressure, 40
van der Waals' equation, 40–44
Non-Newtonian fluids
centrifugal pump, 409–412, 410*f*
characterisation of, 140–142, 141*f*
laminar to turbulent flow, 164–165
rheology and structure of material, 143–144
shear-dependent, 125
shear rate, 124
shear stress, 124
shear-thinning, 125
steady-state shear-dependent behaviour
apparent viscosity, 126–127, 127*f*, 130–131, 131*f*
Bingham-plastic, 133, 134*f*
food products and processes, 131, 132*t*
generalised Bingham equation, 133

Non-Newtonian fluids (*Continued*)
Herschel–Bulkley equation, 133
Ostwald–de Waele law, 129
power-law model, 129
shear stress *vs.* shear rate curve, 127, 127*f*
shear-thickening fluid, 127, 128*f*
shear-thinning fluids approach, 130–131
yield stress, 132, 135*f*
streamline flow
apparent viscosity, 153
Bingham-plastic fluid, 147–155
centre line velocity, 152
general equations, 156–158
generalised Reynolds number, 158–159
laminar-flow, 144
modified Reynolds number, 144
power-law fluids, 144–147, 148*f*, 151, 153–154
rheological equation, 153
velocity-pressure gradient relationships, 159–161
volumetric flowrate, 155
time-dependent behaviour, 125, 135–137, 136*f*
turbulent flow, 161–164
viscoelastic behaviour, 137–140, 138–139*f*
viscoelastic flows, 142–143
yield stress, 125
Notch/weir
definition, 297
rectangular notch, 319–321, 320*f*
triangular notch, 321–322, 321*f*
Nozzle/orifice gas flow
energy balance, 172
fluid flow, 297, 311–312, 311*f*
gas flowrate, 172
isothermal flow, 173–175
nonisothermal flow
critical pressure ratio, 177, 180–182
discharge coefficient, 176
discharge velocity, 176
isentropic conditions, 177–178
mass rate of discharge, 178–182, 179*f*
mass rate of flow, 176
maximum flow conditions, 176–182
velocity of small pressure wave, 177
pressure–volume relation, 172

O

Orifice meter, fluid flow
definition, 297
diameter calculation, 309–310
disadvantage, 304
drilled orifice plate, 304
eccentric orifice plate, 310–311, 310*f*
flow of steam, 307–310
frictional losses, 304
horizontal orifice, 306
incompressible fluid, 304
influencing factors, 302–304, 303*f*
isothermal flow, 305
mass flowrate, 308–309
nonisothermal flow, 306
pressure drop, 309
pressure recovery, 313–315, 314*f*, 315*t*
Reynolds numbers, 305
segmented orifice plate, 310–311
vena contracta, 304
Oscillating-piston meter, 328, 329*f*
Ostwald–de Waele law, 129

P

Piezometer tube, 287
Piezo-resistive effects, 294
Pipeline flow
adiabatic flow
entropy value, 204
equation, 202
Fanno lines, 205
irreversible adiabatic process, 201–202
kinetic energy, 204
mass rate of flow *G*, 203
maximum entropy, 205–206
P and *v* relation, 202
pressure P_2, 203
sonic velocity, 204
energy balance, 190–191
fixed upstream pressure and variable downstream pressure
compressor pressure, 199–200
heat flow, 200–201
maximum flowrate, 196
Reynolds number, 196
vacuum systems, flow of gases, 198–199
values of G, 197, 197*t*, 198*f*
values of G_{max}, 197
isothermal and adiabatic conditions, 189
isothermal flow
average pressure gradients, 194
critical pressure ratio, 194, 196*f*
ideal gas, 191
limiting pressure ratios, 195*t*, 196
maximum flowrate, 193
mean pressure, 192
pressure drop, 192
nonideal gases flow, 206–207
nonisothermal flow, 201
supersonic flow, 189–190
Piston pump, 381–384, 382*f*
Pitot tube, 297–299, 298*f*
Polytetrafluoroethylene (PTFE), 406
Potential energy, 52
Power consumption
high viscosity systems
Bingham plastic fluids, 354
impeller and vessel configuration, 351
impeller speed, 355
k_s value, 352, 352–353*t*
non-Newtonian and Newtonian liquids, 351
power curve, 351, 354*f*
Rushton-type turbine impellers, 355
low viscosity systems
Froude number, 343–344

geometrical arrangement, 346
independent variables, 343
laminar and turbulent zones, 350
power curves, 351
power-law relationship, 344
Power number, 343, 350, 350*t*
Reynolds number, 343–344, 345*f*, 346–347, 349, 349*f*
Power-law fluids, 148*f*, 159
Power-law model, 129, 222
Power requirements
energy balance, 438
gases, 447–449
liquids
fluctuation effect, 445–446
pipe diameter selection, 443–445
streamline flow, 446
turbulent flow, 446–447
Pressure drops
gas-solids mixtures flow
airflow rate, 266, 267*f*
diameter conveying line, 265–266, 266*f*
effect of pipe diameter, 269, 270*f*
free-falling velocity, 267
physical properties, 267, 268*t*
pneumatic conveying, 269, 269*f*
slip velocities, 267, 268*f*
Newtonian liquids flow
allowable water velocity, 82
effect of roughness, 80–89, 80*t*
flow rate, 87
hydraulic gradient *vs.* velocity, 73*f*
kinetic energy, 84–87
liquid flowing, 78–79, 79*f*
pipe diameter, 87
resistance to flow, 75–77
Reynolds number, 74, 81
shear stress, 74–75
turbulent flow, 82–84
Pressure energy, 52
Pressure signal transmission, 292–293, 292–293*f*
Pressure transmitters, 293–295, 294–295*f*
Pressure-volume relationship
compressible fluids, 56–58
incompressible fluids *vs.* independent of pressure, 56
Pumping fluids
centrifugal pump
advantages, 409
cavitation, 404
disadvantages, 409
impeller types, 396, 397*f*
kinetic and pressure energy, 396
non-Newtonian fluids, 409–412, 410*f*
operating characteristics, 402–404, 402–403*f*
radial flow pumps, 397, 397*f*
similarity, 401–402
specific speed, 401
suction head (*see* Suction system)
velocity diagram, 397, 398*f*
virtual head, 398–400, 398*f*
for gas equipment (*see* Gas compressors)
minor losses
flowrate calculation, 452
head loss calculation, 450–456
valves and fittings, 449–450, 450–451*t*
positive-displacement rotary pumps (*see* Gear pump)
power requirements
energy balance, 438
fluctuation effect, 445–446
gases, 447–449
pipe diameter selection, 443–445
streamline flow, 446
turbulent flow, 446–447
reciprocating pump
diaphragm pump, 384–385, 385*f*
metering pumps, 385–387, 386*f*
piston pump, 381–384, 382*f*
plunger/ram pump, 384
vacuum pumps
diffusion pump, 438
ejector flow phenomena, 437, 439*f*
multistage units, 437, 440*f*
sliding vane and liquid ring, 437, 437*f*
steam-jet ejector, 437, 438*f*

R

Reynolds' method, 68–69, 69*f*
Rotameter
coefficients, 317, 317*f*
float and wall, 315–316, 316*f*
flowrate, 317–318

S

Salt injection method, 253
Saturated steam
Centigrade and Fahrenheit units, 488–490*t*
S.I. units, 484–487*t*
Separated flow model, 223
Shock wave
change in entropy, 209
initial rate of discharge, 210–212
Mach number, 208–209
momentum balance, 207
pressure ratio, 210–212
Simple control system, 285, 286*f*
Single-phase liquid mixing, 333
Single-phase pressure drops, 224
Single-screw extruder, 369–370, 369*f*
Solids–liquid mixtures flow
coarse particles (*see* Coarse solids)
coarse suspensions, 240
fine suspensions, 240
homogeneous nonsettling suspensions, 240–243
hydraulic transport, 239–240
Static mixers
pressure drop, 372
quality, 372
twisted-blade type, 370, 370–371*f*
viscous materials, 370–372, 372*f*
Static pressure, 286–287
Steam tables
Abridged callendar steam tables
pressure-enthalpy diagram, 493*f*
saturated steam, 484–490*t*
superheated steam, 491–492*t*

Steam tables (*Continued*)
temperature-entropy diagram, 494*f*
critical constants of gases, 478–479*t*
gas viscosity, 472–473*t*
heats of liquids, 465–466*t*
liquids density, 474–476*t*
liquids viscosity, 474–476*t*
pressure of gases and vapours, 467*t*
surface emissivity, 479–483*t*
thermal conductivities
gases and vapours, 470–471*t*
of liquids, 461–462*t*
vaporisation, 463*t*
water viscosity, 469*t*
Stirred vessels
dynamic similarity, 340–341
flow patterns
agitator offset, 357, 357*f*
disc turbines, 356, 356*f*
double-celled secondary flow, 358–359, 359*f*
flat-bladed turbine, 358, 358*f*
gate-type anchor, 358–359, 358*f*
propeller mixer, 356, 356*f*
single-celled secondary flow, 358–359, 359*f*
vertical baffles, 357, 357*f*
Froude and Weber numbers, 342
geometric similarity, 340
inertial force, 341
kinematic similarity, 340
low and high viscosity systems (*see* Power consumption)
surface tension force, 342
typical configuration and dimensions, 340, 340*f*
viscous forces, 341
Suction system
absolute pressure, 404–405
effect of, 404, 405*f*
efficiency, 406
mechanical seal, 406, 407*f*
NPSH, 404
performance characteristics, 406, 408*f*
vacseal pump, 407–409, 408*f*
Surface emissivity, 479–483*t*

T

Thermal conductivities
gases and vapours, 470–471*t*
of liquids, 461–462*t*
Thermal (heat) units, 8–9
Time-of-flight ultrasonic flowmeter, 325–326
Trouton ratio, 139–140
Turbine flow meters, 328–329, 330*f*
Turbulent mixing, 339
Twin-screw extruders, 369–370, 369*f*
Two-layer model
coefficient of friction, 252
force balance, 251
γ-ray absorption technique, 253
pressure gradients, 253, 254*f*
salt injection method, 253
sensitivity analysis, 253
steady-state force balance, 251
Two-phase gas (vapour)-liquid flow
continuous flow of liquid, 217
critical flow, 225–227
erosion, 231–232
evaluating pressure drop, 224–225
Flow Pattern Map, 217
hold-up, 221–223
horizontal flow, 218–220, 218*f*, 219*t*, 220*f*
non-Newtonian flow, 227–231
separated flow model, 223
streamline/turbulent flow, 217
vertical flow, 218*f*, 220–221, 220*f*
Two-phase pressure drop, 225

U

Units and dimensions
British Engineering System, 2
Buckingham's π theorem
dimensionless groups, 20–23, 20*t*
drop of liquid spreads, 22
m fundamental dimensions, 18
n variables, 18
recurring set, 18–19
scaling laws, 23–25
conversion of, 10–12
dimensional analysis
application of, 14
dimensionless groups, 13–14
dimensionless numbers, 13–14
influence of linear size, 17
mass, length, and time, 15
Reynolds number, 15–16
length and mass dimensions
quantity mass and inertia mass, 27
vector and scalar quantities, 26–27
Newton's law, 2
physical properties, 1–2
physical quantity, 1
standardisation, 1
systems of, 3*t*
British engineering system, 6
cgs system, 4
derived units, 7–8
electrical units, 10
fps system, 5
mks system, 4–5
molar units, 9
noncoherent system, 6–7
thermal (heat) units, 8–9

V

Vacseal pump, 407–409, 408*f*
Vacuum pumps
diffusion pump, 438
ejector flow phenomena, 437, 439*f*
multistage units, 437, 440*f*
sliding vane and liquid ring, 437, 437*f*
steam-jet ejector, 437, 438*f*
van der Waals' equation, 40–41, 43–44
Vaporisation, 463*t*
Vena contracta, 107, 304
Venturi meter, 297, 312–313, 312*f*
Voigt model, 138
Vortex-shedding flowmeters, 325, 325*f*

W

Water viscosity, 469*t*

Printed in the United States
By Bookmasters